CAMBRIDGE LIBRARY COLLECTION

Books of enduring scholarly value

Physical Sciences

From ancient times, humans have tried to understand the workings of
the world around them. The roots of modern physical science go back to
the very earliest mechanical devices such as levers and rollers, the mixing
of paints and dyes, and the importance of the heavenly bodies in early
religious observance and navigation. The physical sciences as we know them
today began to emerge as independent academic subjects during the early
modern period, in the work of Newton and other 'natural philosophers',
and numerous sub-disciplines developed during the centuries that followed.
This part of the Cambridge Library Collection is devoted to landmark
publications in this area which will be of interest to historians of science
concerned with individual scientists, particular discoveries, and advances in
scientific method, or with the establishment and development of scientific
institutions around the world.

Mathematical and Physical Papers

William Thomson, first Baron Kelvin (1824–1907), is best known for
devising the Kelvin scale of absolute temperature and for his work on the
first and second laws of thermodynamics, though throughout his 53-year
career as a mathematical physicist and engineer at the University of Glasgow
he investigated a wide range of scientific questions in areas ranging from
geology to transatlantic telegraph cables. The extent of his work is revealed
in the six volumes of his *Mathematical and Physical Papers*, published from
1882 until 1911, consisting of articles that appeared in scientific periodicals
from 1841 onwards. Volume 3, published in 1890, includes articles from the
period 1858–1890, the majority of which relate to questions around elasticity
and heat, and are accompanied by extensive appendices.

Mathematical and Physical Papers

VOLUME 3

LORD KELVIN

CAMBRIDGE UNIVERSITY PRESS

Cambridge, New York, Melbourne, Madrid, Cape Town,
Singapore, São Paolo, Delhi, Tokyo, Mexico City

Published in the United States of America by Cambridge University Press, New York

www.cambridge.org
Information on this title: www.cambridge.org/9781108029001

© in this compilation Cambridge University Press 2011

This edition first published 1890
This digitally printed version 2011

ISBN 978-1-108-02900-1 Paperback

MATHEMATICAL

AND

PHYSICAL PAPERS.

MATHEMATICAL

AND

PHYSICAL PAPERS

VOLUME III.

ELASTICITY, HEAT, ELECTRO-MAGNETISM.

BY

SIR WILLIAM THOMSON, LL.D., D.C.L., F.R.S.,

PROFESSOR OF NATURAL PHILOSOPHY IN THE UNIVERSITY OF GLASGOW,
AND FELLOW OF ST PETER'S COLLEGE, CAMBRIDGE.

*COLLECTED FROM DIFFERENT SCIENTIFIC PERIODICALS FROM
MAY, 1841, TO THE PRESENT TIME.*

WITH SUPPLEMENTARY ARTICLES WRITTEN FOR THE
PRESENT VOLUME, AND HITHERTO UNPUBLISHED.

LONDON:
C. J. CLAY AND SONS,
CAMBRIDGE UNIVERSITY PRESS WAREHOUSE.
1890

PREFACE TO VOL. III.

In this third volume which consists chiefly of Articles relating to Elasticity and Heat, I have not found it convenient to follow the chronological order of Vols. I. and II.

A printed volume containing my Baltimore Lectures on Molecular Dynamics and the Wave Theory of Light with Appendices, of which a limited edition has already been published in papyro-graph by the Johns Hopkins University, will contain also some later articles relating to those subjects, and will I hope be published soon.

A concluding volume of the present series will I hope contain, in chronological order, all that remain of my Mathematical and Physical Papers.

I take this opportunity of expressing my thanks to Messrs Adam and Charles Black, for their kindness in permitting the Articles on Elasticity and Heat from the Encyclopædia Britannica, to be included in the present volume.

WILLIAM THOMSON.

PETERHOUSE LODGE, CAMBRIDGE.
June 2, 1890.

CONTENTS.

ERRATA.

Page 34, line 17, for "uniform" read "a simple'

" 71, Heading of Col. 7 of Table IV. for "$\dfrac{N}{K}$" read "$\dfrac{K}{N}$"

" 72, line 8 of Table VI. for "Copper" read "Iron, cast"

" " " 9 " " insert "Copper" in Column 1

" 153, " 27, for "cryophorous" read "cryophorus"

" 202, " 5 from foot, for "XII." read "XIII."

" " " 6 " " for "square centimetres"
read "grammes water units per second"

" 205, " 11, delete "not well chosen"

" 226, " 10 of Table XIII. for "·013" read "·0013"

" 228, footnote, for "XCVI." read "XCVIII."

" 396, line 29, insert "half" between "than" and "that"

" 398, " 3 of footnote, for "præcendentium" read "præcedentium"

" 408, " 30, delete comma after "structure"

" 409, " 35, for "point" read "state"

" 410, " 9, for "any" read "every"

" " lines 2 and 3 from foot, for "OK. In their present positions they"
read "OK: and the two assemblages"

" 476, line 6 from foot, delete comma after "force"

" " " 2 " " for "σ" read "−σ"

" 479, " 8, for "15" read "12"

" " " 9, for "·18" read "·225"

" " " 4 from foot of § 13, insert ", and at 1/20 of a centimetre from
the surface, "

" 480, " 16, for "of the" read "if the"

" 481, " 5 of § 18, for "must"
read "may during a part of the time more nearly have been"

" 482, " 1, for "probably" read "possibly"

" 483, " 7, for "18" read "·225"

ERRATA.

p. 499. Last line: *For* 2 *substitute* 4.

p. 499. Line 3 from foot: *For* $=2c$ *substitute* $+\dfrac{1}{r}\dfrac{-du}{dr}=4c.$

p. 499. Line 9 from foot: *After* "shearing" *insert* "together with rate of shearing divided by distance from axis."

MATHEMATICAL AND PHYSICAL PAPERS.

Art. XCII. Elasticity and Heat.

[Reprint, by permission, of articles contributed to the Ninth Edition (1878) of the *Encyclopædia Britannica*.]

PART I. ELASTICITY.

[This article is founded upon, and has incorporated within it, these two papers —" A Mathematical Theory of Elasticity," *Trans. Roy. Soc.* April 24, 1856, and " On the Elasticity and Viscosity of Metals," *Proc. Roy. Soc.* May 18, 1865.]

TABLE OF CONTENTS.

MATHEMATICAL THEORY.

PART I.   *Stresses and Strains.*

PART II.—*On the Dynamical Relation between Stresses and Strains experienced by
an Elastic Solid.*

TABLES.

1. ELASTICITY of matter is that property in virtue of which a body requires force to change its bulk or shape, and requires a continued application of the force to maintain the change, and springs back when the force is removed, and, if left at rest without the force, does not remain at rest except in its previous bulk and shape. The elasticity is said to be perfect, when the body always requires the same force to keep it at rest, in the same bulk and shape and at the same temperature, through whatever variations of bulk, shape, and temperature it be brought. A body is said to possess some degree of elasticity if it requires any force to keep it in any particular bulk or shape. It is convenient to discuss elasticity of bulk and elasticity of shape sometimes separately and sometimes jointly.

2. Every body has some degree of elasticity of bulk. If a body possesses any degrée of elasticity of shape it is called a solid; if it possesses no degree of elasticity of shape it is called a fluid.

3. All fluids possess elasticity of bulk to perfection. Probably so do all homogeneous solids, such as crystals and glasses. It is not probable that any degree of fluid pressure (or pressure acting equally in all directions) on a piece of common glass, or rock crystal, or of diamond, or on a crystal of bismuth, or of copper, or of lead, or of silver, would make it denser after the pressure is removed, or put it into a condition in which at any particular intermediate pressure it would be denser than it was at that pressure before the application of the extreme pressure. Malleable metals and alloys, on the other hand, may have their densities considerably increased and diminished by hammering and by mere traction. By compression between the dies used in coining, the density of gold may be raised from 19·258 to 19·367, and the density of copper from 8·535 to 8·916*; and Mr M'Farlane's experiments quoted below (§ 78), show a piece of copper wire decreasing in density from 8·91 to 8·835 after successive simple tractions, by which its length was increased from 287 centimetres to 317 centimetres, while its modulus of rigidity decreased from 443 to 426 million grammes per square centimetre. Later experiments, recently made for this article by the same experimenter,

---

* *Seventh Annual Report of the Deputy-Master of the Mint*, p. 43, quoting as authority Percy's *Metallurgy of Copper*. London, 1861.

have shown *augmentation* of density from 8·85 to 8·95, produced
by successive tractions which elongated a piece of copper wire
from weighing 16·4 grammes per metre to weighing 13·5 grammes
per metre, the wire having been first annealed by heating it to
redness in sand, and allowing it to cool slowly. Augmentation of
density by traction is a somewhat surprising result, but not altogether so when we consider that the wire had been reduced to
an abnormally small density by the previous thermal treatment
(the "annealing"). The common explanations of these changes
of density in metals, which attributes them to porosity, is probably true; by porosity being understood a porous structure with
such vast numbers of the ultimate molecules in the portions of
the solid substance between pores or interstices that these portions
may be called homogeneous in the sense that a crystal or a liquid
can be called homogeneous (compare § 40 below).

4. The elasticity of shape of many solids is not perfect; it
is not known whether it is perfect for any. It might be expected
to be perfect for glass and rock crystal and diamond and other
hard, brittle, homogeneous substances; but experiment proves
that at all events for glass it is not so, and shows on the contrary
a notable degree of imperfection in the torsional elasticity of glass
fibres. It might be expected that in copper and soft iron and
other plastic metals the elasticity of shape would be very imperfect; experiment shows, on the contrary, that in copper, brass,
soft iron, steel, platinum, provided the distortion does not exceed
a certain limit in each case, elasticity of shape is remarkably
perfect, much more perfect than in glass. It is quite probable
that even in the softer metals—zinc, tin, lead, cadmium, potassium,
sodium, &c.—the elasticity of shape may be as perfect as in the
metals mentioned above, but within narrower limits as to degree
of distortion. Accurate experiment is utterly wanting, to discover
what is the degree of imperfection, if any, of the elasticity of any
metal or alloy, when tested within sufficiently narrow limits of
distortion.

5. The "viscosity of metals" described below (§§ 21—25) does
not demonstrate any imperfectness of elasticity according to the
definition of § 1, which is purely statical. The viscosity of solids
may (for all we yet know by experiment) depend, as does the
viscosity of fluids, upon a resistance varying with the velocity of

the change, and *vanishing when the velocity of the change is zero*, that is to say, when the body is at rest in any configuration ; if so, the elasticity of the substance concerned is perfect within the limits of the experiment in question. If, on the other hand (as the discovery of elastic " fatigue " described in § 30 below seems to indicate may be to some degree the case), the loss of energy from the vibrations in the experiments described, is due to a dependence of the elastic resilient force upon previous conditions of the substance in respect to strain, the " viscosity " would be continuous with a true imperfectness of static elasticity. Here, then, we have a definite question which can be answered by experiment only :—Consider a certain definite stress applied to a solid substance ; as, for example, a certain " couple " twisting a wire or rod ; or a certain weight pulling it out, or compressing it lengthwise ; or a certain weight placed on the middle of a beam supported by trestles under its ends. Let the weight be applied and removed a great many times, and suppose it to be seen that after each application and removal of the stress the body comes to rest in exactly the same configuration as after the previous application or removal of the stress. If now the body be left to itself with the stress removed, and if it be found to remain at rest in the same configuration for minutes, or hours, or days, or years after the removal of the stress, a part of the definition of perfect elasticity is fulfilled. Or, again, if the stress be applied, and kept applied with absolute constancy, and if the body remain permanently in a constant configuration, another item of the definition of perfect elasticity is proved. When any such experiment is made on any metal, unless some of the softer metals (§ 4) is to be excepted, there is certainly very little *if any* change of configuration in the circumstances now supposed. The writer believes, indeed, that nothing of the kind has hitherto been discovered by experiment, provided the stress has been considerably less than that which would break or give a notable permanent twist, or elongation, or bend, to the body ; that is to say, provided the action has been kept decidedly within the limits of the body's elasticity as commonly understood (§§ 7—20 below). Mr J. T. Bottomley, with the assistance of a grant of money from the British Association, has commenced making arrangements for secular experiments on the elasticity of metals, in the tower of the University of Glasgow, to answer this question in respect to permanence or non-permanence through minutes, or hours, or days, or years, or centuries. If several gold wires are hung side by side, one of them bearing the smallest weight that will keep it approximately straight, another wire bearing $\frac{1}{20}$ of the breaking weight, another wire bearing $\frac{2}{20}$ of the breaking weight, and so on; the one of them bearing $\frac{19}{20}$ of the breaking weight will probably, in the course of a few hours or days, show very sensible elongation. Will it go on becoming longer and longer till it breaks, or will the time-curve of its elongation be asymptotic? Even with considerably less than $\frac{19}{20}$ of the breaking weight there will probably be a continually augmenting elongation, but with asymptotic time-curve indicating a limit beyond which the elongation never goes, but which it infinitely nearly reaches in an infinite time. It is not probable that a gold wire stretched by $\frac{1}{10}$ of its present breaking weight, or by $\frac{1}{4}$ of its present breaking weight, or even by $\frac{1}{2}$ of its present breaking weight, would break in a thousand or in a million years. The existence of gold ornaments which have been found in ancient tombs and cities, and have preserved their shapes for thousands of years without running down glacier-wise (as does brittle pitch or sealing-wax in the course of a few years in moderately warm climates), seems to prove that for gold (and therefore leaves no doubt also for many other metals) the time-curve is asymptotic, if indeed there is any slow change of shape at all after the application of a moderate stress well within the limits of elasticity. Egyptian and Greek statues, Etruscan vases, Egyptian obelisks, and other stone monuments with their engraved hieroglyphics, flint implements and boulders, and mountains with the geological evidence we have of their antiquity, prove for stones, and pottery, and rocks of various kinds, a permanence for thousands and millions of years of resistance to distorting stress.

6. The complete fulfilment of the definition of perfect elasticity, is not proved by mere permanence of the extreme configurations assumed by the substance, when a stated amount of the stress is alternately applied and removed. This condition might be fulfilled, and yet the amount of elastic force might be different with the same palpable configuration of the body during gradual augmentation and during gradual diminution of the stress. That it is so in fact is proved by the discovery of viscosity referred to below (§ 31); but it is not yet proved that if, after increasing

the stress to a certain definite amount, the body is brought to rest in the same palpable configuration as before, the amounts of stress required to hold it in this configuration are different in the two cases. If they are, (§ 1) the elasticity is imperfect; if they are not, the elasticity is perfect within the limits of the experiment (compare § 36 below).

7. LIMITS OF ELASTICITY—*Elasticity of Shape.*—The degree of distortion within which elasticity of shape is found is essentially limited in every solid. Within sufficiently narrow limits of distortion every solid shows elasticity of shape to some degree—some solids, to perfection, so far as we know at present. When the distortion is too great, the body either breaks or receives a permanent bend—that is, such a molecular disturbance that it does not return to its original figure when the bending force is removed. If the first notable dereliction from perfectness of elasticity is a breakage, the body is called brittle,—if a permanent bend, plastic or malleable or ductile. The metals are generally ductile; some metals and metallic alloys and compounds of metals with small proportions of other substances, are brittle; some of them brittle only in certain states of temper, others it seems essentially brittle. The steel of before the days of Bessemer and Siemens is a remarkable instance. When slowly cooled from a bright red heat, it is remarkably tough and ductile. When heated to redness and cooled suddenly by being plunged in oil or water or mercury, it becomes exceedingly brittle and hard (glass-hard, as it is called), and to ordinary observation seems incapable of taking a permanent bend (though probably careful observation would prove it not quite so). The definition of steel used to be *approximately pure iron capable of being tempered glass-hard, and again softened to different degrees by different degrees of heat.* Now, the excellent qualities of iron made by Bessemer's and Siemens's processes are called steel, and are reckoned best when *incapable of being tempered glass-hard,* the possibility of brittleness supervening in the course of any treatment which the metal may meet with in its manufacture, being an objection against the use of what was formerly called steel for ship's plates, ribs, stringers, &c., and for many applications of land engineering, even if the material could be had in sufficient abundance.

8. LIMITS OF ELASTICITY (CONTINUED)—*Elasticity of Bulk.* —If we reckon by the amount of pressure, there is probably no

limit to the elasticity of bulk in the direction of increase of pressure for any solid or fluid; but whether continued augmentation produces continued diminution of bulk towards zero without limit, or whether for any or every solid or fluid there is a limit towards which it may be reduced in bulk, but smaller than which no degree of pressure, however great, can condense it, is a question which cannot be answered in the present state of science. Would any pressure, however tremendous, give to gold a density greater than 19·6, or to copper a density greater than 9·0, after the pressure is removed (§ 3 above)? Whether the body be fluid or a continuous non-porous solid, it probably recovers its original density, however tremendously it may have been pressed, and probably shows perfect elasticity of bulk (§ 3 above) through the whole range of positive pressure from zero to infinity, provided the pressure has been equal in all directions like fluid pressure. As for negative pressure, we have no knowledge of what limit, if any, there may be to the amount of force which can be applied to a body pulling its surface out equally in all directions. The question of how to apply the negative pressure is inextricably involved with that of the body's power to resist. The upper part of the mercury of a barometer, adhering to the glass above the level corresponding to the atmospheric pressure, is a familiar example of what is called negative pressure in liquids. Water and other transparent liquids show similar phenomena, one of which is the warming of water above its boiling-point in an open glass or metal vessel varnished with shellac. Attempts to produce great degrees of this so-called negative pressure, are baffled by what seems an instability of the equilibrium which supervenes when the negative pressure is too much augmented. It is a very interesting subject for experimental inquiry to find how high mercury or water or any other liquid can be got to stand above the level corresponding to the atmospheric pressure in a tall hermetically sealed tube, and by how many degrees a liquid can, with all precautions, be warmed above its boiling-point. In each case it seems to be by a minute bubble forming and expanding somewhere at the boundary of the liquid, where it is in contact with the containing vessel, that the possible range of the negative pressure is limited, judging from what we see when we carefully examine a transparent liquid, or the surface of separation between mercury and glass, in any such experiment.

The contrast of the amounts of negative pressure practically obtainable or obtained hitherto in such experiments on liquids (which are at the most those corresponding to the weight of a few metres of the substance), with that obtainable in the case of even the weakest solids, is remarkable; and as for the strongest, consider for instance (§ 22 below) 17 nautical miles of steel pianoforte wire hanging by one end. When a cord, or rod, or wire of any solid substance hangs vertically, the negative pressure (for example, 23,000 atmospheres in the case just cited) in any transverse section is equal to the weight of the part hanging below it. It is an interesting question not to be answered by any experiment easily made or even devised,—How much would the longitudinal pull which can be applied to a cord, rod, or wire without breaking it, be augmented (probably augmented, but possibly diminished) by lateral pull applied all round the sides so as to give equal negative pressure in all directions?

9. LIMITS OF ELASTICITY (CONTINUED)—*Elasticity of Shape for Distortions not Uniform through the Substance, and for Compound Distortions; and Elasticity corresponding to Co-existent Distortion and Change of Bulk:*—

Example 1.—A round wire twisted, or a cylindrical shaft transmitting revolutional motion in machinery, presents, as we shall see (§ 64 below), an instance of simple distortion, but to different degrees in different parts of the substance, increasing from the axis where it is zero, uniformly to the surface where it is greatest.

Example 2.—Elongation of a wire or rod by direct pull, is (§ 23 below) an instance of a compound distortion co-existing with a rarefaction of the substance, both distortion and rarefaction being uniform throughout.

Example 3.—Shortening of a column by end pressure is an instance of a similar compound distortion combined with condensation of the substance, both distortion and condensation being uniform throughout.

Example 4.—Flexure of a round wire or of a bar, or beam, or girder, of any shape of normal section, by opposite bending couples applied at the two ends, is an instance in which one-half of the substance is stretched, and the other half shortened with exactly the same combination of distortions and changes of bulk as in examples 2 and 3. The strain is uniform along the length of the bar, but varies in the cross section in simple proportion to distance from a certain line (§ 62 below) through the centre of gravity of the sectional area, which, in the case of a round bar, is the diameter perpendicular to the plane of curvature.

The limits of elasticity in the cases of these four examples are subjects of vital importance in practical mechanics, and a vast amount of careful and accurate observation and experiment, which has given much valuable practical information regarding them, has been gone through by engineers, in their necessary dealings with questions regarding strength of materials. Still there is great want of definite scientific information on the subject of limits of elasticity generally, and particularly on many elementary questions (§ 21 below), which force themselves upon us when we endeavour to analyze the molecular actions concerned in such cases as the four examples now before us. Some principles of much importance for guidance in practical as well as in theoretical deductions from observations and experiments on this subject, were set forth twenty-nine years ago by Professor James Thomson, in an article published in the *Cambridge and Dublin Mathematical Journal* for November 1848. Nothing is to be gained either in clearness or brevity by any other way of dealing with it than reproducing it *in extenso*. It is accordingly given here, with a few changes made in it with its author's concurrence.

It constitutes the following §§ 10—20.

"*On the strength of materials, as influenced by the existence or non-existence of certain mutual strains* among the particles composing them.* By James Thomson, M.A., College, Glasgow.

10. "My principal object in the following paper is to show that the absolute strength of any material composed of a substance possessing ductility (and few substances, if any, are entirely devoid of this property) may vary to a great extent, according to the state of tension or relaxation in which the particles have been made to exist when the material as a whole is subject to no external strain.

11. "Let, for instance, a round bar of malleable iron, or a piece of iron wire, be made red hot, and then be allowed to cool.

---

* [*Note added Nov.* 1877. More nearly what is now called *stress* than what is now called *strain* is meant by "strain" in this article, which was written before Rankine's introduction of the word stress, and distinct definition of the word strain (see chap. I. of Mathematical Theory below).]

Its particles may now be regarded as being all completely relaxed. Let, next, one end of the bar be fixed, and the other be made to revolve by torsion, till the particles at the circumference of the bar are strained to the utmost extent of which they can admit, without undergoing a permanent alteration in their mutual connexion*. In this condition, equal elements of the cross section of the bar afford resistances proportional to the distances of the elements from the centre of the bar; since the particles are displaced from their positions of relaxation through spaces which are proportional to the distances of the particles from the centre. The couple which the bar now resists, and which is equal to the sum of the couples due to the resistances of all the elements of the section, is that which is commonly assumed as the measure of the torsional strength of the bar. For future reference, this couple may be denoted by L, and the angle through which it has twisted the free end of the bar by Θ.

12. "The twisting of the bar may, however, be carried still farther, and during the progress of this process the outer particles will yield in virtue of their ductility, those towards the interior successively reaching their elastic limits, until, when the twisting has been sufficiently continued, all the particles in the section, except those quite close to the centre, have been strained beyond their elastic limits. Hence, if we suppose† that no change in the hardness of the substance composing the material has resulted from the sliding of its particles past one another, and that there-

* "I here assume the existence of a definite '*elastic limit*,' or a limit within which, if two particles of a substance be displaced, they will return to their original relative positions when the disturbing force is removed. The opposite conclusion, to which Mr Hodgkinson seems to have been led by some interesting experimental results, will be considered at a more advanced part of this paper."

† [*Note added October* 1877. This supposition may be true for some solids; it is certainly not true for solids generally. A piece of copper or of iron taken in a soft and unstrained condition certainly becomes "harder" when strained beyond its first limits of elasticity, that is to say, its limits of elasticity become wider; and a similar result will probably be found in ductile metals generally. Thus the resistance of the outer elements will be greater than those of the inner elements in the case described in the text, until the torsion has been pushed so far as to bring about the greatest hardness in all the elements at any considerable distance from the axis. It may be that before this condition has been attained the hardening of the outer elements will have been overdone, and they may have begun to lose strength, and to have become friable and fissured. The principle set forth in the text is not, however, vitiated by the incorrectness of a supposition introduced merely for the sake of numerical illustration.]

fore all small elements of the section of the bar afford the same resistance, no matter what their distances from the centre may be, it is easy to prove that the total torsional resistance of the bar is $\frac{4}{3}$ of what it was in the former case; or, according to the notation already adopted, it is* now $\frac{4}{3}L$.

13. "If, after this, all external stress be removed from the bar, it will assume a position of equilibrium, in which the outer particles will be strained in the direction opposite to that in which it was twisted, and the inner ones in the same direction as that of the twisting, the two sets of opposite couples thus produced among the particles of the bar balancing one another. It is easy to show that the line of separation between the particles strained in one direction and those in the other is a circle whose radius is $\frac{3}{4}$ of the radius of the bar. The particles in this line are evidently

* "To prove this, let r be the radius of the bar, η the utmost force of a unit of area of the section to resist a strain tending to make the particles slide past one another, or to resist a shearing strain, as it is commonly called. Also, let the section of the bar be supposed to be divided into an infinite number of concentric annular elements,—the radius of any one of these being denoted by x and its area by $2\pi x dx$.

"Now, when only the particles at the circumference are strained to the utmost, and when, therefore, the forces on equal areas of the various elements are proportional to the distances of the elements from the centre, we have $\eta \dfrac{x}{r}$ for the force of a unit of area at the distance of x from the centre. Hence the total tangential force of the element is

$$= 2\pi x dx \cdot \eta \, \frac{x}{r},$$

and the couple due to the same element is

$$= x \cdot 2\pi x dx \cdot \eta \, \frac{x}{r} = 2\pi\eta \, \frac{1}{r} \cdot x^3 dx \, ;$$

and therefore the total couple, which has been denoted above by L, is

$$= 2\pi\eta \, \frac{1}{r} \int_0^r x^3 dx,$$

that is $L = \tfrac{1}{2}\pi\eta r^3 \dots\dots\dots\dots\dots\dots\dots\dots\dots\dots(a).$

Next, when the bar has been twisted so much that all the particles in its section afford their utmost resistance, we have the total tangential force of the element $2\pi x dx \cdot \eta$, and the couple due to the same element

$$= x \cdot 2\pi x dx \cdot \eta = 2\pi\eta \cdot x^2 dx.$$

Hence the total couple due to the entire section is

$$= 2\pi\eta \int_0^r x^2 dx = \tfrac{2}{3}\pi\eta r^3.$$

But this quality is $\frac{4}{3}$ of the value of L in formula (a). That is, the couple which the bar resists in this case is $\frac{4}{3}L$, or $\frac{4}{3}$ of that which it resisted in the former case."

subject to no strain* when no external couple is applied. The
bar with its new molecular arrangement may now be subjected, *as
often as we please†*, to the couple $\frac{4}{5}L$ without undergoing any
farther alteration. Its strength to resist torsion, in the direction
of the couple L has therefore been considerably increased. Its
strength to resist torsion in the opposite direction has, however,
by the same process, been much diminished; for as soon as its
free extremity has been made to revolve backwards through an
angle‡ of $\frac{2}{5}\Theta$ from the position of equilibrium, the particles of
the circumference will have suffered the utmost distortion of
which they can admit without undergoing permanent alteration.
Now, it is easy to prove that the couple required to produce a
certain angle of torsion is the same in the new state of the bar
as in the old§. Hence the ultimate strength of the bar when
twisted backwards is represented by a couple amounting to only
$\frac{2}{5}L$. But, as we have seen, it is $\frac{4}{5}L$ when the wire is twisted
forwards. That is, then, *The wire in its new state has twice as
much strength to resist torsion in one direction as it has to resist it
in the other.*

14. " Principles quite similar to the foregoing, are applicable

* " Or at least they are subject to *no strain of torsion*, either in the one direction
or in the other; though they *may* be subject to a strain of compression or extension in the direction of the length of the bar." [That they are so is proved by
experiments made for the present article by Mr Thomas Gray in October 1877.]
" This, however, does not fall to be considered in the investigation of the text."

† " This statement, if not strictly, is at least extremely nearly true, since from
the experiments made by Mr Fairbairn and Mr Hodgkinson on cast-iron (see
various *Reports of the British Association*), we may conclude that the metals are
influenced only in an extremely slight degree by time. Were the bars composed
of some substance, such as sealing wax, or hard pitch, possessing a sensible amount
of viscidity, the statement in the text would not hold good."

‡ [*Note added October* 1877. This assumes that the limits of elasticity in a
substance which has been already strained beyond its limits of elasticity are equal
on the two sides of the shape which it has when in equilibrium without disturbing
force—a supposition which may be true or may not be true. Experiment is
urgently needed to test it; for its truth or falseness is a matter of much importance
in the theory of elasticity.]

§ " To prove this, let the bar be supposed to be divided into an infinite number
of elementary concentric tubes (like the so-called annual rings of growth in trees).
To twist each of these tubes through a certain angle, the same couple will be
required, whether the tube is already subject to the action of a couple of any
moderate amount in either direction or not. Hence, to twist them all, or, what
is the same thing, to twist the whole bar, through a certain angle, the same couple
will be required whether the various elementary tubes be or be not relaxed, when
the bar as a whole is free from external strain."

in regard to beams subjected to cross strain. As, however, my chief object at present is to point out the existence of such principles, to indicate the mode in which they are to be applied and to show their great practical importance in the determination of the strength of materials, I need not enter fully into their application in the case of cross strain. The investigation in this case closely resembles that in the case of torsion, but is more complicated on account of the different ultimate resistances afforded by any material to tension and to compression, and on account of the numerous varieties in the form of section of beams which for different purposes it is found advisable to adopt. I shall therefore merely make a few remarks on this subject.

15. "If a bent bar of wrought iron or other ductile material be straightened, its particles will thus be put into such a state that its strength to resist cross strain, in the direction towards which it has been straightened, will be very much greater than its strength to resist it in the opposite direction; each of these two resistances being entirely different from that which the same bar would afford were its particles all relaxed when the entire bar is free from external strain. The actual ratios of these various resistances depend on the comparative ultimate resistances afforded by the substance to compression and extension, and also, in a very material degree, on the form of the section of the bar. I may, however, state that in general the variations in the strength of a bar to resist cross strain, which are occasioned by variations in its molecular arrangement, are much greater even than those which have already been pointed out as occurring in the strength of bars subjected to torsion.

16. "What has already been stated is quite sufficient to account for many very discordant and perplexing results which have been arrived at by different experimenters on the strength of materials. It scarcely ever occurs that a material is presented to us, either for experiment or for application to a practical use, in which the particles are free from great mutual strains. Processes have already been pointed out by which we may at pleasure produce certain peculiar strains of this kind These, or other processes producing somewhat similar strains, are used in the manufacture of almost all materials. Thus, for instance, when malleable iron has received its final conformation by the process

termed *cold swaging*, that is, by hammering it till it is cold, the outer particles exist in a state of extreme compression, and the internal ones in a state of extreme tension. The same seems to be the case in cast iron when it is taken from the mould in which it has been cast. The outer portions have cooled first, and have therefore contracted, while the inner ones still continued expanded by heat. The inner ones then contract as they subsequently cool, and thus they, as it were, pull the outer ones together. That is, in the end the outer ones are in a state of compression and the inner ones in the opposite condition.

17. "The foregoing principles may serve to explain the true cause of an important fact observed by Mr Eaton Hodgkinson in his valuable researches in regard to the strength of cast iron (*Report of the British Association* for 1837, p. 362)*. He found, that, contrary to what had been previously supposed, a strain, however small in comparison to that which would occasion rupture, was sufficient to produce a *set*, or permanent change of form, in the beams on which he experimented. Now this is just what should be expected in accordance with the principles which I have brought forward; for if, for some of the causes already pointed out, various parts of a beam previously to the application of an external force have been strained to the utmost, when, by the application of such force, however small, they are still farther displaced from their positions of relaxation, they must necessarily undergo a permanent alteration in their connection with one another, an alteration permitted by the ductility of the material; or, in other words, the beam as a whole must take a set.

18. "In accordance with this explanation of the fact observed by Mr E. Hodgkinson, I do not think we are to conclude with him that 'the maxim of loading bodies within the elastic limit has no foundation in nature.' It appears to me that the defect of elasticity, which he has shown to occur even with very slight strains, exists only when the strain is applied for the first time; or, in other words, that if a beam has already been subjected to a considerable strain, it may again be subjected to any smaller strain in the same direction without its taking a set. It will readily be

---

* For further information regarding Mr Hodgkinson's views and experiments see his communications in the *Transactions of the Sections of the British Association* for the years 1843 (p. 23) and 1844 (p. 25), and a work by him, entitled *Experimental Researches on the Strength and other properties of Cast Iron*, 8vo. 1846.

seen, however, from Mr Hodgkinson's experiments, that the term
'elastic limit,' as commonly employed, is entirely vague, and must
tend to lead to erroneous results.

19. "The considerations adduced seem to me to show clearly
that there really exist *two elastic limits* for any material, between
which the displacements or deflexions, or what may in general be
termed the changes of form, must be confined, if we wish to avoid
giving the material a set, or, in the case of variable strains, if we
wish to avoid giving it a continuous succession of sets which would
gradually bring about its destruction; that these two elastic limits
are usually situated one on the one side and the other on the
opposite side of the position which the material assumes when
subject to no external strain, though they may be both on the
same side of this position of relaxation*; and that they may therefore with propriety be called the *superior* and the *inferior limit* of
the change of form of the material for the particular arrangement
which has been given to its particles; that these two limits are not
fixed for any given material, but that, if the change of form be
continued beyond either limit, two new limits will, by means of
an alteration in the arrangement of the particles of the material, be
given to it in place of those which it previously possessed; and
lastly, that the processes employed in the manufacture of materials
are usually such as to place the two limits in close contiguity with
one another, thus causing the material to take in the first instance
a set from any strain, however slight, while the interval which may
afterwards exist between the two limits, and also, as was before
stated, the actual position assumed by each of them, are determined by
the peculiar strains which are subsequently applied to the material.

20. "The introduction of new, though necessary, elements
into the consideration of the strength of materials may, on the one

* Thus if the section of a beam be of some such form as
that shown in either of the accompanying figures, the one rib
or the two ribs, as the case may be, being very weak in comparison to the thick part of the beam, it may readily occur
that the two elastic limits of deflexion may be situated both
on the same side of the position assumed by the beam when
free from external force. For if the beam has been supported at its extremities and loaded at its middle till the
rib AB has yielded by its ductility so as to make all its parti-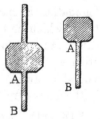
cles exert their utmost tension, and if the load be now gradually removed, the
particles at B may come to be compressed to the utmost before the load has been
entirely removed.

hand, seem annoying from rendering the investigations more complicated. On the other hand, their introduction will really have the effect of obviating difficulties, by removing erroneous modes of viewing the subject, and preventing contradictory or incongruous results from being obtained by theory and experiment. In all investigations, in fact, in which we desire to attain or to approach nearly to truth, we must take facts as they actually are, not as we might be tempted to wish them to be for enabling us to dispense with examining processes which are somewhat concealed and intricate but are not the less influential from their hidden character."

21. Passing now to homogeneous matter (§ 38 below), homogeneously strained (chap. II. of Math. Theory below), we are met by physical questions of great interest regarding limits of elasticity. Supposing the solid to be homogeneously distorted in any particular way to nearly the limit of its elasticity for this kind of distortion, will the limits be widened or narrowed by the superposition of negative or positive pressure equal in all directions producing a dilatation or a condensation? It seems probable that a dilatation would narrow the limits of elasticity, and a condensation widen them. This, however, is a mere guess: experiment alone can answer the question. Take again a somewhat less simple case. A wire is stretched by a weight to nearly its limits of longitudinal elasticity; a couple twisting it is applied to its lower end—Will this either cause the weight to run down and give the wire a permanent set, or break it? Probably,—yes; but experiment only can decide. The corresponding question with reference to a column loaded with a weight *may* have the same answer, but not necessarily so; experiment again is wanting. A wire hanging stretched by a light weight, merely to steady it, is twisted to nearly its limit of torsional elasticity by a couple of given magnitude applied to its lower end ; the stretching weight is increased—Will this cause it to yield to the couple and take a permanent set? Probably,—yes. [Certainly *yes*, for steel pianoforte wire experimented on by Mr M'Farlane to answer this question since it was first put in type for the present article.] If so, then the limits of torsional elasticity of a wire bearing a heavy weight are widened by diminishing or taking off the weight; and no doubt it will follow continuously that a column twisted

by opposing couples at its two ends will have its limits of torsional elasticity widened by the application of forces to its two ends, pressing them towards one another. Experiments to answer these questions would certainly reward the experimenter with definite and interesting results.

22. NARROWNESS OF LIMITS OF ELASTICITY.—*Solids.*—The limits of elasticity of metals, stones, crystals, woods, are so narrow that the distance between any two neighbouring points of the substance never alters by more than a small proportion of its own amount without the substance either breaking or experiencing a permanent set, and therefore the angle between two lines meeting in any point of the substance and passing always through the same matter is never altered by more than a small fraction of the radian*, before the body either breaks or takes a permanent set. By far the widest limits of elasticity hitherto discovered by experiment, for any substance except cork, india-rubber, jellies, are those of steel pianoforte wire. Take, for example, the pianoforte wire at present in use for deep-sea soundings. It is No. 22 of the Birmingham wire gauge, its density is 7·727, it weighs ·034 gramme per centimetre, or 6·298 kilogrammes per nautical mile of 1852·3 metres, and therefore its sectional area and diameter are ·00044 square centimetre and ·0244 centimetre. It bears a weight of 106 kilogrammes, which is equal in weight to about 31 kilometres of its length, and when this weight is alternately hung on and removed the length of the wire varies by $\frac{1}{8}$ of its amount. While this elongation takes place there is a lateral shrinking, as we shall see (§ 47 below), of from $\frac{1}{4}$ to $\frac{3}{10}$ of the same amount.

23. Consider now in the unstrained wire two lines through the substance of the wire at right angles to one another in any plane through or parallel to the axis of the wire in directions equally inclined to this line. When the wire is pulled lengthwise the two vertical angles bisected by the length of the wire become acute, and the other two obtuse by a small difference, as illustrated in the diagram (fig. 2), where the continuous lines represent a portion of the unpulled wire, and the dotted lines the same portion of the wire when pulled. The change in each of the angles would be $\frac{1}{86}$ of the radian in virtue of the elongation were

---

* The *radian* is the angle whose arc is equal to radius; it is equal to 57°·29......

there no lateral shrinking, and about $\frac{1}{330}$ of the radian in virtue
of the lateral shrinking were there no elonga-
tion. The whole change experienced by each of
the right angles is therefore actually (§ 37 below)
$\frac{1}{86} + \frac{1}{330}$, or about $\frac{1}{68}$ of the radian, or 0° ·84.
This is an extreme case. In all other cases of
metals, stones, glasses, crystals, the substance
either breaks or takes a permanent bend,
probably before it experiences any so great
angular distortion as a degree; and except in
the case of steel we may roughly regard the
limits of elasticity as being something between
$\frac{1}{1000}$ and $\frac{1}{100}$ in respect to the linear elongation or contraction, and from $\frac{1}{50}$ of a degree
to half a degree in respect to angular distortion.

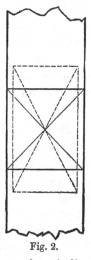

Fig. 2.

24. On the other hand, gelatinous substances, such as india-
rubber and elastic jellies, have very wide limits of elasticity. A
vulcanized india-rubber band, for instance, is capable of being
stretched, again and again, to eight times its length, and returning
always to nearly its previous condition when the stress is removed.
A shape of transparent jelly presents a beautiful instance of great
degrees of distortion with seemingly very perfect elasticity. All
these instances, india-rubber and jellies, show with great changes
of shape but slight changes of bulk. They have, in fact, all, as
nearly as experiment has hitherto been able to determine, the
same compressibility as water.

25. Cork, another body with very wide limits of elasticity
(very imperfect elasticity it is true) is singular, among bodies
seemingly homogeneous to the eye, in its remarkably easy compressibility. It is, in fact, the only seemingly homogeneous solid
which shows to the unaided eye any sensible change of bulk under
any practically applicable forces. A small homogeneous piece
torn out of a cork may, by merely pressing it between the fingers,
be readily compressed to half its bulk, and a large slab of cork
in a Bramah press may be compressed to $\frac{1}{10}$ of its bulk. An
ordinary bottle cork loaded with a small piece of metal presents a
very interesting appearance in an Oersted glass compressing
vessel; first floating, and when compressed to 20 or 30 atmo-

spheres sinking, and shrivelling in bulk very curiously; then on
the pressure being removed, expanding again, but not quite to
previous bulk, and floating up or remaining down according to the
amount of its load.

The divergences, presented by cork and gelatinous bodies,
in opposite directions from the regular elasticity of hard solids
form an interesting subject, to which we shall return later (§ 48
below).

26. *Liquids.*—In respect to liquids, there are no limits of
elasticity so far as regards the magnitude of the positive pressure
applied or conceivably applicable; but in respect to the magnitude
of negative pressure, and in respect to the magnitude of the
change of bulk, whether by negative or positive pressure, there
are probably very decided and not very wide limits. Thus water,
though condensed 1/11·5 of its bulk by 2000 atmospheres in
Perkins'* experiments, corrected roughly for the compression of
his glass "piezometer," which is very nearly at the rate of 1/21,000
per atmosphere found (§ 77 below) more accurately by subsequent
experiments for moderate pressures up to 20 or 30 atmospheres,
may be expected to be compressed by much less than 1/3 of its
volume under a pressure of 7000 atmospheres. How much it
or any other liquid is condensed by a pressure of 10,000 atmospheres, or by 20,000 atmospheres, is an interesting subject for
experimental investigation.

27. *Gases.*—In respect to rarefaction, and in respect to *proportionate* condensation, gases present enormously wider limits
of elasticity than any liquids or solids,—in fact no limit in respect
to dilatation, and in respect to condensation a definite limit only
when the gas is below Andrews's "critical temperature." If the
gas be kept at any temperature above that critical temperature,
it remains homogeneous, however much it be condensed; and
therefore for a fluid above the critical temperature there is, in
respect to magnitude of pressure, no superior limit to its elasticity.
On the other hand, if a fluid be kept at any constant temperature
less than its critical temperature, it remains homogeneous, and
presents an increasing pressure until a certain density is reached;

---

* *Transactions of Royal Society*, June 1826, "On the Progressive Compression
of Water by high degrees of force, with some trials of its effects on other liquids," by
J. Perkins. Communicated by W. H. Wollaston, M.D., V.P.R.S.

when its bulk is further diminished it divides into two parts of less and greater density (the part of less density being called vapour, that of greater density being called liquid, if it is not solid), and presents no further increase of pressure until the vaporous part shrinks to nothing, and the whole becomes liquid (that is to say, homogeneous fluid at the greater of the two densities) or else becomes solid—the question whether the more dense part is liquid or solid depending on the particular temperature below the critical temperature at which the whole substance is kept during the supposed experiment.

28. The thermo-dynamic reasoning of Professor James Thomson, which showed the effect of change of pressure in altering the freezing point of a liquid, leads to analogous considerations regarding the effect of continuous increase or continuous decrease of pressure upon a mass consisting of the same substance, partly in the liquid and partly in the solid state at one temperature. The three cases of transition from gas to liquid, from gas to solid, and from liquid to solid, present us with perfectly definite limits of elasticity,—the only perfectly definite limits of elasticity in nature of which we have any certain knowledge.

29. *Viscosity of Fluids and Solids.*—Closely connected with limits of elasticity, and with imperfectness of elasticity, is viscosity, that is to say, resistance to change of shape depending on the velocity of the change. The full discovery of the viscosity of liquids and gases is due originally to Stokes; and his hypothesis that in fluids the force of resistance is in simple proportion to the velocity of change of shape has been subsequently confirmed by the experimental investigations of Helmholtz, Maxwell, Meyer, Kundt, and Warburg. The definition of a fluid given in § 2 above may, by § 1, be transformed into the following:—A fluid is a body which requires no force to keep it in any particular shape, or—A fluid is a body which exercises no permanent resistance to a change of shape. The resistance to a change of shape presented by a fluid, evanescent as it is when the shape is not being changed (or vanishing when the velocity of the change vanishes), is essentially different from that permanent resistance to change of shape, the manifestation of which in solids constitutes elasticity of shape as defined in § 1 above. Maxwell's admirable kinetic theory of the viscosity of gases points to a full explanation of viscosity,

whether of gases, liquids, or solids, in the consideration of configurations and arrangements of relative motions of molecules,
permanent in a solid under distorting stress, and temporary in
fluids or solids while the shape is being changed, in virtue
of which elastic force in the quiescent solid, and viscous resistance to change of shape in the non-quiescent fluid or solid, are
produced.

30. *Viscosity of Metals and Fatigue of their Elasticity.*—
Experimental exercises performed by students in the physical
laboratory of the university of Glasgow, during the session
1864—65, brought to light some very remarkable and interesting
results, proving a loss of energy in elastic vibrators (sometimes as
much as two or three per cent. of energy lost in the course of a
single vibration in one direction) incomparably greater than anything that could be due to imperfections in their elasticity (§ 1),
and shewing also a very remarkable fatigue of elasticity, according
to which a wire which had been kept vibrating for several hours or
days through a certain range came to rest much quicker when left
to itself than when set in vibration after it had been at rest for
several days and then immediately left to itself. Thus it was
found that the rates of subsidence of the vibrations of the several
wires experimented on were generally much less rapid on the
Monday mornings, when they had been at rest since the previous
Friday, than on other days of the week, or than after several series
of experiments had been made on a Monday. The following statement (§§ 31—34) is extracted from the article, in the *Proceedings
of the Royal Society* for May 18, 1865*, containing some of the
results of these observations.

31. *Viscosity.*—By induction from a great variety of observed
phenomena, we are compelled to conclude that no change of volume
or of shape can be produced in any kind of matter without dissipation of energy. Even in dealing with the *absolutely perfect* elasticity of volume presented by every fluid, and possibly by some
solids—as for instance homogeneous crystals—dissipation of energy
is an inevitable result of every change of volume, because of the
accompanying change of temperature, and consequent dissipation
of heat by conduction or radiation. The same cause gives rise

* Referred to in preliminary note to Part I. The contents of this Paper are
embodied in the present article more particularly in §§ 31-34, 42-43 and 78 below.

necessarily to some degree of dissipation in connection with every change of shape of an elastic solid. But estimates founded on the thermodynamic theory of elastic solids, which I have given elsewhere*, have sufficed to prove that the loss of energy due to this cause is small in comparison with the whole loss of energy observed in many cases of vibration. I have also found, by vibrating a spring alternately in air of ordinary pressure and in the exhausted receiver of an air-pump, that there is an internal resistance to its motions immensely greater than the resistance of the air. The same conclusion is to be drawn from the observation made by Kupffer in his great work on the elasticity of metals, that his vibrating springs subsided much more rapidly in their vibrations than rigid pendulums supported on knife-edges. The subsidence of vibrations is probably more rapid in glass than in some of the most elastic metals, as copper, iron, silver, aluminium†; but it is much more rapid than in glass, marvellously rapid indeed, in some metals (as for instance zinc)‡, and in india-rubber, and even in homogeneous jellies.

32. The *frictional resistance* against change of shape must in every solid be infinitely small when the change of shape is made at an infinitely slow rate, since, if it were finite for an infinitely slow change of shape, there would be infinite rigidity, which we may be sure§ does not exist in nature. Hence there is in elastic solids a *molecular friction* which may be properly called *viscosity of solids*, because, as being an internal resistance to change of shape depending on the rapidity of the change, it must be classed with fluid molecular friction, which by general consent is called *viscosity of fluids*. But, at the same time, it ought to be remarked that the word viscosity, as used hitherto by the best writers, when solids or heterogeneous semi-solid semi-fluid masses are referred to, has not been distinctly applied to molecular friction, especially not to the molecular friction of a highly elastic solid within its limits of high

---

* "On the Thermo-elastic Properties of Solids," *Quarterly Journal of Mathematics*, April, 1855 (Art. XLVIII. Part VII. Vol. I. above).

† We have no evidence that the precious metals are more elastic than copper, iron, or brass. One of the new bronze pennies gives quite as clear a ring as a two-shilling silver piece tested in the usual manner.

‡ Torsional vibrations of a weight hung on a zinc wire subside so rapidly, that it has been found scarcely possible to count more than twenty of them in one case experimented on.

§ Those who believe in the existence of indivisible, infinitely strong and infinitely rigid, very small bodies (finite hard atoms!) deny this.

elasticity, but has rather been employed to designate a property of
slow continual yielding through very great, or altogether unlimited,
extent of change of shape, under the action of continued stress.
It is in this sense that Forbes, for instance, has used the word in
stating that 'viscous theory of glacial motion,' which he demonstrated by his grand observations on glaciers. As, however, he
and many other writers after him have used the words plasticity
and plastic, both with reference to homogeneous solids (such as
wax or pitch even though also brittle, soft metals, &c.) and to
heterogeneous semi-solid semi-fluid masses (as mud, moist earth,
mortar, glacial ice, &c.), to designate the property common to all
those cases of experiencing, under continued stress, either quite
continued and unlimited change of shape, or gradually very great
change at a diminishing (asymptotic) rate through infinite time,
and as the use of the term *plasticity* implies, no more than does
viscosity, any physical theory or explanation of the property, the
word viscosity is without inconvenience left available for the
definition I propose.

33. To investigate the viscosity of metals, I have in the first place
taken them in the form of round wires, and have chosen torsional
vibrations, after the manner of Coulomb, for observation, as being
much the easiest way to arrive at definite results. In every case
one end of the wire was attached to a rigid vibrator with sufficient
firmness (thorough and smooth soldering I find to be always the
best plan when the wire is thick enough); and the other to a
fixed rigid body, from which the wire hangs, bearing the vibrator
at its lower end. I arranged sets of observations to be made for
the separate comparison of the following cases:—

(*a*) The same wire with different vibrators of equal weights,
to give equal stretching-tractions but different moments of inertia
(to test the relation between viscous resistances against motions
with different velocities, through the same range and under the
same stress).

(*b*) The same wire with different vibrators of equal
moments of inertia but unequal weights (to test the effect of
different longitudinal tractions on the viscous resistance to torsion
under circumstances similar in all other respects).

(*c*) The same wire and the same vibrator, but different
initial ranges in successive experiments (to test an effect un-

expectedly discovered, by which the subsidence of vibrations from any amplitude takes place at very different rates according to the immediately previous molecular condition, whether of quiescence or of recurring changes of shape through a wider range).

(d) Two equal and similar wires, with equal and similar vibrators, one of them kept as continually as possible in a state of vibration, from day to day; the other kept at rest, except when vibrated in an experiment once a day (to test the effect of continued vibration on the viscosity of a metal).

34. *Results.*—(a) It was found that the loss of energy in a single vibration through one range was greater the greater the velocity (within the limits of the experiments); but the difference between the losses at low and high speeds was *much less* than it would have been had the resistance been, as Stokes has proved it to be in fluid friction, approximately as the rapidity of the change of shape. The irregularities in the results of the experiments which up to this time I have made, seem to prove that much smaller vibrations (producing less absolute amounts of distortion in the parts of the wires most stressed) must be observed, before any simple law of relation between molecular friction and velocity can be discovered.

(b) When the weight was increased, the viscosity was always at first much increased; but then day after day it gradually diminished and became as small in amount as it had been with the lighter weight. It has not yet been practicable to continue the experiments long enough in any case to find the limit to this variation.

(c) The vibration subsided in aluminium wires much more rapidly from amplitude 20 to amplitude 10, when the initial amplitude was 40, than when it was 20. Thus, with a certain aluminium wire, and vibrator No. 1 (time of vibration one way 1·757 second), the number of vibrations counted were in three trials—

	Vibrations.
Subsidence from 40 initial amplitude to 20	56 64 64
And from 20 (in course of the same experiments) to 10	96 98 96
The same wire and the same vibrator showed—	
Subsidence from 20 initial amplitude to 10 (average of four trials)	112 vibrations.

Again, the same wire, with vibrator No. 2* (time of vibration one way 1·236), showed in two trials—

Vibrations.

Subsidence from 40 initial amplitude to 20 .... 54 52
And continued from 20 to 10... 90 90

Again, same wire and vibrator,—

From initial amplitude 20 to 10 . . 103 (mean of eight trials). This remarkable result suggested the question (d).

(d) In a wire which was kept vibrating nearly all day, from day to day, after several days very much more molecular friction was found than in another kept quiescent except during each experiment. Thus two equal and similar pieces of copper wire were put up about the 26th of April, hanging with equal and similar lead weights, the upper and lower ends of the two wires being similarly fixed by soldering. No. 2 was more frequently vibrated than No. 1 for a few days at first, but no comparison of viscosities was made till May 15. Then

No. 1 subsided from 20 initial range to 10 in 97 vibrations.
No. 2 gave the same subsidence in 77 vibrations.

During the greater part of May 16 and 17, No. 2 was kept vibrating and No. 1 quiescent, and late on May 17 experiments with the following results were made :—

Time per
Vibration.

No. 1 subsided from 20 to 10 after	99	vibrations in	237 secs.,	2·4
,,	,,	,,	98	,,	235 ,,	2·4
,,	,,	,,	98	,,	235 ,,	2·4
No. 2 subsided from 20 to 10 after	58	vibrations in	142 ,,	2·45
,,	,,	,,	60	,,	147 ,,	2·45
,,	,,	,,	57	,,	139 ,,	2·45
,,	,,	,,	60	,,	147 ,,	2·45

[Note of May 27.—No. 1 has been kept at rest from May 17, while No. 2 has been kept oscillating more or less every day till yesterday, May 26, when both were oscillated, with the following results :—

Time per
Vibration.

No. 1 subsided from 20 to 10 after 100 vibrations in 242 secs., 2·42
No. 2   ,,   ,,   44 or 45 vibrations......... 2·495.]

35. The investigation was continued with much smaller degrees of maximum angular distortion, to discover, if possible, the law of the molecular friction, the existence of which was demonstrated by these experiments. Two questions immediately occurred :—

* Of same weight as No. 1, but different moment of inertia.

What is the law of subsidence of range in any single series of oscillations, the vibrator being undisturbed by external force ? and (question (a) of § 33 above) what is the relation between the law of subsidence in two sets of oscillations having different periods, with the same elastic body in the same circumstances of elastic force, as for instance the same or similar metallic wires with equal weights hung upon them, performing torsional oscillations in different times on account of the moments of inertia of the suspended masses being different ?

36. So far as the irregularities depending on previous conditions of the elastic substance allowed any simple law to be indicated, the experimental answer to the first question for degrees of angular distortion much smaller than the palpable limits of elasticity, was the COMPOUND INTEREST LAW, that is to say,—*The diminutions of range per equal intervals of time, or per equal numbers of oscillations, bore a constant proportion to the diminishing range; or, The differences of the logarithms of the ranges were proportional to the intervals of time.*

The only approach to an answer to the second question yet obtained is, that the proportionate losses of amplitude in the different cases are not such as they would be if the molecular resistance were simply proportional to the velocity of change of shape in the different cases. If the molecular friction followed this simple law, the proportionate diminutions of range per period would be inversely as the periods, or per equal intervals of time they would be inversely as the squares of the periods. Instead of the proportion being so, the loss was greater with the longer periods than that calculated, according to the law of square roots, from its amount in the shorter periods. It was in fact as it would be if the result were wholly or partially due to imperfect elasticity, or "elastische Nach-wirkung"—elastic after-working—as the Germans call it (compare § 6 above). To form a rough idea of the results, irrespectively of the ultimate molecular theory (which is to be looked for in the proper extension of Maxwell's kinetic theory of viscosity of gases), consider a perfectly elastic vesicular solid, whether like a sponge with communications between the vesicles, or with each vesicle separately inclosed in elastic solid: imagine its pores and interstices filled up with a viscous fluid, such as oil. Static experiments on such a solid will show perfect elasticity of bulk and shape ; kinetic experiments will show losses of energy such as are really shown by vibrators of india-rubber, jelly, glass, metals, or other elastic homogeneous solids, but more regular, and following more closely the compound interest law for single series, and the law of relation to squares of periods stated above for sets of oscillations in different periods. In short, according to Stokes's law of viscosity of fluids, our supposed vesicular vibrator would follow the law of subsidence of a simple vibrator experiencing a resistance simply proportional to the velocity of its motion, while no such simple law is applicable to the effects of the internal molecular resistance in a vibrating elastic solid.

37. *Hooke's Law.*—A law expressed by Hooke with Latin terseness in the words *Ut tensio sic vis* is the foundation of the mathematical theory of the elasticity of hard solids. By *tensio* here is meant not force (as is generally meant by the English word tension), but an elongation produced by force. In English, then, Hooke's law is that elongation (understood of an elastic solid) is proportional to the force producing it. It is, of course, to be extended continuously from elongation to contraction in respect to the effect, and from pull to push in respect to the cause; and the experiments on which it is founded prove a perfect continuity from a pulling force to a smaller force in the same direction, and from the less force to zero, and from zero of pulling force to different degrees of push or positive pressure, or negative pull. Experimental proof merely of the *continuity of the phenomena through zero of force* suffices to show that, for *infinitely small* positive or negative pulls, positive or negative elongation is simply proportional to the positive or negative pull; or, in other words, positive or negative contraction is proportional to the positive or negative pressure producing it. But now must be invoked minutely accurate experimental measurement to find how nearly the law of simple proportionality holds through finite ranges of contraction and elongation. The answer happily for mathematicians and engineers is *that Hooke's law is fulfilled, as accurately as any experiments hitherto made can tell,* for all metals and hard solids each through the whole range within its limits of elasticity; and for woods, cork, india-rubber, jellies, when the elongation is not more than two or three per cent., or the angular distortion not more than a few hundredths of the radian (or not more than about two or three degrees). The same law holds for the condensation of liquids up to the highest pressures under which their compressi-

bility has hitherto been accurately measured. [A decided but small deviation from Hooke's law has been found, in steel piano-forte wire under combined influence of torsion and longitudinal pull, by Mr M'Farlane in experiments (§ 81 below) made for the present article.]

Boyle's law of the "spring of air" shows that the augmentation of density of a gas is simply proportional to the augmentation of the pressure, through the very wide ranges of pressure through which that law is approximately enough fulfilled. Hence the infinitesimal diminution of volume produced by a given infinitesimal augmentation of pressure varies as the square of the volume, and the *proportionate* diminution of volume (that is to say, the ratio of the diminution of volume to the volume) is proportional to the volume, or inversely proportional to the density. Andrews' experiments on the compressibility of a fluid, such as carbonic acid, at temperatures slightly above the critical temperature, and of gas or vapour, and liquid into which it divides itself at temperatures slightly below the critical temperature, are intensely interesting, not merely in respect to the *natural history* of elasticity, but as opening vistas into the philosophy of molecular action.

We cannot expect to find any law of simple proportionality between stress and change of dimensions, or proportionate change of dimensions, in the case of any elastic or semi-elastic " soft " solids, such as cork on the one hand or india-rubber or jellies on the other, when strained to large angular distortions, or to large proportionate changes of dimensions. The exceedingly imperfect elasticity of all these solids, and the want of definiteness of the substance of many of them, renders accurate experimenting un-available for obtaining any very definite or consistent numerical results; but it is interesting to observe roughly the forces required to produce some of the great strains of which they are capable without any total break down of elastic quality ; for instance, to hang weights successively on an india-rubber band and measure the elongations. This any one may readily do, and may be sur-prised to find the enormous increase of resistance to elongation presented by the attenuated band before it breaks.

38. *Homogeneousness defined.*—A body is called homogeneous when any two equal, similar parts of it, with corresponding lines parallel and turned towards the same parts, are undistinguishable

from one another by any difference in quality. The perfect fulfilment of this condition, without any limit as to the smallness of
the parts, though conceivable, is not generally regarded as probable, for any of the real solids or fluids known to us, however
seemingly homogeneous. It is held by all naturalists that there
is a *molecular structure*, according to which, in *compound* bodies
such as water, ice, rock-crystal, &c., the constituent substances
lie side by side, or are arranged in groups of finite dimensions, and
even in bodies called *simple* (that is those not known to be
chemically resolvable into other substances) there is no ultimate
homogeneousness. In other words, the prevailing belief is that
every kind of matter with which we are acquainted has a more
or less *coarse-grained* texture; whether having visible molecules
(as great masses of solid brick-work or stone-building, or as
natural sandstone or granite-rocks) or having molecules too small
to be directly visible or measurable, but not *undiscoverably* small
(as seemingly homogeneous metals, or continuous crystals, or
liquids, or gases),—really, it is to be believed, of dimensions to
be accurately determined in future advances of science. Practically
the definition of *homogeneousness* may be applied on a very large
scale to masses of building or to coarse-grained conglomerate rock,
or on a more moderate scale to blocks of common sandstone,
or on a very small scale to seemingly homogeneous metals*; or
on a scale of extreme, undiscovered fineness, to vitreous bodies,
continuous crystals, solidified gums, as india-rubber, gum-arabic,
&c., and fluids.

39. *Isotropic and Æolotropic Substances defined.* The substance of a homogeneous solid is called *isotropic*† when a spherical
portion of it, tested by any physical agency, exhibits no difference
in quality however it is turned. Or, which amounts to the same,
a cubical portion, cut from any position in an isotropic body,
exhibits the same qualities relatively to each pair of parallel
faces. Or two equal and similar portions cut from *any* positions
in the body, not subject to the condition of parallelism (§ 38
above), are undistinguishable from one another. A substance

---

* Which, however, we know, as proved by Deville and Van Troost, are porous
enough at high temperatures to allow very free percolation of gases. Helmholtz and
Root find percolation of platinum by hydrogen at ordinary temperature (*Berl.
Sitzungsbericht*).

† Thomson and Tait's *Natural Philosophy*, § 676.

which is not isotropic, but exhibits differences of quality in different directions, is called *æolotropic*. The remarks of § 38 above relative to homogeneousness in the aggregate, and the supposed ultimately heterogeneous texture of all substances, however seemingly homogeneous, indicate corresponding limitations and non-rigorous practical interpretations of isotropy and æolotropy.

40. *Isotropy and Æolotropy of different sets of properties.*— The substance of a homogeneous solid may be isotropic in one quality or class of qualities, but æolotropic in others. Or a transparent substance may transmit light at different velocities in different directions through it (that is, be *doubly-refracting*), and yet a cube of it may (and does in many natural crystals) show no sensible difference in its absorption of white light transmitted across it perpendicularly to any of its three pairs of faces. Or (as a crystal which exhibits *dichroism*) it may be sensibly æolotropic relatively to the absorption of light, but not sensibly double-refracting, or it may be dichroic and doubly-refracting, and yet it may conduct heat equally in all directions. Still, as a rule, a homogeneous substance which is æolotropic for one quality, must be more than infinitesimally æolotropic for every quality which has directional character admitting of a corresponding æolotropy.

41. *Moduluses of Elasticity.*—A modulus of Elasticity is the number obtained by dividing the number expressing a stress* by the number expressing the strain† which it produces. A modulus is called a principal modulus when the stress is such that it produces a strain of its own type.

(1) An isotropic solid has two principal moduluses—a *modulus of compression* and a *modulus of rigidity*.

(2) A crystal of the cubic class (fluor-spar, for instance) has three principal moduluses,—*one modulus of compression* and *two rigidities*.

(3) An æolotropic solid having (what *no natural crystal* has, but what a *drawn wire* has) perfect isotropy of physical qualities relative to all lines perpendicular to a certain axis of its substance has three principal moduluses,—*two* determinable from its different compressibilities along and perpendicular to the axis, or from one compressibility and the "Young's modulus" (§ 42 below) of an

---

* Mathematical Theory, below, chap. I.   † *Ibid.*

axial bar of the substance, or determinable from two compressibilities; and *one rigidity* determinable by measurement of the
torsional rigidity of a round axial bar of the substance.

(4) A crystal of Iceland spar has four principal moduluses,—
three like those of case (3), and another rigidity depending on
(want of complete circular symmetry, and) possession of *triple*
symmetry of form, involving sextuple elastic symmetry, round
the crystalline axis.

(5) A crystal of the rectangular parallelepiped (or "tesseral")
class has six distinct principal moduluses which, when the directions of the principal axes are known, are determinable by six
single observations,—three, of the three (generally unequal) compressibilities along the three axes; and three, of the three rigidities
(no doubt generally unequal) relatively to the three simple distortions of the parallelepiped, in any one of which one pair of
parallel rectangular faces of the parallelepiped become oblique
parallelograms.

(6) An æolotropic solid generally has six principal moduluses*,
which, when a piece of the solid is presented without information,
and without any sure indication from its appearance of any particular axis or axes of symmetry of any kind, require just twenty-
one independent observations for the determination of the fifteen
quantities specifying their types, and the six numerical values of
the moduluses themselves.

42. *"Young's Modulus,"* or *Modulus of Simple Longitudinal
Stress.*—Thomas Young called *the modulus of elasticity* of an elastic
solid the amount of the end-pull or end-thrust required to produce
any infinitesimal elongation or contraction of a wire, or bar, or
column of the substance, multiplied by the ratio of its length to
the elongation or contraction. In this definition the definite
article is clearly misapplied. There are, as we have seen, two
moduluses of elasticity for an isotropic solid,—one measuring
elasticity of bulk, the other measuring elasticity of shape. An
interesting and instructive illustration of the confusion of ideas so
often rising in physical science from faulty logic is to be found in
"An Account of an Experiment on the Elasticity of Ice: By
Benjamin Bevan, Esq., in a letter to Dr Thomas Young, Foreign
Sec. R. S." and in Young's "Note" upon it, both published in the

* Mathematical Theory, chap. XVI.

Transactions of the Royal Society for 1826.   Bevan gives an interesting account of a well-designed and well-executed experiment on the flexure of a bar, 3·97 inches thick, 10 inches broad, and 100 inches long, of ice on a pond near Leighton Buzzard (the bar remaining attached by one end to the rest of the ice, but being cut free by a saw along its sides and across its other end), by which he obtained a fairly accurate determination of "the modulus of ice* (his result was 21,000,000 ft.);" and says that he repeated the experiment in various ways on ice bars of various dimensions, some remaining attached by one end, others completely detached, and found results agreeing with the first as nearly "as the admeasurement of the thickness could be ascertained." He then proceeds to compare "the modulus of ice" which he had thus found with "the modulus of water," which he quotes from Young's *Lectures* as deduced from Canton's experiments on the compressibility of water.   Young in his "Note" does *not* point out that the two moduluses were essentially different, and that *the modulus of his definition*, the modulus determinable from the flexure of a bar, is essentially zero for every fluid.   We now call "Young's modulus" the particular modulus of elasticity defined as above by Young, and so avoid all confusion.

43. *Modulus of Rigidity.*—The "modulus of rigidity" of an isotropic solid is the amount of tangential stress divided by the deformation it produces,—the former being measured in units of force per unit of the area to which it is applied in the manner indicated by the annexed diagram (fig. 3), and the latter by the variation of each of the four right angles reckoned in fraction of the radian. By drawing either diagonal of the square in the diagram we see that the distorting stress represented by it gives rise to a normal traction on every surface of the substance perpendicular to the square and parallel to one of its diagonals, and an equal normal *pressure* on every surface of the solid perpendicular to the square and parallel to the other diagonal; and that the amount of

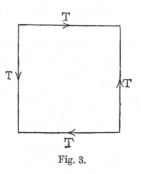

Fig. 3.

* See table of Moduluses, § 77, below.

each of these normal forces* per unit of area is equal to the
amount per unit area of the tangential forces which the diagram
indicates. The corresponding† geometrical proposition, also easily
proved, is as follows : A strain compounded of a simple *extension*
in one set of parallels, and a simple contraction of equal amount
in any other set perpendicular to those, is the same as a simple
shear in either of the two sets of planes cutting the two sets of
parallels at 45°, and the numerical value of this shear or simple
distortion is equal to *double* the amount of the elongation or
contraction, each reckoned per unit of length.

Hence we have another definition of "modulus of rigidity"
equivalent to the preceding :—The modulus of rigidity of an
isotropic substance is the amount of normal traction or pressure
per unit of area, divided by *twice* the amount of elongation in
the direction of the traction or of contraction in the direction of
the pressure, when a piece of the substance is subjected to a
stress producing uniform distortion.

44‡. *Conditions fulfilled in Elastic Isotropy.*—To be elastically
isotropic, a spherical or cubical portion of any solid, if subjected
to uniform normal pressure (positive or negative) all round, must,
in yielding, experience no deformation, and therefore must be
equally compressed (or dilated) in all directions. But, further, a
cube cut from any position in it, and acted on by *tangential* or
distorting stress in planes parallel to two pairs of its sides, must
experience simple deformation, or "shear" parallel to either
pair of these sides, unaccompanied by condensation or dilatation§,
and the same in amount for all the three ways in which a stress

* The directions of these forces are called the "axes" of the stress. The corresponding directions in the corresponding strain are called the axes of the strain.
   † Mathematical Theory, below, chap. VI.
   ‡ This, with several of the following sections, 44—51, is borrowed, with but
slight change, from the second edition of Thomson and Tait's *Natural Philosophy*
(§§ 679-694), by permission of the authors.
   § It must be remembered that the changes of figure and volume we are concerned
with are so small that the principle of superposition is applicable ; so that if
any distorting stress produced a condensation, an opposite distorting stress would
produce a dilatation, which is a violation of the isotropic condition. But it is
possible that a shearing stress may produce, in a truly isotropic solid, condensation or dilatation in proportion to the square of its value : and it is probable that
such effects may be sensible in india-rubber, or cork, or other bodies susceptible of
great deformations or compressions with persistent elasticity.

may be thus applied to any one cube, and for different cubes taken from any different positions in the solid. Hence the elastic quality of a perfectly elastic, homogeneous, isotropic solid is fully defined by two elements,—its resistance to distortion and its resistance to compression. The first has been already considered (§ 43). The second is measured by the amount of uniform pressure in all directions per unit area of its surface required to produce a stated very small compression. The numerical measure of the second is the compressing pressure divided by the diminution of the bulk of a portion of the substance which, when uncompressed, occupies the unit volume. It is sometimes called the "*elasticity of bulk*," or sometimes the "*modulus of bulk-elasticity*," sometimes the *resistance to compression*. Its reciprocal, or the amount of compression on unit of volume divided by the compressing pressure, or, as we may conveniently say, the compression per unit of volume per unit of compressing pressure, is commonly called the *compressibility*.

45. *Strain produced by a single Longitudinal Stress (subject of Young's Modulus).* Any stress whatever may* be made up of simple longitudinal stresses. Hence, to find the relation between any stress and the strain produced by it, we have only to find the strain produced by a single longitudinal stress, which, for an isotropic solid, we may do at once thus:—A simple longitudinal stress P is equivalent to a uniform dilating tension $\frac{1}{3}P$ in all directions, compounded with two distorting stresses, each equal to $\frac{1}{3}P$, and having a common axis in the line of the given longitudinal stress, and their other two axes any two lines at right angles to one another and to it. The diagram (fig. 4), drawn in a plane through one of these latter lines and the former, sufficiently indicates the synthesis,—the only forces not shown being those perpendicular to its plane.

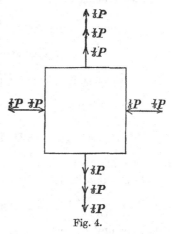

Fig. 4.

* Mathematical Theory, below, chap. VIII.

Hence if n denote the *rigidity*, and k the *modulus of compression*, or the *modulus of bulk-elasticity* (being the same as the reciprocal of the compressibility), the effect will be an equal dilatation in all directions, amounting, per unit of volume, to

$$\frac{\frac{1}{3}P}{k} \quad\text{..(1),}$$

compounded with two equal distortions, each amounting to

$$\frac{\frac{1}{3}P}{n} \quad\text{..(2),}$$

and having (§ 43, footnote) their axes in the directions just stated for the axes of the distorting stresses.

46.   The dilatation and two shears thus determined may be conveniently reduced to simple longitudinal strains by following the indications of § 43, thus:—

The two shears together constitute an elongation amounting to $\frac{1}{3}P/n$ in the direction of the given force P, and equal contraction amounting to $\frac{1}{6}P/n$ in all directions perpendicular to it.   And the cubic dilatation $\frac{1}{3}P/k$ implies a lineal dilatation, equal in all directions, amounting to $\frac{1}{9}P/k$.   On the whole, therefore, we have

$$\left.\begin{array}{l} \text{linear elongation} = P\left(\dfrac{1}{3n} + \dfrac{1}{9k}\right)\text{, in the direction} \\[4pt] \text{of the applied stress, and} \\[4pt] \text{linear contraction} = P\left(\dfrac{1}{6n} - \dfrac{1}{9k}\right)\text{, in all directions} \\[4pt] \text{perpendicular to the applied stress.} \end{array}\right\}...(3).$$

47.   Hence " Young's Modulus " $= \dfrac{9nk}{3k + n}$, and when the ends of a column, bar, or wire of isotropic material are acted on by equal and opposite forces, it experiences a lateral lineal contraction equal to $\dfrac{3k - 2n}{2(3k + n)}$ of the longitudinal dilatation, each reckoned as usual per unit of lineal measure.   One specimen of the fallacious mathematics referred to in chap. XVI. of the Mathematical Theory below is a celebrated conclusion of Navier's and Poisson's, that the ratio of lateral contraction to elongation by pull without transverse force is 1/4.   This would require the rigidity to be 3/5 of the resistance to compression, for all solids; which was first shown

to be false by Stokes* from many obvious observations, proving enormous discrepancies from it in many well-known bodies, and rendering it most improbable that there is any approach to a constancy of ratio between rigidity and resistance to compression in any class of solids. Thus clear elastic jellies and india-rubber present familiar specimens of isotropic homogeneous solids which while differing very much from one another in rigidity ("stiffness"), are probably all of very nearly the same compressibility as water, which is about $\frac{1}{21000}$ per atmosphere. Their resistance to compression, measured by the reciprocal of this, is obviously many hundred times the absolute amount of the rigidity of the stiffest of those substances. A column of any of them, therefore, when pressed together or pulled out, within its limits of elasticity, by balancing forces applied to its ends (or an india-rubber band when pulled out), experiences no sensible change of volume, though very sensible change of length. Hence the proportionate extension or contraction of any transverse diameter must be sensibly equal to half the longitudinal contraction or extension; and such substances may be practically regarded as incompressible elastic solids in interpreting all the phenomena for which they are most remarkable. Stokes gave reasons for believing that metals also have in general greater resistance to compression, in proportion to their rigidities, than according to the fallacious theory, although for them the discrepancy is very much less than for the gelatinous bodies. This probable conclusion was soon experimentally demonstrated by Wertheim, who found the ratio of lateral to longitudinal change of lineal dimensions, in columns acted on solely by longitudinal force, to be about $\frac{1}{3}$ for glass and brass; and by Kirchhoff, who, by a well-devised experimental method, found ·387 as the value of that ratio for brass, and ·294 for iron. For copper it is shown to lie between ·226 and ·441, by experiments quoted below (§§ 78, 81), measuring the torsional and longitudinal rigidities of copper wires.

48. All these results indicate rigidity *less* in proportion to the compressibility than according to Navier's and Poisson's theory.

---

* "On the Friction of Fluids in Motion, and the Equilibrium and Motion of Elastic Solids," *Trans. Camb. Phil. Soc.*, April 1845. See also *Camb. and Dub. Math. Journ.*, March 1848 [or Vol. I. of "*Mathematical and Physical Papers*," by G. G. Stokes].

And it has been supposed by many naturalists who have seen the necessity of abandoning that theory as inapplicable to ordinary solids, that it may be regarded as the proper theory for an ideal *perfect solid*, and as indicating an amount of rigidity not quite reached in any real substance, but approached to in some of the most rigid of natural solids (as for instance, iron). But it is scarcely possible to hold a piece of cork in the hand without perceiving the fallaciousness of this last attempt to maintain a theory which never had any good foundation. By careful measurements on columns of cork of various forms (among them, cylindrical pieces cut in the ordinary way for bottles), before and after compressing them longitudinally in a Bramah's press, we have found that the change of lateral dimensions is insensible both with small longitudinal contractions and return dilatations, within the limits of elasticity, and with such enormous longitudinal contractions as to $\frac{1}{6}$ or $\frac{1}{8}$ of the original length. It is thus proved decisively that cork is much more rigid, while metals, glass, and gelatinous bodies are all less rigid, in proportion to resistance to compression, than the supposed "perfect solid;" and the practical invalidity of the theory is experimentally demonstrated. By obvious mechanism of jointed bars a solid may be designed which shall swell laterally when pulled, and shrink laterally when compressed, in one direction, and which shall be homogeneous in the same sense (§ 40 above) as crystals and liquids are called homogeneous.

49. *Modulus of Simple Longitudinal Strain.*—In §§ 45, 46 above, we examined the effect of a simple longitudinal stress in producing elongation in its own direction, and contraction in lines perpendicular to it. With stresses substituted for strains, and strains for stresses, we may apply the same process to investigate the longitudinal and lateral tractions required to produce a simple longitudinal strain, (that is, an elongation in one direction, with no change of dimensions perpendicular to it) in a rod or solid of any shape.

Thus a simple longitudinal strain e, is equivalent to a cubic dilatation e without change of figure (or lineal dilatation $1/3.e$ equal in all directions), and two distortions consisting each of dilatation $1/3.e$ in the given direction and contraction $1/3.e$ in each of two directions perpendicular to it and to one another. To produce the cubic dilatation e alone requires (§ 44 above) a normal traction ke equal in all directions. And, to produce either of the distortions simply, since the measure (§ 43 above) of each is $2/3.e$, requires a distorting stress equal to $n \times \frac{2}{3}e$, which consists of tangential tractions each equal to this amount, positive (or drawing outwards) in the line of the given elongation, and negative (or pressing inwards) in the perpendicular direction. Thus we have in all

$$\left.\begin{array}{l} \text{normal traction} = (k + \tfrac{4}{3}n)e, \text{ in the direction of the} \\ \qquad \text{given strain, and} \\ \text{normal traction} = (k - \tfrac{2}{3}n)e, \text{ in every direction per-} \\ \qquad \text{pendicular to the given strain} \end{array}\right\} \dots (4).$$

Hence the modulus of simple longitudinal strain is $k + \frac{4}{3}n$.

50. *Weight-Modulus and Length of Modulus.*—Instead of reckoning moduluses in units of force per unit of area, it is sometimes convenient to express them in terms of the weight of unit bulk of the solid. A modulus thus reckoned, or, as it is called by some writers, the length of the modulus, is of course found by dividing the weight-modulus by the weight of the unit bulk. It is useful in many applications of the theory of elasticity, as, for instance, in this result, which is proved in the elementary dynamics of waves in an elastic solid or fluid (see chap. XVII. of the Mathematical Theory, below):—the velocity of transmission of longitudinal* vibrations (as of sound) along a bar or cord, or of waves of simple distortion, or of simple longitudinal extension and contraction in a homogeneous isotropic solid, or of sound waves in a fluid, is equal to the velocity acquired by a body in falling from a height equal to half the length of the proper modulus† for the case;—that is, the Young's modulus $\left(\dfrac{9kn}{3k+n}\right)$

---

* It is to be understood that the vibrations in question are so much spread out through the *length* of the body that inertia does not sensibly influence the transverse contractions and dilatations which (unless the substance have in this respect the peculiar character presented by cork, § 48 above) take place along with them.

† In §§ 73—76 below we shall see that changes of shape and bulk produced by the varying stresses cause changes of temperature which, in ordinary solids, render the velocity of transmission of longitudinal vibrations sensibly greater than that calculated by the rule stated in the text, if we use the *static modulus* as understood from the definition there given; and it will be shown how to take into account the thermal effect by using a definite *static modulus*, or *kinetic modulus*, according to the circumstances of any case that may occur.

for the first case, the modulus of rigidity (n) for the second, the modulus of simple longitudinal strain ($k + \frac{4}{3}n$) for the third, the modulus of compression k for the fourth. Remark that for air the static "length-modulus of compression" at constant temperature is the same as what is often technically called the "height of the homogeneous atmosphere."

51. In reckoning moduluses there must be a definite understanding as to the unit in terms of which the force is measured, which may be either the *kinetic unit* or the *gravitation unit* for a specified locality, that is, the weight in that locality of the unit of mass. Experimenters have usually stated their results in terms of the gravitation unit, each for his own locality,—the accuracy hitherto attained being scarcely in any cases sufficient to require corrections for the different intensities of gravity in the different places of observation.

The most useful and generally convenient specification of the modulus of elasticity of a substance is in grammes-weight per square centimetre. This has only to be divided by the specific gravity of the substance to give the *length of the modulus*. British measures, however, being still unhappily sometimes used in practical and even in scientific statements, we too often meet with reckonings of the modulus in pounds per square inch, or per square foot, in tons per square inch, or of the length of the modulus in feet or in British statute miles.

A reckoning sometimes used in some British treatises on mechanics and in practical statements is pounds per square inch. The modulus thus stated must be divided by the weight of 12 cubic inches of the solid, or by the product of its specific gravity into ·4335*, to find the length of the modulus in feet.

---

* This decimal being the weight in pounds of 12 cubic inches of water. The one great advantage of the French metrical system is that the mass of the unit volume (1 cubic centimetre) of water at its temperature of maximum density (3·945° C.) is unity (1 gramme) to a sufficient degree of approximation for almost all practical purposes. (Professor W. H. Miller, of Cambridge, concludes, from a very trustworthy comparison of standards by Kupffer, of St Petersburg, that the weight of a cubic decimetre of water at temperature of maximum density is 1000·013 grammes.) Thus, according to this system, the density of a body and its specific gravity mean one and the same thing; whereas on the British no-system the density is expressed by a number found by multiplying the specific gravity by one number or another, according to the choice of a cubic inch, cubic foot, cubic yard, or cubic mile, that is made for the unit of volume; and the grain, scruple, gun-

To reduce from pounds per square inch to grammes per square centimetre, multiply by (453·59/6·45) 70·31, or divide by ·014223. French engineers generally state their results in kilogrammes per square metre, and so bring them to more convenient numbers; being 1/100,000 of the inconveniently large numbers expressing moduluses in grammes weight per square centimetre, but it is much better to reckon in millions of grammes per square centimetre.

51′ The same statements as to units, reducing factors, and nominal designations, are applicable to the bulk-modulus of any elastic solid or fluid, and to the rigidity (§ 44 above) of an iso-tropic body; or, in general, to any one of the 21 moduluses in the expressions (Mathematical Theory below, chaps. XIII.—XVI.) for stresses in terms of strains, or to the reciprocal of any one of the 21 moduluses in the expressions (Mathematical Theory below, chaps. XIII.—XVI.) for strains in terms of stresses, as well as to the modulus defined by Young.

51″. The convenience, for residents on the Earth, of the length-reckoning of moduluses is illustrated by the theorems stated at the end of § 50 above, and others analogous to it as follows :—

(1) The velocity of propagation of a wave of distortion in an isotropic homogeneous solid, is equal to the velocity acquired by a body in falling through a height equal to half the length-modulus of rigidity.

(2) The velocity of the other kind of wave possible in an isotropic homogeneous solid, that is to say a wave analogous to that of sound, is equal to the velocity acquired by a body falling through a height equal to half the length-modulus for simple longi-tudinal strain (compare § 42 above); just as the Young's modulus is reckoned for simple stress. The modulus for simple longitudinal strain may be found by enclosing a rod or bar of the substance in an infinitely rigid, perfectly smooth and frictionless tube fitting

maker's drachm, apothecary's drachm, ounce Troy, ounce avoirdupois, pound Troy, pound avoirdupois, stone (Imperial, Ayrshire, Lanarkshire, Dumbartonshire), stone for hay, stone for corn, quarter (of a hundredweight), quarter (of corn), hundred-weight, or ton, that is chosen for unit of mass. It is a remarkable phenomenon, belonging rather to moral and social than to physical science, that a people tending naturally to be regulated by common sense should voluntarily condemn themselves, as the British have so long done, to unnecessary hard labour in every action of common business or scientific work relating to measurement, from which all the other nations of Europe have emancipated themselves.

it perfectly all round, and then dealing with it as the rod with its sides all free is dealt with for finding the Young's modulus. Of course it is understood that the ideal tube, which gives positive normal pressure when the two ends of the elastic rod within it are pressed together, must be supposed to give the negative normal pressure, or the normal traction, required to prevent lateral shrinkage, when the two ends of the wire are pulled asunder. (Compare § 47 above.)

(3) The velocity of sound in a liquid is the velocity a body would acquire in falling through a height equal to half the length-modulus of compression.

(4) The Newtonian velocity of sound (that is to say, the velocity which sound would have in air if the pressure in the course of the vibration varied simply according to Boyle's law, without correction for the heat of condensation, and the cold of rarefaction) is equal to the velocity a body would acquire in falling through half the height of the homogeneous atmosphere for the actual temperature of the air whatever it may be. ("The Height of the Homogeneous Atmosphere" is a short expression commonly used to designate the depth that an ideal incompressible liquid of the same density as air must have, to give by its weight the same pressure at the bottom as the actual pressure of the air at the supposed temperature and density.)

(5) The velocity of a long wave* in water of uniform depth, supposed incompressible, is the velocity a body would acquire in falling through a height equal to half the depth.

(6) The velocity of propagation of a transverse pulse in a stretched cord is equal to the velocity acquired by a body falling through a height equal to half the length of a quantity of cord amounting in weight to the stretching force.

52. "Resilience" is a very useful word, introduced about forty years ago (when the *doctrine of energy* was beginning to become practically appreciated) by Lewis Gordon, first Professor of Engineering in the University of Glasgow, to denote the quantity of work that a spring (or elastic body) gives back, when strained

* A "Long wave" is a technical expression in the theory of waves in water, used to denote a wave of which the length is a large multiple (20 or 30 or more) of the depth.

to some stated limit and then allowed to return to the condition in which it rests when free from stress. The word " resilience ", used without special qualification, may be understood as meaning *extreme resilience*, or the work given back by the spring after being strained to the extreme limit within which it can be strained again and again without breaking or taking a permanent set. In all cases for which Hooke's law of simple proportionality between stress and strain holds, the resilience is obviously equal to the work done by a constant force of half the amount of the extreme force, acting through a space equal to the extreme deflection.

53. When force is reckoned in " gravitation measure," resilience per unit of the spring's mass is simply the height to which the spring itself, or an equal weight, could be lifted against gravity by an amount of work equal to that given back by the spring returning from the stressed condition.

54. Let the elastic body be a long homogeneous cylinder or prism with flat ends (a bar as we may call it for brevity), and let the stress for which its resilience is reckoned be *positive* normal pressures on its ends. The resilience per unit mass is equal to the greatest height from which the bar can fall with its length vertical, and impinge against a perfectly hard frictionless horizontal plane without suffering stress beyond its limits of elasticity. For in this case (as in the case of the direct impact of two equal and similar bars meeting with equal and opposite velocities, discussed in Thomson and Tait's *Natural Philosophy*, § 303), the kinetic energy of the translational motion preceding the impact is, during the first half of the collision, wholly converted into potential energy of elastic force, which during the second half of the collision is wholly reconverted into kinetic energy of translational motion in the reverse direction. During the whole time of the collision the stopped end of the bar experiences a constant pressure, and at the middle of the collision the whole substance of the bar is for an instant at rest in the same state of compression as it would have permanently if in equilibrium under the influence of that pressure and an equal and opposite pressure on the other end. From the beginning to the middle of the collision the compression advances at a uniform rate through the bar from the stopped end to the free end. Every particle of the bar which the compression has not reached continues moving uniformly with the velocity of the whole

before the collision, until the compression reaches it, when it instantaneously comes to rest. The part of the bar which at any instant is all that is compressed, remains at rest till the corresponding instant in the second half of the collision.

55. From our preceding view of a bar impinging against an ideal perfectly rigid frictionless plane, we see at once all that takes place in the real case of any rigorously direct longitudinal collision between two equal and similar elastic bars with flat ends. In this case the whole of the kinetic energy which the bodies had before collision reappears as purely translational kinetic energy after collision. The same would be approximately true of any two bars, provided the times taken by a pulse of simple longitudinal stress to run through their lengths are equal. Thus if the two bars be of the same substance, or of different substances having the same value for Young's modulus, the lengths must be equal, but the diameters may be unequal. Or if the Young's modulus be different in the two bars, their lengths must (Math. Theory, below, chap. XVII.) be inversely as the square roots of its values. To all such cases the laws of "collision between two perfectly elastic bodies," whether of equal or unequal masses, as given in elementary dynamical treatises, are applicable. But in every other case part of the translational energy which the bodies have before collision is left in the shape of vibrations after collision, and the translational energy after collision is accordingly less than before collision. The losses of energy observed in common elementary dynamical experiments on collision between solid globes of the same substance are partly due to this cause. If they were wholly due to it they would be independent of the substance, when two globes of the same substance are used. They would bear the same proportion to the whole energy in every case of collision between two equal globes, or again, in every case of collision between two globes of any stated proportion of diameters, provided in each case the two which collide are of the same substance; but the proportion of translational energy converted into vibrations would not be the same for two equal globes as for two unequal globes. Hence when differences of proportionate losses of energy are found in experiments on different substances, as in Newton's on globes of glass, iron, or compressed wool, this must be due to imperfect elasticity of the material. It is to be expected that careful experiments

upon hard well-polished globes striking one another with such
gentle forces as not to produce even at the point of contact any
stress approaching to the limit of elasticity, will be found to give
results in which the observed loss of translational energy can be
almost wholly accounted for by vibrations remaining in the globes
after collision.

56. *Examples of Resilience.—Example* 1. *Longitudinal resilience of a wire, rod, or column subjected to end-pull or thrust,
and free all round its sides.* Let M be the Young's modulus, in
units of force per unit of cross-sectional area, and let ϵ be the extreme elastic elongation or shortening: so that if l be the length,
and A the cross-sectional area, the force required to produce this
change of length is equal to $MA\epsilon$, and the actual elongation or
shortening is ϵl. Hence (§ 52 above) the resilience is equal to
$\frac{1}{2}\epsilon^2 MAl$: and therefore the resilience per unit of volume is equal
to $\frac{1}{2}M\epsilon^2$.

Example 2. *The resilience per unit bulk of a homogeneous
simple shear in an isotropic solid,* is similarly found to be $\frac{1}{2}n\delta^2$;
where n is the rigidity modulus (§ 43 above) and δ is the extreme
elastic shear.

Example 3. *Torsional resilience of a round tube, or round
solid rod, or wire.* The torsional rigidity (§ 64 below) is (§ 65
below) equal to

$$\tfrac{1}{2}\pi\left(r^2 - r'^2\right)\left(r^2 + r'^2\right)n;$$

where r and r' are the radii of the outer and inner cylindrical
boundaries. Hence the angle turned through by one end of the
tube with a couple G, applied to it, while the other end is held
fixed is equal to

$$Gl/\tfrac{1}{2}\pi\left(r^2 - r'^2\right)\left(r^2 + r'^2\right)n,$$

and r/l of this is the shear in the matter contiguous with the
outer boundary of the tube. Hence if, as before, δ denote the
elastic permanent shear, we have

$$G = \tfrac{1}{2}\pi nr^{-1}\left(r^2 - r'^2\right)\left(r^2 + r'^2\right)\delta.$$

The work done in producing the supposed amount of twist from
zero being $\frac{1}{2}Glr^{-1}\delta$, is therefore equal to

$$\tfrac{1}{4}\pi lr^{-2}\left(r^2 - r'^2\right)\left(r^2 + r'^2\right)n\delta^2,$$

which is therefore the whole resilience of the supposed piece of matter. The volume of the piece of matter is $\pi l (r^2 - r'^2)$ and therefore the resilience per unit of volume is

$$\tfrac{1}{4} n \left(r^2 + r'^2 \right) \delta^2 / r^2.$$

Example 3′. For the case of an infinitely thin tube—$(r - r')$ infinitely small—the preceding expression for the resilience per unit volume becomes $\tfrac{1}{2} n \delta^2$; which agrees with the case of Example 2, as it clearly ought to do.

Example 3″. For the case of a round solid rod or wire, the expression becomes $\tfrac{1}{4} n \delta^2$; which is just half the resilience per unit volume of a body strained throughout to the amount δ of simple shear.

Example 4. Comparing Examples (1) and (3″) we see that the torsional resilience of a round solid rod or wire is

$$\tfrac{1}{2} n \delta^2 / M \epsilon^2, \text{ of its longitudinal resilience.}$$

By §§ 47 and 48 above, we see that for india-rubber, Young's modulus is approximately equal to $3n$; while for cork, it is approximately equal to $2n$. For all other natural solids it is probably between these limits, and for solids fulfilling Navier's and Poisson's conclusions, referred to in § 47 above, Young's modulus is equal to two-and-a-half times the modulus of rigidity. Taking then as a rough average $M/n = 2\tfrac{1}{2}$, we find $\tfrac{1}{5} \delta^2 / \epsilon^2$ for the ratio of the torsional to the longitudinal resilience of a round solid rod or wire. This is a very important conclusion with reference to the theory of Coulomb's torsion balance and the theory of spiral springs, §§ 67—72 below.

Comparing this with Experimental Example (1) below, according to Example (3″) above, we see that the extreme shear in the surface parts of the wire, when it is twisted as far as it can be without giving it any permanent set, is only $\sqrt{(5 \times 1\cdot3)/179}$, or about $\tfrac{1}{5}$ of the elongation produced in the same wire when pulled with the greatest force which can be applied to it without producing any permanent elongation.

Experimental Example (1).—In respect to simple longitudinal pull, the extreme resilience of steel piano-forte wire of the gauge and quality referred to in § 22 above (calculated by multiplying the breaking weight into half the elongation produced by it according to the experimental data of § 22) is 6162 metre-grammes (gravitation measure) per ten metres of the wire. Or, whatever the length of the wire, its resilience is equal to the work required to lift its weight through 180 metres.

Experimental Example 2.—The torsional resilience of the same wire, twisted in either direction as far as it can be without giving it any notable permanent set, I have found by experiment to be equal to the work required to lift its weight through 1·3 metres.

Experimental Example 3.—The extreme longitudinal resilience of a vulcanized india-rubber band weighing 12·3 grammes I have found to be equal to the work required to lift its weight through 1200 metres. The result was obtained by stretching the india-rubber band by gradations of weights up to the breaking weight, representing the results by aid of a curve, and measuring its area to find the integral work given back by the spring after being stretched by a weight just short of the breaking weight.

The figures given in Table I. below show, conveniently for comparison, the longitudinal and the torsional resiliences and the rigidities and the Young's moduluses of several different substances, india-rubber and metals, and of different specimens of the same metal; obtained from the recorded results of experiments which have been made at various times in the Physical Laboratory of Glasgow University. The torsional resiliences are calculated from observations of the stretching, within the limits of elasticity, of spiral springs formed from the substance to be tested. They show, as is to be expected, very great differences for the same metal; due of course to differences of temper.

TABLE I. RESILIENCES AND MODULUSES.

Substance.	Torsional resilience in cms.	Longitudinal Resilience in cms.	Rigidity-modulus in grammes weight per square cm.	Young's modulus in grammes weight per square cm.
India-rubber band ...	....	120000	....	....
Piano-forte steel wire	130 to 1203	17620	834×10^6	2049×10^6
Platinoid ,,	271 to 1580	1693	476×10^6	1222×10^6
German silver ,,	168	514	567×10^6	1082×10^6
Brass ,,	860 to 940	728	350×10^6	1001×10^6
Delta metal ,,	750 to 1250	2708	332 to 363×10^6	715 to 1016×10^6
Phosphor bronze ,,	....	4904	....	477×10^6
,, ,, ,,	....	3545	....	573×10^6
,, ,, ,,	....	1842	....	951×10^6
Silicium ,, ,,	....	3166	....	626×10^6
Manganese ,, ,,	....	2998	....	622×10^6

57. *Flexure of a Beam or Rod.*—In the problem of simple flexure a bar or uniform rod or wire, straight when free from stress, is kept in a circular form by equal opposing couples properly applied to its ends. The parts of the bar on the convex side of the circle must obviously be stretched longitudinally, and those on the concave side contracted longitudinally, by the flexure. It is not obvious, however, what are the conditions affecting the lateral shrinkings and swellings of ideal filaments into which we may imagine the bar divided lengthwise. Earlier writers had assumed without proof that each filament, bent as it is in its actual position in the

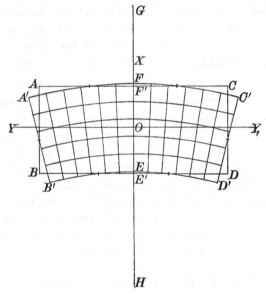

Fig. 5.

bar, is elongated or contracted by the same amount as it would be if it were detached, and subjected to the same end pull or end compression with its sides quite free to shrink or expand, but they had taken no account of the lateral shrinking or swelling which the filament must really experience in the bent bar. The subject first received satisfactory mathematical investigation from St Venant[*]. He proved that the old supposition is substantially correct, with the important practical exception of the flat spring

---

[*] *Mémoires des Savants Etrangers*, 1855, "De la Torsion des Prismes, avec des considérations sur leur Flexion, &c."

referred to in § 59 below. His theory shows that, in fact, if we imagine the whole rod divided parallel to its length into infinitesimal filaments, each of these shrinks or swells laterally with sensibly the same freedom as if it were separated from the rest of the substance and subjected to end pull or end compression, lengthening or shortening it in a straight line to the same extent as it is really lengthened or shortened in the circular arc which it becomes in the bent rod. He illustrates the distortion of the cross section by which these changes of lateral dimensions are necessarily accompanied in the annexed diagram (fig. 5), in which either the whole normal section of a rectangular beam, or a rectangular area in the normal section of a beam of any figure, is represented in its strained and unstrained figures, with the central point O common to the two. The flexure is in planes perpendicular to YOY_1, and is concave upwards (or towards X),—G, the centre of curvature, being in the direction indicated, but too far away to be included in the diagram. The straight sides AC, BD, and all straight lines, parallel to them, of the unstrained rectangular area, become concentric arcs of circles concave in the opposite direction, their centre of curvature H being (§§ 47, 48 above) for rods of india-rubber or gelatinous substance, or of glass or metal, from 2 to 4 times as far from O on one side as G is on the other. Thus the originally plane sides AC, BD of a rectangular bar become anticlastic* surfaces, of curvatures $\dfrac{1}{\rho}$ and $\dfrac{-\sigma}{\rho}$, in the two principal sections, if σ denote the ratio of lateral shrinking to longitudinal extension. A flat rectangular, or a square, rod of india-rubber [for which σ amounts (§ 47 above) to very nearly $\frac{1}{2}$, and which is susceptible of very great amounts of strain without utter loss of corresponding elastic action] exhibits this phenomenon remarkably well.

58. *Limits to the bending of Rods or Beams of hard solid substance.*—For hard solids, such as metals stones glasses woods ivory vulcanite papier-maché, elongations and contractions to be within the limits of elasticity must generally (§ 23 above) be less than $\frac{1}{100}$. Hence the breadth or thickness of the bar in the plane

---

* See Thomson and Tait's *Natural Philosophy*, second edition, Vol. I. Part I. § 128.

of curvature must generally be less than $\frac{1}{100}$ of the radius of curvature in order that the bending may not break it, or give it a permanent bend, or strain it beyond its "limits of elasticity."

59. *Exceptional case of Thin flat Spring, too much bent to fulfil conditions of* § 57.—St Venant's theory shows that a farther condition must be fulfilled if the ideal filaments are to have the freedom to shrink or expand as explained in § 57 above. For unless the *breadth* AC of the bar (or diameter perpendicular to the plane of flexure) be very small in comparison with the mean proportional between the radius OH and the thickness AB, the distances from Y, Y_1 to the corners A', C', would fall short of the half thickness, OE, and the distances to B', D', would exceed it, by differences comparable with its own amount. This would give rise to sensibly less and greater shortenings and stretchings in the filaments towards the corners than those supposed in the ordinary calculation of flexural rigidity (§ 61 below), and so vitiate the result. Unhappily, mathematicians have not hitherto succeeded in solving, possibly not even tried to solve, the beautiful problem thus presented by the flexure of a broad very thin band (such as a watch spring) into a circle of radius comparable with a third proportional to its thickness and its breadth.

60. But, provided the radius of curvature of the flexure is not only a large multiple of the *greatest* diameter, but also of a third proportional to the diameters in and perpendicular to the plane of flexure; then, however great may be the ratio of the greatest diameter to the least, the preceding solution is applicable; and it is remarkable that the necessary distortion of the normal section (illustrated in the diagram of § 57 above) does not sensibly impede the free lateral contractions and expansions in the filaments, even in the case of a broad thin lamina (whether of precisely rectangular section, or of unequal thicknesses in different parts).

61. *Flexural Rigidities of a Rod or Beam.*—The couple required to give unit curvature in any plane to a rod or beam is called its flexural rigidity for curvature in that plane. When the beam is of circular cross section and of isotropic material, the flexural rigidity is clearly the same, whatever be the plane of flexure through the axis, and the plane of the bending couple coincides with the plane of flexure. It might be expected that a round bar of æolotropic material, such as a wooden rod with the

annual woody layers sensibly plane and parallel to a plane through its axis, would show different flexural rigidities in different planes,—in the case of wood, for example, different according as the flexure is in a plane parallel or perpendicular to the annual layers. This is not so, however; on the contrary, it is easy to show, by an extension of St Venant's theory, that in the case of the wooden rod the flexural rigidity is equal in all planes through the axis, and that the plane of flexure always agrees with the plane of the bending couple; and to prove generally that the flexure of a bar of æolotropic substance, composed it may be of longitudinal filaments of heterogeneous materials, is precisely the same as if it were isotropic; and that its flexural rigidities are calculated by the same rule from its Young's modulus, provided that the æolotropy is not such as (§ 81 below) to give rise to alteration of the angle between the length and any diameter perpendicular to the length when weight is hung on the rod, or on any longitudinal filament cut from it. Excluding then all cases in which there is any such oblique æolotropy, we have a very simple theory for the flexure of bars of any substance, whether isotropic or æolotropic, and whether homogeneous or not homogeneous through the cross section.

62. *Principal Flexural Rigidities and Principal Planes of Flexure of a Beam.*—The flexural rigidity of a rod is generally not equal in different directions, and the plane of flexure does not generally coincide with the plane of the bending couple. Thus a flat ruler is much more easily bent in a plane perpendicular to its breadth than in the plane of its breadth; and if we apply opposing couples to its two ends in any plane through its axis not either perpendicular or parallel to its breadth, it is obvious that the plane in which the flexure takes place will be more inclined to the plane of the breadth than to the plane of the bending couple. Very elementary statical theory, founded on St Venant's conclusions of § 57 above, shows that, whatever the shape and the distribution of matter in the cross section of the bar, there are two planes at right angles to one another, such that if the bar be bent in either of these planes the bending couple will coincide with the plane of flexure. These planes are called principal planes of flexure, and the rigidities of the bar for flexure in these planes are called its principal flexural rigidities. When the principal flexural rigidities are known, the flexure of the bar in any plane oblique to the

principal planes is readily found by supposing it to be bent in one
of the principal planes and simultaneously in the other, and
calculating separately the couples required to produce these two
component flexures. The positions of the principal planes of
flexure, the relative flexural rigidities, and the law of elongation
and contraction in different parts of the cross section, are found
according to the following simple rules :—

(1) Imagine an infinitely thin plane disc of the same shape
and size as the cross section, loaded with matter in simple proportion to the Young's modulus in different parts of the cross section.
Let the quantity of matter per unit area on any point of the disc
be equal to the Young's modulus on the corresponding point of
the rod when the material is heterogeneous : on the other hand,
when the material is homogeneous it is more convenient to call
the quantity of matter unity per unit area of the disc. Considering
different axes in the plane of the disc through its centre of inertia,
find the two principal axes of greatest and least moments of
inertia, and find the moments of inertia round them.

(2) In whatever plane the bar be bent, it will experience
neither elongation nor contraction in the filament which passes
through the centres of inertia of the cross sections found according
to rule (1), nor in the diameter of the cross section perpendicular
to the plane of flexure.

(3) Thus all the particles which experience neither elongation
nor contraction lie in a surface cutting the plane of flexure
perpendicularly through the centres of inertia of the cross sections.
All the material on the outside of this cylindrical surface is
elongated, and all on the interior is contracted, in simple proportion
to distance from it : the amount of the elongation or contraction
being in fact equal to distance from this neutral surface divided
by the radius of its curvature.

(4) Hence it is obvious that the portions of the solid on the
two sides of any cross section must experience mutual normal
force, pulling them from one another in the stretched part, and
pressing them towards one another in the condensed part, and that
the amount of this negative or positive normal pressure per unit
of area must be equal to the Young's modulus at the place
multiplied into the ratio of its distance from the neutral line of
the cross section to the radius of curvature.

The sum of these positive and negative forces over the whole area of the cross section is zero in virtue of condition (2). Their resultant couple has its axis perpendicular to the plane of curvature when this line is either of the principal axes (3) of the cross section; and its moment is clearly equal to the moment of inertia of the material disc (1) divided by the radius of curvature. Hence the principal flexural rigidities are simply equal to the principal moments of inertia of this disc; and the principal flexural planes are the planes through its principal axes and the length of the bar; or taking the quantity of matter per unit area of the disc unity for the case of a homogeneous bar, we have the rule, that the principal rigidities are equal to the product of the Young's modulus into the principal moments of inertia of the cross sectional areas, and the principal planes of flexure are the longitudinal planes through the principal axes of this area.

63. *Law of Torsion.*—One of the most beautiful applications of the general equations of internal equilibrium of an elastic solid hitherto made, is that of M. de St Venant to "the torsion of prisms*." In this work the mathematical methods invented by Fourier for the solution of problems regarding conduction of heat have been most ingeniously and happily applied by St Venant to the problem of torsion. To reproduce St Venant's mathematical investigation here would make this article too long (it occupies 227 quarto pages of the *Mémoires des Savants Étrangers*); but a statement of some of the chief results is given (§§ 65—72 below), not only on account of their strong scientific interest, but also because they are of great practical value in engineering; and the reader is referred to Thomson and Tait's *Natural Philosophy*, §§ 700—710, for the proofs and for further details regarding results, but much that is valuable and interesting is only to be found in St Venant's original memoir.

64. *Torsion Problem stated and Torsional Rigidity defined.* —To one end of a long, straight prismatic rod, wire, or solid or hollow cylinder of any form, a given couple is applied in a plane perpendicular to the length, while the other end is held fast: it is required to find the degree of twist produced, and the distribution of strain and stress throughout the prism. The moment of

---

* *Mémoires des Savants Etrangers*, 1855, "De la Torsion des Prismes, avec des considérations sur leur Flexion, &c."

the couple divided by the amount of the twist per unit length is called the torsional rigidity of the rod or prism. This definition is founded simply on the extension of Hooke's law to torsion discovered experimentally by Coulomb, according to which a rod or wire when twisted within its limits of torsional elasticity exerts a reactive couple in simple proportion to the angle through which one end is turned relatively to the other. The internal conditions to be satisfied in the torsion problem are that the resultant action between the substance on the two sides of any normal section is a couple, in the normal plane, equal to the given couple. This problem has not hitherto been attacked for æolotropic solids. Even such a case as that of the round wooden rod (§ 61 above) with annular layers sensibly parallel to a plane through its length, will, when twisted, experience a distribution of strain complicated much by its æolotropy. The following statements of results are confined to rods of isotropic material.

65. *Torsion of Circular Cylinder.*—For a solid or hollow circular cylinder, the solution (given first, we believe, by Coulomb) obviously is that each circular normal section remains unchanged in its own dimensions, figure, and internal arrangement (so that every straight line of its particles remains a straight line of unchanged length), but is turned round the axis of the cylinder through such an angle as to give a uniform *rate of twist* equal to the applied couple divided by the product of the moment of inertia of the circular area (whether annular or complete to the centre) into the modulus of rigidity of the substance.

For, if we suppose the distribution of strain thus specified to be actually produced, by whatever application of stress is necessary, we have, in every part of the substance, a simple shear parallel to the normal section, and perpendicular to the radius through it. The elastic reaction against this requires, to balance it (§ 43 above), a simple distorting stress consisting of forces in the normal section, directed as the shear, and others in planes through the axis, and directed parallel to the axis. The amount of the shear is, for parts of the substance at distance r from the axis, equal obviously to τr, if τ be the rate of twist reckoned in radians per unit of length of the cylinder. Hence the amount of the tangential force in either set of planes is $n\tau r$ per unit of area, if n be the rigidity of the substance. Hence there is no force between parts of the substance lying on the two sides of any element of any circular cylinder coaxal with the bounding cylinder or cylinders; and consequently no force is required on the cylindrical boundary to maintain the supposed state of strain. And the mutual action between the parts of the substance on the two sides of any normal plane section consists of force in this plane, directed perpendicular to the radius through each point, and amounting to $n\tau r$ per unit of area. The moment of this distribution of force round the axis of the cylinder is (if $d\sigma$ denote an element of the area) $n\tau \iint d\sigma r^2$, or the product of $n\tau$ into the moment of inertia of the area round the perpendicular to its plane through its centre, which is therefore equal to the moment of the couple applied at either end.

66. *Prism of any shape constrained to a Simple Twist.*— Farther, it is easily proved that if a cylinder or prism of any shape be compelled to take exactly the state of strain above specified (§ 65), with the line through the centres of inertia of the normal sections, taken instead of the axis of the cylinder, the mutual action between the parts of it on the two sides of any normal section will be a couple of which the moment will be expressed by the same formula, that is, the product of the rigidity, into the rate of twist, into the moment of inertia of the section round its centre of inertia. But for any other shape of prism than a solid or symmetrical hollow circular cylinder, the supposed state of strain requires, besides the terminal opposed couples, force parallel to the length of the prism, distributed over the prismatic boundary, in proportion to the distance PE along the tangent,

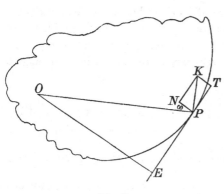

from each point of the surface, to the point in which this line is cut by a perpendicular to it from O the centre of inertia of the normal section. To prove this let a normal section of the prism be represented in the annexed diagram (fig. 6). Let PK, representing the shear at any point P, close to the prismatic

Fig. 6.

boundary, be resolved into PN and PT along the normal and tangent respectively. The whole shear PK being equal to τr, its component PN is equal to $\tau r \sin \omega$ or $\tau . PE$. The corresponding component of the required stress is $n\tau$. PE, and involves equal forces in the plane of the diagram, and, in the plane through TP, perpendicular to it, each amounting to $n\tau . PE$ per unit of area.

An application of force equal and opposite to the distribution thus found over the prismatic boundary, would of course alone produce in the prism, otherwise free, a state of strain which, compounded with that supposed above, would give the state of strain actually produced by the sole application of balancing couples to the two ends. The result, it is easily seen, consists of an increased twist, together with a warping of naturally plane normal sections, by infinitesimal displacements perpendicular to themselves, into certain surfaces of anticlastic curvature, with equal opposite curvatures. In bringing forward this theory, St Venant not only pointed out the falsity of the supposition admitted by several previous writers, and used in practice fallaciously by engineers, that Coulomb's law holds for other forms of prism than the solid or hollow circular cylinder, but he discovered fully the nature of the requisite correction, reduced the determination of it to a problem of pure mathematics, worked out the solution for a great variety of important and curious cases, compared the results with observation in a manner satisfactory and interesting to the naturalist, and gave conclusions of great value to the practical engineer.

67. "*Hydrokinetic Analogue to Torsion Problem**.—We take advantage of the identity of mathematical conditions in St Venant's torsion problem, and in a hydrokinetic problem first solved a few years earlier by Stokes†, to give the following statement, which will be found very useful in estimating deficiencies in torsional rigidity below the amount calculated from the fallacious extension of Coulomb's law :"—

"Conceive a liquid of density n completely filling a closed infinitely light prismatic box of the same shape within as the

---

* Extracted from Thomson and Tait's *Natural Philosophy*, Vol. I. Part II. §§ 704, 705.

† "On some cases of Fluid Motion."—*Camb. Phil. Trans.* 1843; or [*Mathematical and Physical Papers*, by G. G. Stokes, Vol. I.].

given elastic prism and of length unity, and let a couple be applied to the box in a plane perpendicular to its length. The *effective* moment of inertia of the liquid* will be equal to the correction by which the torsional rigidity of the elastic prism, calculated by the false extension of Coulomb's law, must be diminished to give the true torsional rigidity."

" Farther, the actual *shear* of the solid, in any infinitely thin plate of it between two normal sections, will at each point be, when reckoned as a differential sliding (§ 43 above), parallel to their planes, equal to and in the same direction as the velocity of the liquid relatively to the containing box."

68. *Solution of Torsion Problem.*—To prove these propositions and investigate the mathematical equations of the problem, the process followed in Thomson and Tait's *Natural Philosophy*, § 706, is first to show that the conditions of § 64 above are verified by a state of strain compounded of (1) a simple twist round the line through the centres of inertia, and (2) a distortion of each normal section by infinitesimal displacements perpendicular to its plane; then find the interior and surface equations to determine this warping; and lastly, calculate the actual moment of the couple to which the mutual action between the matter on the two sides of any normal section is equivalent.

69. St Venant's treatise abounds in beautiful and instructive graphical illustrations of his results, from which the following are selected †:—

(1) *Elliptic Cylinder.*—The plain and dotted curvilineal arcs are (fig. 7) " contour lines " (*coupes topographiques*) of the section as warped by torsion; that is to say, lines in which it is cut by a series of parallel planes, each perpendicular to the axis. The arrows indicate the direction of rotation in the part of the prism *above* the plane of the diagram.

(2) *Contour lines for St Venant's " étoile à quatre points arrondis."*—This diagram (fig. 8) shows the contour lines, in all

* "That is, the moment of inertia of a rigid solid which, as will be proved in Vol. II., may be fixed within the box, if the liquid be removed, to make its motions the same as they are with the liquid in it."

† A full mathematical consideration of these cases will be found in Thomson and Tait's *Natural Philosophy*, 2nd Edition, Vol. I. Part II. § 707 (A) and (B); see also § 708.

respects as in case (1), for the case of a prism having for section

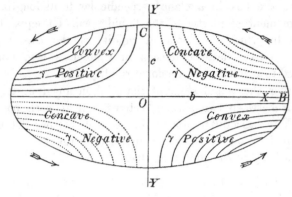

Fig. 7.

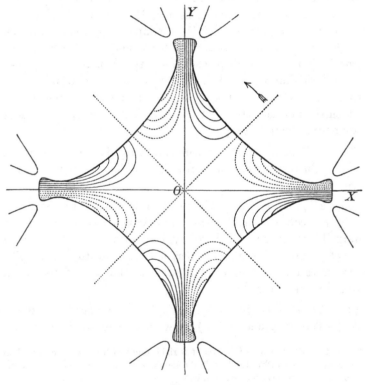

Fig. 8.

the figure indicated. The portions of curve outside the continuous closed curve are merely indications of mathematical extensions irrelevant to the physical problem.

(3) *Contour lines,* shown as in case (1), *of normal section of triangular prism, as warped by torsion* (fig. 9).

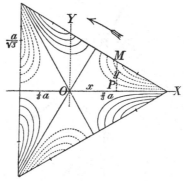

Fig. 9.

(4) *Contour lines of normal section of square prism as warped by torsion* (fig. 10).

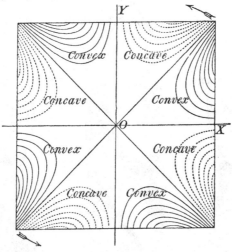

Fig. 10.

(5) *Diagram of St Venant's curvilinear squares for which the torsion problem is algebraically solvable.*—This diagram (fig. 11) shows the series of lines represented by the equation

$$x^2 + y^2 - a(x^4 - 6x^2y^2 + y^4) = 1 - a,$$

with the indicated values for a. It is remarkable that the values

$a = 0.5$ and $a = -\frac{1}{2}(\sqrt{2}-1)$ give similar but not equal curvilinear squares (hollow sides and acute angles), one of them turned through half a right angle relatively to the other.

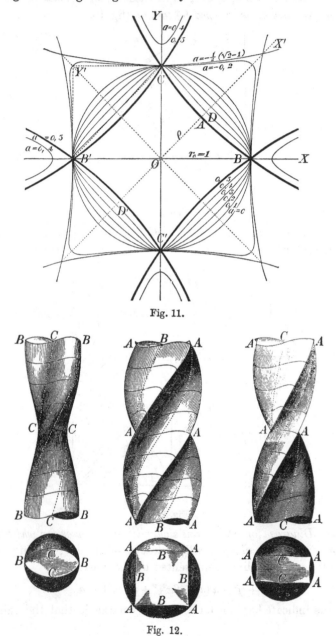

Fig. 11.

Fig. 12.

(6), (7) and (8). *Elliptic, square, and flat rectangular bars twisted* (Fig. 12). These are shaded drawings, showing the appearances presented by elliptic, square, and flat rectangular bars under exaggerated torsion, as may be realised with such a substance as India-rubber.

70. *Torsional Rigidity less in proportion to sum of principal Flexural Rigidities than according to false extension* (§ 66 above) *of Coulomb's Law.*—Inasmuch as the moment of inertia of a plane area about an axis through its centre of inertia perpendicular to its plane, is obviously equal to the sum of its moments of inertia round any two axes through the same point at right angles to one another in its plane, the fallacious extension of Coulomb's law, referred to in § 66 above, would make the torsional rigidity of a bar of any section, equal to the product of the ratio of the modulus of rigidity to the Young's modulus, into the sum of its flexural rigidities (§§ 61—2 above) in any two planes at right angles to one another through its length. The true theory, as we have seen (§§ 67—8 above), always gives a torsional rigidity less than

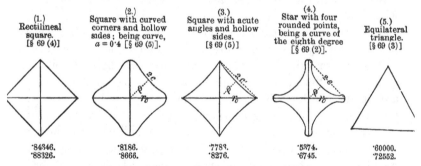

(1.) Rectilineal square. [§ 69 (4)]	(2.) Square with curved corners and hollow sides; being curve, $a = 0\cdot4$ [§ 69 (5)].	(3.) Square with acute angles and hollow sides. [§ 69 (5)]	(4.) Star with four rounded points, being a curve of the eighth degree [§ 69 (2)].	(5.) Equilateral triangle. [§ 69 (3)]
·84346. ·88326.	·8186. ·8666.	·7783. ·8276.	·5374. ·6745.	·60000. ·72552.

Fig. 13. Diagrams showing torsional rigidities.

this. How great the deficiency may be expected to be in cases in which the figure of the section presents projecting angles, or considerable prominences (which may be imagined from the hydrokinetic analogy given in § 67 above), has been pointed out by M. de St Venant, with the important practical application, that strengthening ribs, or projections (see, for instance, (4) of the annexed diagrams, fig. 13), such as are introduced in engineering to give stiffness to beams, have the reverse of a good effect when *torsional* rigidity or strength is an object, although they are truly of great value in increasing the flexural rigidity, and giving strength to bear ordinary strains, which are always more or less flexural. With remarkable ingenuity and mathematical skill he has drawn beautiful illustrations of this important practical principle from his algebraic and transcendental solutions. Thus, for an equilateral triangle, and for the rectilinear and three curvilinear squares shown in the diagrams (fig. 13), he finds for the torsional rigidities the values stated. The number immediately below the diagram indicates in each case the fraction which the true torsional rigidity is of the old fallacious estimate (§ 66 above),—the latter being the product of the rigidity of the substance into the moment of inertia of the cross section round an axis perpendicular to its plane through its centre of inertia. The second number indicates in each case the fraction which the torsional rigidity is of that of a solid circular cylinder of the same sectional area.

71. *Places of greatest Distortion in Twisted Prisms.*—M. de St Venant also calls attention to a conclusion from his solutions which to many may be startling, that in his simpler cases the places of greatest distortion are those points of the boundary which are nearest to the axis of the twisted prism in each case, and the places of least distortion those farthest from it. Thus in the elliptic cylinder the substance is most strained at the ends of the smaller principal diameter, and least at the ends of the greater. In the equilateral triangular and square prisms there are longitudinal lines of maximum strain through the middles of the sides. In the oblong rectangular prism there are two lines of greater maximum strain through the middles of the longer pair of sides, and two lines of less maximum strain through the middles of the shorter pair of sides. The strain is, as we may judge from the hydrokinetic analogy (§ 67 above), excessively small, but not evanescent, in the projecting ribs of a prism of the figure shown in (2) of § 69 above. It is quite evanescent infinitely near the angle, in the triangular and rectangular prisms, and in every other case, as (5) of § 69 above, in which there is a finite angle, whether acute or obtuse, projecting outwards. This reminds us of a general remark we have to make, although consideration of space may oblige us to leave it without formal proof.

72. *Strain at Projecting Angles, evanescent; at Re-entrant Angles, infinite; Liability to Cracks proceeding from Re-entrant*

Angles, or any places of too sharp concave curvature.—A solid of
any elastic substance, isotropic or æolotropic, bounded by any
surfaces presenting projecting edges or angles, or re-entrant angles
or edges, however obtuse, cannot experience any finite stress or
strain in the neighbourhood of a *projecting* angle (trihedral, poly-
hedral, or conical); in the neighbourhood of an edge, can only
experience simple longitudinal stress parallel to the neighbouring
part of the edge; and generally experiences infinite stress and
strain in the neighbourhood of a *re-entrant* edge or angle; when
influenced by any distribution of force, exclusive of surface tractions
infinitely near the angles or edges in question. An important
application of the last part of this statement is the practical rule,
well known in mechanics, that every re-entering edge or angle
ought to be rounded, to prevent risk of rupture, in solid pieces
designed to bear stress. An illustration of these principles is
afforded by the concluding example of torsion in Thomson and
Tait's *Natural Philosophy*, § 707; in which we have the complete
mathematical solution of the torsion problem for prisms of fan-
shaped sections, such as the annexed forms (fig. 14). The solution

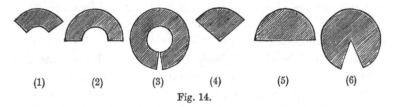

(1)   (2)   (3)   (4)   (5)   (6)

Fig. 14.

shows that when the solid is continuous from the circular cylindrical surface to its axis, as in (4), (5), (6), the strain is zero or
infinite according as the angle between the bounding planes of
the solid is less than or greater than two right angles as in cases
(4) and (6) respectively.

73. *Changes of Temperature produced by Compressions or
Dilatations of a Fluid and by Stresses of any kind in an Elastic
Solid.*—From thermodynamic theory* it is concluded that cold is
produced whenever a solid is strained by opposing, and heat
when it is strained by yielding to, any elastic force of its own,

---

* On " Thermo-elastic Properties of Matter," in *Quarterly Journal of Mathematics*,
April 1855 (republished in *Phil. Mag.* 1877, second half year). [Art. XLVIII. Part VII.
Vol. I. above.]

the strength of which would diminish if the temperature were raised; but that, on the contrary, heat is produced when a solid is strained against, and cold when it is strained by yielding to, any elastic force of its own, the strength of which would increase if the temperature were raised. When the strain is a condensation or dilatation, uniform in all directions, a fluid may be included in the statement. Hence the following propositions :—

(1) A cubical compression of any elastic fluid or solid in an ordinary condition causes an evolution of heat; but, on the contrary, a cubical compression produces cold in any substance, solid or fluid, in such an abnormal state that it would contract if heated while kept under constant pressure. Water below its temperature of maximum density (3°·9 Cent.) is a familiar instance. (See table of § 76 below.)

(2) If a wire already twisted be suddenly twisted further, always, however, within its limits of elasticity, cold will be produced; and if it be allowed suddenly to untwist, heat will be evolved from itself (besides heat generated externally by any work allowed to be wasted, which it does in untwisting). It is assumed that the torsional rigidity of the wire is diminished by an elevation of temperature, as the writer of this article had found it to be for copper, iron, platinum, and other metals (compare § 78 below).

(3) A spiral spring suddenly drawn out will become lower in temperature, and will rise in temperature when suddenly allowed to draw in. [This result has been experimentally verified by Joule ("Thermodynamic Properties of Solids," *Trans. Roy. Soc.*, 1858 and "Scientific Papers," Vol. I. pp. 413—473) and the amount of the effect found to agree with that calculated, according to the preceding thermodynamic theory, from the amount of the weakening of the spring which he found by experiment.]

(4) A bar or rod or wire of any substance with or without a weight hung on it, or experiencing any degree of end thrust, to begin with, becomes cooled if suddenly elongated by end pull or by diminution of end thrust, and warmed if suddenly shortened by end thrust or by diminution of end pull; except abnormal cases in which with constant end pull or end thrust elevation of temperature produces shortening; in every such case pull or diminished thrust produces elevation of temperature, thrust or diminished pull lowering of temperature.

(5) An india-rubber band suddenly drawn out (within its limits of elasticity) becomes warmer; and when allowed to contract, it becomes colder. Any one may easily verify this curious property by placing an india-rubber band in slight contact with the edges of the lips, then suddenly extending it—it becomes very perceptibly warmer: hold it for sometime stretched nearly to breaking, and then suddenly allow it to shrink—it becomes quite startlingly colder, the cooling effect being sensible not merely to the lips but to the fingers holding the band. The first published statement of this curious observation is due to Gough (*Memoirs of the Literary and Philosophical Society of Manchester*, 2nd series, vol. I. p. 288), quoted by Joule in his paper on " Thermodynamic Properties of Solids" (*Trans. Roy. Soc.*, 1858 and " Scientific Papers," vol. I. pp. 413—473). The thermodynamic conclusion from it is that an india-rubber band, stretched by a constant weight of sufficient amount hung on it, must, when heated, pull up the weight, and, when cooled, allow the weight to descend: this Gough, independently of thermodynamic theory, had found to be actually the case. The experiment any one can make with the greatest ease, by hanging a few pounds weight on a common india-rubber band, and taking a red-hot coal in a pair of tongs, or a red-hot poker, and moving it up and down close to the band. The way in which the weight rises when the red-hot body is near, and falls when it is removed, is quite startling. Joule experimented on the amount of shrinking per degree of temperature, with different weights hung on a band of vulcanized india-rubber, and found that they agreed closely with the amounts calculated by thermo-dynamic theory* from the heating effects of pull, and cooling effects of ceasing to pull, which he had observed in the same piece of india-rubber.

74. The thermodynamic theory gives one formula† by which the change of temperature in every such case may be calculated when the other physical properties are known:—

---

* On " Thermo-elastic Properties of Matter," in *Quarterly Journal of Mathematics*, April 1855 (republished in *Phil. Mag.* Jan. 1878), [Art. XLVIII. Part VII. Vol. I. above.].

† " Dynamical Theory of Heat" [Art. XLVIII. Vol. I. above] (§ 49), *Trans. R.S.E.*, March 1851, " Thermo-elastic Properties of Matter," [Art. XLVIII. Part VII. Vol. I. above], and " On the Alteration of Temperatures accompanying changes of Pressure in Fluids," *Proc. R. S.*, June 1857; *Phil. Mag.* June Suppl. 1858 [Appendix to Part II. of present Article].

$$\theta = \frac{tep}{JK\rho} \; ;$$

where θ denotes the elevation of temperature produced by the sudden application of a stress p;

t, the temperature of the substance on the absolute thermodynamic scale*, the change of temperature θ being supposed to be but a very small fraction of t;

e, the geometrical effect (expansion or other strain) produced by an elevation of temperature of one degree when the body is kept under constant stress;

K, the specific heat of the substance per unit mass under constant stress;

ρ, the density;

and J, Joule's equivalent (taken as 42400 centimetres).

In using the formula for a fluid, p must be normal pressure equal in all directions. For a solid it may be normal pressure on a set of parallel planes, or tangential traction on one or other of the two sets of mutually perpendicular parallel planes which (§ 43 above) experience tangential traction when the body is subjected to a simple distorting stress; or, quite generally, p may be the proper numerical reckoning (Mathematical Theory, chap. x. below) of any stress, simple or compound. When p is pressure uniform in all directions, e must be expansion of bulk, whether the body expands equally in all directions or not. When p is pressure perpendicular to a set of parallel planes, e must be expansion in the direction opposed to this pressure, irrespectively of any change of shape not altering the distance between the two planes of the solid perpendicular to the direction of p. When p is a simple tangential stress, reckoned as in § 43 above, e must be the change, reckoned in fraction of the radian, of the angle, infinitely nearly a right angle, between the two sets of parallel planes in either of which there is the tangential traction denoted by p. In each of these cases p is reckoned simply in units of force per unit of area. Quite generally p may be any stress, simple or compound, and

---

* "Dynamical Theory of Heat" [Art. XLVIII. Vol. I. above], Part vi. §§ 97, 100, *Trans. R. S. E.*, May 1854. According to the scale there defined on thermodynamic principles, independently of the properties of any particular substance, t is found, by Joule and Thomson's experiments, to agree very approximately with temperature centigrade, with 274° added.

e must be the component (Math. Th., chaps. VIII. and IX. below) relatively to the type of p, of the strain produced by an elevation of temperature of one degree when the body is kept under constant stress. The constant stress for which K and e are reckoned ought to be the mean of the stresses which the body experiences with and without p. Mathematically speaking, p is to be infinitesimal, but practically it may be of any magnitude moderate enough not to give any sensible difference in the value of either K or e, whether the "constant stress" be with p or without p, or with the mean of the two; thus for air p must be a small fraction of the whole pressure, for instance a small fraction of one atmosphere for air at ordinary pressure; for water or watery solutions of salts or other solids, for mercury, for oil, and for other known liquids p may, for all we know, amount to twenty atmospheres or one hundred atmospheres without transgressing the limits for which the preceding formula is applicable. When the law of variation of K and e with pressure is known, the differential formula is readily integrated to give the integral amount of the change of temperature produced by greater stresses than those for which the differential formula is applicable. For air and other permanent gases Boyle's law of compression, and Charles's law of thermal expansion, supply the requisite data with considerable accuracy up to twenty or thirty atmospheres. The result is expressed by the following formulas, showing the relations between temperature, pressure and volume when a gas experiences condensation or dilatation; without gain or loss of heat across the walls of the containing vessel :—

$$\frac{t'}{t} = \left(\frac{V}{V'}\right)^{k-1} \quad \dots\dots\dots\dots\dots\dots\dots\dots (1),$$

$$\frac{P'}{P} = \left(\frac{V}{V'}\right)^{k} \quad \dots\dots\dots\dots\dots\dots\dots\dots\dots(2),$$

and therefore $\qquad \dfrac{t'}{t} = \left(\dfrac{P'}{P}\right)^{\frac{k-1}{k}} \quad \dots\dots\dots\dots\dots\dots\dots\dots(3)\,;$

where k denotes the ratio of the thermal capacity, pressure constant, to the thermal capacity, volume constant, of the gas, a number which thermodynamic theory proves to be approximately constant for all temperatures and densities, for any fluid approximately fulfilling Boyle's and Charles's laws;

P, V, and t the initial pressure, volume, and temperature of the gas;

P', V', and t' the altered pressure, volume, and temperature of the gas.

For the case of $V' - V$ a small fraction of V the formula gives

$$t' - t = (k-1)\left(\frac{V' - V}{V}\right) t \quad \dots\dots\dots\dots\dots\dots(4).$$

It is by an integration of this formula that (1) is obtained.

For common air the value of k is very approximately 1·41. Thus if a quantity of air be given at 15° C. ($t = 289°$) and the ordinary atmospheric pressure, and if it be compressed gradually to $\frac{1}{32}$ of its initial volume, or dilated to 32 times its initial volume, and perfectly guarded against gain or loss of heat from or to without, its temperature in several different cases, chosen for example, will be according to the following table of differences of temperature above the primitive temperature, calculated by (1).

TABLE II.   EFFECTS OF PRESSURE ON TEMPERATURE.

Air given at temperature 15° Cent. (289° absolute).

Value of $\frac{P'}{P}$	$\frac{V}{V'}$	Elevation of temperature produced by compression.	Value of $\frac{P'}{P}$	$\frac{V}{V'}$	Lowering of temperature produced by dilatation.
2·66	2	95°	·3763	$\frac{1}{2}$	71°
7·07	4	221	·1416	$\frac{1}{4}$	125
18·77	8	389	·0533	$\frac{1}{8}$	166
49·87	16	612	·0200	$\frac{1}{16}$	196
132·50	32	911	·0075	$\frac{1}{32}$	219

But we have no knowledge of the effect of pressures of several thousand atmospheres in altering the expansibility or specific heat in liquids, or in fluids which at less heavy or at ordinary pressures are " gases."

75.   When change of temperature, whether in a solid or a fluid is produced by the application of a stress, the corresponding modulus of elasticity will be greater in virtue of the change of temperature than what may be called the static modulus defined as above, on the understanding that the temperature if changed by the stress is brought back to its primitive degree before the

measurement of the strain is performed. The modulus calculated on the supposition that the body, neither losing nor gaining heat during the application of the stress and the measurement of its effect, retains the whole change of temperature due to the stress, will be called for want of a better name the kinetic modulus, because it is this which must (as in Laplace's celebrated correction of Newton's calculation of the velocity of sound) be used in reckoning the elastic forces concerned in waves and vibrations in almost all practical cases. To find the ratio of the kinetic to the static modulus remark that $e\theta$, according to the notation of § 74 above, is the diminution of the strain due to the change of temperature θ. Hence if M denote the static modulus (§ 41 above), the strain actually produced by it when the body is not allowed either to gain or lose heat is $\dfrac{p}{M} - e\theta$, or, with θ replaced by its value according to the formula of § 74,

$$\frac{p}{M} - e\,\frac{tep}{JK\rho}.$$

Dividing p by this expression we find for the kinetic modulus

$$M' = \frac{1}{\dfrac{1}{M} - \dfrac{te^2}{JK\rho}}\,.$$

Hence

$$\frac{M'}{M} = \frac{1}{1 - \dfrac{te^2 M}{JK\rho}}\,.$$

76. For any substance, fluid or solid, it is easily proved without thermodynamic theory, that

$$\frac{M'}{M} = \frac{K}{N};$$

where K denotes the thermal capacity of a stated quantity of the substance under constant stress, and N its thermal capacity under constant strain (or thermal capacity when the body is prevented from change of shape or change of volume). For permanent gases, and generally for fluids approximately fulfilling Boyle's and Charles's laws as said above, k is proved by thermodynamic theory to be approximately constant. Its value for all gases for which it has been measured differs largely from unity, and probably also for liquids generally (except water near its temperature of maximum density).

On the other hand, for solids whether the stress considered be uniform compression in all directions, or of any other type, the value of $\dfrac{M'}{M}$ or $\dfrac{K}{N}$ differs but very little from unity; and both for solids and liquids it is far from constant at different temperatures (in the case of water it is zero at 3°·9 Cent., and varies as the square of the difference of the temperature from 3° 9 C., at all events for moderate differences from this critical temperature, whether above or below it). The following tables (III. and IV.) show the value of $\dfrac{M'}{M}$ or $\dfrac{K}{N}$, and the value of θ by the formula of § 74 above, for different fluid and solid substances at the temperature 15° Cent. (289° absolute scale). The first table (A) is for compression uniform in all directions; the second (B), necessarily confined to solids, is for the stress dealt with in "Young's Modulus," that is, normal pressure (positive or negative) on one set of parallel planes, with perfect freedom to expand or contract in all directions in these planes. A wire or rod pulled longitudinally is a practical application of the latter.

TABLE III. THERMODYNAMIC TABLE A.

Pressure equal in all directions—Ratio of Kinetic to Static Bulk-Modulus. Temperature 15° C. (289° absolute) J=42400 centimetres.

Substance.	Density.	Thermal Capacity per unit mass $=K$.	Expansibility $=e$.	Elevation of Temperature produced by a pressure of one gramme per square centimetre $=\dfrac{te}{JK\rho}$.	Static Bulk-Modulus in grammes per square centimetre $=M$.	Deduced value of $\dfrac{M'}{M}$ or $\dfrac{K}{N}$ $=\left(1-\dfrac{te^2M}{JK\rho}\right)^{-1}$
Air.............	·001225	·2375	·00346	·08106	1033	1·408
Distilled water......	1·000	1·000	·00016	·0000011	$22·63 \times 10^6$	1·004
Alcohol......	·795	·6148	·00106	·0000148	$11·4 \times 10^6$	1·217
Ether	·7005	·5157	·00155	·0000292	$8·07 \times 10^6$	1·577
Mercury	13·56	·0330	·00018	·00000274	$552·5 \times 10^6$	1·028
Glass, flint .	2·942	·1770	·000026	·000000340	423×10^6	1·004
Brass,drawn	8·471	·09391	·0000545	·000000466	1063×10^6	1·028
Iron	7·677	·1098	·0000395	·000000319	1485×10^6	1·019
Copper	8·843	·0949	·0000545	·000000443	1717×10^6	1·043

TABLE IV. THERMODYNAMIC TABLE B.

*Pressure parallel to one direction in a solid—Ratio of Kinetic to Static Young's
Modulus. Temperature 15° C. (289° absolute).*

Substance.	Density $=\rho$.	Thermal Capacity per unit mass $= K$.	Expansibility $= e$.	Lowering of Temperature produced by a pull of one gramme per square centimetre $= \dfrac{te}{JK\rho}$.	Static Young's Modulus in grammes per square centimetre $= M$.	Deduced value of $\dfrac{M'}{M}$ or $\dfrac{N}{K}$ $= \left(1 - \dfrac{te^2M}{JK\rho}\right)^{-1}$.
Zinc	7·008	·0927	·0000249	·000000308	873×10^6	1·0080
Tin	7·404	·0514	·000022	·000000394	417×10^6	1·00362
Silver	10·369	·0557	·000019	·000000224	736×10^6	1·00315
Copper	8·933	·0949	·000018	·000000145	1245×10^6	1·00325
Lead	11·215	·0293	·000029	·000000602	177×10^6	1·00310
Glass.........	2·942	·177	·0000086	·000000113	614×10^6	1·000600
Iron	7·553	·1098	·000013	·000000107	1861×10^6	1·00259
Platinum ...	21·275	·0314	·0000086	·0000000778	1704×10^6	1·00129

77. *Experimental Results.*—The following tables (v. vi. and
vii.) show determinations of moduluses of compression, of rigidity-
moduluses, and of Young's modulus, by various experimenters and
various methods. It will be seen that the Young's moduluses
obtained by Wertheim by vibrations, longitudinal or transverse, are
generally in excess of those which he found by static extension;
but the differences are enormously greater than those due to the
heating and cooling effects of elongation and contraction (§ 76),
and are to be certainly reckoned as errors of observation. It is
probable that his moduluses determined by static elongation are
minutely accurate; the discrepancies of those found by vibrations
are probably due to imperfections of the arrangements for carrying
out the vibrational method.

78. A question of great importance in the physical theory
of the elasticity of solids, "What changes are produced in the
moduluses of elasticity by permanent changes in its molecular condition?" has occupied the attention, no doubt, of every
"naturalist" who has studied the subject, and valuable contributions to its answer by experiment had been given by Wertheim
and other investigators, but solely with reference to Young's
modulus. In 1865 an investigation of the effect on the torsional
rigidity of wires of different metals, produced by stretching them

longitudinally beyond their limits of elasticity, was commenced in
the physical laboratory of the university of Glasgow in its old

TABLE V.   MODULUSES OF COMPRESSIBILITY.

Substance.	Moduluses of compressibility in grammes per square centimetre.	Temperature.	Authority.
Distilled water.....................	$22 \cdot 63 \times 10^6$	15°	⎫
Alcohol	$12 \cdot 4 \ \times 10^6$	0°	Amaury and
,, ...	$11 \cdot 4 \ \times 10^6$	15°	Descamps,
Ether.................................	$9 \cdot 5 \ \times 10^6$	0°	⎬ Comptes Rendus,
,, ...	$8 \cdot 07 \times 10^6$	14°	tome xvii. p. 1564
Bisulphide of carbon	$16 \cdot 3 \ \times 10^6$	14°	(1869).
Mercury	$552 \cdot 5 \ \times 10^6$	15°	⎭
Glass.................................	423×10^6	...	Everett's *Illustrations of the Centimetre-Gramme-Second System of Units.*
Another specimen	354×10^6	...	
Steel	1876×10^6	...	
Iron, wrought	1485×10^6	...	
Copper...............................	1717×10^6	...	
Brass, different specimens..⎱	Mean 1063×10^6	⎱ ... ⎰	Wertheim, *Ann. de Chim.*, 1848.
Tuff	55×10^6	...	Gray & Milne.
Clay rock	96×10^6	...	Quar. Jour. Geol. Soc. 1883.

TABLE VI.   RIGIDITY-MODULUSES.

Substance.	Modulus of Rigidity in grammes per square centimetre.	Authority.
Glass, different specimens............	Mean 150×10^6	Wertheim, *Annales de Chimie*, 1848.
Brass, different specimens............	Mean 350×10^6	
Glass..................................	243×10^6	⎫
Another specimen	240×10^6	
Brass, drawn	373×10^6	Everett's *Ill. of the Centimetre-Gramme-Second System of Units.*
Steel	834×10^6	
Iron, wrought	785×10^6	
Copper	542×10^6	
	456×10^6	⎭
German Silver.........................	496×10^6	Gray, *Trans. R. S. E.* 1880.
Tuff	102×10^6	Gray & Milne. Quar. Jour. Geol. Soc. 1883.
Clay rock	177×10^6	
Cocoon silk fibre......................	13×10^6	T. Gray.

buildings.   The following description of experiments and table (VIII.)
of results is extracted from the paper "On the Elasticity and
Viscosity of Metals," already quoted (§ 30 above), with reference to
viscosity and fatigue of elasticity.

"To determine rigidities by torsional vibrations, taking advantage of an obvious but most valuable suggestion made to me by Dr Joule, I used as vibrator in each case a thin cylinder of sheet brass, turned true outside and inside (of which the radius of gyration must be, to a very close degree of approximation, the arithmetic mean of the radii of the outer and inner cylindrical surfaces)*, supported by a thin flat rectangular bar, of which the square of the radius of gyration is one-third of the square of the distance from the centre to the corner. The wire to be tested passed perpendicularly through a hole in the middle of the bar, and was there firmly soldered. The cylinder was tied to the middle of the bar by light silk thread so as to hang with its axis vertical. Each wire, after having been suspended and stretched with just force enough to make it as nearly straight as was necessary for accuracy, was vibrated. Then it was stretched by hand (applied to the cross bar soldered to its lower end), and vibrated again, and stretched again, and so on till it broke." The experiments were performed with great care and accuracy by Mr Donald M‘Farlane. "The results, as shown in the accompanying table, (VIII.) were most surprising."

The highest and lowest rigidities found for copper in the table are as follows:

Highest rigidity 473×10^6, being that of a wire which had been softened by heating it to redness and plunging it into water, and which was found to be of density 8·91.

Lowest rigidity $393\cdot4 \times 10^6$, being that of a wire which had been rendered so brittle by heating it to redness, surrounded by powdered charcoal in a crucible, and letting it cool very slowly, that it could scarcely be touched without breaking it, and which had been found to be reduced in density by this process to as low as 8·674. The wires used were all commercial specimens—those of copper being all, or nearly all, cut from hanks supplied by the Gutta Percha Company, having been selected as of high electric conductivity, and of good mechanical quality, for submarine cables.

It ought to be remarked that the change of molecular condition produced by permanently stretching a wire or solid cylinder of metal is certainly a change from a condition which, if origin-

---

* It is exactly the square root of the mean of their squares.

TABLE VII. MODULUSES

Substance.	Density.	Young's Modulus.		Tenacity in 10^4 Grammes per square centimetre.
		In 10^6 Grammes per square centimetre.	Length Modulus: in 10^6 cms.	
Iron or Steel..................	...	About 2100	About 274	...
Wood	...	105 to 280	122 305	...
Stone	...	About 350	About 152	...
Slate	...	910 to 1120	...	...
Ice	...	...	640	...
Brass, cast.....................	...	645	...	127
„ wire	...	1001	...	343
Bronze, or gun metal	...	696	...	252
Copper, cast	...	...	...	134
„ sheet	...	...	...	211
„ bolts	...	...	...	253
„ wire..................	...	1195	...	422
Iron, cast	...	984 to 1610	...	94 to 204
„ wrought, plates.......	...	...	...	359
„ „ bars and bolts	...	2040	...	422 to 492
Steel, plates	...	...	...	563
„ bars.....................	...	2040 to 2953	...	703 to 914
Lead, sheet	...	51	...	23
Tin, cast	...	...	...	32
Zinc	...	...	...	49 to 56
Ash	...	113	...	120
Beech...........................	...	95	...	81
Birch	...	116	...	105
Cedar of Lebanon	...	34	...	80
Fir, red pine...................	...	118	...	91
Spruce	...	113	...	87
Fir, larch	...	79	...	68
Mahogany......................	...	88	...	105
Oak, European................	...	103	...	105
Sycamore	...	73	...	91
Teak, Indian	...	169	...	105
Lead, cast......................	11·215	177	16	22
„	...	198	...	...
„	...	199	...	...
Tin, cast	7·404	417·2	56	41·6
„	...	464	...	...
Cadmium, drawn.............	8·665	542	63	...
„	...	609	...	...
Gold, drawn	18·514	813	44	266 to 284
„	...	864	...	...

AND STRENGTHS.

Length of Modulus of Rupture in centimetres (or Tenacity in terms of Weight of Unit-Bulk). In 10⁴ cms.	Extreme Elastic Elongation.	Resilience per cubic centimetre in centimetre-grammes.	Resilience per Unit Mass in centimetres.	Authority.	Method of Determination.
...	...	...	...	Dr T. Young.	Probably flexure, (Young's *Works*, vol. ii. p. 133).
...	...	...	...	,,	
...	...	...	...	,,	
...	...	...	...	Rankine's "Rules and Tables."	
...	...	...	...	Bevan.	Flexure (see § 42).
...	·00198	1256	...	Rankine's "Rules and Tables."	
...	·00344	5905	...		
...	·00362	4562	...	,,	
...	...	...	...	,,	
...	...	...	...	,,	
...	...	...	...	,,	
...	·0036	7480	...	,,	
...	·00116	879	...	,,	
...	...	...	...	,,	
...	·00224	5120	...	,,	
...	...	...	...	,,	
...	·00324	1310	...	,,	
...	·00451	518	...	,,	
...	...	...	...	,,	
...	...	...	...	,,	
...	·0106	6370	...	,,	
...	·00853	3455	...	,,	
...	·00905	4752	...	,,	
...	·0235	9410	...	,,	
...	·00771	3510	...	,,	
...	·0077	3347	...	,,	
...	·00861	2927	...	,,	
...	·0120	6265	...	,,	
...	·0102	5352	...	,,	
...	·0125	5670	...	,,	
...	·00621	3262	...	,,	
...	·0012	...	12	Wertheim.	By direct elong.
...	...	...	...	,,	,, trans. vibr.
...	...	...	...	,,	,, longitud. ,,
...	·001	...	28	,,	,, trans. ,,
...	...	...	...	,,	,, longitud. ,,
...	...	...	...	,,	,, trans. ,,
...	...	...	...	,,	,, long. ,,
15	·0034	...	250	,,	,, direct elong.
...	...	...	...	,,	,, trans. vibr.

TABLE VII.

Substance.	Density.	Young's Modulus.		Tenacity in 10⁴ grammes per square centimetre.
		In 10⁶ Grammes per square centimetre.	Length Modulus: in 10⁸ cms.	
Gold, drawn	...	860	...	...
Silver, drawn	10·369	736	71	296
,, ...	...	782	...	...
,, ...	...	758	...	...
Zinc, common drawn	7·008	873	124	158
,, ...	...	879	...	...
,, ...	...	955	...	...
Palladium	11·35	1175	104	272
,, ...	...	1239	...	...
Copper, drawn	8·933	1245	139	410
,, ,, ...	...	1251	...	...
,, ,, ...	...	1254	...	...
,, annealed............	8·936	1052	118	316
,, ,, ...	...	1183	...	...
,, ,, ...	...	1254	...	...
Platinum, wire, fine........	21·166	1593	75	350
,, ,, ...	...	1618	...	...
,, ,, medium...	21·275	1704	...	...
,, ,, ...	...	1715	...	...
,, ,, ...	...	1716	...	...
,, ,, thick	21·259	1581	...	...
,, ,, ...	...	1616	...	...
Iron wire, common	7·553	1861	246	625 to 651
Steel, cast, drawn............	7·717	1955	...	838
,, ,, ...	...	1825	...	...
,, ,, ...	...	1982	...	...
Steel wire, English drawn. {	7·718	1881	244	859 to 991
	...	2071	...	...
,, ,, ,,	...	1944	...	...
Steel wire, common, tempered blue	7·420	1804	243	...
English steel, pianoforte wire	7·727	2049	265	2362
Copper wire	8·9	1150 to 1200	...	...
German silver	8·8	1380	157	
Silk fibre	...	72	...	...
Granite	2·63	416	159	17·1
Marble	2·70	400	148	13·7
Tuff	2·28	189	83	13·1
Clay rock	2·68	329	123	25·2
Slate	2·74	686	250	41·7

(*Continued*).

Length of Modulus of Rupture in centimetres (or Tenacity in terms of Weight of Unit-Bulk). In 10^4 cms.	Extreme Elastic Elongation.	Resilience per cubic centimetre in centimetre-grammes.	Resilience per Unit Mass in centimetres.	Authority.	Method of Determination.
...	...	...	...	Wertheim.	By long. vibr.
28	·0041	...	575	,,	,, direct elong.
...	...	...	...	,,	,, trans. vibr.
...	...	...	...	,,	,, long.　,,
23	·0018	...	204	,,	,, direct elong.
...	...	...	...	,,	,, trans. vibr.
...	...	...	...	,,	,, long.　,,
23	·0023	...	277	,,	,, direct elong.
...	...	...	...	,,	,, trans. vibr.
46	·0033	...	756	,,	,, direct elong.
...	...	...	...	,,	,, trans. vibr.
...	...	...	...	,,	,, long.　,,
35	·003	...	531	,,	,, direct elong.
...	...	...	...	,,	,, trans. vibr.
...	...	...	...	,,	,, long.　,,
17	·0022	...	187	,,	,, trans.　,,
...	...	...	...	,,	,, long.　,,
...	...	...	...	,,	,, direct elong.
...	...	...	...	,,	,, trans.　,,
...	...	...	...	,,	,, long.　,,
...	...	...	...	,,	,, trans. vibr.
...	...	...	...	,,	,, long.　,,
85	·0034	...	1450	,,	,, direct elong.
108	...	...	...	,,	,,　,,　,,
...	...	...	...	,,	,, trans. vibr.
...	...	...	...	,,	,, long.　,,
125	·0050	...	2945	,,	,, direct elong.
...	...	...	...	,,	,, trans. vibr.
...	...	...	...	,,	,, long.　,,
...	...	...	...	,,	,, direct elong.
306	·0115	...	17600	D. M'Farlane.	,,　,,　,,
...	...	...	...	,,	,,　,,　,,
...	...	...	...	T. Gray.	
...	...	...	...	T. Gray.	
...	·00055	470	...	T. Gray & Milne.	
...	·00060	411	...	,,	
...	·00062	406	...	,,	
...	·00054	680	...	,,	
...	·00062	1293	...	,,	

TABLE VIII.   EFFECT OF PERMANENT LONGITUDINAL STRETCHING
ON TORSIONAL RIGIDITY.

Substance.	Length of Wire in centimetres. l.	Volume in cubic centimetres. V.	Density. ρ.	Moment of Inertia of Vibrator Wk^2.	Time of Vibration one way or (half period) in seconds. T.	Rigidity in 10^b grammes weight per square centimetre $\dfrac{2\pi^3l^3Wk^2}{gT^2V^2}$.
Aluminium[1]...	60·3	1·1845	2·764	31771	1·14	241
Zinc[2]..........	304·9	2·351	7·105	31896	4·31	359·6
Brass............	237·7	...	...	...	4·76	410·3
,, ...	248·3	...	...	...	5·456	354·8
,, ...	261·9	1·703	8·398	...	5·96	350·1
Copper	2435·0	15·30	8·91	38186	16·375	448·7
,, ...	,,	,,	,,	61412	20·77	448·4
Copper[3]........	214·4	1·348	8·864	31771	5·015	433·0
,, ...	,,	,,	,,	61412	6·982	431·8
Copper[4]........	143·7	·9096	8·674	...	3·381	393·4
Copper[5]........	286·8	...	...	20612	4·245	442·9
,, •••......	291	...	...	,,	4·375	435·6
,, ...	293	...	...	,,	4·417	436·2
,, ...	296·1	...	...	,,	4·500	433·8
,, ...	300·0	...	...	,,	4·588	434·0
,, ...	303·4	...	...	,,	4·646	437·8
,, ...	309·3	...	...	,,	4·833	428·6
,, ...	313·2	...	...	,,	4·931	427·5
,, ...	317·4	1·962	8·835	,,	5·040	425·9
Copper[6]........	315·6	...	...	31771	8·155	442·3
,, ...	235·5	...	...	,,	9·425	432·2
,, ...	251·9	·827	8·872	,,	10·463	428·6
Copper[7]........	253·2	1·580	8·91	...	5·285	472·9
,, ...	262·8	...	...	...	5·640	464·3
,, ...	270·4	...	...	...	5·910	460·4
,, ...	278·7	...	...	...	6·20	458·5
,, ...	287·9	...	...	...	6·5325	455·0
,, ...	297·5	...	...	...	6·8195	451·0
,, ...	308·8	...	...	...	7·3075	448·9
Copper[8]........	256·5	1·6145	8·90	...	4·2226	463·5
,, ...	267·9	...	...	...	4·5625	453·3
,, ...	280·1	...	...	...	4·915	446·2
,, ...	292·2	...	...	...	5·240	445·5
,, ...	301·9	...	...	...	5·532	438·2
Soft Iron[9]......	316·8	...	...	...	6·655	791·4
,, ...	322·1	...	...	...	6·88	778·3
,, ...	335·1	...	...	...	7·301	779·0
,, ...	347·4	...	...	...	7·768	766·6
,, ...	366·0	1·357	7·657	...	8·455	756·0
Platinum	39·4	·1745	20·805	20612	2·05	622·25
Gold	65·9	·1825	19·8	10902	...	281
Silver	75·7	·1185	10·21	10967	...	270

Remarks on Table VIII.

[1] Only forty vibrations from initial arc of convenient amplitude could be counted. Had been stretched considerably before this experiment.

[2] So viscous that only twenty vibrations could be counted. Broke in stretching.

[3] A piece of the preceding stretched.

[4] The preceding made red-hot in a crucible filled with powdered charcoal and allowed to cool slowly, became very brittle : a part of it with difficulty saved for the experiment.

[5] Another piece of the long (2435 centims.) wire; stretched by successive simple tractions.

[6] A finer gauge copper wire ; stretched by successive tractions.

[7] A finer gauge copper wire, softened by being heated to redness and plunged in water. A length of 260 centimetres cut from this, suspended, and elongated by successive tractions.

[8] Another piece of 260 centimetres cut from the same, and similarly treated.

[9] One piece, successively elongated by simple tractions till it broke.

ally isotropic, becomes æolotropic as to some qualities[*], and that the changed conditions may therefore be presumed to be æolotropic as to elasticity.
If so, the rigidities corresponding to the direct and
diagonal distortions (indicated by No. 1 and No. 2,
in fig. 15) must in all probability become different
from one another when a wire is permanently
stretched, instead of being equal as they must be
when its substance is isotropic. It becomes,
therefore, a question of extreme interest to
find whether rigidity No. 2 is not *increased* by
this process, which, as is proved by the experiments above described, diminishes, to a very remarkable

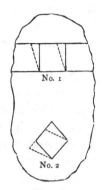

Fig. 15.

degree, the rigidity No. 1. The most obvious experiment,
and indeed the only practicable experiment adapted to answer
this question, for a wire or round bar is that of Cagniard-
Latour, in which an accurate determination of the difference produced in the *volume* of the substance is made by applying and
removing longitudinal traction within its limits of elasticity. With
the requisite apparatus, which must be much more accurate than
that of Cagniard-Latour, a most important and interesting investigation might be made. The results, along with an accurate
determination of the Young's modulus for the particular case, give
(§ 47 above) the modulus of compression, and the rigidity No. 2.
Regnault suggested the use of hollow instead of solid cylinders, to be
subjected to longitudinal pull, and (after the manner of the bulb
and tube of a thermometer) a capillary tube to aid in measuring
changes of volume of the hollow; and Wertheim, adopting this
excellent suggestion, obtained seemingly very accurate results for
brass and glass, which are given in the tables (V. and VI.) of § 77
above.

79. The following tables (IX. and X.) show the effects of
differences of temperature on the rigidity-modulus, and modulus
of compressibility of various substances.

[Note of February 19, 1886. The original paper contained
under this paragraph, a table of results regarding the effect of
changes of temperature on the Young's Modulus, extracted from

---

[*] For example, see paper "On Electrodynamic Qualities of Metals," *Trans. Roy.
Soc.* Feb. 1856 [Art. XCI. Vol. II. above].

Wertheim's "Mémoires" on Elasticity, *Ann. de Chim. et Phys.*, tom. xii. (1884), page 443; but on looking into the matter just now, I see that the results must be very far wrong, and therefore I do not reproduce the table.

An experiment made in the Cavendish Laboratory, by Messrs Macleod and Clarke (see *Phil. Trans.* Vol. 171, Part i. 1880), the object of which was to ascertain the change in the period of a steel tuning fork, due to change of temperature, enables us to calculate the corresponding change in the Young's Modulus. They find that the period augments at the rate of 11.0×10^{-5} per degree centigrade of elevation of temperature. Now the linear expansion of steel is 1.2×10^{-5} per degree, and therefore the period of the steel fork would diminish at the rate of $.6 \times 10^{-5}$ per degree if there were no change of the Young's Modulus. The amount of the augmentation of period due to diminution of the Young's Modulus must therefore be 11.6×10^{-5}. The proportionate diminution of the Young's Modulus must be twice this amount, because, for the same linear dimensions, the period is inversely as the square root of the Young's Modulus. We conclude that the Young's Modulus diminishes at the rate of 23.2×10^{-5} per degree centigrade of elevation of temperature. W. T.]

The change in the rigidity-modulus produced by change of temperature was investigated by F. Kohlrausch and Francis E. Loomis*. They found that it is expressed by the formula $n = n_0 (1 - \alpha t - \beta t^2)$, where n_0 denotes the value of the rigidity-modulus at $0°$ C., n its value at temperature t, and α, β, co-efficients the values of which for iron, copper, and brass are as follows:—

TABLE IX.—VALUES OF COEFFICIENTS IN KOHLRAUSCH'S FORMULA FOR CHANGE OF RIGIDITY-MODULUS DUE TO CHANGE OF TEMPERATURE.

	α	β
Iron	0·000447	0·00000012
Copper	0·000520	0·00000028
Brass	0·000428	0·00000136

* *Pogg. Ann.* Bd. cxli. 1870, pp. 350—366; or *Amer. Jour. Science*, Vol. L. 1870, pp. 481—503.

TABLE X.   EFFECT OF CHANGE OF TEMPERATURE ON MODULUS
OF COMPRESSIBILITY OF WATER, ALCOHOL, AND ETHER*.

Temp. Cent.	Modulus of compressibility in million grammes weight per square centimetre.			Authority.
	Water.	Alcohol.	Ether.	
0°	20·6	12·4	9·5	For water,
1·5	20·2	...	...	Grassi, *Ann.*
4·1	20·7	...	...	*de Chim.*, tome
10·8	21·5	...	...	xxxi. (1851).
13·4	21·6	...	...	
14·0	...	...	8·07	For ether and
15·0	...	11·4	...	alcohol,
18·0	22·4	...	...	Amaury and
25·0	22·6	...	...	Descamp,
34·0	22·8	...	...	*Comptes Ren-*
43·0	23·3	...	...	*dus*, tome xvii.
53·0	23·5	...	...	p. 1564 (1869).

80. *Tempering soft iron by long-continued stress.*—Preliminary
experiments by Mr J. T. Bottomley towards the investigation
promised in § 5 above have discovered a very remarkable property
of soft iron wire respecting its ultimate tensile strength. Eight
different specimens, tested by the gradual application of more and
more weight within ten minutes of time in each case until the
wire broke, bore from 43½ to 46 lbs. (average 45·2) just before breaking, with elongations of from 17 per cent. to 22 per cent. Another
specimen left with 43 lbs. hanging on it for 24 hours, and then
tested by the gradual addition of weights during 25 minutes till
it broke, bore 49¼ lbs. before breaking, with elongation of 15 per
cent. Another left for 3 days 11 hours 40 minutes with 43 lbs. hanging on it, and then tested by the gradual addition of weights during
34 minutes till it broke, bore 51½ lbs. just before breaking, with
elongation of 14·4 per cent. Another specimen of the same wire
was set up with 40 lbs. hanging on it on the 5th of July, 1877, on
the 6th of July 3 lbs. were added, on the 9th 1½ lb. more, and on
the 10th ¾ lb. more, making in all on this date 45¼ lbs. Thence-
forward day by day, with occasional intervals of two days or three
days, the weight was increased first by half a pound at a time, and
latterly by a quarter of a pound at a time, until on the 3rd of
September the wire broke with 57¼ lbs. (elongation not recorded).

* The modulus seems to be a minimum near the temperature of maximum
density.

This gradual addition of weight therefore had increased the tensile strength of the metal by 26·7 per cent.!

81. *Experiments made for this article.*—There are many subjects in the theory of elasticity regarding which information, to be obtained by experiment only, is greatly wanted. Several of these have been pointed out above (§ 21), and while this article was being put in type, experiments were made in the physical laboratory of the University of Glasgow with a view of answering some of the questions proposed. Mr Donald M'Farlane, besides making the experiments referred to in §§ 3 and 21 above, investigated the effects of applying different amounts of pull to a steel pianoforte wire which had been twisted to nearly its limits of elasticity, and which was kept twisted by means of a couple. The results proved a deviation from Hooke's law by showing a diminution of the torsional rigidity, of about 1·6 per cent., produced by hanging a weight of 112 lbs. on the wire. Of this 1·2 per cent. is accounted for by elongation, and by shrinkage of the diameter, leaving ·4 per cent. of diminution of the rigidity-modulus.

It was also found that when the wire was twisted far beyond its limits of elasticity, and then freed from torsional stress, a weight hung on it caused it to untwist slightly. When the weight was removed and reapplied again and again, the lower end of the wire always turned in the same direction as the permanent twist when the weight was removed, and in the opposite direction when it was applied. This result shows the development of æolotropic quality in the substance of the wire, according to which a small cube cut from any part of it far out from the axis, with two sides of the cube parallel to the length, and the other two pairs of sides making angles of 45° with the length, would show different compressibilities in the directions perpendicular to the last-mentioned pairs of sides.

Another very interesting result, discovered in the course of these experiments, was that when a length of five metres of the steel wire, with a weight of 39 lbs. hung upon it, was twisted to the extent of 95 turns, it became gradually elongated to the extent of $\frac{1}{1600}$ of the length of the wire; when farther twisted it began to shorten till, when 25 turns more had been given (in all 120 turns), the weight had risen from its lowest position through

nearly $\frac{1}{6000}$ of the length of the wire, so that the previous elongation had been diminished by about $\frac{1}{4}$ of its amount.

Experiments were also made by Mr Andrew Gray and Mr
Thomas Gray for the purpose of determining the effects of various
amounts of permanent twist, in altering the rigidity-modulus and
the Young's modulus of wires of copper, iron, and steel. A copper
wire, of 3·15 metres length and ·154 centimetre diameter, No. 17
B.W.G., which had a rigidity-modulus of 442 million grammes per
square centimetre to begin with, was found to have 420 after 10
turns, showing a diminution in the modulus of $\frac{1}{20}$ of its own
amount. The diminution went on rapidly until 100 turns of permanent twist had been given, when the modulus was as low as
385. The diminution of the modulus continued with further
twist, but very slowly, up to 1225 turns, when the modulus was
found to be 371, showing a diminution to the extent of 1/6 of its
original value! There was little farther change until 1400 turns
had been given, when the modulus began to increase. At 1525
turns its value was 373, and at 1625 it was 377. Twenty turns
more broke the wire before the torsional elasticity had been again
determined.

A piece of iron wire of nearly the same length (about three
metres), but of smaller diameter (·087 centimetre), showed continued
diminution of torsional rigidity as far as 1350 turns of permanent
twist, when the diminution had amounted to 14 per cent. of the
primitive value; 36 turns more broke the wire before another
determination of torsional rigidity had been made.

The steel pianoforte wire also showed a diminution of torsional
rigidity with permanent twist, and (as did the copper wire) showed
first a diminution and then a slight augmentation. The amount
of the diminution in the steel wire was enormously greater than
the surprisingly great amount which had been discovered in the
copper wire, and the ultimate augmentation was considerably
greater in the steel than what it had been in the copper before
rupture. Thus after 473 turns of permanent twist the torsional
modulus had diminished from 751 million grammes per square
centimetre to 414! 95 more turns of permanent twist augmented
the rigidity from 414 to 430, and when farther twisted the wire
broke before another observation had been made. The vibrator
used in these experiments was a cylinder of lead weighing 56 lbs.,

which was kept hanging on the wire while it was being twisted, and in fact during the whole of about 100 hours from the beginning of the experiment till the wire broke, except on two occasions for a few minutes, while the top fastening which had given way was being resoldered. The period of vibration was augmented from 39·375 seconds to 51·9 seconds by the twist. The wire took the twist very irregularly, some parts not beginning to show signs of permanent twist till near the end of the experiment.

In two specimens of copper wire of the same length and gauge as those described above, the Young's modulus was found to be increased 10 per cent by 100 turns of permanent twist.

Five metres of the steel pianoforte wire, bearing a weight of 39 lbs., was, in one of Mr M'Farlane's experiments, twisted 120 turns, and then allowed to untwist, and $38\frac{1}{4}$ turns came out, leaving the wire in equilibrium with $81\frac{3}{4}$ turns of permanent twist. Its Young's modulus was then found not to differ as much as $\frac{1}{2}$ per cent. from the value it had before the wire was twisted.

MATHEMATICAL THEORY OF ELASTICITY*.

PART I.—ON STRESSES AND STRAINS†.

CHAPTER I.—*Initial Definitions and Explanations.*

DEF. I. A stress is an equilibrating application of force to a body.

COR. The stress on any part of a body in equilibrium will thus signify the force which it experiences from the matter touching that part all round, whether entirely homogeneous with itself, or only so across a portion of its bounding surface.

DEF. 2. A strain is any definite alteration of form or dimensions experienced by a solid.

* The substance of Chaps. I.—XVI. of this part of the present article was read before the Royal Society on April 24, 1856, and was published in the *Transactions* for that year. Chapter XVII., containing the mathematical theory of Waves in an æolotropic or isotropic elastic solid, is new.

† These terms were first definitively introduced into the Theory of Elasticity by Rankine, and have been found very valuable in writing on the subject. It will be seen that there is a slight deviation from Rankine's definition of the word "stress." It is here applied to the direct action experienced by a body from the matter around it, and not, as proposed by him, to the elastic reaction of the body equal and opposite to that action.

Examples.—Equal and opposite forces acting at the two ends of a wire or rod of any substance constitute a stress upon it. A body pressed equally all round—for instance, any mass touched by air on all sides—experiences a stress. A stone in a building experiences stress if it is pressed upon by other stones, or by any parts of the structure, in contact with it. Any part of a continuous solid mass, simply resting on a fixed base, experiences stress from the surrounding parts in consequence of their weight. The different parts of a ship in a heavy sea experience stresses from which they are exempt when the water is smooth.

If a rod of any substance become either longer or shorter, it is said to experience a strain. If a body be uniformly condensed in all directions it experiences a strain. If a stone, a beam, or a mass of metal in a building, or in a piece of framework, becomes condensed or dilated in any direction, or bent, or twisted, or distorted in any way, it is said to experience a strain, to become strained, or often in common language, simply "to strain." A ship is said to "strain" if in launching, or when working in a heavy sea, the different parts of it experience relative motions.

CHAPTER II.— *Homogeneous Stresses and Homogeneous Strains.*

DEF. 1. A stress is said to be homogeneous throughout a body when equal and similar portions of the body, with corresponding lines parallel, experience equal and parallel pressures or tensions on corresponding elements of their surfaces.

COR. 1. When a body is subjected to any homogeneous stress, the mutual tension or pressure between the parts of it on two sides of any plane, amounts to the same per unit of surface as that between the parts on the two sides of any parallel plane; and the former tension or pressure is parallel to the latter.

DEF. 2. A strain is said to be homogeneous throughout a body, or the body is said to be homogeneously strained, when equal and similar portions, with corresponding lines parallel, experience equal and similar alterations of dimensions.

COR. 1. All the particles of the body in parallel planes remain in parallel planes, when the body is homogeneously strained in any way.

Examples.—A long uniform rod, if pulled out, or a pillar loaded with a weight, will experience a uniform strain, except near its ends, where there will be a sensible heterogeneousness of the strain, because of the end attachments, or other circumstances preventing the ends from shrinking or expanding laterally to the same extent as the middle does.

A piece of cloth held in a plane, and distorted so that a warp and woof, instead of being perpendicular to one another, become two sets of parallels cutting one another obliquely, experiences a homogeneous strain. The strain is heterogeneous as to intensity, from the axis to the surface of a cylindrical wire under torsion, and heterogeneous as to direction in different positions in a circle round the axis.

CHAPTER III.—*On the Distribution of Force in a Stress.*

THEOREM.—In every homogeneous stress there is a system of three rectangular planes, each of which is perpendicular to the direction of the mutual force between the parts of the body on its two sides.

For let $P(X)$, $P(Y)$, $P(Z)$ denote the components, parallel to X, Y, Z, any three rectangular lines of reference, of the force

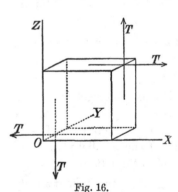

experienced per unit of surface at any portion of the solid bounded by a plane parallel to (Y, Z); $Q(X)$, $Q(Y)$, $Q(Z)$, the corresponding components of the force experienced by any surface of the solid parallel to (Z, X); and $R(X), R(Y), R(Z)$, those of the force at a surface parallel to (X, Y). Now, by considering the equilibrium of a cube of the solid with faces parallel to the planes of reference (fig. 15), we

Fig. 16.

see that the couple of forces $Q(Z)$ on its two faces perpendicular to Y is balanced by the couple of forces $R(Y)$ on the faces perpendicular to Z. Hence we must have

$$Q(Z) = R(Y).$$

Similarly it is seen that

$$R(X) = P(Z)$$

and

$$P(Y) = Q(X).$$

For the sake of brevity, these pairs of equal quantities (being tangential forces respectively perpendicular to X, Y, Z) may be denoted by $T(X)$, $T(Y)$, $T(Z)$.

Consider a tetrahedral portion of the body (surrounded it may be with continuous solid) contained within three planes A, B, C, through a point O, parallel to the planes of the pairs of lines of reference, and a fourth plane K cutting these at angles α, β, γ respectively; so that as regards the areas of the different sides we shall have

$$A = K \cos \alpha, \quad B = K \cos \beta, \quad C = K \cos \gamma.$$

The forces actually experienced by the sides A, B, C have nothing to balance them except the force actually experienced by K. Hence those three forces must have a single resultant, and the force on K must be equal and opposite to it. If, therefore, the force on K per unit of surface be denoted by F and its direction cosines by l, m, n, we have

$$F \cdot K \cdot l = P(X) A + T(Z) B + T(Y) C,$$
$$F \cdot K \cdot m = T(Z) A + Q(Y) B + T(X) C,$$
$$F \cdot K \cdot n = T(Y) A + T(X) B + R(Z) C,$$

and, by the relations between the cases stated above, we deduce

$$Fl = P(X) \cos \alpha + T(Z) \cos \beta + T(Y) \cos \gamma,$$
$$Fm = T(Z) \cos \alpha + Q(Y) \cos \beta + T(X) \cos \gamma,$$
$$Fn = T(Y) \cos \alpha + T(X) \cos \beta + R(Z) \cos \gamma.$$

Hence the problem of finding (α, β, γ), so that the force $F(l, m, n)$ may be perpendicular to it, will be solved by substituting $\cos \alpha$, $\cos \beta$, $\cos \gamma$ for l, m, n in these equations. By the elimination of $\cos \alpha$, $\cos \beta$, $\cos \gamma$ from the three equations thus obtained, we have the well-known cubic determinantal equation, of which the roots, necessarily real, lead, when no two of them are equal, to one and only one system of three rectangular axes having the stated property.

DEF. The three lines thus proved to exist for every possible homogeneous stress are called its axes. The planes of their pairs are called its Normal Planes; the mutual forces between parts of the body separated by these planes, or the forces on portions of the bounding surface parallel to them, are called the Principal Tensions.

COR. 1. The Principal Tensions of the stress are the roots of the determinant cubic referred to in the demonstration.

COR. 2. If a stress be specified by the notation $P(X)$, &c., as explained above, its Normal Planes are the principal planes of the surface of the second degree whose equation is

$$P(X)X^2 + Q(Y)Y^2 + R(Z)Z^2 + 2T(X)YZ$$
$$+ 2T(Y)ZX + 2T(Z)XY = 1,$$

and its Principal Tensions are equal to the reciprocals of the squares of the lengths of the semi-principal axes of the same surface, (quantities which are negative of course for the principal axis or axes which do not cut the surface, when the surface is a hyperboloid of one or of two sheets).

COR. 3. The ellipsoid whose equation, referred to the rectangular axes of a stress, is

$$(1 - 2eF)X^2 + (1 - 2eG)Y^2 + (1 - 2eH)Z^2 = 1,$$

where F, G, H denote the principal tensions, and e any infinitely small quantity, *represents the stress*, in the following manner :—

From any point P in the surface of the ellipsoid draw a line in the tangent plane, half-way towards the point where this plane is cut by a perpendicular to it through the centre; and from the end of the first-mentioned line draw a radial line to meet the surface of a sphere of unit radius concentric with the ellipsoid. The tension at this point of the surface of a sphere of the solid is in the line from it to the point P; and its amount per unit of surface is equal to the length of that infinitely small line, divided by e.

COR. 4. Any stress is fully specified by six quantities, viz., its three principal tensions (F, G, H), and three angles (θ, ϕ, ψ) or three numerical quantities equivalent to the nine direction cosines specifying its axes.

CHAPTER IV.—*On the Distribution of Displacement in a Strain.*

PROP. 1. In every homogeneous strain any part of the solid bounded by an ellipsoid remains bounded by an ellipsoid.

For all particles of the solid in a plane remain in a plane, and two parallel planes remain parallel. Consequently every system of conjugate diametral planes of an ellipsoid of the solid, retain the property of conjugate diametral planes with reference to the altered curve surface containing the same particles. This altered surface is therefore an ellipsoid.

PROP. 2. There is a single system (and only a single system, except in the cases of symmetry) of three rectangular planes for every homogeneous strain, which remain at right angles to one another in the altered solid.

DEF. 1. These three planes are called the normal planes of the strain, or simply the strain-normals. Their lines of intersection are called the axes of the strain. The elongations of the solid per unit of length along these axes, or perpendicular to these planes, are called the Principal Elongations of the strain.

Remark. The preceding propositions and definitions are not limited to infinitely small strains, but are applicable to whatever extent the body may be strained.

PROP. 3. If a body, while experiencing an infinitely small strain, be held with one point fixed, and the normal planes of the strain parallel to three fixed rectangular planes through the point O, a sphere of the solid of unit radius having this point for its centre becomes, when strained, an ellipsoid whose equation, referred to the strain-normals through O, is

$$(1 - 2x)\,X^2 + (1 - 2y)\,Y^2 + (1 - 2z)Z^2 = 1,$$

if x, y, z denote the elongations of the solid per unit of length, in the directions respectively perpendicular to these three planes; and the position, on the surface of this ellipsoid, attained by any particular point of the solid, is such that if a line be drawn in the tangent plane, half-way to the point of intersection of this plane with a perpendicular from the centre, a radial line drawn through its extremity cuts the primitive spherical surface in the primitive position of that point.

COR. 1. For every stress, there is a certain infinitely small strain, and conversely, for every infinitely small strain, there is a certain stress, so related that if, while the strain is being acquired, the centre and the strain-normals through it are unmoved, the absolute displacements of particles belonging to a spherical surface of the solid represent, in intensity (according to a definite convention as to units for the representation of force by lines) and in direction, the force (reckoned as to intensity, in amount per unit of area) experienced by the enclosed sphere of the solid, at the different parts of its surface, when subjected to the stress.

COR. 2.   Any strain is fully specified by six quantities, viz., its three principal elongations, and three angles (θ, ϕ, ψ), or nine direction cosines, equivalent to three independent quantities specifying its axes.

DEF. 2.   A stress and an infinitely small strain related in the manner defined in Cor. 1 of Prop. 3 above, are said to be of the same type.   The ellipsoid by means of which the distribution of force over the surface of a sphere of unit radius is represented in one case, and by means of which the displacements of particles from the spherical surface are shown in the other, may be called the geometrical type of either.

COR.   Any stress- or strain-type is fully specified by *five* quantities, viz., two ratios between its principal strains or elongations and three quantities specifying the angular position of its axes.

CHAPTER V.   *Conditions of Perfect Concurrence between Stresses and Strains.*

DEF. 1.   Two stresses are said to be coincident in direction, or to be perfectly concurrent, when they only differ in absolute magnitude.   The same relative designations are applied to two strains differing from one another only in absolute magnitude.

COR.   If two stresses or two strains differ by one being reverse to the other, they may be said to be negatively coincident in direction, or to be directly opposed or directly contrary to one another.

DEF. 2.   When a homogeneous stress is such, that the normal component of the mutual force between the parts of the body on the two sides of any plane whatever through it, is proportional to the augmentation of distance between the same plane and another parallel to it and initially at unit distance, due to a certain strain experienced by the same body, the stress and the strain are said to be perfectly concurrent; also to be coincident in direction. The body is said to be yielding directly to a stress applied to it, when it is acquiring a strain thus related to the stress; and in the same circumstances, the stress is said to be working directly on the body, or to be acting in the same direction as the strain.

COR. 1.　Perfectly concurrent stresses and strains are of the same type.

COR. 2.　If a strain is of the same type as the stress, its reverse will be said to be negatively of the same type, or to be directly opposed to the strain.　A body is said to be working directly against a stress applied to it when it is acquiring a strain directly opposed to the stress; and in the same circumstances, the matter round the body is said to be yielding directly to the reactive stress of the body upon it.

CHAPTER VI.—*Orthogonal Stresses and Strains.*

DEF. 1.　A stress is said to act right across a strain, or to act orthogonally to a strain, or to be orthogonal to a strain, if work is neither done upon nor by the body in virtue of the action of the stress upon it while it is acquiring the strain.

DEF. 2.　Two stresses are said to be orthogonal when either coincides in direction with a strain orthogonal to the other.

DEF. 3.　Two strains are said to be orthogonal when either coincides in direction with a stress orthogonal to the other.

Examples.—(1) A uniform cubical compression, and any strain involving no alteration of volume, are orthogonal to one another.

(2) A simple extension or contraction in parallel lines unaccompanied by any transverse extension or contraction, that is, " a simple longitudinal strain," is orthogonal to any similar strain in lines at right angles to those parallels.

(3) A simple longitudinal strain is orthogonal to a " simple tangential strain*" in which the sliding is parallel to its direction or at right angles to it.

(4) Two infinitely small simple tangential strains in the same plane†, with their directions of sliding mutually inclined at an angle of 45°, are orthogonal to one another.

---

* That is, a homogeneous strain in which all the particles in one plane remain fixed, and other particles are displaced parallel to this plane.

† " The plane of a simple tangential strain," or the plane of distortion in a simple tangential strain, is a plane perpendicular to that of the particles supposed to be held fixed, and parallel to the lines of displacement of the others.

(5) An infinitely small simple tangential strain is orthogonal to every infinitely small simple tangential strain, in a plane either parallel to its plane of sliding or perpendicular to its line of sliding.

Chapter VII.—*Composition and Resolution of Stresses and of Strains.*

Any number of simultaneously applied homogeneous stresses are equivalent to a single homogeneous stress which is called their resultant. Any number of superimposed homogeneous strains are equivalent to a single homogeneous resultant strain. Infinitely small strains may be independently superimposed; and in what follows it will be uniformly understood that the strains spoken of are infinitely small, unless the contrary is stated.

Examples.—(1) A strain consisting simply of elongation in one set of parallel lines, and a strain consisting of equal contraction in a direction at right angles to it, applied together, constitute a single strain, of the kind which that described in Example (3) of the preceding chapter (VI.) is when infinitely small, and is called a plane distortion, or a simple distortion. It is also sometimes called a simple tangential strain, and when so considered, its plane of sliding may be regarded as either of the planes bisecting the angles between planes normal to the lines of the component longitudinal strains.

(2) Any two simple distortions in one plane may be reduced to a single simple distortion in the same plane.

(3) Two simple distortions not in the same plane have for their resultant a strain which is a distortion unaccompanied by change of volume, and which may be called a compound distortion.

(4) Three equal longitudinal elongations or condensations in three directions at right angles to one another are equivalent to a single dilatation or condensation equal in all directions. The single stress equivalent to three equal tensions or pressures in directions at right angles to one another is a negative or positive pressure equal in all directions.

(5) If a certain stress, or infinitely small strain, be defined (Chapter III. Cor. 3, or Chapter IV.) above by the ellipsoid

$$(1 + A) X^2 + (1 + B) Y^2 + (1 + C) Z^2 + D YZ + EZX + FXY = 1,$$

and another stress or infinitely small strain by the ellipsoid

$$(1 + A')X^2 + (1 + B')Y^2 + (1 + C')Z^2 + D'YZ + E'ZX + F'XY = 1,$$

where A, B, C, D, E, F, &c., are all infinitely small, their resultant stress or strain is that represented by the ellipsoid

$$(1 + A + A')X^2 + (1 + B + B')Y^2 + (1 + C + C')Z^2 + (D + D')YZ$$
$$+ (E + E')ZX + (F + F')XY = 1.$$

CHAPTER VIII.—*Specification of Strains and Stresses by their Components according to chosen Types.*

PROP. Six stresses or six strains of six distinct arbitrarily chosen types may be determined to fulfil the condition of having a given stress or a given strain for their resultant, provided those six types are so chosen that a strain belonging to any one of them cannot be the resultant of any strains whatever belonging to the others.

For, just six independent parameters being required to express any stress or strain whatever, the resultant of any set of stresses or strains may be made identical with a given stress or strain by fulfilling six equations among the parameters which they involve; and therefore the magnitudes of six stresses or strains belonging to the six arbitrarily chosen types may be determined, if their resultant be assumed to be identical with the given stress or strain.

COR. Any stress or strain may be numerically specified in terms of numbers expressing the amounts of six stresses or strains of six arbitrarily chosen types which have it for their resultant.

Types arbitrarily chosen for this purpose will be called types of reference. The specifying elements of a stress or strain will be called its components according to types of reference. The specifying elements of a strain may also be called its co-ordinates, with reference to the chosen types.

Examples.—(1) Six strains in each of which one of the six edges of a tetrahedron of the solid is elongated while the others remain unchanged, may be used as types of reference for the specification of any kind of strain or stress. The ellipsoid representing any one of those six types will have its two circular sections parallel to the faces of the tetrahedron which do not contain the stretched side.

(2) Six strains consisting, any one of them, of an infinitely small alteration either of one of the three edges, or of one of the three angles between the faces, of a parallelepiped of the solid, while the other five angles and edges remain unchanged, may be taken as types of reference, for the specification of either stresses or strains. In some cases, as for instance in expressing the probable elastic properties of a crystal of Iceland spar, it might possibly be convenient to use an oblique parallelepiped for such a system of types of reference; but more frequently it will be convenient to adopt a system of types related to the deformations of a cube of the solid.

CHAPTER IX.—*Orthogonal Types of Reference.*

DEF. A normal system of types of reference is one in which the strains or stresses of the different types are all six mutually orthogonal (fifteen conditions). A normal system of types of reference may also be called an orthogonal system. The elements specifying, with reference to such a system, any stress or strain, will be called orthogonal components or orthogonal co-ordinates.

Examples.—(1) The six types described in Example (2) of Chapter VIII. are clearly orthogonal, if the parallelepiped referred to is rectangular. Three of these are simple longitudinal extensions, parallel to the three sets of rectangular edges of the parallelepiped. The remaining three are plane distortions parallel to the faces, their axes bisecting the angles between the edges. They constitute the system of types of reference uniformly used hitherto by writers on the theory of elasticity.

(2) The six strains in which a spherical portion of the solid is changed into ellipsoids having the following equations—

$$(1 + A)X^2 + Y^2 + Z^2 = 1,$$
$$X^2 + (1 + B)Y^2 + Z^2 = 1,$$
$$X^2 + Y^2 + (1 + C)Z^2 = 1,$$
$$X^2 + Y^2 + Z^2 + DYZ = 1,$$
$$X^2 + Y^2 + Z^2 + EZX = 1,$$
$$X^2 + Y^2 + Z^2 + FXY = 1,$$

are of the same kind as those considered in the preceding example, and therefore constitute a normal system of types of reference. The resultant of the strains specified, according to those equations, by the elements A, B, C, D, E, F, is a strain in which the sphere becomes an ellipsoid whose equation—see above, Chapter VII. Ex. (5)—is

$$(1 + A)X^2 + (1 + B)Y^2 + (1 + C)Z^2 + DYZ + EZX + FXY = 1.$$

(3)* A compression equal in all directions (I.), three simple distortions having their planes at right angles to one another and their axes† bisecting the angles between the lines of intersection of these planes (II.), (III.), (IV.), any simple or compound distortion consisting of a combination of longitudinal strains parallel to those lines of intersections (V.), and the distortion (VI.), constituted from the same elements which is orthogonal to the last, afford a system of six mutually orthogonal types which will be used as types of reference below in expressing the elasticity of cubically isotropic solids. (Compare Chapter X. Example 7 below.)

CHAPTER X.—*On the Measurement of Strains and Stresses.*

DEF. 1. Strains of any types are said to be to one another in the same ratios as stresses of the same types respectively, when any particular plane of the solid acquires, relatively to another plane parallel to it, motions in virtue of those strains which are to one another in the same ratios as the normal components of the forces between the parts of the solid on the two sides of either plane due to the respective stresses.

DEF. 2. The magnitude of a stress and of a strain of the same type are quantities, which, multiplied one by the other, give the work done on unity of volume of a body acted on by the stress while acquiring the strain.

COR. 1. If x, y, z, ξ, η, ζ denote orthogonal components of a certain strain, and if P, Q, R, S, T, U denote components, of the same type respectively, of a stress applied to a body while acquiring that strain, the work done upon it per unit of its volume will be

$$Px + Qy + Rz + S\xi + T\eta + U\zeta.$$

---

* This example, as well as (7) of Chapter X., (5) of Chapter XI., and the example of Chapter XII., are intended to prepare for the application of the theory of Principal Elasticities to cubically and spherically isotropic bodies, in Part II., Chapter XV.

† The "axes of a simple distortion" are the lines of its two component longitudinal strains.

COR. 2. The condition that two strains or stresses specified by $(x, y, z, \xi, \eta, \zeta)$ and $(x', y', z', \xi', \eta', \zeta')$, in terms of a normal system of types of reference, may be orthogonal to one another is

$$xx' + yy' + zz' + \xi\xi' + \eta\eta' + \zeta\zeta' = 0.$$

COR. 3. The magnitude of the resultant of two, three, four, five or six mutually orthogonal strains or stresses is equal to the square root of the sum of their squares. For if P, Q, &c., denote several orthogonal stresses, and F the magnitude of their resultant; and x, y, &c., a set of proportional strains of the same types respectively, and r the magnitude of the single equivalent strain, the resultant stress and strain will be of one type, and therefore the work done by the resultant stress will be Fr. But the amounts done by the several components will be Px, Qy, &c., and therefore

$$Fr = Px + Qy + \&c.$$

Now we have, to express the proportionality of the stresses and strains,

$$\frac{P}{x} = \frac{Q}{y} = \&c. = \frac{F}{r}.$$

Each member must be equal to

$$\frac{P^2 + Q^2 + \&c.}{Px + Qy + \&c.};$$

and also equal to

$$\frac{Px + Qy + \&c.}{x^2 + y^2 + \&c.}.$$

Hence, $\quad \dfrac{F}{r} = \dfrac{P^2 + Q^2 + \&c.}{Fr}$, which gives $F^2 = P^2 + Q^2 + \&c.$,

and $\quad \dfrac{F}{r} = \dfrac{Fr}{x^2 + y^2 + \&c.}$, which gives $r^2 = x^2 + y^2 + \&c.$

COR. 4. A definite stress of some particular type chosen arbitrarily may be called unity; and then the numerical reckoning of all strains and stresses becomes perfectly definite.

DEF. 3. A uniform pressure or tension in parallel lines, amounting in intensity to the unit of force per unit of area normal to it, will be called a stress of unit magnitude, and will be reckoned as positive when it is tension, and negative when pressure.

Examples.—(1) Hence the magnitude of a simple longitudinal strain, in which lines of the body parallel to a certain direction experience elongation to an extent bearing the ratio κ to their original dimensions, must be called κ.

(2) The magnitude of the single stress equivalent to three simple pressures in directions at right angles to one another each unity is $-\sqrt{3}$; a uniform compression in all directions of unity per unit of surface is a negative stress equal to $\sqrt{3}$ in absolute value.

(3) A uniform dilatation in all directions, in which lineal dimensions are augmented in the ratio $1 : 1 + x$, is a strain equal in magnitude to $x\sqrt{3}$; or a uniform "cubic expansion" E is a strain equal to $\dfrac{E}{\sqrt{3}}$.

(4) A stress compounded of unit pressure in one direction and an equal tension in a direction at right angles to it, or which is the same thing, a stress compounded of two balancing couples of unit tangential tensions in planes at angles of 45° to the direction of those forces, and at right angles to one another amounts in magnitude to $\sqrt{2}$.

(5) A strain compounded of a simple longitudinal extension x and a simple longitudinal condensation of equal absolute value, in a direction perpendicular to it, is a strain of magnitude $x\sqrt{2}$; or, which is the same thing (if $\sigma = 2x$), a simple distortion such that the relative motion of two planes at unit distances parallel to either of the planes bisecting the angles between the two planes mentioned above is a motion σ parallel to themselves, is a strain amounting in magnitude to $\dfrac{\sigma}{\sqrt{2}}$.

(6) If a strain be such that a sphere of unit radius in the body becomes an ellipsoid whose equation is

$$(1 - A)X^2 + (1 - B)Y^2 + (1 - C)Z^2 - DYZ - EZX - FXY = 1,$$

the values of the component strains corresponding, as explained in Example (2) of Chap. IX. above, to the different coefficients respectively, are

$$\tfrac{1}{2}A, \quad \tfrac{1}{2}B, \quad \tfrac{1}{2}C, \quad \frac{D}{2\sqrt{2}}, \quad \frac{E}{2\sqrt{2}}, \quad \frac{F}{2\sqrt{2}}.$$

For the components corresponding to A, B, C are simple longitudinal strains, in which diameters of the sphere along the axes of co-ordinates become elongated from 2 to $2 + A$, $2 + B$, $2 + C$

respectively; D is a distortion in which diameters in the plane YOZ, bisecting the angles YOZ and $Y'OZ$, become respectively elongated and contracted from 2 to $2 + \frac{1}{2}D$, and from 2 to $2 - \frac{1}{2}D$; and so for the others. Hence, if we take $x, y, z, \xi, \eta, \zeta$ to denote the magnitudes of six component strains, according to the orthogonal system of types described in Examples (1) and (2) of Chap. IX. above, the resultant strain equivalent to them will be one in which a sphere of unit radius in the solid becomes an ellipsoid whose equation is

$$(1 - 2x)X^2 + (1 - 2y)Y^2 + (1 - 2z)Z^2 - 2\sqrt{2}(\xi YZ + \eta ZX + \zeta XY) = 1,$$

and its magnitude will be

$$\sqrt{(x^2 + y^2 + z^2 + \xi^2 + \eta^2 + \zeta^2)}.$$

(7) The specifications, according to the system of reference used in the preceding Example, of the unit strains of the six orthogonal types defined in Example (3) of Chap. IX. are respectively as follows :—

	x	y	z	ξ	η	ζ
(I.)	$\frac{1}{\sqrt{3}}$	$\frac{1}{\sqrt{3}}$	$\frac{1}{\sqrt{3}}$	0	0	0
(II.)	0	0	0	1	0	0
(III.)	0	0	0	0	1	0
(IV.)	0	0	0	0	0	1
(V.)	l	m	n	0	0	0
(VI.)	l'	m'	n'	0	0	0

where l, m, n, l', m', n' denote quantities fulfilling the following conditions :—

$$l^2 + m^2 + n^2 = 1,$$
$$l + m + n = 0,$$
$$ll' + mm' + nn' = 0,$$
$$l'^2 + m'^2 + n'^2 = 1,$$
$$l' + m' + n' = 0.$$

(8) If $(1 - 2eP)X^2 + (1 - 2eQ)Y^2 + (1 - 2eR)Z^2$
$$- 2e\sqrt{2}(SYZ + TZX + UXY) = 1,$$

be the equation of the ellipsoid representing a certain stress, the amount of work done by this stress, if applied to a body while acquiring the strain represented by the equation in the preceding example (7), will be

$$Px + Qy + Rz + S\xi + T\eta + U\zeta.$$

Cor. Hence, if variables X, Y, Z be transformed to any other set (X', Y', Z') fulfilling the condition of being the co-ordinates of the same point, referred to another system of rectangular axes, the coefficients x, y, z, &c., $x_{,}$, $y_{,}$, $z_{,}$, &c., in two homogeneous quadratic functions of three variables,

$$(1 - 2x) X^2 + (1 - 2y) Y^2 + (1 - 2z) Z^2 - 2\sqrt{2}(\xi YZ + \eta ZX + \zeta XY)$$

and

$$(1 - 2x_{,}) X^2 + (1 - 2y_{,}) Y^2 + (1 - 2z_{,}) Z^2 - 2\sqrt{2}(\xi_{,} YZ + \eta_{,} ZX + \zeta_{,} XY),$$

and the corresponding coefficients x', y', z', &c., $x_{,}'$, $y_{,}'$, $z_{,}'$, &c., in these functions transformed to x', y', z', will be so related that

$$x'x_{,}' + y'y_{,}' + z'z_{,}' + \xi'\xi_{,}' + \eta'\eta_{,}' + \zeta'\zeta_{,}' = xx_{,} + yy_{,} + zz_{,} + \xi\xi_{,} + \eta\eta_{,} + \zeta\zeta_{,};$$

or the function $xx_{,} + yy_{,} + zz_{,} + \xi\xi_{,} + \eta\eta_{,} + \zeta\zeta_{,}$ of the coefficients is an "invariant" for linear transformations fulfilling the conditions of transformation from one to another set of rectangular axes. Since $x + y + z$ and $x_{,} + y_{,} + z_{,}$ are clearly invariants also, it follows that $AA_{,} + BB_{,} + CC_{,} + 2DD_{,} + 2EE_{,} + 2FF_{,}$ is an invariant function of the coefficients of the two quadratics

$$AX^2 + BY^2 + CZ^2 + 2DYZ + 2EZX + 2FXY$$

and $$A_{,}X^2 + B_{,}Y^2 + C_{,}Z^2 + 2D_{,}YZ + 2E_{,}ZX + 2F_{,}XY,$$

which it is easily proved to be by direct transformation.

This is the simplest form of the algebraic theorem of invariance with which we are concerned.

CHAPTER XI.—*On Imperfect Concurrences of Two Stress or Strain Types.*

DEF. 1. The concurrence of any stresses or strains of two stated types, is the proportion which the work done, when a body of unit volume experiences a stress of either type while acquiring a strain of the other, bears to the product of the numbers measuring the stress and strain respectively.

Cor. 1. In orthogonal resolution of a stress or strain, its component of any stated type, is equal to its own amount multiplied

by its concurrence with that type; or the stress or strain of a
stated type which, along with another or others orthogonal to it,
have a given stress or strain for their resultant, is equal to the
amount of the given stress or strain reduced in the ratio of its
concurrence with that stated type.

COR. 2. The concurrence of two coincident stresses or strains
is unity; or a perfect concurrence is numerically equal to unity.

COR. 3. The concurrence of two orthogonal stresses and strains
is zero.

COR. 4. The concurrence of two directly opposite stresses or
strains is -1.

COR. 5. If x, y, z, ξ, η, ζ are orthogonal components of any
strain or stress r, its concurrences with the types of reference are
respectively

$$\frac{x}{r}, \frac{y}{r}, \frac{z}{r}, \frac{\xi}{r}, \frac{\eta}{r}, \frac{\zeta}{r},$$

where

$$r = (x^2 + y^2 + z^2 + \xi^2 + \eta^2 + \zeta^2)^{\frac{1}{2}}.$$

COR. 6. The mutual concurrence of two stresses or strains is

$$ll' + mm' + nn' + \lambda\lambda' + \mu\mu' + \nu\nu',$$

if l, m, n, λ, μ, ν denote the concurrences of one of them with six
orthogonal types of reference, and l', m', n', λ', μ', ν' those of the
other.

COR. 7. The most convenient *specification of a type* for strains
or stresses, being in general a statement of the components, according to the types of reference, of a unit strain or stress of the
type to be specified, becomes a statement of its concurrences with
the types of reference when these are orthogonal.

Examples—(1) The mutual concurrence of two simple longitudinal strains or stresses, inclined to one another at an angle θ, is
$\cos^2\theta$.

(2) The mutual concurrence of two simple distortions in the
same plane, whose axes are inclined at an angle θ to one another,
is $\cos^2\theta - \sin^2\theta$, or $2\sin(45° - \theta)\cos(45° - \theta)$.

Hence the components of a simple distortion δ, along two rectangular axes in its plane, and two others bisecting the angle between these, taken as axes of component simple distortions, are

$$\delta\,(\cos^2\theta - \sin^2\theta) \quad \text{and} \quad \delta\,.\,2\sin\theta\cos\theta$$

respectively, if θ be the angle between the axis of elongation in the given distortion and in the first component type.

(3) The mutual concurrence of a simple longitudinal strain and a simple distortion is

$$\sqrt{2}\,.\,\cos\alpha\,\cos\beta,$$

if α and β be the angles at which the direction of the longitudinal strain is inclined to the lines bisecting the angles between the axes of the distortion; it is also equal to

$$\frac{1}{\sqrt{2}}\,(\cos^2\phi - \cos^2\psi),$$

if ϕ and ψ denote the angles at which the direction of the longitudinal strain is inclined to the axes of the distortion.

(4) The mutual concurrence of a simple longitudinal strain and of a uniform dilatation is $\dfrac{1}{\sqrt{3}}$.

(5) The specifying elements exhibited in Example (7) of the preceding Chapter (X.), are the concurrences of the new system of orthogonal types described in Example (3) of Chap. IX. with the ordinary system, as given in Examples (1) and (2), Chap. IX.

CHAPTER XII.—*On the Transformation of Types of Reference for Stresses or Strains.*

To transform the specification $(x,\,y,\,z,\,\xi,\,\eta,\,\zeta)$ of a stress or strain with reference to one system of types, into $(x_1,\,x_2,\,x_3,\,x_4,\,x_5,\,x_6)$ with reference to another system of types. Let $(a_1,\,b_1,\,c_1,\,e_1,\,f_1,\,g_1)$ be the components, according to the original system, of a unit strain of the first type of the new system; let $(a_2,\,b_2,\,c_2,\,e_2,\,f_2,\,g_2)$ be the corresponding specification of the second type of the new system; and so on. Then we have, for the required formulæ of transformation—

$$x = a_1 x_1 + a_2 x_2 + a_3 x_3 + a_4 x_4 + a_5 x_5 + a_6 x_6,$$
$$y = b_1 x_1 + b_2 x_2 + b_3 x_3 + b_4 x_4 + b_5 x_5 + b_6 x_6,$$
$$\dotfill$$
$$\dotfill$$
$$\xi = g_1 x_1 + g_2 x_2 + g_3 x_3 + g_4 x_4 + g_5 x_5 + g_6 x_6.$$

Example.—The transforming equations to pass from a specification $(x, y, z, \xi, \eta, \zeta)$ in terms of the system of reference used in Examples (6) and (7), Chapter X., to a specification $(\sigma, \xi, \eta, \zeta, \varpi, \omega)$ in terms of the new system described in Example (3) of Chapter IX., and specified in Example (7) of Chapter X., are as follows:

$$x = \frac{1}{\sqrt{3}}\,\sigma + l\varpi + l'\omega,$$

$$y = \frac{1}{\sqrt{3}}\,\sigma + m\varpi + m'\omega,$$

$$z = \frac{1}{\sqrt{3}}\,\sigma + n\varpi + n'\omega,$$

$$\xi = \xi,\ \eta = \eta,\ \zeta = \zeta;$$

where, as before stated, l, m, n, l', m', n' are quantities fulfilling the conditions

$$l^2 + m^2 + n^2 = 1,$$
$$l + m + n = 0,$$
$$l'^2 + m'^2 + n'^2 = 1,$$
$$l' + m' + n' = 0,$$
$$ll' + mm' + nn' = 0.$$

PART II.—ON THE DYNAMICAL RELATIONS BETWEEN STRESSES AND STRAINS EXPERIENCED BY AN ELASTIC SOLID.

CHAPTER XIII.—*Interpretation of the Differential Equation of Energy.*

In a paper on the Thermo-elastic Properties of Matter, published in the first number of the *Quarterly Mathematical Journal*, April 1855, and republished in the *Philosophical Magazine*, 1877, second half year [Art. XLVIII., Part VII., Vol. I. above], it was proved, from general principles in the theory of the Transformation of Energy, that the amount of work (w) required to reduce an elastic solid, kept at a constant temperature, from one stated condition of internal strain to another depends solely on these two conditions, and not at all on the cycle of varied states through which the body may have been made to pass in effecting the change, provided always there has been no failure in the elasticity under any of the strains it has experienced. Thus for a homogeneous solid homogeneously strained, it appears that w is a function of six independent variables x, y, z, ξ, η, ζ, by which the condition of the solid as to strain is specified. Hence to strain the body to the infinitely small extent expressed by the variation from $(x, y, z, \xi, \eta, \zeta)$ to $(x + dx, y + dy, z + dz, \xi + d\xi, \eta + d\eta, \zeta + d\zeta)$, the work required to be done upon it is

$$\frac{dw}{dx} dx + \frac{dw}{dy} dy + \frac{dw}{dz} dz + \frac{dw}{d\xi} d\xi + \frac{dw}{d\eta} d\eta + \frac{dw}{d\zeta} d\zeta.$$

The stress which must be applied to its surface to keep the body in equilibrium in the state $(x, y, z, \xi, \eta, \zeta)$ must therefore be such that it would do this amount of work if the body, under its action, were to acquire the arbitrary strain dx, dy, dz, $d\xi$, $d\eta$, $d\zeta$; that is, it must be the resultant of six stresses:—one orthogonal to the five strains dy, dz, $d\xi$, $d\eta$, $d\zeta$, and of such a magnitude as to do the work $\frac{dw}{dx} dx$ when the body acquires the strain dx; a second orthogonal to dx, dz, $d\xi$, $d\eta$, $d\zeta$, and of such a magnitude as to do the work $\frac{dw}{dy} dy$ when the body acquires the strain dy; and so on. If a, b, c, f, g, h denote the respective concurrences of these six stresses, with the types of reference used in the specification $(x, y, z, \xi, \eta, \zeta)$ of the strains, the amounts of the six stresses which fulfil those conditions will (Chapter XI.) be given by the equations

$$P = \frac{1}{a} \frac{dw}{dx}, \qquad Q = \frac{1}{b} \frac{dw}{dy}, \qquad R = \frac{1}{c} \frac{dw}{dz},$$

$$S = \frac{1}{f} \frac{dw}{d\xi}, \qquad T = \frac{1}{g} \frac{dw}{d\eta}, \qquad U = \frac{1}{h} \frac{dw}{d\zeta};$$

and the types of these component stresses are determined by being orthogonal to the fives of the six strain-types, wanting the first, the second, &c., respectively.

Cor. If the types of reference used in expressing the strain of the body constitute an orthogonal system, the types of the component stresses will coincide with them, and each of the concurrences will be unity. Hence the equations of equilibrium of an elastic solid referred to six orthogonal types are simply

$$P = \frac{dw}{dx}, \qquad Q = \frac{dw}{dy}, \qquad R = \frac{dw}{dz},$$

$$S = \frac{dw}{d\xi}, \qquad T = \frac{dw}{d\eta}, \qquad U = \frac{dw}{d\zeta}.$$

CHAPTER XIV.—*Reduction of the Potential Function, and of the Equations of Equilibrium, of an Elastic Solid to their Simplest Forms.*

If the condition of the body from which the work denoted by w is reckoned, be that of equilibrium under no stress from without, and if $x, y, z, \xi, \eta, \zeta$ be chosen each zero for this condition, we shall have, by Maclaurin's theorem,

$$w = H_2(x, y, z, \xi, \eta, \zeta) + H_3(x, y, z, \xi, \eta, \zeta) + \&c.,$$

where H_2, H_3, &c., denote homogeneous functions of the second order, third order, &c., respectively. Hence $\dfrac{dw}{dx}$, $\dfrac{dw}{dy}$, &c., will each be a linear function of the strain co-ordinates, together with functions of higher orders derived from H_3, &c. But experience shows (§ 37, above) that, within the elastic limits, the stresses are very nearly, if not quite, proportional to the strains they are capable of producing; and therefore H_3, &c., may be neglected, and we have simply

$$w = H_2(x, y, z, \xi, \eta, \zeta).$$

Now in general there will be twenty-one terms, with independent coefficients, in this function; but by a choice of types of reference, that is, by a linear transformation of the independent variables, we may, in an infinite variety of ways, reduce it to the form

$$w = \tfrac{1}{2}(Ax^2 + By^2 + Cz^2 + F\xi^2 + G\eta^2 + H\zeta^2).$$

The equations of equilibrium then become

$$P = \frac{A}{a}x, \quad Q = \frac{B}{b}y, \quad R = \frac{C}{c}z,$$

$$S = \frac{F}{f}\xi, \quad T = \frac{G}{g}\eta, \quad U = \frac{H}{h}\zeta,$$

the simplest possible form under which they can be presented. The interpretation can be expressed as follows.

PROP. An infinite number of systems of six types of strains or stresses, exist in any given elastic solid, such that, if a strain of any one of those types be impressed on the body, the elastic reaction is balanced by a stress orthogonal to the five others of the same system.

CHAPTER XV.—*On the Six Principal Strains of an Elastic Solid.*

To reduce the twenty-one coefficients of the quadratic terms in the expression for the potential energy to six, by a linear transformation, we have only fifteen equations to satisfy; while we have thirty disposable transforming coefficients, there being five independent elements to specify a type, and six types to be changed. Any further condition expressible by just fifteen independent equations may be satisfied, and makes the transformation determinate. Now the condition that six strains may be mutually orthogonal is expressible by just as many equations as there are different pairs of six things, that is, fifteen. The well-known algebraic theory of the linear transformation of quadratic functions shows for the case of six variables—(1) that the six coefficients in the reduced form are the roots of a "determinant" of the sixth degree necessarily real; (2) that this multiplicity of roots leads determinately to one, and only one, system of six types fulfilling the prescribed conditions, unless two or more of the roots are equal to one another, when there will be an infinite number of solutions and definite degrees of isotropy among them; (3) that there is no equality between any of the six roots of the determinant in general, when there are twenty-one independent coefficients in the given quadratic.

PROP. 1.   Hence a single system of six mutually orthogonal types, may be determined for any homogeneous elastic solid, so that its potential energy when homogeneously strained in any way, is expressed by the sum of the products of the squares of the components of the strain, according to those types, respectively multiplied by six determinate coefficients.

DEF.   The six strain-types thus determined are called the Six Principal Strain-types of the body.

The concurrences of the stress-components used in interpreting the differential equation of energy, with the types of the strain-co-ordinates, in terms of which the potential function of elasticity is expressed, being perfect when these constitute an orthogonal system, each of the quantities denoted above by a, b, c, f, g, h, is unity when the six principal strain-types are chosen for the co-ordinates. The equations of equilibrium of an elastic solid may therefore be expressed as follows :—

$$P = Ax, \quad Q = By, \quad R = Cz,$$
$$S = F\xi, \quad T = G\eta, \quad U = H\zeta,$$

where $x, y, z, \xi, \eta, \zeta$ denote strains belonging to the six Principal Types, and P, Q, R, S, T, U the components according to the same types, of the stress required to hold the body in equilibrium when in the condition of having those strains. The amount of work that must be spent upon it per unit of its volume, to bring it to this state from an unconstrained condition, is given by the equation

$$w = \tfrac{1}{2}(Ax^2 + By^2 + Cz^2 + F\xi^2 + G\eta^2 + H\zeta^2).$$

DEF. The coefficients A, B, C, F, G, H are called the six Principal Elasticities of the body.

The equations of equilibrium express the following propositions :—

PROP. 2. If a body be strained according to any one of its six Principal Types, the stress required to hold it so is directly concurrent with the strain.

Examples.—(1) If a solid be cubically isotropic in its elastic properties, as crystals of the cubical class probably are, any portion of it will, when subject to a uniform positive or negative normal pressure all round its surface, experience a uniform condensation or dilatation in all directions. Hence a uniform condensation is one of its six principal strains. Three plane distortions with axes bisecting the angles between the edges of the cube of symmetry are clearly also principal strains, and since the three corresponding principal elasticities are equal to one another, any strain whatever compounded of these three is a principal strain. Lastly, a plane distortion whose axes coincide with any two edges of the cube, being clearly a principal distortion, and the principal elasticities corresponding to the three distortions of this kind being equal to one another, any distortion compounded of them is also a principal distortion.

Hence the system of orthogonal types treated of in Examples (3) Chap. IX., and (7) Chap. X., or any system in which, for (II.), (III.), and (IV.) of Example (7) Chap. X., any three orthogonal strains compounded of them are substituted, constitutes a system of six Principal Strains in a solid cubically isotropic. There are only three distinct Principal Elasticities for such a body, and these are—(A) its modulus of compressibility, (B) its rigidity against diagonal distortion in any of its principal planes (three equal elasticities), and (C) its rigidity against rectangular distortions of a cube of symmetry (two equal elasticities).

(2) In a perfectly isotropic solid, the rigidity against all distortions is equal. Hence the rigidity (B) against diagonal distortion must be equal to the rigidity (C) against rectangular distortion, in a cube; and it is easily seen that if this condition is fulfilled for one set of three rectangular planes for which a substance is isotropic, the isotropy must be complete. The conditions of perfect or spherical isotropy are therefore expressed in terms of the conditions referred to in the preceding example, with the farther condition $B = C$.

A uniform condensation in all directions, and any system whatever of five orthogonal distortions, constitute a system of six Principal Strains in a spherically isotropic solid. Its Principal Elasticities are simply its Modulus of Compressibility and its Rigidity.

PROP. 3. Unless some of the six Principal Elasticities be equal to one another, the stress required to keep the body strained otherwise than according to one or other of six distinct types is oblique to the strain.

PROP. 4. The stress required to maintain a given amount of strain is a maximum or a maximum-minimum, or a minimum, if it is of one of the six Principal Types.

COR. If A be the greatest and H the least of the six quantities A, B, C, F, G, H, the principal type to which the first corresponds, is that of a strain *requiring a greater stress to maintain it* than any other strain of equal amount; and the principal type to which the last corresponds, is that of a strain which *is maintained by a less stress* than any other strain of equal amount in the same body. The stresses corresponding to the four other principal strain-types have each the maximum-minimum property in a determinate way.

PROP. 5. If a body be strained in the direction of which the concurrences with the principal strain-types are l, m, n, λ, μ, ν, and to an amount equal to r, the stress required to maintain it in this state will be equal to Ωr, where

$$\Omega = (A^2l^2 + B^2m^2 + C^2n^2 + F^2\lambda^2 + G^2\mu^2 + H^2\nu^2)^{\frac{1}{2}},$$

and will be of a type of which the concurrences with the principal types are respectively

$$\frac{Al}{\Omega},\ \frac{Bm}{\Omega},\ \frac{Cn}{\Omega},\ \frac{F\lambda}{\Omega},\ \frac{G\mu}{\Omega},\ \frac{H\nu}{\Omega}.$$

PROP. 6.   A homogeneous elastic solid, crystalline or non-crystalline, subject to magnetic force or free from magnetic force, has neither right-handed, nor left-handed, nor any dipolar properties dependent on elastic forces simply proportional to strains.

COR.   The elastic forces concerned in the luminiferous vibrations of a solid or fluid medium possessing the right- or left-handed property, whether axial or rotatory, such as quartz crystal, or tartaric acid, or solution of sugar, either depend on the heterogeneousness, or on the magnitude, of the strains experienced.

Hence as they *do not depend on the magnitude of the strain,* they *do depend on its heterogeneousness through the portion of a medium containing a wave.*

COR.   There cannot possibly be any characteristic of elastic forces simply proportional to the strains, in a homogeneous body, corresponding to certain peculiarities of crystalline form which have been observed,—for instance corresponding to the plagihedral faces discovered by Sir John Herschel to indicate the optical character, whether right-handed or left-handed, in different specimens of quartz crystal,—or corresponding to the distinguishing characteristics of the crystals of the right-handed and left-handed tartaric acids, obtained by M. Pasteur from racemic acid, —or corresponding to the dipolar characteristics of form said to have been discovered in electric crystals.

CHAPTER XVI.—*Application of Conclusions to Natural Crystals.*

It is easy to demonstrate that a body, homogeneous when regarded on a large scale, may be constructed to have twenty-one arbitrarily prescribed values for the coefficients in the expression for its potential energy in terms of any prescribed system of strain co-ordinates.   This proposition was first enunciated in the paper on the Thermo-elastic Properties of Solids, published April 1855, in the *Quarterly Mathematical Journal* [Art. XLVIII. Part VII. Vol. I. above] alluded to above. We may infer the following.

PROP. A solid may be constructed to have arbitrarily pre-scribed values for its six Principal Elasticities, and an arbitrary orthogonal system of six strain-types, specified by fifteen inde-pendent elements, for its principal strains: for instance, five arbitrarily chosen systems of three rectangular axes, for the normal axes of five of the Principal Types; those of the sixth consequently, in general, distinct from all the others and determi-nate; and the six times two ratios between the three stresses or strains of each type, also determinate. The fifteen equations expressing (Chap. VI. above) the mutual orthogonality of the six types, determine the twelve ratios for the six types, and the three quantities specifying the axes of the sixth type in the particular case here suggested: or generally the fifteen equations determine fifteen out of the thirty quantities (viz. twelve ratios and eighteen angular coordinates) specifying six Principal Types.

COR. There is no reason for believing that natural crystals do not exist for which there are six unequal Principal Elasticities, and six distinct strain-types for which the three normal axes con-stitute six distinct sets of three Principal rectangular axes of elasticity.

It is easy to give arbitrary illustrative examples regarding Principal Elasticities: also, to investigate the principal strain-types and the equations of elastic force referred to them or to other natural types, for a body possessing the kind of symmetry as to elastic forces that is possessed by a crystal of Iceland spar, or by a crystal of the "tesseral class," or of the included "cubical class." Such illustrations and developments, though proper for a students' text-book of the subject, are unnecessary here.

For applications of the Mathematical Theory of Elasticity to the question of the earth's rigidity and elasticity as a whole, and to the equilibrium of elastic solids in general, which are beyond the scope of the present article, the reader is referred to Thomson and Tait's *Treatise on Natural Philosophy* (Second Edition), §§ 740, 832 to 848, and Appendix C.

CHAPTER XVII.—*Plane Waves in a Homogeneous Æolotropic Solid.*

A plane wave in a homogeneous elastic solid, is a motion in which every line of particles, in a plane parallel to one fixed plane, experiences simply a motion of translation—but a motion differing from the motions of particles in planes parallel to the same. Let OX, OY, OZ be three fixed rectangular axes; OX perpendicular to the wave front (as any of the parallel planes of moving particles referred to in the definition is called), and OY, OZ *in* the wave front. Let $x + u$, $y + v$, $z + w$ be the coordinates at time t of a particle, which, if the solid were free from strain, would be at (x, y, z). The definition of wave motion amounts simply to this, that u, v, w are functions of x and t.

The *strain* of the solid (Chap. VII. above) is the resultant of a simple longitudinal strain in the direction OX, equal to du/dx, and two differential slips dv/dx, dw/dx, parallel to OY and OZ, constituting simple distortions of which the numerical magnitudes (Chap. X. above) are

$$\frac{dv}{dx}\sqrt{2}, \quad \text{and} \quad \frac{dw}{dx}\sqrt{2}.$$

Put then

$$\frac{du}{dx} = \xi, \quad \frac{dv}{dx}\sqrt{2} = \eta, \quad \frac{dw}{dx}\sqrt{2} = \zeta \dots\dots\dots\dots (1);$$

and let W denote the work per unit of bulk required to produce the strain represented by this notation. We have (Chap. XIII. above)

$$W = \tfrac{1}{2}(A\xi^2 + B\eta^2 + C\zeta^2 + 2D\eta\zeta + 2E\zeta\xi + 2F\xi\eta)\dots\dots\dots(2),$$

where A, B, C, D, E, F denote moduluses of elasticity of the solid. Let p, q, r denote the three components of the traction per unit area of the wave front. We have (Chap. XIV. above)

$$\left.\begin{aligned} p &= A\xi + F\eta + E\zeta \\ q\sqrt{\tfrac{1}{2}} &= F\xi + B\eta + D\zeta \\ r\sqrt{\tfrac{1}{2}} &= E\xi + D\eta + C\zeta \end{aligned}\right\} \dots\dots\dots\dots\dots\dots (3).$$

Now let ξ, η, ζ be taken such that

$$\left.\begin{aligned} A\xi + F\eta + E\zeta &= M\xi \\ F\xi + B\eta + D\zeta &= M\eta \\ E\xi + D\eta + C\zeta &= M\zeta \end{aligned}\right\} \dots\dots\dots\dots\dots (4)$$

the determinantal cubic gives three real positive values for M, and with M equal to any one of these values, equations (4) determine the ratios $\xi : \eta : \zeta$. Hence when the solid is strained in any one of the three ways thus determined, we have

$$p = M\frac{du}{dx}, \qquad q = M\frac{dv}{dx}, \qquad r = M\frac{dw}{dx} \quad\dots\dots\dots\dots (5).$$

The three components of the whole force due to the tractions on the sides of an infinitely small parallelepiped $(\delta x, \delta y, \delta z)$ of the solid, are clearly

$$\frac{dp}{dx}\,\delta x\delta y\delta z, \qquad \frac{dq}{dx}\,\delta x\delta y\delta z, \quad\text{and}\quad \frac{dr}{dx}\,\delta x\delta y\delta z \dots\dots\dots (6);$$

and therefore, if ρ be its density, and consequently $\rho\,\delta x\delta y\delta z$ its mass, the equations of its motion are

$$\rho\frac{d^2u}{dt^4} = \frac{dp}{dx}, \qquad \rho\frac{d^2v}{dt^2} = \frac{dq}{dx}, \qquad \rho\frac{d^2w}{dx^2} = \frac{dr}{dx} \dots\dots\dots (7).$$

These, putting for p, q, r their values by (5), become

$$\rho\frac{d^2u}{dt^2} = M\frac{d^2u}{dx^2}, \qquad \rho\frac{d^2v}{dt^2} = M\frac{d^2v}{dx^2}, \qquad \rho\frac{d^2w}{dt^2} = M\frac{d^2w}{dx^2} \dots\dots (8).$$

And by (4) and (1) we have

$$\left.\begin{array}{l} Au + (Fv + Ew)\sqrt{2} = Mu \\ Fu + (Bv + Dw)\sqrt{2} = Mv\sqrt{2} \\ Eu + (Dv + Cw)\sqrt{2} = Mw\sqrt{2} \end{array}\right\} \dots\dots\dots\dots\dots\dots (9).$$

Let M_1, M_2, M_3 be the three roots of the determinantal cubic, and b_1, c_1; b_2, c_2; b_3, c_3; the corresponding values of the ratios $\dfrac{v}{u}$, $\dfrac{w}{u}$ determined by (9). The complete solution of (8), subject to (9), is

$$\left.\begin{array}{l} u = u_1 + u_2 + u_3, \\ v = b_1u_1 + b_2u_2 + b_3u_3, \\ w = c_1u_1 + c_2u_2 + c_3u_3; \end{array}\right.$$

where

$$\left.\begin{array}{l} u_1 = f_1\left(x + t\sqrt{\dfrac{M_1}{\rho}}\right) + F_1\left(x - t\sqrt{\dfrac{M_1}{\rho}}\right), \\[2mm] u_2 = f_2\left(x + t\sqrt{\dfrac{M_2}{\rho}}\right) + F_2\left(x - t\sqrt{\dfrac{M_2}{\rho}}\right), \\[2mm] u_3 = f_3\left(x + t\sqrt{\dfrac{M_3}{\rho}}\right) + F_3\left(x - t\sqrt{\dfrac{M_3}{\rho}}\right), \end{array}\right\} \dots\dots (10),$$

f_1, F_1, f_2, F_2, f_3, F_3 denoting arbitrary functions. Hence we conclude that there are three different wave-velocities,

$$\sqrt{\frac{M_1}{\rho}}, \quad \sqrt{\frac{M_2}{\rho}}, \quad \sqrt{\frac{M_3}{\rho}}$$

and three different modes of waves, determined by equations (9).

Waves in an Isotropic Solid.—If the solid be isotropic, we have

$$\left. \begin{array}{l} B = C \\ D = E = F = 0 \\ M_1 = A, \quad M_2 = M_3 = B \end{array} \right\} \quad \dots\dots\dots\dots\dots\dots (11).$$

Hence, instead of three different waves with different velocities, we have just two,—a wave (like that of sound in air or other elastic fluid) in which the motions are perpendicular to the wave front, and the other (like the waves of light in an isotropic medium) in which the motions are parallel to the wave front.

Waves in an Incompressible Solid (Æolotropic or Isotropic).— If the solid be incompressible, we have $A = \infty$, and u must be zero.

Hence $\qquad\qquad W = B\eta^2 + C\zeta^2 + 2D\eta\zeta \quad \dots\dots\dots\dots\dots (12)$,

and by a determinantal quadratic, instead of cubic, we find two wave-velocities and two wave-modes, in each of which the motion is parallel to the wave front. In the case of isotropy the two wave velocities are equal.

It is to be noticed that M_1, M_2, M_3 in the preceding investigation are not generally true "Principal moduluses," but special moduluses corresponding to the particular plane chosen for the wave front. In the particular case of isotropy, however, the equal moduluses M_2, M_3 of (11) are principal moduluses, being each equal to *the* modulus of rigidity, but M_1 is a mixed modulus of compressibility and rigidity—not a principal modulus. In the case of incompressibility, the two moduluses found from the determinantal quadratic by the process indicated above are not principal moduluses generally, because the distortions by the differential motions of planes of particles parallel to the wave front, must generally give rise to tangential stresses orthogonal to them, which do not influence the wave motion.

PART II. HEAT.

[This Paper contains the whole of the article Heat as it appeared in the
Encyclopaedia Britannica, excepting a Mathematical Appendix which has
been already reprinted in this collection of my Papers as Article LXXII.
Vol. II. (above). I have added as an Appendix five short papers bearing
on this subject, three of them by myself and two of them by my former
Assistant in the Physical Laboratory of the University of Glasgow,
Mr Donald Macfarlane. W. T.]

TABLE OF CONTENTS.

HEAT is a property of matter which first became known to us by one of six very distinct senses.

1. *Sense of Heat.*—The sense of touch, as commonly meant, has two distinct objects—force and heat. If a person stretches out his hand till it meets anything solid, or holds it out while something solid is placed upon it, he experiences a sensation of force. He perceives resistance to the previous motion of his hand in one case, in the other case the necessity of resisting to prevent his hand from being forced downwards; the immediate object of this perception in each case is force*. But there is another very

* The sense of smoothness and roughness to which physiologists have sometimes given the special name "tactile sense" is as clearly a sense of force as is what they call the muscular sense. The sense of roughness is a sense of force at places of application distributed over the skin of the finger, while in the muscular sense of force the place of application on a larger scale is distinguished by the position of the hand perceiving the force. The internal mechanism of tissue and nerves in one case and that of muscles in the other, through which the perception of places of application of force is obtained, are no doubt different, but the thing perceived is essentially the same—force—the complete discrimination of which involves magnitude of the force, its place of application, and its direction.

distinct sensation, that of heat or cold, which he may or may not perceive in either of those cases, and which he may also perceive, still by what is commonly called the sense of touch, in other cases even when no sense of force is also experienced. Thus, in the first case, if the solid be a fixed can of warm water, or of iced water, the person perceives a sense of heat or of cold; and, supposing him to have performed the operation with his eyes shut, his mind is informed by the double sense of touch that his hand has met with a hot fixed body or a cold fixed body: in the other case he may perceive that a hot heavy solid, or a cold heavy solid, has been laid upon his hand. But if he dips his hand gently into a can of water, or if he holds it towards a fire, or if he exposes it to a gentle current of air, or waves it about through the air, he perceives heat or cold without any accompanying sense of force.

The earliest scientific thoughts respecting these sensations of heat and cold must have led to the true conclusion that there is some property of external matter on which the sensations depend, and a little advance into the natural philosophy of the subject has suggested and proved that this property is also possessed by the living body, and that the sensation of heat or cold in the hand, in the instances referred to above, depends on the change produced in the hand in respect to this property by a change of circumstances which preceded the sensation. We now call heat the property of matter concerned in these sensations, and temperature a certain variable quality of matter varying according to its temporary condition in respect to heat.

In the strictest modern scientific language (compare § 3 below) the word heat is used to denote something communicable from one body or piece of matter to another, and temperature a definite variable quality of matter, varying generally in any particular piece of matter when heat is communicated to it or taken from it, varying also as we shall see (§§ 8 and 9 below) in consequence of operations which can take place within the body itself, or which may be performed upon it from without, but which cannot be described as communication of heat to it or drawing off of heat from it.

2. *Latent Heat.*—There are exceptional cases in which temperature does not vary in a mass of matter when heat is communicated to it from, or taken from it to, external matter. For instance, when the body is ice at the melting point, heat com-

municated to it does not raise its temperature; or if the body be
water at the freezing point with ever so small a piece of ice in it,
heat taken from it does not cause its temperature to fall; or if
the whole mass considered be ice and water well mixed, heat may
be either communicated to it or taken from it without altering
its temperature; or if the body be water at the boiling point in
the open air, heat very slowly communicated to it in however
great quantities does not raise its temperature sensibly, but causes
it to disappear by evaporation from its surface; or
if the body be steam in a cylinder with a little
water in the bottom and with a frictionless piston
above it for roof (fig. 1), under atmospheric
pressure, heat taken from it very slowly does not
cool it until the whole steam has become condensed into water, and heat communicated to it
very slowly does not warm it until the whole
water has become evaporated into steam; or if
the body be ice (or frozen water), in place of the
liquid water of the last case, and if the pressure
on the upper side of the piston, instead of atmospheric pressure of about 1033 grammes per square
centimetre (14 7 lb. per square inch), be anything
less than $\frac{1}{30}$th of a gramme per square centimetre,
the same statement will still apply with "ice"
substituted for water. Black's celebrated doctrine of latent heat
is merely the declaration of a class of phenomena of which the
preceding illustrations sufficiently indicate the character. Modern
mysticism has been much exercised in respect to the terms sensible
heat and latent heat, whether in decrying them, or in continuing
to use them, but with aggravating haziness instead of the clear
wrongness of the old doctrine. It has become of late years somewhat the fashion to decry the designation of latent heat, because
it had been very often stated in language involving the assumption
of the materiality of heat*. Now that we know heat to be a

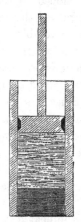

Fig. 1.

* A hundred years ago those deeper philosophers who in their judgment antici-
pated, or tended to anticipate, what we now know to be the true theory of the nature
of heat, had indeed good grounds to be jealous of even the phrase *latent heat*.
Maxwell says—"It is worthy of remark that Cavendish, though one of the greatest
chemical discoverers of his time, would not accept the phrase *latent heat*. He
prefers to speak of the generation of heat when steam is condensed, a phrase incon-
sistent with the notion that heat is matter, and objects to Black's term as relating

mode of motion, and not a material substance, the old "impressive,
clear, and wrong" statements regarding latent heat, evolution and
absorption of heat by compression, specific heats of bodies and
quantities of heat possessed by them, are summarily discarded.
But they have not yet been generally enough followed by equally
clear and concise statements of what we now know to be the
truth. A combination of impressions surviving from the old
erroneous notions regarding the nature of heat, with imperfectly
developed apprehension of the new theory, has somewhat liberally
perplexed the modern student of thermodynamics with questions
unanswerable by theory or experiment, and propositions which
escape the merit of being false by having no assignable meaning.
There is no occasion to give up either " sensible heat " or " latent
heat"; and there is a positive need to retain the term latent heat,
because if it were given up a term would be needed to replace it,
and it seems impossible to invent a better. Heat given to a
substance and warming it, is said to be sensible in the substance.
Heat given to a substance and not warming it, is said to become
latent. These designations express with perfect clearness the
relation of certain material phenomena to our sensory perception
of them. Thus when heat given to a quantity of water warms
it, the heat becomes sensible to a hand held in the water. When
a basin of warm water and a basin of water and ice are placed
side by side, a hand dipped first in one and then in the other
perceives the heat. If now the warm water be poured into the
basin of ice and water, and stirred for a few seconds of time
(unless there is enough of warm water to melt all the ice), the
hand perceives no warmth; on the contrary, it perceives that the
temperature is the same as it was in the basin of ice and water
at the beginning. Thus the heat which was sensible in the basin
of warm water has ceased to be sensible in the water that was in

Sensible heat and latent heat.

' to an hypothesis depending on the supposition that the heat of bodies is owing to
their containing more or less of a substance called the matter of heat; and, as
I think Sir Isaac Newton's opinion that heat consists in the internal motion of the
particles of bodies much the most probable, I chose to use the expression *heat is
generated*' (*Phil. Trans.*, 1783, quoted by Forbes). We shall not now be in danger
of any error if we use latent heat as an expression meaning neither more nor less
than this:—

DEFINITION.—*Latent heat is the quantity of heat which must be communicated to
a body in a given state in order to convert it into another state without changing its
temperature.*"—Maxwell's *Theory of Heat*, pp. 72, 73.

that basin, and has not become sensible in the other.   It is therefore well said to have become latent.

CALORIMETRY.

3. *Calorimetry by Latent Heat.*—The doctrine of latent heat
leads us very smoothly to a most important measurement in
thermal science, the measurement of quantities of this wonderful
property of matter which we call heat; and this without our
knowing anything of what the nature of heat is,—whether it be
a subtle elastic fluid, or a state of motion, or possibly some
modification of matter related to action of force.   Without, in
the first place, admitting into our minds any definite idea as to
the nature of heat, we may agree to measure quantities of heat
by quantities of ice melted into water without change of temperature.   Thus if a kilogramme of ice is melted by a large
quantity of water at a lukewarm temperature, or by a comparatively small quantity of very hot water, the same quantity of
heat has certainly gone from the warm water to the ice in each
case, supposing that the result in each case is the ice and warm
water left all in a state of ice-cold water.   The measurement of
quantities of heat, whether thus by the melting of ice, or by any
other means, received the name of "Calorimetry," when the essence
of heat was supposed to be a fluid, and this fluid called caloric.
The name calorimetry is still by general consent retained to
designate measurement of quantities of heat, as distinguished from
thermometry, or the measurement of temperature (§§ 10–67 below).
As long as the truth or falsity of the materialistic hypothesis
seemed an open question, the word "Caloric" was held to imply
the materiality of heat.   Thus Davy, after discussing some of the
fundamental dogmas of the "Calorists," as he called them, and
describing his own experiments, which proved beyond all doubt
the falsity of their fundamental hypothesis that heat is matter,
varied the statement of his conclusion by saying, "or caloric does
not exist."   While accepting Davy's conclusion, however, we need
not accept this way of stating it; and as most of our best modern
writers still use the word calorimetry, and as French writers have,
in comparatively recent times, introduced the word "calorie" to
designate a unit quantity of heat, it is decidedly convenient still
to retain the name *caloric* to denote definitely the measureable
essence of heat.   This is convenient scientifically as tending to

give precision to language and ideas respecting the two classes of measurement, calorimetry and thermometry; and it has the advantage of leaving the more popular word *heat* available for that somewhat lax general usage, from which we cannot altogether displace it; in which it may sometimes mean high temperature, as when we speak of great heat, or summer heat, or blood heat; sometimes a measureable quantity of heat, as in the term latent heat; and sometimes a branch of study or science dealing with the transference of heat by conduction and radiation, as in the title of Fourier's great work *Théorie analytique de la Chaleur;* or the whole province of science concerned with heat, including calorimetry and thermometry, and conduction and radiation of heat, and generation of heat, and dynamical relations of heat, as in English titles of separate books such as Dixon's, Balfour Stewart's, and Maxwell's, or of chapters or divisions of larger treatises, such as even the present article.

4. *Calorimetry by Melting of Ice.*—Calorimetry was first practised by means of the melting of ice as explained above, and the first thermal unit, or unit quantity of heat, or " Calorie," although not then called calorie, was the quantity of heat required to melt unit weight of ice. This, for example, is the unit on which Fourier founds his reckoning illustratively when he explains the fundamental principles of his theory of the conduction of heat. Ice seems to have been first used for calorimetry by Wilcke, a Swede. For the systematic application of this method for the measurement of quantities of heat in various physical inquiries

Laplace and Lavoisier's ice-calorimeter.

Laplace and Lavoisier constructed an instrument, the first to which the name of calorimeter was applied, and described it in the memoirs of the French Academy of Sciences for 1780*. Though in the hands of Laplace and Lavoisier it gave good results, it had a great inconvenience, which with less careful and less scientific experimenters might lead to great inaccuracies, on account of the water adhering by capillary attraction to the broken ice, instead of draining away from it completely and showing exactly how much ice had been melted. To avoid this evil Sir John Herschel suggested that, instead of draining away the water from the ice, the water and ice should all be kept together, and the whole bulk measured. The diminution of bulk of the whole

---

* The instrument itself is preserved in the *Conservatoire des Arts et Métiers* in Paris. It is described and explained in Maxwell's *Theory of Heat*, chap. 3.

thus gives an accurate measurement of the quantity of ice melted, because ice melting into water comes to occupy just 91·675 per cent. of its original volume.   This suggestion is admirably carried out by Bunsen[*] in his ice-calorimeter, an instrument possessing also other novel features of remarkable beauty and scientific interest.   It is particularly valuable for the measurement of small quantities of heat.   Its inventor, for example, by means of it succeeded in making satisfactory determinations of the specific heats of some of those rarer metals, such as Indium, of which only a few grammes have been obtained.

(margin: Bunsen's ice-calorimeter.)

5. *Calorimetry by the Evaporation of Water.*—By another application of Black's doctrine of latent heat, the evaporation of water may be used for calorimetry with great advantage in many scientific investigations.   It is used generally in engineering practice, particularly for testing the heating power of different qualities of coal and the economy of various forms of furnaces.   The thermal unit, which presents itself naturally in this system, is the quantity of heat required to evaporate unit weight of water when the pressure of the atmosphere as measured by the barometer is of some conventional standard amount, such as that called one atmosphere, or one atmo, being that for which the barometer, with its mercury column at zero centigrade (or the temperature at which ice melts), stands at 76 centimetres in the latitude of Paris, 48° 50′[†], or at

$$76 \times \frac{(1 + ·00531 \sin^2 48° 50')}{1 + ·00531 \sin^2 l}$$

in any latitude l.   This thermal unit is, according to Regnault's observations, equal to 6·8 times the ice-calorimetric unit.

6. *Thermometric Calorimetry.*—The most prevalent mode of calorimetry in scientific investigation has been hitherto, however,

---

[*] Pogg. *Ann.*, Sept. 1870, and *Phil. Mag.*, 1871 ; Maxwell's *Theory of Heat*, p. 61.

[†] This is chosen because all the most accurate experimental determinations depending on a conventional standard for atmospheric pressure, such as measurements of thermal expansions and specific heats of gases, of latent heat of melting solids in terms of a calorimetric unit depending on the centigrade thermometric scale, of latent heats of vapours, and thermal expansions of mercury and glass, and comparisons of mercury and air thermometers, are those of Regnault, and were made in Paris and calculated and given to the world according to an arbitrary standard atmosphere corresponding to 76 centimetres of mercury there.

neither that by the melting of ice, nor that by evaporation of water, nor indeed by any mode founded on the doctrine of latent heat at all. It has been founded on the elevation of temperature produced in water by the communication to it of the heat to be measured; and, for the sake of distinction from calorimetry by latent heat or otherwise, it may be called thermometric calorimetry. We can only consider it now in anticipation, as we have not yet reached the foundation of any thermometric scale; but even now we can see that, if in any way we fix upon any two particular determinate temperatures, the quantity of water warmed from the lower to the higher of them by the heat to be measured is a perfectly definite measure for the quantity of this heat. The two temperatures chosen for thermometric calorimetry are those marked 0° and 100° on the centigrade scale. The first of these we can understand at present, being the temperature at which ice melts under ordinary atmospheric pressure. The second is fully defined in §§ 35, 37, 51, 67 below. The quantity of heat required to raise unit mass of water (1 kilogramme, or 1 gramme, or 1 milligramme, or 1 lb., as the case may be) from zero to 1° C. is called the thermal unit centigrade, and sometimes, especially by French writers, the "calorie."

7. *Comparison of Calorimetric Units.* — Observations by Prevostaye and Desians, and by Regnault, on the latent heat of fusion of ice, show it to be 79·25 thermal units centigrade, a result differing but little from Black's original determination, which made it 142 thermal units Fahr.,—this being equal to 78·9 thermal units centigrade. Thus if one kilogramme of ice be put into 79¼ kilogrammes of water at 1° C., and left till the whole is melted (the process may be accelerated by not too violent stirring, § 9 below), the result will be 80¼ kilogrammes of water at 0° C.

Regnault's experiments on the latent heat of steam show that the quantity of heat required to convert into steam unit mass of water at the boiling temperature, under standard atmospheric pressure (§ 5 above), is 536·5 thermal units centigrade. This number, which is no doubt very accurate, differs but little from Watt's final result, 960 thermal units Fahr. (equal to 533·3 thermal units centigrade), obtained by him, in a repetition in 1781 of experiments which he had commenced in 1765 at the invitation of Black, whose pupil he was.

8. *Dynamical Calorimetry.—Preliminary regarding the Nature of Heat.*—From the dawn of science till the close of the eighteenth century two rival hypotheses had been entertained regarding the nature of heat, each with more or less of plausibility, but neither on any sure experimental basis:—one that heat consisted of a subtle elastic fluid permeating through the pores or interstices among the particles of matter, ·like water in a sponge ; the other that it was an intestine commotion among the particles or molecules of matter. In the year 1799 Davy, in his first published work entitled *An Essay on Heat, Light, and Combinations of Light**, conclusively overthrew the former of these hypotheses, and gave good reason for accepting as true the latter, by his celebrated experiment of converting ice into water by rubbing two pieces of ice together, without communicating any heat from surrounding matter. A few years earlier Rumford had been led to the same conclusion, and had given very convincing evidence of it in his observation of the great amount of heat produced in the process of boring cannon in the military arsenal at Munich, and the experimental investigation on the excitation of heat by friction† with which he followed up that observation. He had not, however, given a perfect logical demonstration of his conclusion, nor even quite a complete experimental basis on which it could be established with absolute certainty. According to the materialistic doctrine it would have been held that the heat excited by the friction was not *generated*‡, but was *produced*, squeezed out, or let flow out like honey from a broken honeycomb, from those parts of the solid which were cut or broken into small fragments, or rubbed to powder in the frictional process. If this were true, the very small fragments or powder would contain much less heat in them than an equal mass of continuous solid of the same substance as theirs. But unhappily the caloristic doctrine, besides its fundamental hypothesis, which we now know to be wrong, had given an absurd and illogical test for quantity of

---

* Published in 1799 in *Contributions to Physical and Medical Knowledge, principally from the West of England*, collected by Thomas Beddoes, M.D., and republished in Dr Davy's edition of his brother's collected works, vol. ii., London, 1836.

† "An Enquiry concerning the Source of Heat which is excited by Friction" (*Phil. Trans.*, abridged, vol. xviii. p. 286; see also vol. i. of "The Complete Works of Count Rumford" published by *The American Academy of Arts and Sciences*, Boston, 1870).

‡ Compare quotation from Cavendish, footnote, § 2 above.

heat in a body, of which a not altogether innocuous influence still survives in our modern name "specific heat;" and Rumford actually, in trying to disprove the materialistic doctrine, was baffled by this sophism. That is to say, he measured the specific heat or "capacity for heat" of the powder, and he found that the powder took as much heat to warm it to a certain degree as did an equal mass of the continuous solid, and from this he concluded that the powder did not contain less heat than the continuous solid at the same temperature. This conclusion is so obviously unwarranted by the premises that it is difficult to imagine how Rumford could have for a moment put forward the "capacity for heat" experiment as proving it, or could have rested in the conclusion without a real proof, or at least the suggestion of a real proof. All that Rumford's *argument* proved was that the fundamental hypothesis of the "calorists," and their other altogether gratuitous doctrine of equality of "specific heat" as a test for equality of whole quantities of heat in matter, could not be both true; and any one not inclined to give up the materialistic hypothesis might have cheerfully abandoned the minor doctrine, and remained unmoved by Rumford's argument. If Rumford had but melted a quantity of the powder (or dissolved it in an acid), and compared the heat which it took with that taken by an equal weight of the continuous solid, he would have had no difficulty in proving that the enormous quantity of heat which he had found to be excited by the friction had not been squeezed, or rubbed, or pounded, out of the solid matter, but was really brought into existence, and therefore could not be a material substance. He might even, without experiment, have pointed out that, if the materialistic doctrine were true, it would follow that sufficiently long-continued pounding of any solid substance by pestle and mortar, whether by hand or by aid of machinery, would convert it into a marvellous powder possessing one or other of two properties about equally marvellous. Either the smallest quantity of it thrown into an acid would constitute a freezing mixture of unlimited intensity,—the longer it had been pounded, the more intense would be its frigorific effect on being dissolved,—or the powder would be incapable of being warmed by friction, because it had already parted with all the heat which friction could rub out of it. The real effect of Rumford's argument seems to have been to salve the intellectual consciences of those who were not inclined to give up the materialistic doctrine, and to save them from the trouble of reading through Rumford's paper and thinking for themselves, by which they would have seen that his philosophy was better than his logic, and would inevitably have been forced to agree with him in his conclusion. It is remarkable that Davy's logic, too, was at fault, and on just the same point as Rumford's, but with even more transparently logical fallaciousness, because his argument is put in a more definitely logical form.

"Let heat be considered as matter, and let it be granted that the temperature of bodies cannot be increased unless their capacities are diminished from some cause, or heat added to them from some bodies in contact"!!

*　　　*　　　*.　　　*　　　*　　　*

"*Experiment II.*—I procured two parallelepipedons of ice*, of the temperature of 29°, 6 inches long, 2 wide, and 2/3 of an inch thick; they were fastened by wires to two bars of iron. By a peculiar mechanism their surfaces were placed in contact and kept in a continued and violent friction for some minutes. They were almost entirely converted into water, which water was collected and its temperature ascertained to be 35° after remaining in an atmosphere of a lower temperature for some minutes. The fusion took place only at the plane of contact of the two pieces of ice, and no bodies were in friction but ice. From this experiment it is evident that ice by friction is converted into water, and, according to the supposition, its capacity is diminished; but it is a well-known fact that the capacity of water for heat is much greater than that of ice, and ice must have an absolute quantity of heat added to it before it can be converted into water. Friction consequently does not diminish the capacities of bodies for heat."— Davy's *Essay on Heat, Light, and Combinations of Light*, pp. 10—12.

[Delete from "and, according to the supposition," to "greater than that of ice" inclusive; and delete the lame and impotent conclusion stated in the last eleven words. The residue constitutes an unanswerable demonstration of Davy's negative proposition that heat is not matter.]

9. *Joule's Dynamical Equivalent of Heat.*—It is remarkable that, while Davy's experiment alone sufficed to overthrow the

---

* "The result of the experiment is the same if wax, tallow, resin, or any substance fusible at a low temperature be used; even iron may be fused by collision, as is evident by the first experiment."

hypothesis that heat is matter, and Rumford's, with the addition
of just a little consideration of its relations to possibilities or probabilities of inevitable alternatives, in effect did the same, fifty years
passed before the scientific world became converted to their conclusion,—a remarkable instance of the tremendous efficiency of bad
logic in confounding public opinion and obstructing true philosophic
thought. Joule's great experiments from 1840 to 1849*, creating
new provinces of science in the thermodynamics of electricity, and
magnetism, and electro-chemistry, recalled attention to Davy's
and Rumford's doctrine regarding the nature of heat, and supplied
several fresh proofs, each like Davy's absolutely in itself complete
and cogent, that heat is not a material substance, and each advancing with exact dynamical measurement on the way pointed
out by Rumford in his measurement of the quantity of heat
generated in a certain time by the action of two horses not urged
to overwork themselves. The full conversion of the scientific
world to the kinetic theory of heat took place about the middle of
this century, and was no doubt an immediate consequence of
Joule's work, although Rumford's and Davy's demonstrative experiments, and the ingenious and penetrating speculations of Mohr,
and Séguin, and Mayer, and the experimental thermodynamic
measurements of Colding, all no doubt contributed to the result.

* List of titles of, and references to, papers by Dr James Prescott Joule, F.R.S.:—
"Description of an Electromagnetic Engine," *Sturgeon Ann. Electr.* ii. 1838, pp.
122—123. "On the Production of Heat by Voltaic Electricity," *Roy. Soc. Proc.*,
iv., 1840, pp. 280—282. "On the Heat evolved by Metallic Conductors of Electricity, and in the Cells of a Battery during Electrolysis," *Phil. Mag.* xix., 1841,
pp. 260—277. "On the Electric Origin of the Heat of Combustion," *Brit. Assoc.
Report*, 1842 (Pt. 2), p. 31. "On the Electrical Origin of Chemical Heat," *Phil.
Mag.*, xxii., 1843, pp. 204—208. "On the Calorific Effects of Magneto-electricity,
and on the Mechanical Value of Heat," *Phil. Mag.*, xxiii., 1843, pp. 263—276,
347—355, 435—443. "On the Changes of Temperature produced by the Rarefaction
and Condensation of Air," *Roy. Soc. Proc.*, v., 1844, pp. 517—518. "On the
Mechanical Equivalent of Heat," *Brit. Assoc. Report*, 1845 (Pt. 2), p. 31. "On the
Existence of an Equivalent Relation between Heat and the ordinary forms of
Mechanical Power," *Phil. Mag.*, xxvii., 1845, pp. 205—207. "On the Heat evolved
during the Electrolysis of Water (1843)," *Manchester Phil. Soc. Mem.*, vii., 1846,
pp. 87—113. "On a new Theory of Heat," *Manchester Phil. Soc. Mem.*, vii., 1846,
pp. 111—112. "On the Theoretical Velocity of Sound," *Phil. Mag.*, xxxi., 1847, pp.
114—115. "On the Mechanical Equivalent of Heat as determined by the Friction
of Fluids," *Phil. Mag.*, xxxi., 1847, pp. 173—176. "On the Mechanical Equivalent
of Heat, and on the Constitution of Elastic Fluids," *Brit. Assoc. Report*, 1848 (Pt. 2),
pp. 21—22. "On Shooting Stars," *Phil. Mag.*, xxxii., 1848, pp. 349—351. [See
Joule's collected papers, vol. i. (Physical Society of London), 1844.]

Each of the several subjects of thermodynamic measurement undertaken by Joule gave him a means of estimating the quantity of work required to generate a certain quantity of heat; but after several years of trials he was led to prefer to all others the direct method of simply stirring a quantity of water by a paddle, and measuring the quantity of heat produced by a measured quantity of work; and this method he has accordingly used in all his experiments for the purpose of determining the "dynamical equivalent of heat" from the year 1845 to the present time. By this he found his final result of 1849*, which was 772 Manchester foot-pounds for the quantity of work required to warm by 1° Fahr., at any temperature between 55° and 61° Fahr., 1 lb of water weighed in vacuum. In 1870 he commenced work for a fresh determination of the dynamical equivalent of heat at the request of the British Association, and the result was communicated to the Royal Society † about the end of 1877, with the following preface :—

"The committee of the British Association on standards of electrical resistance having judged it desirable that a fresh determination of the mechanical equivalent of heat should be made by observing the thermal effects due to the transmission of electrical currents through resistances measured by the unit they had issued, I undertook experiments with that view, resulting in a larger figure (782·5, *Brit. Assoc. Report,* Dundee, 1867, p. 522) than that which I had obtained by the friction of fluids (772·6, *Phil. Trans.* 1850, p. 82).

"The only way to account for this discrepancy was to admit the existence of error either in my thermal experiments or in the unit of resistance. A committee, consisting of Sir William Thomson, Professor P. G. Tait, Professor Clerk Maxwell, Professor B. Stewart, and myself, were appointed at the meeting of the British Association in 1870; and with the funds thus placed at my disposal I was charged with the present investigation, for the purpose of giving greater accuracy to the results of the direct method."

The result of this final investigation of Joule s is 772·43 Manchester foot-pounds for the quantity of heat required to warm from 60° to 61° Fahr. a pound of water weighed in vacuum,

---

* Joule "On the Mechanical Equivalent of Heat," *Philosophical Transactions of Royal Society* for 1850.

† " New Determination of the Mechanical Equivalent of Heat," by James Prescott Joule, *Phil. Trans. of Royal Society* for 1878, pp. 365—383.

which is about $\frac{1}{20}$th per cent. greater than the result of 1849 expressed in the same terms. According to Regnault's measurements* of the thermal capacity of water at different temperatures
from 0° to 230°C., it must be about ·08 per cent. greater at 60° Fahr.
than at 32°. According to this, Joule's thermodynamic result
would be 771·81 Manchester foot-pounds, for the work required
to warm a pound of water from 32° to 33° Fahr., or 1389·26 to
warm a pound of water from 0° to 1° C. Reducing 1389·26 feet
to metres, we have 423·437 metres. At Paris the force of gravity
is about $\frac{4}{100}$ per cent. less than in Manchester. Hence for about
the middle of France and the southern latitudes of Germany,
Joule's result, according to the ordinary reckoning of French and
German engineers, may be stated as 423·5 kilogramme-metres
for the amount of work required to warm 1 kilogramme of water
from 0° to 1°C. The force of gravity at Manchester is 981·34
dynes (centimetres per second per second). Multiplying 423·437
by this, we find accordingly 41,553,000 centimetre-dynes, or "ergs,"
for the amount of work in C.G.S. measure required to warm
1 gramme of water from 0° to 1° C.

THERMOMETRY.

10. *Preliminary for Thermometry.*—*Sense of Heat* (*resumed
from* § 1 *above*).—The sense of heat and cold is not simply dependent on the temperature of the body touched. If a person
takes a piece of iron, or a stone, or a piece of wood, or a ball of
worsted, or a quantity of finely carded cotton-wool, or of eider
down, in his hand, or touches an iron column, or a stone wall, or
a wooden beam, or a mass of wool or of down, he will perceive
the iron cold, the stone cold, but less cold than the iron, the
wood but slightly cold—much less cold than the stone, the wool
or down decidedly warm.

We now know that if all the bodies before being touched
were near one another in similar exposure, they must have been
at the same temperature, and from the iron and stone being felt
cold we know that this mean temperature is lower than the
temperature of the hand. Each of the bodies touched must at the
first instant have taken some heat from the hand, and therefore,
if the perception were quick enough, all at the very instant of being
touched would have seemed cold to the sense. The iron by its

* *Relation des Expériences*, vol. i. p. 748, Paris, 1847.

high thermal conductivity (§§ 76, 78, 80 below) keeps drawing off heat from the hand and lowering its temperature, till after many seconds of time an approximately permanent temperature is reached, which may be considerably lower than the temperature of the hand before contact, but somewhat higher than the previous temperature of the iron, because of the internal furnace generating heat in the hand. A similar result, but in less time and with less ultimate lowering of temperature of the hand, takes place when stone is touched. When wood is touched its comparatively small conductivity (§ 76 below) allows its surface to be warmed again after the first few seconds, sometimes to a higher temperature than that of the hand before contact; and thus, if the sensation could be perfectly remembered, it would be perceived that the wood was first felt to be cold, and afterwards to be warm. This latter warmth is rendered very perceptible by first holding the hand in contact with a piece of wood, as for instance a mahogany table, for a considerable time, half a minute or more, and then suddenly removing it; a sense of cold is immediately perceived in consequence of the exposure of the hand to the air. The foot is similarly sensitive. If, after holding a bare foot for some time in the air, it be placed on a varnished wooden floor, the floor is perceived to be cold, and if, after standing some time with it pressed to the floor, the foot be suddenly lifted, the air now seems cold by contrast. If a person walks with bare feet on a wooden floor, a continued sense of cold is experienced; and if, immediately after doing so, he sits down, and holds his feet in the air, the air seems to be warm by contrast. The same sensations are perceived even on a carpeted floor, but much less markedly than on a plain wooden floor, and much less markedly on a plain wooden floor than on a varnished wooden floor, and much less markedly on a varnished wooden floor than on a stone floor. In the case of touching soft wool, or finely carded cotton-wool, or eider down, the first instantaneous sensation of cold is scarcely if at all perceived, and that which first provokes consciousness is the subsequent heating; and it is very startling to find a body which we know to be ice-cold on a frosty day feeling positively warm to the first consciously perceptible sensation after it is touched. In this case the small thermal conductivity or great thermal resistance of the substance is such that heat is carried off by it from the hand slower than it was carried off by radiation and aerial convection

(§§ 70, 71 below) before contact; and thus, after the first momentary cooling of the hand by the initial cataract of heat from it to the cold body touched, in a small fraction of a second of time a higher temperature is attained by the hand than it had before contact.

11. *Sense of Temperature.*—The sense of heat is in reality a somewhat delicate thermal test when properly used. Even an unskilled hand alternately dipped into two basins of water will, as we have found by experiment, detect a difference of temperature of less than a quarter of a degree centigrade; and there can be no doubt that bath and hospital attendants, and persons occupied with hot liquors in various manufactures, such as dyeing, can detect much smaller differences of temperature than that, and, what is still more remarkable, can remember permanently sensations of absolute temperature sufficiently to tell within less than a degree centigrade that the temperature of a bath, or a poultice, or dyeing liquor is " blood heat," or " fever heat," or some other definite temperature to which they have been accustomed.

12. *Thermometry by Sense of Heat—with arbitrary Centigrade Scale deduced from Mixtures of Hot and Cold Water.*—Without knowing anything of the nature of heat we might found a complete system of thermometry on the mixing of hot and cold water with no other thermoscope (§ 13 below) than our sense of heat, if we had but two definite constant temperatures of reference. These in practical thermometry are supplied by the melting-point of ice and the temperature of steam from water boiling in air at a definite pressure (the " atmo " or standard atmosphere, § 5 above). Thus, suppose perfectly abundant supplies of iced water and of water at the boiling temperature to be available, and suppose it to be desired to measure the temperature of a river, or lake, or sea. Take measured quantities of the boiling and of the ice-cold water, and mix them by trial until, tested by the hand, the mixture is found to have the same temperature as that of the mass of water of which the temperature is to be determined. Suppose, for example, the mixture giving the required temperature to consist of 86·6 parts by weight of ice-cold water, and 13·4 parts by weight of boiling water; the required temperature is 13·4 on a perfectly definite scale of thermometry in which the temperature of ice-cold water is called zero, that of boiling water 100, and other temperatures are reckoned according to the law of proportion of mixtures of water in the manner

An arbitrary centigrade scale founded on assuming the specific heat (§ 68) of water to be constant.

indicated by the example, and defined generally in § 31 below·
For temperatures within the range of sensibility of the hand this
method would give more accurate results than many common
thermometers sold by instrument makers for ordinary popular
purposes. It may be relied upon for absolute accuracy within
$\frac{2}{10}$ths of a degree centigrade, provided the mixing of hot and cold
water is performed with sufficiently large quantities of water, and
with all proper precautions to obtain in that part of the process
all the accuracy obtainable by the living thermoscope.

We shall see (§ 25 below) that with the most accurate mercury
or air thermometers, made for scientific investigation and carefully
tested, absolute determinations can scarcely be depended upon
within $\frac{1}{10}$th of a degree centigrade. The method of mixtures
with only the sensory thermoscope is not limited to the range of
temperature directly perceptible with unimpaired sensibility; but
when the temperature to be tested is beyond this range an indirect
method may be followed, as thus:—

A large quantity of water too warm for the hand is to be
tested. Mix it with say twice its weight of ice-cold water, this
giving a convenient temperature for the hand; then find by trial
what proportions of ice-cold and boiling water give a mixture of
the same temperature as tested by the hand; suppose these
proportions to be 26·2 of boiling water and 73·8 of ice-cold water.
The temperature of the mixture is by definition 26·2, and on
the same principle the required temperature is three times this,
or 78·6.

This system of thermometry is, however, strictly limited to
the range between the freezing and boiling points of water, for
we do not at present consider the possibilities of obtaining and
using thermometrically quantities of water below the freezing
point and above the boiling point. It is described here, not only
because it is very instructive in respect to the principles of thermometry, but because it is in point of fact the thermometric
method used through a large range of processes not only in the
arts but in scientific investigation. In many cases the hand is
a more convenient and easy test than a common mercury thermometer, and it has just about the same sensibility; the commonest
thermometers in popular use being in fact scarcely to be read to a
quarter of a degree centigrade. In respect to accuracy a common
cheap thermometer, though perhaps a degree or two wrong in its

absolute indications, may still be used as an accurate indicator of equality of temperatures just as is the hand in the method of mixtures.

In many cases the hand is more convenient than the thermometer, in other cases the thermometer is more convenient than the hand, but in many cases the thermometer is applicable when the hand is not. When the quantities of water tested are abundant, the hand is always the quicker test, but there must be abundance of water to allow it to be satisfactorily and accurately applicable.

THERMOSCOPES DIFFERENTIAL AND INTRINSIC.

Differential Thermoscopes essentially continuous.—Intrinsic Thermoscopes discontinuous and continuous.—Single and Multiple Intrinsic Thermoscopes (discontinuous).—Continuous Intrinsic Thermoscopes.

Differential thermoscopes.

13. A thermoscope is an indicator of temperature. A differential thermoscope is a thermoscope which shows difference, or tests equality, of simultaneous temperatures in two places. Its action is essentially continuous, depending on difference of temperature between the two places, and showing zero continuously when the temperatures of the two places are varied, provided they are kept exactly equal. Every kind of differential thermoscope, and of continuous intrinsic thermoscope, must be founded on some property of matter continuously varying with the temperature, as density of a fluid under constant pressure, pressure of a fluid in constant volume, volume of the liquid part of a whole mass of liquid and solid kept in constant volume*, steam-pressure of a solid or liquid†, shape or density of an elastic solid under constant stress, stress of an elastic solid in a constant state of strain, viscosity of a fluid, electric current in a circuit of two metals with their junctions at unequal temperatures, electric resistance of a conductor, magnetic moment of a steel or loadstone magnet.

Continuous thermoscopes.

Examples:—(1) Leslie's differential air thermometer; (2) steam-pressure differential thermometers (§§ 39—44 below); (3) Joule's hydraulic and pneumatic differential thermoscopes (*Mem.*

---

* This is the principle of the ordinary mercury or spirit thermometer.
† For definition of steam, see § 17 below.

Chem. Soc., vol. III. p. 201 ; *Proc. Lit. and Phil. Soc. Manchester*,
vol. III. p. 73; *Ibid.*, vol. VII. p. 35) (Joule's *Papers*, vol. I. pp. 535,
573); (4) viscosity differential thermoscope (*Proc. R.S.E.*, April 19,
1880); (5) thermo-electric differential thermometer; (6) Siemens
electric resistance differential thermometer; (7) thermo-magnetic
differential thermometer (see *Proc. R.S.E.*, April 19, 1880).

14. *Intrinsic Thermoscopes.*—An intrinsic thermoscope is
an instrument capable of indicating one definite temperature or
several definite temperatures, or all temperatures within the range
of the instrument, whatever it may be—the temperature or temperatures indicated being intrinsically determined by the constitution of the instrument, and indicated by some recognizable
feature of the instrument which changes discontinuously or continuously, as the case may be, and which is always the same when
the instrument is brought back again and again to the same temperature, whatever changes it may have experienced in the
intervals.   Discontinuous intrinsic thermoscopes show only a
limited number of temperatures.   A continuous intrinsic thermoscope shows any temperature whatever throughout the range of
efficiency of the instrument, ideally any temperature whatever,
though in practice every thermoscope is limited, some with both
inferior and superior limit, as the mercury thermometer by the
freezing of mercury at about − 39° C., and the bursting pressure of
mercury-steam a little above + 360° C.; others with only a superior
limit, as metallic thermoscopes, whether thermo-elastic, or thermo-
electric, or electric-resistance, or thermo-magnetic, by the melting
of their substances at very high temperatures, or, in the case of
the thermo-magnetic instrument, by the total or partial loss of its
magnetism at some temperature much below the melting point of
its substance.   A continuous intrinsic thermoscope, when applied
to a body whose temperature is changing, shows continuously every
variation of temperature within its range of efficiency.

15. *Discontinuous Intrinsic Thermoscopes.*—A single intrinsic
thermoscope is a thermoscope which shows whether the temperature of the body to which it is applied is higher or lower than
some one definite temperature depending on the intrinsic quality
of the instrument.

Examples :—(1) a piece of ice, or of wax, or of fusible metal;
(2) an apparatus for boiling water or other liquid under a perfectly

constant pressure; (3) an apparatus for boiling water under the natural atmospheric pressure, and a barometer to measure exactly what the pressure is at the time.

A multiple intrinsic thermoscope might be made by preparing a graduated series of metallic alloys, numbering them in the order of their melting points, and arranging them together conveniently for use*. The temperature might be reckoned numerically, according to the number of the alloys that melt, when the whole series is exposed to the temperature to be tested. This discontinuous numerical reckoning of temperature is perfectly analogous to the Birmingham reckoning of wires and sheet metals by numbered gauges. Ideally it may be made infinitely nearly continuous by making a series of alloys with fine enough gradation of composition, but the method is in its essence discontinuous. It is useful for many special applications in science and in the arts, as for instance in that very fundamental one (§ 12 above) of giving one of the fixed points in the ordinary thermometric scale, the "freezing point"; also in a form of safety valve for boilers or hot-water pipes, in which a plug fixed by solder is released by the melting of the solder when the temperature reaches a certain limit; also an exceedingly useful guard against overheating in the flue of a stove, by which a stopper is allowed to fall by the melting of a leaden support, and stop the draught, before the temperature reaches the highest limit judged permissible.

16. *Continuous Intrinsic Thermoscopes.*—Continuity of indication requires, as said in § 13 above, choice of some property or properties of matter varying continuously with temperature, such as those enumerated in § 13 above. A continuous intrinsic thermoscope must have a feature, depending on the chosen property of matter, which shall vary with perfect continuity when the temperature is gradually changed, and shall always be the same when the instrument is brought to the same temperature again and again, whatever variation of temperature it may have experienced in the intervals. The accuracy of an intrinsic thermo-

* [Note of *March* 26, 1885. A multiple intrinsic thermoscope has actually been made, according to this plan, by Mr J. J. Coleman, and was exhibited by him at a meeting of the Philosophical Society of Glasgow on 23rd January, 1884 (*Proc. Phil. Soc.* vol. xv. p. 94). Mr Coleman has used in this instrument a graduated series of paraffins which become solid at certain definite temperatures between 40° F. and 100° F., and mixtures of glycerine and water for the temperatures from − 35° F. to 30° F. W. T.]

meter, whether discontinuous or continuous, depends upon permanence of quality of the material and of the mechanical constitution
of the instrument, according to which the recognized feature shall
always be very accurately the same for the same temperature.
The sensibility or delicacy of a continuous intrinsic thermometer
depends upon the recognisability of change in its indicating feature
with very small change of temperature.

17. The property of matter chosen as the foundation of
almost all ordinary continuous intrinsic thermoscopes in common
use is interdependence of the density, the temperature, and the
pressure of a fluid. The only other thermoscopes which can be
said to be in common use at all are "metallic thermometers"; these
depend upon the change of shape of a rigid elastic solid under
a stated stress, or on the change of shape of a compound solid,
composed of two elastic solids of different substances melted or
soldered together. For the present we confine our attention to the
former and much larger class of instruments. The general type of
all those instruments, except the steam-pressure thermometer
(§§ 39—46 below), is a glass measure, measuring the bulk of
a fluid. To give the requisite practical sensibility to the measurement, the glass, except for the case of the constant-pressure gas
thermometer (§§ 64—67 below) and of the steam-pressure thermometer, is made of a shape which may be generally described as a
bottle with a long narrow neck. The body of the bottle, which
may either be spherical or of an elongated form, is called the bulb,
and the neck is called the tube or stem (stem we shall most frequently call it, to obviate ambiguities without circumlocutions).
The thermometric fluid may be all liquid, as mercury, or oil, or alcohol, or ether, or glycerine and water ; or it may be all gas, as common
air, or hydrogen, or carbonic acid; or it may be partly liquid and
partly steam (steam being a name which we shall invariably use to Steam,
designate the less dense portion of a fluid substance at one tem- definition of.
perature and pressure throughout, and in equilibrium, in two portions of different densities). This last case is different from the
two preceding, in respect to the character of the thermometric indication : the whole volume of the thermometric substance may be
changed from that of all liquid to that of all steam, without
changing the temperature or the pressure, and the pressure cannot
be changed without changing the temperature, provided the substance is kept in the double condition of part liquid and part

steam; in other words, in this case the pressure depends upon the temperature alone and is independent of the volume. In the steam-pressure thermometer, therefore, there is no delicate measuring of volume of the thermometric substance, and the vessel containing it is not in the shape of bulb and stem; but the instrument consists essentially of a means of measuring the pressure of the thermometric substance, with a test that it is really in the twofold condition of part liquid and part steam, whether by seeing it through a glass containing-vessel, or by a proper hydraulic appliance for ascertaining that the pressure is not altered by rarefaction or condensation when the temperature is kept constant. Realized thermometers of this species, quite convenient for many practical purposes, with steam of sulphurous acid, of water, and of mercury, to serve for different ranges of temperature, from below − 30° C. to above + 520° C., are described in §§ 39—44 below.

Mano-metric thermo-metry.

18. In respect to general convenience for large varieties of uses, whether for scientific investigation, or for the arts, or for ordinary use, liquid thermometers are generally and with good reason preferred; but the general preference of either mercury or spirits of wine for the liquid, which is so much the rule, is not (§ 20 below) so clearly reasonable. For ordinary uses in which the thermometer has to be moved about and placed in various positions, gas thermometers are much less convenient, because they require essentially an accurate measurement of pressure, and generally for this purpose a column of liquid. But when the thermometer is to be kept always in one position, as for instance when it is devoted to testing the temperature of the air indoors or out of doors, Amonton's air or gas thermometer is really as convenient and as easily read as any liquid thermometer can be : but even it, simple as it is, involves a triple division of the hermetically sealed space, with three different conditions of occupation,—one part occupied by the thermometric substance, another by the pressure-measuring

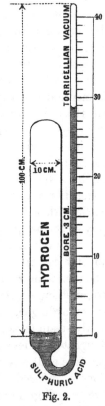

Fig. 2.

vapourless liquid*, and the third vacuous : and it is by so much the less simple than the liquid thermometer, that in the liquid thermometer the enclosed space is divided into only two parts, one occupied by the thermometric liquid, and the other by its steam, with or without some admixture of common air. For accuracy the air or gas thermometer is superior, we might almost say incomparably superior, to the mercury thermometer, and, though in a much less degree, still decidedly superior to even the most accurate liquid thermometer, on account of the imperfect constancy of the glass containing-vessel.

19. If we were quite sure of the bulk measurement given by the glass bulb and tube, liquid thermometers would be quite as

---

* An instrument closely resembling that shown in the drawing (fig. 2), but with common air instead of hydrogen, was made for the writer of this article, by Casella, about fifteen years ago, and has been used for illustrations in the natural philosophy class in Glasgow University ever since. It is probably an exceedingly accurate air thermometer. When it was set up in the new lecture-room after the migration to its present locality in 1870, the tube above the manometric liquid column was cleared of air. To do this the instrument must be held in such a sloping position with the closed end of the tube down, as to allow the bubble of air always found in it to rise and burst in the bulb. If now the instrument is placed in its upright position, the liquid refuses to leave the top of the tube, and it would remain filling the tube (probably, for ever?) if left in that position. No violence of knocking which has been ventured to try to bring it down has succeeded. To bring it down a bubble of air must be introduced. The bubble must be very small, so that the pressure of the air which fills it may become insensible when this air expands into the space of the tube left above the manometric column after it descends to its proper thermo-scopic position. Special experiments made for this article in September 1879 showed that in the nine years during which the instrument had remained undis-turbed in the lecture-room a very sensible quantity of air (enough to render the temperature indication about 35° C. too low) had leaked from the bulb through the sulphuric acid into the tube above the liquid column. This air was eliminated, and the instrument again set up for use, an operation completed in a minute at any time if need be. Some careful experiments were then made by Mr Macfarlane to ascertain if the pressure of vapour or gas from the sulphuric acid, in the tube, was sensible, with a happily decisive result in the negative. The bulb was kept at a very constant temperature by cold water ; the uppermost few centimetres of the liquid column, and the whole of the tube above it, were heated to about 100° C. by steam blown through a glass jacket-tube, fitted round it for the purpose. The height of the manometric column remained sensibly unchanged! Further experiments must be made to ascertain whether or not there is enough of variation of absorption of the air by the sulphuric acid with variation of temperature, and enough of the conse-quent variation of pressure in the bulb, to vitiate sensibly the thermometric use of the instrument. If, as seems improbable, the answer to this question be unhappily affirmative, a satisfactory negative might be found by substituting hydrogen for common air.

accurate as gas thermometers. For there is no difficulty in giving any required degree of sensibility to the instrument by making the bulb large enough; and the quality of the liquid itself, hermetically sealed in glass, may be regarded as being as constant as anything we know of in the material world. But, alas for thermometry, the glass measure is not constant! In fact, glass is a substance of very imperfect elasticity (Part I, § 4, above); and it is found that the bulb of a thermometer is not always of the same volume at the same temperature, but that, on the contrary, it experiences uncertain changes exceedingly embarrassing in thermometry. In the course of a few months after a thermometer is filled and sealed, the bulb generally shrinks by some uncertain amount of from $\frac{1}{40000}$ to $\frac{1}{20000}$ of its bulk, sometimes even in the course of years to almost $\frac{1}{10000}$. This has been discovered by a gradual rising of the freezing point in new mercury thermometers, generally as much as from $\frac{1}{4}°$ to $\frac{1}{2}°$, sometimes to as much as 1° C., which corresponds to a shrinkage of $1\cdot8/10^4$, as the bulk-expansion of mercury is, when its temperature is raised from 0° to 100°C. (Table II. below) $1/55\cdot1$, or $\cdot01815$, of its bulk at 0° C. After a few months or a few years this progressive shrinkage ceases to be sensible; but if the thermometer at any time is exposed to the temperature of boiling water or any higher temperature, an abrupt sub-permanent enlargement of the bulb is produced, and the freezing point, if tested for by placing the thermometer in ice and water, is found to be lowered; then again for weeks or months or years there is a gradual shrinkage, as shown by a gradual rising of the freezing point when the thermometer is tested again and again by placing it in ice and water. A very·delicate mercury thermometer, which has been kept for years at ordinary atmospheric pressures when out of use, and never when in experimental use exposed to any temperature higher than about 30° C., or much lower than the freezing point, becomes very constant, and probably may never show any change of as much as $\frac{1}{10}$ of a degree C. in its freezing point or in its indication at any other absolutely definite temperature, within some such range as from − 20° or − 10° C. to + 30° or + 40° C. But the abrupt and irregular changes, produced by exposing the thermometer to temperatures much above or much below some such limited range as that, constitute a very serious difficulty in the way of accurate thermometry by the mercury-in-glass thermometer. Although the greatest care has been bestowed by Regnault, Joule, and all other accurate thermo-metric experimenters to avoid error from this cause, we have still but little definite information as to its natural history in thermo-meters of different qualities of glass, different shapes of bulb, or differently constructed in respect to processes of glass-blowing, boiling the mercury, and sealing the stem. We do not even know whether the excess of the atmospheric pressure outside the bulb, over the pressure due to mercury and Torricellian vacuum inside, is influential sensibly, or to any considerable degree, in producing the gradual initial shrinkage. If it were so we might expect that the effect of heating the thermometer up to 100° C. or more at any time would be rather to produce an accelerated shrinkage for the time than what it is found to be, which is a return towards the original larger volume, followed by gradual shrinkage from day to day and week to week afterwards. A careful comparison between two thermometers constructed similarly in all respects, except sealing one of them with Torricellian vacuum and the other with air above the mercury, would be an important contribution to knowledge of this subject, interesting, not only in respect to ther-mometry, but also to that very fundamental question of physical science, the imperfect elasticity of solids (see Part I, § 4 above).

20. The error of a thermometer due to irregular shrinkages and enlargements of the bulb, is clearly the less the greater is the expansion of the thermometric fluid with the given change of tem-perature. By the investigation of § 30 below we can calculate exactly how much the error is for any stated amount of abnormal change of bulk in the bulb. But it is enough at present to remark that for different liquids in the same or in similar bulbs the errors are very nearly in the inverse proportions of the expansions of the liquids. Now (Table III. below) in being warmed from 0° to 1° C. alcohol expands 6 times as much as mercury, methyl butyrate 7 times, and sulphuric ether $8\frac{1}{2}$ times. Hence if irregular changes of bulk of the bulb leave, as they probably do in practice, an uncertainty of $\frac{2}{10}$ths of a degree in respect to absolute temperature by the best possible mercury-in-glass thermometers used freely at all temperatures from the lowest up to 100° C., the uncertainty from this cause will be reduced to $\frac{1}{30}$th of a degree by using alcohol, or $\frac{1}{35}$th by using methyl butyrate instead of mercury; it may therefore, in a glass thermometer of alcohol or of the methyl buty-rate, be considered as practically annulled (§ 19 above) after a few

weeks or months have passed, and the first main shrinkage is over.

21.   An alcohol-in-glass thermometer is easily made strong enough to bear a temperature of 100° C., as this gives by the pressure of the vapour an internal bursting pressure of not quite an atmosphere and a quarter in excess of the atmospheric pressure outside.   The boiling point of methyl butyrate (Table III. below) is 103°·5 C.; a thermometer of it may therefore be used for temperatures considerably above 100° C., but how much above we cannot tell, as we have not experiments as yet on the pressure of its vapour at temperatures above its boiling point.   The pressure of vapour of sulphuric ether (Table VIII. below) is too great to allow a thermometer of this liquid sealed in glass to be used much above 60° or 70° C., but for low temperatures it makes a very valuable thermometer.   It was used in 1850 by the author in finding by experiment the lowering of the freezing point of water, predicted theoretically by Prof. J. Thomson in 1849 (*Trans. R. S. E.*), and gave a sensibility of 128 divisions to 1° C.   Glass thermometers with ether, or chloroform (whose expansion is about 4 per cent. greater than that of ether), were used by Joule and the author in experiments* on changes of temperature experienced by bodies moving through air, in one of which the sensibility was as great as 330 scale divisions to the 1° C.   All these liquids, and many others in the modern chemist's repertory of oils and ethers and alcohols, besides the superior sensibility which they give by their greater expansions, have a great advantage over mercury for some thermometric purposes in their smaller specific gravity.   This allows the bulb to be larger, with less liability to break or to give disturbed readings through distortion by the weight of the contained liquid.

22.   Liquids which wet the glass have another great advantage over mercury in their smaller capillary attraction and in the constancy of their 180° angle of contact with the glass, instead of the much greater absolute intensity of capillary attraction in the mercury, and its very variable angle of contact, averaging about 45° when the mercury is rising, and about as much as 90° when it is falling.   On account of these variations the bulb of the mercury thermometer is subjected to abrupt variations of pressure when the mercury is rising or falling.   The greatest and least pressures

---

* *Phil. Trans.* for 1860, p. 325 [Art. XLIX. Pt. III. Vol. I. above].

due to this cause are experienced when the angle of contact is respectively least and greatest, and differ by the pressure due to a vertical column of mercury equal in height to the difference of depressions of mercury in a capillary tube of the same bore as the thermometer stem, when the angle of contact is changing from one to the other of the supposed extreme values. Hence the mercury in a thermometer rises and falls by jerks very noticeable in a delicate thermometer when looked at with a lens of moderate magnifying power, or even with the naked eye. Dr Joule informs us that this defect is much greater in some thermometers than in others, and that he believes it is greatly owing to the tube being left unsealed for too long a time after the introduction of the mercury, (by which it is to be presumed something of a film of oxide of mercury is left on the glass, to reappear on the surface of the mercury when it sinks as it cools after the sealing of the end). In Joule's own thermometers not the smallest indication has ever been detected of what he calls "this untoward phenomenon, which is calculated to drive an observer mad, if he discovers it towards the close of a series of careful experiments." Their admirable quality in this respect is no doubt due to the great care taken by the maker, Mr Dancer, under Joule's own instructions, to have the mercury and the interior of the bulb and tube thoroughly clean, and to guard it from exposure to any " matter in its wrong place" until completion of the sealing. But no amount of care could possibly produce a mercury thermometer of moderate dimensions moving otherwise than by jerks of ever so many divisions, if its stem were of fine enough bore to give anything approaching to two or three hundred divisions to the centigrade degree.

23.   One chief objection to the use of alcohol or other volatile liquid, for the thermometric substance in ordinary glass thermometers, is the liability to distillation of some of the liquid into the stem and head reservoir, unless the glass above the level of the liquid be kept at least as warm as the liquid. On this account a spirit thermometer is not suitable for being plunged into a space warmer than the surrounding atmosphere with the stem simply left to take the temperature to which it comes in the circumstances. But whether for elaborate experimental use, or for the most ordinary thermometric purposes, there is little difficulty in arranging to keep the part of the stem which is above the liquid

surface somewhat warmer than at the liquid surface, and this suffices absolutely to prevent the evil of distillation. The only other objection of any grave validity* against the use of highly expansive liquids instead of mercury, is the difficulty of allowing for the expansion of the liquid in the stem, if it is not at the same temperature as the bulb. With the same difference of temperatures in different parts of the instrument, the error on this account is clearly in simple proportion to the expansibility of the liquid; and therefore, the residual error due to want of perfect accuracy in the data for the allowance will, generally speaking, be greater with the more expansive than with the less expansive liquid. But in every case in which the bulb and stem can all conveniently be kept at one temperature, a thermometer having for its thermometric substance some highly expansive oil or alcohol or ether, or other so-called organic liquid of perfectly permanent chemical constitution, cannot but be much more accurate and sensitive than the mercury thermometer, which has hitherto been used almost exclusively in thermometric work of the highest rank. We shall see (§§ 62 and 64—68 below) that the ultimate standard for thermometry, according to the absolute thermodynamic scale (§ 34 below), is practically attained by the use of hydrogen or nitrogen gas as the thermometric substance, but that for ordinary use a gas thermometer can scarcely be made as convenient as one in which the thermometric substance is a liquid. For practical thermometry of the most accurate kind it seems that the best plan will be to use as ordinary working standard thermometers highly sensitive thermometers constructed of some chosen " organic " liquid, and graduated according to the absolute thermodynamic scale, by aid of the thermodynamically corrected air thermometer (§ 62 below)

* There is one other objection which, though often stated as very grave against the thermometric use of any other liquid than mercury, we do not admit to be so. It is that when the temperature is rapidly sinking, before becoming stationary, a little of the liquid lags behind the descending free surface, detained on the glass, and, trickling slowly down to rejoin the main column, must be waited for before the stationary temperature can be correctly read. We believe that if a fairly inviscid (or mobile) liquid such as alcohol or ether or butyrate of oxide of methyl be used, there will be practically *no time lost* from this cause, and certainly no accuracy lost when proper care is taken by the observer. The observer must be on his guard against a possibly false steadiness, through the falling of temperature being *momentarily* balanced in its effect on the free surface by the trickling down of liquid from the glass above, when the free surface is still above, or, it may be, has gone down to a little below, the true position for the final temperature.

used as ultimate standard of reference. The great convenience of the mercury thermometer in respect to freedom from liability to distillation, and smallness of error on account of difference of temperature between the bulb and stem, renders it the most convenient for a large variety of scientific and practical purposes in which the most minute accuracy or the most extreme sensibility is not required.

24. Without any *thermodynamic* reason for preferring air to mercury as thermometric fluid, Regnault preferred it for two very good reasons. (1) Its expansion is 20 times that of mercury and 160 times the cubical expansion of glass, and therefore with air the error due to irregularity in the expansion of the glass is 20 times smaller than with mercury, and small enough to produce no practical defalcation from absolute accuracy in thermometry, as he found by elaborate and varied trials. So far as this is concerned, some highly expansive organic liquids would answer nearly as well as air for thermometric fluid, and would have the advantage of giving a thermometer much more easily used. (2) For an ultimate standard of reference air has the advantage over organic liquids generally, that different samples of it taken at different times, or in different parts of the world, and purified of water and carbonic acid* by well-known and easily practised processes, are sufficiently uniform to give thermometric results between which the accordance is practically perfect, provided the thermometric plan according to which the different samples are used is the same, or as approximately the same as is easily secured in practice. Two plans for the thermometric use of air naturally present themselves :—
(I.) augmentation of volume of air kept in constant pressure ; and
(II.) augmentation of pressure of air kept in constant volume. Regnault tried both plans, but found that he could only arrange his apparatus to give good results by the second, and on it therefore he founded what he called his "normal air thermometer." For the sake of perfect definiteness he chose, as the density of the air in his normal thermometer, the density which air has when at

Regnault's normal air thermometer.

---

* Henceforth, to avoid circumlocutions, the unqualified word "air" will be used to denote atmospheric air taken in any part of the world, and deprived of carbonic acid and whatever vapour of water it may have contained, by aid of hydrated lime, or caustic potash, or some other suitable reagent for removing the carbonic acid, and quicklime, or chloride of calcium, or sulphuric acid, or phosphoric acid, for removing the water.

the temperature of melting ice and under the pressure of one atmo. He adopted the centigrade scale in respect to the marking of the freezing and boiling points by 0° and 100°; and the principle which he assumed for the reckoning of other temperatures was to call equal those differences of temperature for which differences of pressure of the air in his normal thermometer are equal. Thus he was led to a definition of temperature expressed by the following formula :—

$$t = 100 \frac{p - \Pi}{p_{100} - \Pi} \quad \dots\dots\dots\dots\dots\dots(1_a),$$

where Π denotes the pressure one atmo, and p and p_{100} the pressures of the air of the normal thermometer at the temperatures denoted by t and by 100 respectively, the latter being the temperature of steam issuing from water boiling under the pressure of one atmo. By the most accurate observations which he could make Regnault found for his "normal air" $p_{100} = 1\cdot3665 \times \Pi$. Hence his thermometric formula becomes

$$t = 100 \frac{p - \Pi}{\cdot3665 \times \Pi} = 272\cdot85 \left(\frac{p}{\Pi} - 1\right) \quad \dots\dots\dots(1_b).$$

Regnault's comparison of thermometers.

25. Regnault compared with his normal air thermometer other thermometers on the same plan of constant volume, but with air at other than the normal density of one atmo, and with other gases than air; also air and gas thermometers on the plan of constant *apparent* volume as measured in a glass bulb and stem; also a thermometer founded simply on the dilatation of mercury; also thermometers of mercury in different kinds of glass, each graduated on the glass stem with divisions corresponding to exactly equal volumes of the bore; also *overflowing thermometers* (thermomètres à déversement), in which a bulb with a short piece of fine stem was perfectly filled with mercury at 0° and the quantity of mercury expelled by the high temperature to be measured was weighed, instead of being volumetrically measured by divisions of a long stem as in the ordinary thermometer.

The whole of this thermometric investigation is full of scientific interest, and abounds with results of great practical value in respect even of the minutest details of Regnault's work. It will be found fully described in the first of his three volumes, entitled *Relation des Expériences entreprises par ordre de Monsieur le Ministre des Travaux Publics et sur la proposition de la Commission centrale des Machines à Vapeur pour Déterminer les Principales Lois et Données Numériques qui entrent dans le Calcul des Machines à Vapeur*, which were published at Paris in 1847, 1862, and 1870. Here we can but state some of the most important of the general conclusions :—

(1) The air thermometers with pressure at $0°$ of from 44 to 149 centimetres of mercury agreed perfectly with the normal air thermometer calculated according to the same formula (1_b), and nearly the same numerical coefficient $272·85$. A slightly larger value $272·98$ (or $·0036632^{-1}$) gave the best agreement for the 44 centimetres pressure, and the somewhat smaller value $272·7$ (or 003667^{-1}) for the pressure 149 centimetres.

(2) The hydrogen gas thermometer, with pressure one atmo at $0°$, and with its indications calculated according to formula (1_b) but with a different numerical coefficient*, agreed perfectly with the normal air thermometer from $0°$ to $325°$.

(3) The carbonic acid gas thermometer with pressure 46 centimetres at $0°$, and its indications calculated with the coefficient $271·59$ ($·003682^{-1}$), agreed perfectly with the normal air thermometer from $0°$ to $308°$).

(4) The carbonic acid gas thermometer with pressure 74 centimetres (or nearly 1 atmo) at $0°$, calculated with the coefficient $270·64$ ($·003695^{-1}$) to make it agree with the normal air thermometer at $100°$, gave numbers somewhat too large for all temperatures from $200°$ to $323°$. The difference seemed to rise to a maximum at about $180°$, when it was about $\frac{1}{3}°$, and to diminish so as to be only about $\frac{1}{10}°$ at the highest temperatures of the comparison. Two sulphurous acid gas thermometers, with pressures 59 centimetres and 75 centimetres at $0°$, calculated with coefficients $263·6$ ($·003794^{-1}$) and $261·4$ ($·003825^{-1}$) respectively to make them agree at $100°$ with the normal air thermometer, each gave numbers too small for the higher temperatures by differences increasing gradually from $\frac{1}{2}°$ at $140°$ to $3°$ at $320°$.

* Instead of the 003665 of his normal air thermometer, Regnault states that for his hydrogen thermometer he used ·003652 (which would make the coefficient in formula (1_b) be 273·82 instead of 272·85). But this must surely be a mistake, as he found ·0036678 for the "coefficient of dilatation" of hydrogen calculated from its increase of pressure in constant volume, and ·0036613 for the coefficient of dilatation observed directly for hydrogen under constant pressures of from 1 to 4 atmos (*Expériences*, Vol. I. pp. 78, 80, 91, 115, 116), and he nowhere speaks of having found any smaller value than ·003661 for hydrogen.

(5) Air and gas thermometers, calculated according to differences of pressure of the gas kept at the same *apparent* volume, (that is to say, with the bounding mercury column at a constant mark on the glass stem of the thermometer) gave numbers too small at the higher temperatures by differences gradually increasing up to $2\frac{1}{4}°$ at 350° in the case of *Choisi le Roi* crystal, a hard glass without lead, and to as much as $3\frac{1}{4}°$ in the case of ordinary glass.

In connexion with these observations Regnault remarks that the greatest cause of uncertainty in his air thermometry is the allowance for expansion of the glass. It was only by most carefully made special experiments* on each particular bulb and tube, to determine its expansion throughout the range for which it was to be used, that he succeeded in obtaining the great accuracy which we find in his results, according to which the probable error, whether by his normal air thermometer, or by other air or gas thermometers of those stated above to agree with it perfectly, was not more than from ·1 to ·15 of a degree for any temperature up to 350°.

(6) The mercury-in-glass thermometers which Regnault generally used for comparison with his normal air thermometer were overflowing thermometers, because he found that with such he could more easily obtain the very minute accuracy at which he aimed than with the ordinary volumetric thermometers; but the formula by which he calculated temperature from the overflowing thermometer, was adapted to give exactly the same result as would have been obtained by the ordinary thermometer with divisions on the stem corresponding to equal volumes of the bore. It must be remembered, however, that this perfect agreement between the volumetric and overflowing thermometers, would not be found unless the expansion of the bulb and tube were uniform and isotropic throughout.

(7) The general results of Regnault's comparisons of mercury thermometers with his normal air thermometer, were given by

---

* These experiments were made by finding the weight of mercury contained in each bulb and tube at several different temperatures throughout the range through which it was to be used, and thence calculating the bulks according to the density of mercury for the different temperatures found by his independent investigation of the absolute dilatation of mercury by the hydrostatic method, this method being independent of the expansion of the containing glass or other solid.

himself in a diagram of curves from which the accompanying is copied on a reduced scale (fig. 3). It shows that at a temperature of 320° the independent mercury thermometer stands at 329·8°, the thermometer of mercury in Choisi le Roi crystal at 327·25°, and the thermometer of mercury in ordinary glass at 321·8°; and that the independent mercury thermometer and the mercury in Choisi le Roi crystal stand 10° higher than the normal air thermometer at the temperatures by it of 323° and 345° respectively.

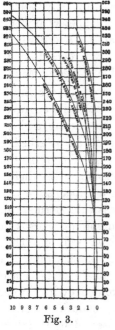

Absolute expansion of mercury thermometer.

26. The curve for the independent mercury thermometer is merely Regnault's graphic representation of his experiments on the absolute expansion of mercury (*Relation des Expériences*, Vol. I. p. 328). It shows that the addition of bulk given to the same mass of mercury under constant pressure by elevation of temperature is for the same difference of temperatures as indicated by his normal air thermometer regularly greater and greater the higher the temperature.

27. It is interesting to see by the diagram that at the high temperatures all the mercury thermometers keep nearer to the air thermometer than does the independent mercury thermometer, and that

Fig. 3.

the mercury in ordinary soft glass keeps much nearer to the air thermometer than does the mercury in the hard Choisi le Roi glass. We infer that, still reckoning temperature by the air thermometer, we have regular augmentation of expansion at the high temperatures in all the different glasses, each greater than the augmentation of expansion of mercury, and that this augmentation is greater in the soft ordinary glass than in the hard Choisi le Roi glass, being in the ordinary glass great enough to overcompensate in the resulting thermometric indication the augmenting expansion of the mercury from 100° to 245°; while above 245° in the ordinary glass thermometer, and at all temperatures above 100° in the Choisi le Roi thermometer, the compensation is only partial. Between 0° and 100° the independent

10—2

mercury thermometer stands regularly lower than the air thermometer by as great a difference as ·35° at 50°, where it
is a maximum. The curves for the mercury-in-glass thermometers are not shown between 0° and 100°, but it is
clear from the diagram that the Choisi le Roi thermometer
must, like the independent mercury thermometer, ·stand
lower than the air thermometer, but by a smaller difference,
probably only about ·2° at 50°; and the ordinary glass
thermometer higher than the air thermometer from 0° to
100° by a difference which may be ·2° or ·3° at 50°. This
last inference from the diagram is confirmed by Regnault's
table of results facing page 227 of his first volume.

28. In the best modern thermometers the graduations
are actually engraved on the glass; but in most popular
thermometers, and in many for scientific investigation, they
are on an attached scale of wood, or ivory, or brass, or
paper. Some of the best popular thermometers are the
German bath thermometers, in which the graduation is
on a paper scale guarded by being enclosed in a wide
glass tube hermetically sealed round the stem and over the
bulb of the glass which contains the mercury in the manner
shown in fig 4. The graduation is clearer and more easily
read in this kind of thermometer than in any other. The
complete protection of the paper scale against damp and damage,
afforded by its hermetically sealed glass envelope, gives a perennially enduring quality to this form of thermometer*, such as is
possessed by no others except those graduated on the glass; and
the lightness of the paper renders its proper attachment to the
inner stem, by gum or otherwise, thoroughly trustworthy, when
once well done by the maker of the instrument. For scientific
purposes the paper scale was too cheap, and common, and
good, to satisfy the ideas of those instrument makers who in
Germany and France substituted the heavy graduated slab of

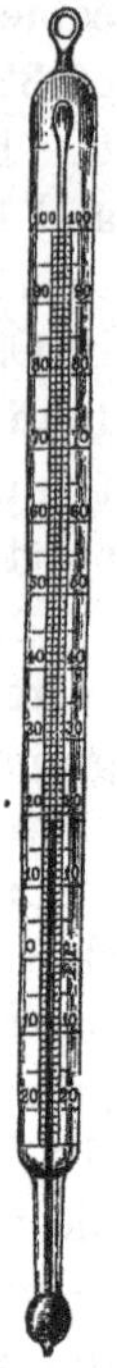

Fig. 4.

---

* Provided it is never exposed to "browning" temperatures (or temperatures
high enough to produce partially destructive distillation of the paper). Instrument
makers ignoring this caution have actually made it with graduation extending to
such temperatures for kitchen use. The result is that it gets injured to the extent
of partially browning the hermetically sealed paper, and befogging the inner surface
of the glass envelope, by applying it to test the temperature of melted fat in
cooking. For this purpose the simple scientific thermometer with graduation on
the glass stem is proper.

opal glass for the paper, while still adhering to the bath thermometer pattern in hermetically enclosing this scale in an
outer containing glass tube,—very unnecessarily, as the
glass scale, unlike the paper scale, does not require any
such protection.

Ordinary scientific thermometer of mercury in glass.

This is now, however, a thing of the past. At the present time all high-class scientific thermometers are graduated
on the glass of the stem, without any attached scale of
other material. Except in respect to ease of reading the
indications, this simplest form is, both for popular and for
scientific purposes, superior even to the German bath thermometer with hermetically sealed paper scale; and this will
be the form intended (Fig. 5) when we speak of a mercury
thermometer, or a spirit thermometer, or a liquid thermometer, without any special qualification.

29. *Properties of Matter concerned in Liquid Thermometers.*—The indications of the liquid thermometer depend
not only upon the expansion of the liquid with heat; they
are seriously modified by the expansion experienced also by
the containing solid. The instrument in fact consists of a
glass measure measuring the bulk of a liquid. If the bulk
of the hollow space in the glass and the bulk of the liquid
expand by the same amount, the apparent bulk of the liquid
as thus measured will remain unchanged. Now, supposing
the glass to be perfectly homogeneous and isotropic (see
Part I. above; §§ 38, 39, and chap. I. of Mathematical
Theory), and the bulb to be free from internal stress, the
glass will, when warmed uniformly, expand equally in all
directions, and the volume of the hollow space will be altered
in the same ratio as the volume of the glass itself. Hence
the indications of the thermometer depend on a difference
between the expansion of the glass and the expansion of
the liquid.

30. To define exactly the indications of a thermometer founded on the expansion of a fluid, let the volume
of the bore of the stem between two consecutive divisions
be called for brevity a *degree-measure*. The degree measure

Fig. 5.

is habitually made as nearly as possible equal throughout the
scale in the best mercury-in-glass thermometers; and, as we shall
see (§ 62 below), it ought to be so in an air-thermometer to give

indications agreeing with the absolute thermodynamic scale, nearly enough for the most accurate practical thermometry. But in practical spirit-thermometers the divisions are made to correspond as nearly as may be to degrees of a standard mercury or air thermometer, and the degree measures are therefore (Table II. below) larger and larger from the lower to the upper end of the scale. For the purpose, however, of comparing the thermometric performances of different liquids, we shall suppose the degree-measure to be of equal volume throughout the scale in each case.

Let N be the number of degree-measures contained in the volume of the bulb and stem up to the point marked zero on the scale; and let D_t denote the volume, at any temperature t, of the degree-measure reckoned in absolute units of volume. The volume of the bulb and stem up to zero will be ND_t. On the supposition of perfect isotropy and freedom from stress in the glass, N will be independent of the temperature and D_t/D_o will be the ratio of the volume of any portion of the glass at temperature t to its volume at the temperature called zero, if D_o denote the volume of the degree-measure when the glass is at this zero temperature. Let now L_t and L_o denote the volumes of the whole liquid in a thermometer at the two temperatures t and 0; we have $L_o = ND_o$. And if s be the number of scale divisions marking the place of the liquid surface in the thermometer tube, we have $L_t = (N + s) D_t$. Hence $L_t/L_o = (1 + s/N) D_t/D_o$. Hence $s = N \left(\dfrac{L_t/L_o}{D_t/D_o} - 1 \right)$. Hence, if E_t denote augmentation of bulk of the liquid, and $E_t{}'$ augmentation of bulk of each degree-measure of the stem, when the temperature is raised from 0 to t, each reckoned in terms of the bulk at zero temperature, we have

$$s = N \left(\frac{1 + E_t}{1 + E_t{}'} - 1 \right) = N \frac{E_t - E_t{}'}{1 + E_t{}'}.$$

This is the formula for the ordinary liquid thermometer. It is also applicable to the constant-pressure air thermometer, in which, with proper instrumental means to keep the pressure constant, air is allowed to expand or contract with elevation or depression of temperature, and its volume is measured in a properly shaped glass measuring vessel. We may arbitrarily determine to take s as the numeric for the temperature which is

indicated by any one particular thermometer of this kind, for
instance, a methyl butyrate thermometer, or an alcohol thermometer, or a mercury or an air thermometer. But if $s = t$ for any
one individual thermometer, it cannot be *exactly* so for any other.
In the first advances towards accurate thermometry it was taken
so for *the* mercury-in-glass thermometer, and by general consent
it was continued so until it was found (§ 25 above) that different
mercury-in-glass thermometers, each made with absolute accuracy,
differ largely in their reckonings of temperature.

31. *Numerical Thermometry.*—In § 12 above, a perfectly
definite and very simple basis for numerical thermometry was
described, not as having been adopted in practice, but as an
illustration of a very general principle upon which reckoning of
temperature may be done in numbers. The principle is this.
Two definite temperatures depending on properties of some particular substance or substances are first fixed upon and marked
by two arbitrary numbers,—as, for instance, the temperature of
melting ice marked zero, and the temperature of steam issuing
from boiling water under atmospheric pressure of exactly one
atmo, marked 100. Then any intermediate temperature t is
obtained by taking t parts of water at 100° and $(100 - t)$ parts at 0°
and mixing them together. As said in § 12 above this method is
limited to temperatures at which liquid water can be obtained,
and therefore practically it is only applicable between the melting
point of ice and the boiling point of water, under ordinary
atmospheric pressure.

Thermo-
meter
defined
to make
constant
the spe-
cific
heat of
water.

32. Any other liquid of permanent chemical constitution
might be used instead of water, as the thermometric substance in
thermometry founded on mixtures; so even might a powdered
solid. Oil, if used instead of water, would have the advantage
of being available for higher temperatures; but want of perfect
definiteness and constancy of chemical constitution is a fatal dis-
qualification for it as the fundamental thermometric substance for
thermometry by mixtures. Liquid mercury might be used with
the advantage of being available for both higher and lower temperatures than water, through a much wider whole range indeed
than either water or oil. For use as thermometric substance for
the method of mixtures both water and mercury, in the conditions
of approximate purity in which they are easily obtained in

Thermo-
metry
by mix-
tures.

abundance, have a paramount advantage over all other liquids in near enough approximation to perfect definiteness and constancy of constitution, to give practically perfect thermometric results.

Different equally permissible thermometric assumptions;

33. In §§ 12, 24, 25, 30, 31, and 32 above, several distinct definitions of numerical reckoning of temperature have been given. In each of these the differences of temperature which are to be called equal are defined specially, and this is the essence of the thermometric scale in each case (the marking of 0° and 100° for the "freezing" and "boiling" points being common to all as a matter of practical usage, and not an essential of the thermometric principle in any case). Thus in §§ 12, 31, and 32 above differences of temperature are called equal, which are produced by the communication of equal quantities of heat to a given quantity of the particular thermometric substance chosen—water, for example, or mercury; in other words (§ 68 below), this thermometric system is chosen so as to make the specific heat of a particular thermometric substance the same for all temperatures. Again in § 24 above differences of temperature are called equal, for which the differences of pressure are equal in air of the particular density which air has if its pressure is one atmo when its temperature is "freezing." This is Regnault's "normal thermometry." In § 25

each founded on a particular property of a particular substance.

(1), (2), (3), and (4) above, other reckonings of temperature differing essentially from this, though, as Regnault's experiments proved, by but very small differences, are given simply by the substitution of air of other than Regnault's normal density, and of other gases than air, for the air of Regnault's normal thermometer. In § 25 (5) above, a thermometry founded on a complex coefficiency of change of pressure and volume of a gas and change of volume of some one particular glass vessel is defined and compared with Regnault's normal thermometry; and in § 25 (6) and (7) above, the same is done for the ordinary mercury-in-glass thermometer, which depends on a coefficiency of glass and mercury leading to the reckoning of temperature defined in § 30 above. Again in §§ 20 to 23 above is indicated a system of thermometry founded on the absolute dilatation of some fluid, such as mercury or alcohol or butyrate of oxide of methyl or other permanent liquid or air, at some constant pressure, such as one atmo, with equal differences of temperature defined as those which give

equal dilatations of the particular substance chosen as the thermometric fluid.

34. Each of all these different definitions of temperature
is founded on some particular property of a particular substance.
A thermometer graduated to fulfil one of the definitions for one
particular substance would not agree with another thermometer
graduated according to the same definition for another substance,
or according to some of the other definitions. A much more satis- Prelim-
factory foundation for thermometry is afforded by thermodynamic thermo-
science, which gives us a definition of temperature depending on dynamic
certain thermodynamic properties of matter, in such a manner tion of
that if a thermometer is graduated according to it from observation tempera-
of one class of thermal effects in one particular substance, it will ture.
agree with a thermometer graduated according to the same
thermodynamic law from the same class of effects in any other
substance, or from the same or from some other class of effects
in another substance. Thus we have what is called the absolute
thermodynamic scale. This scale is now in modern thermal
science the ultimate scale of reference for all thermometers of
whatever kind (§ 67 below). It is defined in §§ 35 and 37 below
after the following preliminary. A piece of matter which we shall
call the "thermometric body" or "thermometric substance" must
be given, and at each instant it must be throughout at one
temperature, whatever operations we perform upon it. For simplicity we shall suppose it to be of one substance throughout. It
may be all solid, or it may be partly solid and the remainder
gaseous (as the contents of a wholly frozen cryophorous or any
other form of closed vessel full of ice and vapour of water,
but with no air); or it may at one particular temperature in the
course of its use be partly solid and partly liquid and partly
gaseous (as the contents of a partially frozen cryophorus); or
it may be partially liquid and partially gaseous (as the contents of an unfrozen cryophorus or of a "philosopher's hammer");
or it may be all liquid; or it may be all gas; or it may
be all fluid at a temperature above the Andrews "critical
temperature." If it be all solid it may be under any homogeneous
stress (Part I. above, *Mathematical Theory*, Part I. Chap. I.); but in
any case we suppose for simplicity the stress to be homogeneous
throughout, and therefore if the thermometric body be partly solid
and partly fluid, the stress in the solid as well as in the fluid must

be *uniform pressure in all directions*. To avoid excluding the case of all solid from our statements, we shall use generally the word *stress*, which will mean normal pressure reckoned in number of units of force per unit of area in every case in which the whole or any part of the thermometric body is fluid, and will denote this or any other possible stress when the thermometric body is all solid.

First
definition:—
by ratio
of two
tempera-
tures.

Carnot's
thermo-
dynamic
cycle
gener-
alized.

35. (1) Alter the bulk or shape of the thermometric sub-stance till it becomes warmer to any desired degree. (2) Keeping it now at this higher temperature, alter bulk or shape farther, and generate the heat which the substance takes to keep its tempera-ture constant, by stirring water, or a portion of the substance itself, if it is partly fluid, and measure the quantity of work spent in this stirring. (3) Bring it back towards its original bulk and shape till it becomes cooled to its original temperature. (4) Keep-ing it at this temperature, reduce it to its original bulk and shape, carrying off, by a large quantity of water, the heat which it must part with to prevent it from becoming warmed. Find by a special experiment how much work must be done to give an equal amount of heat to an equal amount of water by stirring. Then the ratio of the first measured quantity of work to the second is the ratio of the higher temperature to the lower on the absolute thermodynamic scale.

36. The following is equivalent to § 35, and is more con-venient for analytical use. It is derived from § 35 by supposing the first and third operations to be so small that the ratio defined as the ratio of the two temperatures is infinitely nearly unity, and conversely § 35—our first form of definition of absolute tempe-rature—may be derived from the second, which is to be now given, by passing through a finite range of temperature by successive infinitesimal steps, and applying the second definition to each step.

37. Let the thermometric body be infinitesimally warmed by stirring a portion or the whole of itself if it be partially or wholly fluid, or by stirring a quantity of fluid in space around it if it be all solid; and during the process let the stress upon the body be kept unchanged. The body expanding or contracting or changing its shape with the heat, as the case may be, does work upon the surrounding material by which its stress is main-tained. Find the ratio of the amount of work thus done to the

amount of work spent in the stirring. For brevity we shall call this the work-ratio. Again, let the stress be infinitesimally in- creased, the thermometric body being now for the time enclosed in an impermeable envelope so that it may neither gain nor lose heat. It will rise (or fall)* in temperature in virtue of the augmentation of stress. *The ratio of this infinitesimal elevation* *of temperature to the whole absolute temperature is equal to the* *work-ratio multiplied into the ratio of the infinitesimal augmenta-* *tion of stress to the whole stress.*

38. To show how our definition of absolute temperature is to be applied in practice take the following examples. Example 1. —Any case in which the thermometric substance is part in one condition and the remainder in another of different density, as part solid and part vapour, or part solid and part liquid, or part liquid and part steam. In this last case, as explained above (§ 34 above), we suppose the stress to be uniform pressure in all directions.

Let p be its amount, and let t be the absolute temperature corresponding to this pressure. Let σ be the ratio of the density of the rarer to that of the denser portion, ρ the density of the rarer portion, and $J\kappa$ the quantity of work required to generate the heat taken to convert unit mass of the substance from the lower to the higher condition (κ being the "latent heat" of transi-tion from the lower to the higher condition per unit mass of the substance, and J the dynamical equivalent of the thermal unit in which κ is measured). The work done by the substance in passing from the denser to the rarer condition per unit volume of the latter is $p(1-\sigma)$, and the amount of work required to generate the heat taken in doing so is $J\rho\kappa$. Hence the work-ratio of our second definition is

$$\frac{p(1-\sigma)}{J\rho\kappa} \quad\dots\dots\dots\dots\dots\dots\dots\dots\dots \text{(1)}.$$

Let now the pressure be increased by an infinitely small quantity dp, and, the substance being still in the two conditions but of uniform temperature throughout, let dt be the corresponding rise in temperature. We have by the definition (§ 37)

---

* In the case of fall the elevation of temperature is to be regarded as negative; and in this case the "work-ratio" is negative also.

$$\frac{dt}{t} = \frac{p\,(1-\sigma)}{J\rho\kappa}\frac{dp}{p} = \frac{1-\sigma}{J\rho\kappa}\,dp.$$

Hence
$$\frac{1}{t}\frac{dt}{dp} = \frac{1-\sigma}{J\rho\kappa} \quad\dotfill (2).$$

Hence by integration

$$\left.\begin{array}{l} \log\dfrac{t}{t_0} = \displaystyle\int_{p_0}^{p}\dfrac{(1-\sigma)\,dp}{J\rho\kappa} \\[3mm] t = t_0 \epsilon^{\displaystyle\int_{p_0}^{p}\frac{(1-\sigma)dp}{J\rho\kappa}} \end{array}\right\} \quad\dotfill (3).$$

or

Water-steam thermometer.   39.   Fig. 6 represents a thermometer constructed to show absolute temperature on the plan of Example 1, § 38 above, realised for the case of water and vapour of water as thermometric substance. The containing vessel consists of a tube with cylindric bulb like an ordinary thermometer; but, unlike an ordinary thermometer, the tube is bent in the manner shown in the drawing. The tube may be of from 1 to 2 or 3 millims. bore, and the cylindrical part of the bulb of about ten times as much. The length of the cylindrical part of the bulb may be rather more than 1/100 of the length of the straight part of the tube. The contents, water and vapour of water, are to be put in and the glass hermetically sealed to enclose them, with the utmost precautions to obtain pure water as thoroughly freed from air as possible, after better than the best manner of instrument makers in making cryophoruses and water hammers. The quantity of water left in at the sealing must be enough to fill the cylindrical part of the bulb and the horizontal branch of the tube. When in use the straight part of the tube must be vertical with its closed end up, and the part of it occupied by the manometric water-column must be kept at a nearly enough definite temperature by a surrounding glass jacket-tube of iced water. This glass jacket-tube is wide enough to allow little lumps of ice to be dropped into it from its upper end, which is open. By aid of an india-rubber tube connected with its lower end, and a little movable cistern, as shown in the drawing, the level of the water in the jacket is kept from a few inches above to a quarter of an inch below that of the interior manometric column. Thus, by dropping in lumps of ice so as always to keep some unmelted ice floating in the water of the jacket, it is easy to keep the temperature of the top

of the manometric water-column exactly at
the freezing temperature. As we shall see
presently, the manometric water below its
free surface may be at any temperature
from freezing to 10° C. above freezing without more than 1/40 per cent. of hydrostatic
error. The temperature in the vapour-space
above the liquid column may be either freezing or anything higher. It ought not to be
lower than freezing, because, if it were so,
vapour would condense as hoar frost on the
glass, and evaporation from the top of the
liquid column would either cryophoruswise
freeze the liquid there, or would cool it below
the freezing point.

40. The chief object of keeping the
top of the manometric column exactly at
the freezing point is to render perfectly
definite and constant the steam pressure
in the space above it.

A second object of considerable importance when the bore of the tube is so
small as one millimetre is to give constancy
to the capillary tension of the surface of the
water. The elevation by capillary attraction
of ice-cold water in a tube of one millimetre
bore is about 7 millims. The constancy of
temperature provided by the surrounding
iced water will be more than sufficient to
prevent any perceptible error due to in-
equality of this effect. To avoid error from
capillary attraction the bore of the tube
ought to be very uniform, if it is so small
as one millimetre. If it be three millimetres

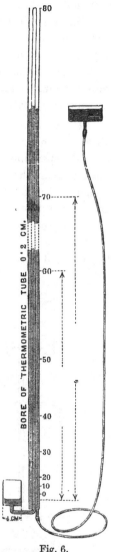

Fig. 6.

or more, a very rough approach to uniformity would suffice.

A third object of the iced-water jacket, and one of much more
importance than the second is to give accuracy to the hydrostatic
measurement by keeping the density of the water throughout the
long vertical branch definite and constant. But the density of
water at the freezing point is only 1/40 per cent. less than the

maximum density, and is the same as the density at 8° C.; and
therefore when 1/40 per cent. is an admissible error on our thermometric pressure, the density will be nearly enough constant with
any temperature from 0° to 10° C. throughout the column. But
on account of the first object mentioned above the very top of the
water column must be kept with exceeding exactness at the
freezing temperature.

41.   In this instrument the "thermometric substance" (§ 34
above) is the water and vapour of water in the bulb, or more
properly speaking the portions of water and vapour of water infinitely near their separating interface. The rest of the water
is merely a means of measuring hydrostatically the fluid pressure
at the interface. When the temperature is so high as to make
the pressure too great to be conveniently measured by a water
column, the hydrostatic measurement may be done, as shown in
the annexed drawing (fig. 7), by a mercury column in a glass tube,

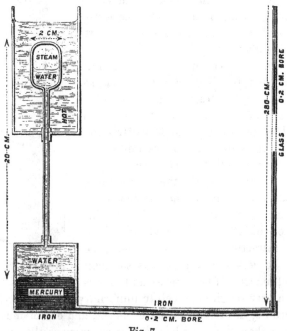

Fig. 7.

surrounded by a glass water-jacket not shown in the drawing, to
keep it very accurately at some definite temperature so that the
density of the mercury may be accurately known.

The simple form of steam thermometer represented with figured dimensions in fig. 6 will be very convenient for practical use for temperatures from freezing to 60° C. Through this range the pressure of water-steam, reckoned in terms of the balancing column of water of maximum density, increases (Table V.) from 6·25 to 202·3 centimetres; and for this therefore a tube of a little more than 2 metres will suffice. From 60° to 140° C. the pressure of steam now reckoned in terms of the length of a balancing column of mercury at 0° increases from 14·88 to 271·8 centimetres; and for this a tube of 280 centimetres may be provided. For higher temperatures a longer column, or several columns, as in the multiple manometer, or an accurate air pressure-gauge, or some other means, such as a very accurate instrument constructed on the principle of Bourdon's metallic pressure-gauge, may be employed, so as to allow us still to use water and vapour of water as thermometric substance.

42. At 230° C., the superior limit of Regnault's high-pressure steam experiments, the pressure is 27·53 atmos, but there is no need for limiting our steam thermometer to this temperature and pressure. Suitable means can easily be found for measuring with all needful accuracy much higher pressures than 27 atmos. But at so high a temperature as 140° C., vapour of mercury measured by a water column, as shown in the diagram (fig. 8), becomes available for purposes for which one millimetre to the degree is a sufficient sensibility. The mercury-steam-pressure thermometer, with pressure measured by water-column, of dimensions shown in the drawing, serves from 140° to 280° C., and will have very ample sensibility through the upper half of its scale. At 280° C. its sensibility will be about 4¾ centimetres to the degree! For temperatures above 280° C. sufficient sensibility for most purposes is obtained by substituting mercury for water in that simplest form of steam thermometer shown in fig. 6, in which the pressure of the steam is measured by a column of the liquid itself kept at a definite temperature. When the liquid is mercury there is no virtue in the particular temperature 0° C., and a stream of water as nearly as may be of atmospheric temperature will be the easiest as well as the most accurate way of keeping the mercury at a definite temperature. As the pressure of mercury steam is at all ordinary atmospheric temperatures quite imperceptible to the hydrostatic test when mercury itself is the balancing

(marginal note:) High-pressure steam thermometer.

liquid, that which was the chief reason for fixing the temperature
at the interface between liquid and vapour at the top of the
pressure-measuring column when the balancing liquid was water
(§ 40 above) has no weight in the present case; but, on the other
hand, a much more precise definiteness than the ten degrees

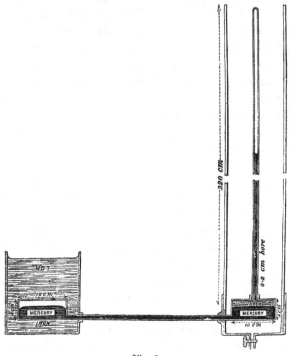

Fig. 8.

latitude allowed in the former case for the temperature of the
main length of the manometric column is now necessary. In fact,
a change of temperature of 2·2° C. in mercury at any atmospheric
temperature produces about the same proportionate change of density as is produced in water by a change of temperature from 0° to
10° C., that is to say, about 1/20 per cent.; but there is no difficulty in keeping, by means of a water jacket, the mercury column
constant to some definite temperature within a vastly smaller
margin of error than 2·2° C., especially if we choose for the definite
temperature something near the atmospheric temperature at the
time, or the temperature of whatever abundant water supply may
be available. If the vertical tube for the pressure-measuring

mercury column be 830 centimetres long, the simple mercury-
steam thermometer may be used up to 520° C., the highest temperature reached by Regnault in his experiments (Table V. below)
on mercury-steam. By using an iron bulb and tube for the part
of the thermometer exposed to the high temperature, and for the
lower part of the measuring column to within a few metres of its
top, with glass for the upper part to allow the mercury to be seen,
a mercury-steam-pressure thermometer can with great ease be
made which shall be applicable for temperatures giving pressures
up to as many atmospheres as can be measured by the vertical
height available. The apparatus may of course be simplified by
dispensing with the Torricellian vacuum at the upper end of the
tube, and opening the tube to the atmosphere, when the steam-
pressure to be measured is so great that a rough and easy barometer observation gives with sufficient accuracy the air-pressure
at the top of the measuring column. The easiest, and not necessarily in practice the least accurate, way of measuring very high
pressures of mercury-steam will be by enclosing some air above
the cool pressure-measuring column of mercury, and so making
it into a compressed-air pressure-gauge, it being understood that
the law of compression of the air under the pressures for which
it is to be used in the gauge is known by accurate independent
experiments such as those of Regnault on the compressibility of
air and other gases.

43. The water-steam thermometer may be used, but somewhat precariously, for temperatures below the freezing point, because water, especially when enclosed and protected as the portion
of it in the bulb of our thermometer is, may be cooled many degrees
below its freezing point without becoming frozen : but, not to
speak of the uncertainty or instability of this peculiar condition
of water, the instrument would be unsatisfactory on account of insufficient thermometric sensibility for temperatures more than
two or three degrees below the freezing point. Hence, to make
a steam thermometer for such temperatures some other substance
than water should be taken, and none seem better adapted
for the purpose than sulphurous acid, which, in the apparatus
represented with figured dimensions in the accompanying diagram
(fig. 9), makes an admirably convenient and sensitive thermometer for temperatures from + 20° to something far below

$-30°$ C., as we see from the results of Regnault's measurements (Table VIII. below).

44. To sum up, we have in §§ 39—43 above, a complete series of steam-pressure thermometers, of sulphurous acid, of water, and of mercury, adapted to give absolutely definite and highly sensitive thermometric indications throughout the wide range from something much below $-30°$ to considerably above $520°$ of the centigrade scale. The graduation of the scales of these thermometers to show absolute temperature is to be made by calculation from formula (3) of § 38 above, when the requisite experimental data, that is to say, the values of σ and $\rho\kappa$ for different values of p throughout the range for which each substance is to be used as thermometric fluid are available. Hitherto these requisites have not been given by direct experiment for any one of the three substances with sufficient accuracy for our thermometric purpose through any range whatever. Water, naturally, is the one for which the nearest approach to the requisite information has been obtained. For it Regnault's experiments have given, no doubt with great accuracy, the values of p (the steam pressure) and of κ (the latent heat of steam per unit mass) for all temperatures reckoned by his normal air thermometer, which we now regard merely as an arbitrary scale of temperature, through the range from $-30°$ to $+230°$. If he, or any other experimenter, had given us with similar accuracy through the same range, the

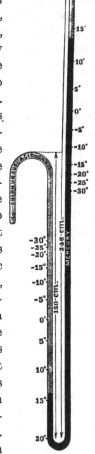

Fig. 9.

values of ρ (the density of steam) and σ (the ratio of the density of steam to the density of water in contact with it), for temperatures reckoned on the same arbitrary scale, we should have all the data from experiment required for the graduation of our water-steam thermometer to absolute thermodynamic scale. For it is to be remarked that all reckoning of temperature is eliminated from the second member of formula (3) of § 38 above, and that in our use of it Regnault's normal thermometer has merely been referred to for the values of $\rho\kappa$ and of $1-\sigma$, which correspond to stated values of p. The arbitrary constant of integration, t_0, is truly arbitrary. It will be convenient to give it such a value that the difference of values of t between the freezing point of water and the temperature for which p is equal to one atmo shall be 100, as this makes it agree with the centigrade scale in respect to the difference between the numbers measuring the temperatures which on the centigrade scale are marked 0° and 100° C. We shall see (§ 56 below) that indirectly by means of experiments on hydrogen gas this assignation of the arbitrary constant of integration would give 273 for the absolute temperature 0° C., and 373 for that of 100° C. Meantime, as said above, we have not the complete data from direct experiments even on water-steam for graduating the water-steam thermometer; but on the other hand we have, from experiments on air and on hydrogen and other gases, data which allow us to graduate indirectly any continuous intrinsic thermoscope (§ 19 above) according to the absolute scale; and we shall see that by thus indirectly graduating the water-steam thermometer, we learn the density of steam at different temperatures more accurately than it has hitherto been made known by any direct experiments on water-steam itself.

45. Merely viewed as a continuous intrinsic thermoscope, the steam thermometer, in one or other of the forms described above to suit different parts of the entire range from the lowest temperatures to temperatures somewhat above 520° C., is no doubt superior, in the conditions for accuracy specified in § 16 above, to every other thermoscope of any of the different kinds hitherto in use; and it may be trusted more surely for accuracy than any other as a thermometric standard, when once it has been graduated according to the absolute scale, whether by practical experiments on steam, or indirectly by experiments on air or other gases. In fact the use of steam-pressure, measured in definite units of pressure, as a thermoscopic effect in the steam thermometer, is simply a continuous extension to every temperature of the principle already practically adopted for fixing the temperature which is called 100° on the centigrade scale; and it stands on precisely the same theoretical footing as an air thermometer, or a mercury-in-glass thermometer, or an alcohol thermometer, or a methyl butyrate thermometer, in respect to the graduation of its scale according to absolute temperature. Any one intrinsic thermoscope

11—2

may be so graduated ideally by thermodynamic experiments on the substance itself without the aid of any other thermometer or any other thermometric substance; but the steam-pressure thermometer has the great practical advantage over all others except the air thermometer, that these experiments are easily realizable with great accuracy; instead of being, though ideally possible, hardly to be considered possible as a practical means of attaining to thermodynamic thermometry. In fact, for water-steam it is only the most easily obtained of experimental data, the measurement of the density of the steam at different pressures, that has not already been actually obtained by direct experiment. Whether or not, when this lacuna has been filled up by direct experiments, the data from water-steam alone may yield more accurate thermodynamic thermometry than we have at present from the hydrogen or nitrogen gas thermometer (§§ 64—69 below), we are unable at present to judge. But when once we have the means, directly from itself, or indirectly from comparison with hydrogen or nitrogen or air thermometers, of graduating once for all a sulphurous acid steam thermometer, a water-steam thermometer, or a mercury-steam thermometer, that is to say, when once we have a table of the absolute thermodynamic temperatures corresponding to the different steam pressures of the substances sulphurous acid, water, and mercury, we have a much more accurate and more easily reproducible standard than either the air or gas thermometer of any form, or the mercury thermometer, or any liquid thermometer, can give. In fact, the series of steam thermometers for the whole range from the lowest temperatures can be reproduced with the greatest ease in any part of the world by a person commencing with no other material than a piece of sulphur and air to burn it in*, some pure water, and some pure mercury, and with no other apparatus than can be made by a moderately skilled glass-blower, and with no other standard of physical measurement of any kind than an accurate linear measure. He may assume the force of gravity to be that calculated for his latitude, with the ordinary rough allowance for his elevation above the sea, and his omission to measure with higher accuracy the actual force of gravity in his locality

---

* Practically, the best ordinary chemical means of preparing sulphurous acid, as from sulphuric acid by heating with copper, might be adopted in preference to burning sulphur.

can lead him into no thermometric error which is not in-comparably less than the inevitable errors in the reproduction and use of the air thermometer, or of mercury or other liquid thermometers. In temperatures above the highest for which mercury-steam pressure is not too great to be practically available, nothing hitherto invented but Deville's air thermometer with hard porcelain bulb suited to resist the high temperature is available for accurate thermometry. Deville's accurate pyro-meter.

46. We have given the steam thermometer as our first example of thermodynamic thermometry, because intelligence in thermodynamics has been hitherto much retarded, and the student unnecessarily perplexed, and a mere quicksand has been given as a foundation for thermometry, by building from the beginning on an ideal substance called perfect gas, with none of its properties realized rigorously by any real substance, and with some of them unknown, and utterly unassignable, even by guess. But after having been moved by this reason to give the steam-pressure thermometer as our first theoretical example, we have been led into the preceding carefully detailed examination of its practical qualities, and we have thus become convinced that though hitherto used in scientific investigations only for fixing the "boiling point," and (through an inevitable natural selection) by practical en-gineers for knowing the temperatures of their boilers by the pressures indicated by the Bourdon gauge, it is destined to be of great service both in the strictest scientific thermometry and as a practical thermometer for a great variety of useful applications.

47. *Example* 2 (including *Example* 1, § 38 above).—Any case in which the stress is uniform pressure in all directions.

Let p and v denote the pressure and volume. The condition of the substance (single, double, or triple, as the case may be) is determinate when p and v are given, and it will therefore be spoken of shortly as the condition (p, v). Let e be the energy which must be communicated to the substance to bring it from any conveniently defined zero condition (p_0, v_0) to any condition whatever (p, v). Remark that e is a function of the two inde-pendent variables p, v to be found by experiment, and that the finding of it by experiment is a perfectly determinate practical problem, which can be carried out without the aid of any thermo-scope, and without any consideration whatever relating to temperature. We shall see in fact that accurate practical solutions of it for many different substances have been obtained by experiment. The absolute temperature t is also a function of p and v to be also determined by experiment, according to the equivalent definitions of §§ 35 and 37 above. Let heat be communicated to the substance so as to cause its volume to increase by dv, the pressure being kept constant. The energy of the body will be augmented by

$$\frac{de}{dv} \cdot dv.$$

At the same time the body in expanding and pressing out the matter around it does work to the extent of

$$p \cdot dv \dotfill (4).$$

Hence the whole work required to generate the heat given to it amounts to

$$\left(\frac{de}{dv} + p\right) dv \dotfill (5).$$

Hence the ratio of (4) to (5), or

$$\frac{p}{\dfrac{de}{dv} + p} \dotfill (6),$$

is the "work-ratio" of § 37 above. Hence by the definition

$$\frac{Dt}{t} = \frac{dp}{p} \frac{p}{\dfrac{de}{dv} + p} = \frac{dp}{\dfrac{de}{dv} + p} \dotfill (7),$$

Operation called adiabatic compression by Rankine. where Dt denotes the change of temperature produced by augmenting the pressure by dp, and at the same time preventing the substance from either giving heat to or taking heat from the surrounding matter. To express this last condition analytically, let dv be the augmentation of volume (negative, of course, if dp be positive) which it implies. The work done on the substance by the pressure from without is $-pdv$, and the energy of the substance is augmented by just this amount, because of the condition to be expressed. Hence

$$\frac{de}{dv} dv + \frac{de}{dp} dp = -pdv \dotfill (8);$$

whence
$$dv = \frac{\dfrac{de}{dp}}{\dfrac{de}{dv} + p}\, dp \dotfill (9).$$

But
$$Dt = \frac{dt}{dp}\, dp + \frac{dt}{dv}\, dv,$$

and so we have

$$Dt = \left(\frac{dt}{dp} - \frac{\dfrac{dt}{dv}\dfrac{de}{dp}}{\dfrac{de}{dv} + p} \right) dp \dotfill (10).$$

Eliminating Dt/dp from this by (7) we find

$$t = \left(p + \frac{de}{dv} \right) \frac{dt}{dp} - \frac{de}{dp}\frac{dt}{dv} \dotfill (11).$$

48. This is a linear partial differential equation of the first order for the determination of t, supposing, as we do for the present, that e is a known function of p and v. The following graphical illustration of the well-known analytical process for finding the complete solutions of such equations shows exactly how much towards determination of temperature can be done with no other data from experiment than the values of $\dfrac{de}{dv}$ and $\dfrac{de}{dp}$ as functions of p and v, and what additional information is required to fully determine t.

First remark that (11) is the condition that $\dfrac{1}{t}$ be a factor rendering $\left(p + \dfrac{de}{dv} \right) dv + \dfrac{de}{dp}\, dp$ a complete differential of a function[*] of two independent variables p and v. Let ϕ be this function,—that is to say, let ϕ be such that

Called thermodynamic function by Rankine, called entropy by Clausius and subsequent writers.

---

[*] This function is of great importance in practical thermodynamics: multiplied by t_0, it is equal to the excess of the energy of the substance above its motivity. Motivity (defined in paper "On Thermodynamic Motivity," *Phil. Mag.* May, 1879 [Art. L. Vol. I. above]) is the amount of work obtainable by letting the substance pass from the state (p, v) in which it is given to the zero condition (p_0, v_0), without either taking in heat from or giving out heat to matter at any other temperature than t_0.

$$\frac{d\phi}{dv} = \frac{1}{t}\left(p + \frac{de}{dv}\right), \text{ and } \frac{d\phi}{dp} = \frac{1}{t}\frac{de}{dp}$$

or

$$t = \frac{p + \dfrac{de}{dv}}{\dfrac{d\phi}{dv}} = \frac{\dfrac{de}{dp}}{\dfrac{d\phi}{dp}} \qquad \Bigg\} \quad \dots\dots\dots (12).$$

Then every solution of the differential equation

$$\frac{dp}{dv} = -\frac{p + \dfrac{de}{dv}}{\dfrac{de}{dp}} \quad \dots\dots\dots\dots (13)$$

Called adiabatics by Rankine, isentropics by Willard Gibbs. Complete diagram of adiabatics drawn.

renders ϕ constant, and conversely, every series of values of p and v which renders ϕ constant, constitutes a solution of (13). Now this differential equation may be solved graphically by taking p and v as rectangular coordinates of a point in a plane, and drawing the whole series of curves which satisfy it as follows. Commence with any point and calculate for its values of p and v the value of the second member of (13). Draw through this point an infinitesimal line in the direction of the tangent to the curve given by the value so found for $\frac{dp}{dv}$. With the altered values of p and v corresponding to the other end of this infinitesimal line, calculate a fresh value of $\frac{dp}{dv}$, and continue the curve in the slightly altered direction thus found, and so on. Take another point anywhere infinitesimally near this curve but not in it, and draw by a similar process the curve through it satisfying the equation. Take a third point infinitely near this second curve, and draw through it a third curve satisfying (13), and so on till the whole area of values p, v, possible for the substance in question, is filled with a series of curves one of which passes through, or infinitely nearly through, every point of the area. Assign arbitrarily a particular value of ϕ to each of these curves; then graphically find $\frac{d\phi}{dv}$ and $\frac{d\phi}{dp}$ for any or every value of p and v. Then either of the two second forms of equation (12) gives us explicitly a value of t for any values whatever of p and v.

Completed solution of
49. The solution for t thus obtained involves the arbitrary assumption of a particular value of ϕ for each one of the series of

curves which we have determinately traced. Hence, to render t differential equation for t with one arbitrary function. wholly determinate, something more must be given than e as a function of p and v. Now the only thing that can be given respecting temperature for any particular substance before we have a thermometric scale is the relation subsisting between p and v when the temperature is constant. This relation can, with merely a single-temperature-thermoscope (§ 15 above), in addition to dynamical instruments, be determined for some one particular temperature; and this, if e be known for every value of p and v, is the only additional knowledge required for the Arbitrary function of the analytical solution determined by experiment. determination of t for every value of p and v. For let $p = f(v)$ be the relation between p and v for some one particular temperature, t_0. If by this we eliminate p from (12) we find

$$\frac{d\phi}{dv} = \frac{1}{t_0}\left[f(v) + \frac{de}{dv} \right] \quad\ldots\ldots\ldots\ldots\ldots(14);$$

where $\frac{de}{dv}$, when $p = f(v)$, becomes a known function of v alone. Hence by integration we find

$$\phi = \frac{1}{t_0}[F(v) + C] \quad\ldots\ldots\ldots\ldots\ldots(15),$$

where F denotes a known function and C an arbitrary constant. Completed determination of absolute temperature from results of experiment. Now trace the curve $p = f(v)$ on our diagram. It must generally cut every one of the previously drawn determinate series of curves. Hence equation (15), with two arbitrarily assigned constants t_0 and C, gives determinately the value of ϕ for every one of the diagram of curves, and thus ϕ is determined for every value of p and v. Either of equations (12) then gives t determinately as a function of p and v, with only the value t_0 arbitrary. The information from experiment, regarding the properties of the thermometric substance, on which this determination is founded, consists of a knowledge of the relation between p and v for any one temperature, and of the value of $e - e_0$ for all values of p, v, (e_0 denoting the unknown value of e for some particular values $p_0 v_0$). Although, theoretically, this information is attainable by purely dynamical operations and measurements, with no other thermal guidance or test than that afforded by a single-temperature-intrinsic-thermoscope (§ 15 above), the whole of it has not in fact been explicitly obtained for any one substance. But less than the whole of it suffices to make a perfect absolute thermometer of any given substance.

50. For this purpose it is not necessary to find t for all values of p and v: it is enough to know it for all values mutually related in any manner convenient for thermometric practice. For example, if we could find t for every value of v with p constant at some one particular chosen value,—this would give a "constant pressure" absolute thermometer. Or again, if we find t for every value of p with v kept constant,—this would give us a "constant volume" absolute thermometer. Let us now examine into the restricted dynamical and thermoscopic investigations upon any particular substance, which will suffice to allow us to make of it a standard absolute thermometer of one or other of these species.

51. *Dynamical and thermoscopic investigation required to graduate, according to the absolute scale, a constant-pressure thermometer of any particular fluid.*—Let a large quantity of fluid be given, and let proper mechanical means be taken to cause it to flow slowly and uniformly through a pipe, in one short length of which there is a fixed porous plug. If, as is the case with common air, nitrogen, oxygen, carbonic acid, and no doubt many other gases, the fluid leaves the plug cooler than it enters it, let there be a paddle in the stream flowing from the plug, and let this paddle be turned so as to stir the fluid and cause the temperature, when the rapids are fairly past and the eddies due to the stirring subsided, to be the same as in the stream flowing towards the plug. When, as in the case of hydrogen and of all ordinary fluids, the stream flows away from the plug warmer than it entered it, let a uniform stream of water be kept flowing in a separate canal outside the tube round a portion of it in which the internal flow is from the plug, and by this means let the temperature of the internal fluid be brought to equality with that which it had on entering the plug. By a separate thermodynamic experiment find how much work would have to be spent in stirring the external stream of water by a paddle to warm it as much as it is warmed by conduction from the internal fluid across the separating tube. Returning now to the internal fluid flowing towards and from the plug, let $p + \delta p$ be the pressure in the steady stream approaching the disturbed region, and p the pressure in the steady stream flowing from the disturbed region; and let δw be the

quantity of work done by the paddle per unit of mass of the
fluid passing by, reckoned positive in the first case, that namely
in which the paddle compensates a cooling effect experienced in
passing through the porous plug. In the second case $- \delta w$ (in
this case a positive quantity) must denote the work done by
the paddle upon the supposed external stream of water in the
separate thermodynamic experiment. It is to be reckoned per
unit mass of the internal fluid, irrespectively of the rate of flow
of the external water. Let t denote the temperature of the fluid
according to the thermodynamic scale, and let δt denote the
infinitely small change of temperature which it must experience
to produce an infinitesimal expansion from volume v to volume
$v + \delta v$ under constant pressure. We have

$$\frac{v}{t}\frac{dt}{dv} = \cfrac{1}{1 + \cfrac{1}{v}\cfrac{\delta w}{\delta p}} \dots\dots\dots\dots\dots(16).$$

Proof.—Let $v + \delta v$, v, and $e + \delta e$, e, be respectively the amounts
of the volume and of the energy of the fluid per unit mass, in
the tranquil stream before and after passing the disturbed region.
The work done by an ideal piston pressing the fluid in towards
the disturbed region is $(p + \delta p)(v + \delta v)$, and the work done by the
emergent stream upon an ideal piston moving before it is pv, each
reckoned per unit of mass, of the fluid. The whole work done
on the fluid per unit mass by these ideal pistons is $p\delta v + v\delta p$;
add to this δw done by the paddle, and we find that, on the whole,
an amount of work equal to $p\delta v + v\delta p + \delta w$ is done on the fluid
in passing through the disturbed region. Hence e exceeds $e + \delta e$
by this amount; that is to say,

$$- \delta e = p\delta v + v\delta p + \delta w\dots\dots\dots\dots\dots(17).$$

Now the paddle and plug together act so as to render the temperatures equal in the tranquil streams at pressures p and $p + \delta p$.
But if there were change of temperature its analytical expression
would be

$$\delta t = \frac{dt}{dv}\delta v + \frac{dt}{dp}\delta p\dots\dots\dots\dots\dots(18).$$

Hence δv and δp are so proportioned as to make this vanish.
That is to say, we have

$$\delta v = -\frac{\dfrac{dt}{dp}}{\dfrac{dt}{dv}}\,\delta p \dots\dots\dots\dots\dots\dots(19);$$

and we have

$$\delta e = \frac{de}{dv}\,\delta v + \frac{de}{dp}\,\delta p;$$

hence (17) divided by δp becomes

$$\frac{de}{dv}\cdot\frac{\dfrac{dt}{dp}}{\dfrac{dt}{dv}} - \frac{de}{dp} = p\,\frac{-\dfrac{dt}{dp}}{\dfrac{dt}{dv}} + v + \frac{\delta w}{\delta p}\dots\dots\dots\dots(20).$$

Using this in (11) we find

$$\frac{t}{\dfrac{dt}{dv}} = v + \frac{\delta w}{\delta p}\dots\dots\dots\dots\dots\dots\dots(21).$$

Dividing (21) by v, and taking the reciprocal of both members, we have the equation (16) which was to be proved.

52. Now if for any particular fluid at some one given pressure p, with infinitesimal excess δp above this pressure for the higher pressure in the thermodynamic experiment, we find neither heating nor cooling effect in passing through the porous plug, the paddle has nothing to do; that is, $\delta w = 0$. If, with always the same pressure p, but with different values of v, that is to say, with the fluid given at different temperatures, but with pressures infinitely nearly the same, we always find the same result, $\delta w = 0$, it follows from (16) that for this particular fluid at the particular pressure of the experiment, and for all the temperatures of the experiment, we have

$$\frac{v}{t}\frac{dt}{dv} = 1\dots\dots\dots\dots\dots\dots\dots\dots(22).$$

Hence by integration

$$t = Cv\dots\dots\dots\dots\dots\dots\dots\dots\dots(23).$$

Hence we infer that with this fluid for thermometric substance, with the particular pressure of the experiment, and throughout the range of temperatures for which experiment has given us $\delta w = 0$, absolute temperature is shown on a scale graduated and numbered in simple proportion to the whole volume of the fluid.

53.　If the thermodynamic test repeated for the same fluid at different pressures gives still the same result, we have, for all pressures and temperatures within the range for which the supposed result $\delta w = 0$ has been found by the experiment,

$$t = f(p)\ v \dots\dots\dots\dots\dots\dots\dots\dots(24)\ ;$$

where $f(p)$ denotes the quantity which depends only on the pressure of the fluid and is independent of its density.

54.　Joule and Thomson's experiments on the thermal effects of fluids in motion* showed that for pressures of from one to five or six atmos hydrogen gas, common air, nitrogen, oxygen, and carbonic acid, all somewhat approximately fulfil the condition of passing through the porous plug without change of temperature,—hydrogen much more approximately, carbonic acid much less approximately, than any of the others. Hence we infer that absolute temperature is somewhat approximately proportional to the volume of the fluid, if any one of these gases be used as the thermometric fluid in a constant-pressure thermometer. We shall presently see that the requisite correction of this statement for the case of hydrogen is so small as to be almost within the limits of accuracy of the most accurate thermometric usage.

55.　In the case of common air, nitrogen, oxygen, and carbonic acid, the experiments showed a slight cooling effect upon the fluid in passing through the porous plug ; in the case of hydrogen, a much smaller heating effect. According to the rigorous dynamical form of our statement of § 51 above, we have no right to measure these heating and cooling effects on any scale of temperature, as we have not yet formed a thermometric scale. And it is interesting to remark that in point of fact the thermodynamic experiment described in that section involves the use of a differential thermoscope (§ 13 above) and not of any intrinsic thermoscope at all ; and in respect to this requisite it may be contrasted with the thermodynamic investigation of § 49 previously, which involved the use, not of any continuous thermoscope, but only of a single-temperature intrinsic thermoscope (§ 14 above). Now, instead of reckoning on any thermometric scale the cooling effect or the heating effect of passage through the plug, we have to measure the quantity of work (δw) required to annul it, in the case of the

* *Transactions Royal Society*, June 1853, June 1854, June 1860, and June 1862 [Art. XLIX. Vol. I. above].

majority of gases; and in the case of hydrogen, instead of reckoning on any thermometric scale the heating effect, we have to
measure $-\delta w$ as explained in § 51. The experiments as actually
made by Joule and Thomson simply gave the cooling effects and
heating effects shown by mercury thermometers in the tranquil
stream towards and from the plug; but the very thermometers
that were used had been used by Joule in his original experiments
determining the dynamical equivalent of heat, and again in his
later experiments by which for the first time the specific heat of
air at constant pressure was measured with sufficient accuracy for
our present purpose. Hence by putting together different experiments which had actually been made with those thermometers of
Joule's, the operation of measuring δw, at all events for the case of
air, was virtually completed. Thus according to our present view
the mercury thermometers are merely used as a step in aid of the
measurement of δw, and their scales may be utterly arbitrary,
provided we know the quantity of work required to raise unit mass
of any of the fluids concerned through the particular differences of
temperature actually shewn by the thermometers in the Joule and
Thomson experiment. The best way of doing this of course is to
take advantage of the best measurements, that is to say, Regnault's,
of the thermal capacity of air at constant pressure, and then to
calculate according to Joule's own measurement the dynamical
equivalent of the heat required to warm water through one degree
of his own thermometers.

56. Let K be the thermal capacity, pressure constant, of the
fluid experimented on, J the dynamical equivalent of the thermal
unit, and δt the cooling effect (reckoned negative when the effect
is *rise* of temperature), as measured by Joule's thermometers. We
have

$$\delta w = JK\,\delta t \dots\dots\dots\dots\dots\dots\dots(25).$$

Hence (16) becomes

$$\frac{v}{t}\frac{dt}{dv} = \frac{1}{1 + \dfrac{JK}{v}\dfrac{\delta t}{\delta p}} \ \dots\dots\dots\dots\dots\dots(26).$$

The experiment showed δt to be simply proportional to δp not
merely for an infinitesimal difference of pressures but for pressures
up to 5 or 6 atmos. For the case of hydrogen* the heating effect

* Joule and Thomson, *Transactions Royal Society*, June 1860 [Art. XLIX. Vol. I.
above].

observed amounted, per 100 inches of mercury, to ·100 of a degree
centigrade at temperatures of 4° or 5° centigrade, and to ·155 of a
degree centigrade at temperatures of from 89° to 93° centigrade.
The investigation was not carried out in sufficient detail to give
any law of variation of this effect with temperature, and it was not
even absolutely proved to be greater for the higher than for the
lower temperature. In the circumstances we may take the mean
of the results for the higher and lower temperatures, say ·13 per
100 inches of mercury, or ·039 per atmo. Hence if Π denotes the
force per unit of area in the pressure called "one atmo," we have

$$\frac{\delta t}{\delta p} = -\frac{\cdot039}{\Pi},$$

$$\frac{v}{t}\frac{dt}{dv} = \frac{1}{1 - \cdot039 JK/\Pi v} \qu\dotsquad\dots\dots\dots\dots(27).$$

Hence

$$\frac{dt}{t} = \frac{dv}{v - \cdot039 JK/\Pi},$$

which gives by integration

$$t = C(v - \cdot039 JK/\Pi)\dots\dots\dots\dots\dots(28).$$

The arbitrary constant C depends on the unit adopted for temperature. Let this be such that the difference of temperature between
freezing and boiling is 100 (which will make our arbitrary scale
agree with the ordinary centigrade scale in respect to the difference
between these two temperatures). Denote now by t_0 the absolute
temperature corresponding to 0° C. The absolute temperature
corresponding to 100° C. will be $t_0 + 100$. Denote also by v_0 and
v_{100}, for the same two temperatures, the bulks of unit mass of
hydrogen at any constant pressure within the limits of Joule and
Thomson's experiments, say, from one to five or six atmos. Then
by dividing the value of each member of (28) for 0° C. by the
difference of its values for 0° and 100°, we find

$$\frac{t_0}{100} = \frac{v_0 - \cdot039 JK/\Pi}{v_{100} - v_0} \quad\dots\dots\dots\dots\dots(29).$$

Hence

$$t_0 = \frac{100}{E}(1 - \cdot039 JK/\Pi v_0)\dots\dots\dots(30);$$

where E denotes the expansion of hydrogen, pressure constant,
from 0° to 100° C. in terms of its volume at 0°, that is to say,

$$E = \frac{v_{100} - v_0}{v_0} \quad\dots\dots\dots\dots\dots\dots(31).$$

Let V_0 denote what the volume would be at 0° C. if the pressure were Π instead of the actual pressure p. We have

$$t_0 = \frac{100}{E}\left(1 - \frac{V_0}{v_0}\cdot 039JK/\Pi V_0\right) \dots\dots\dots(32).$$

Regnault finds (*Expériences*, vol. ii. p. 122) that the value of $K/\Pi V_0$ for hydrogen agrees within 1/2 per cent. with its value for common air; and for common air he finds $K = \cdot 238$. Thus with 423·5 for the value of J in metres (§ 9 above) we find $JK = 100\cdot79$ metres. And Regnault's observations on the density of air give for ΠV_0 (or the height of the homogeneous atmosphere at 0° C.) 7990 metres. Hence for common air, and therefore also for hydrogen, $JK/\Pi V_0 = \cdot0126$; and thus (32) becomes

$$t_0 = \frac{100}{E}\left(1 + c\,\frac{V_0}{v_0}\right) \dots\dots\dots\dots(33),$$

with $c = -\cdot00049$ for hydrogen. For this gas expanding under constant pressure of one atmo Regnault found (*Expériences*, vol. i. p. 80) $E = \cdot36613$, which gives $\frac{100}{E} = 273\cdot13$. Hence (33), with $v_0 = V_0$, gives

$$t_0 = 273\cdot00 \dots\dots\dots\dots\dots(34);$$

that is to say:—

57. We conclude from Regnault's observations on the expansion of hydrogen from 0° to 100° C. under a constant pressure of one atmo, and from the small heating effect discovered in Joule and Thomson's experiments on the forcing of hydrogen through a porous plug, that the absolute temperature of melting ice is 273·00°, if the unit or degree of absolute temperature is so chosen as to make the difference one hundred between the temperatures of melting ice and of water with steam at one atmo of pressure.

58. An almost identical number for that most important physical constant, the absolute temperature of melting ice, is obtained from observations on common air, and a not very different number from observations on carbonic acid, the only two gases besides hydrogen for which Regnault (*Expériences*, vol. i. p. 90) measured the expansion under constant pressure, and for which Joule and Thomson made their experiment on the thermal effect of passage through a porous plug. For each of these two gases the thermal effect observed was a lowering of temperature, and was found

to vary at different temperatures very nearly in the inverse proportion of the square of temperature C., by mercury thermometer, with 273 added. Hence nearly enough for use in the small term of the denominator of (26) we have, for air and carbonic acid,

$$\frac{\delta t}{\delta p} = A \left(\frac{273}{t}\right)^2 \frac{p}{\Pi} \quad \ldots\ldots\ldots\ldots\ldots\ldots\ldots(35),$$

where t denotes as before absolute temperature, and A the amount of the cooling effect per atmo of difference of pressures on the two sides of the plug, at the temperature of melting ice. The values of A found for common air and carbonic acid are ·275 and 1·388. Regnault (*Expériences*, vol. ii p. 126) finds $JK/\Pi V_0$ greater for carbonic acid than for common air in the ratio of 1·39 to 1 on the average of temperatures from 0° to 210°. But he found also that the specific heat of carbonic acid varies greatly with the temperature; and, taking the mean of the values which he finds for it at 0° and 100° (*Expér.* vol. ii. p. 130), as the proper mean for our present purpose we find for $JK/\Pi V_0$, a value 1·29 times its value

Name of Gas.	Expansion at one atmo according to Regnault. E.	Proper mean[1] cooling-effect of forcing through porous plug per atmo according to Joule and Thomson. M.	Uncorrected estimate of absolute temperature of melting ice. $\dfrac{100}{E}$.	Correction calculated from cooling-effect. $\dfrac{100}{E} \times \dfrac{JK}{\Pi V_0} M$.	Resulting estimate of absolute temperature of melting ice. t_0.
Hydrogen	·36613	− 0°·039	273·13	− 0°·13	273·00
Air..............	·36706	+ 0°·208	272·44	+ 0°·70	273·14
Carbonic acid .	·37100	+ 1°·005	269·5	+ 4°·4	273·9

for common air. From these experimental results we find by the mathematical process below (§ 61) still the same approximate formula (33), but with $c = + ·0026$ for common air and $c = + ·0163$ for carbonic acid. At constant pressure of one atmo Regnault's measurements gave $E = ·36706$ for common air, and $E = ·3710$ for carbonic acid; and dividing 100 by these decimals we find respectively 272·44 and 269·5. The corrections on these numbers by formula (33) to give the absolute temperature of freezing are accordingly + ·70 and + 4·4, and the corresponding estimates for the required absolute temperature are 273·14 and 273·9. Bringing

[1] Investigated in § 61 below.

together the results in the three cases, we see them conveniently in the preceding table.

The close agreement of the results from hydrogen and common air is very satisfactory, and it is interesting to see it brought about with so large a correction calculated from the Joule and Thomson effect. It is also interesting to see the sevenfold larger correction of nearly 5° bringing so nearly the same result from the 1 per cent. larger expansion of carbonic acid. The ⅓ per cent. discrepance which remains between the results from carbonic acid and from hydrogen is not satisfactory, and requires explanation, particularly when we remark that, of five measurements by Regnault (*Expériences*, vol. I. p. 84) of the expansion of carbonic acid under constant pressure of one atmo, all lie within $\frac{1}{5\cdot7}$ per cent. of the mean number 3710 which he has given, and we have taken, as his result.

Notwithstanding that the Joule and Thomson correction is so much greater for common air than for hydrogen, the result from common air is probably the most trustworthy of the three, because both Regnault's experiments and Joule and Thomson's were probably more accurate for air than for either of the other two gases. The true result to one place of decimals may therefore be considered as most probably being 273·1°, but the probability that it is nearer 273·1° than 273·0° is scarcely enough to make it worth while to use in any ordinary thermodynamic calculations any other number than 273°, which is exactly that found from hydrogen.

59. The real meaning of our result 273·1° for the absolute temperature of melting ice, expressed without any choice of degrees or units for temperature, is that the ratio of the temperature at which vapour of water has a pressure of one atmo to the temperature at which ice melts is 373·1/273·1. Still another way of saying the same thing, this time eliminating all numerical reckoning of temperature, is as follows :—

Determination of duty of perfect engine with source and refrigerator at those temperatures.

FOR EVERY HUNDRED UNITS OF HEAT CONVERTED INTO WORK BY A PERFECT THERMODYNAMIC ENGINE, 373·1 ARE TAKEN FROM THE SOURCE, AND 273·1 REJECTED TO THE REFRIGERATOR, IF THE TEMPERATURE OF THE SOURCE BE THAT AT WHICH STEAM OF WATER HAS A PRESSURE OF ONE ATMO, AND THE TEMPERATURE OF THE REFRIGERATOR THAT AT WHICH ICE MELTS.

60. *Integration of differential equation* (26), § 56 above, *between volume and absolute temperature for a gas, derived from the Joule and Thomson experiment.*

Returning to § 56 we may write equation (26) as follows :—

$$t\frac{dv}{dt} - v = JK\frac{\delta t}{\delta p} \dots\dots\dots\dots\dots\dots(36).$$

For each of the five gases experimented on, namely, common air, oxygen, nitrogen, carbonic acid, hydrogen, the experiment showed that, for all pressures up to five or six atmos, $\delta t/\delta p$ was sensibly independent of the pressure, but that it varied very considerably with the temperature. Hence, if we put $\delta t/\delta p = \theta/\Pi$, θ, which will thus denote the cooling effect per atmo of differential pressure, is a function of the temperature, and is independent of the whole pressure. With this notation (36) becomes

$$t\frac{dv}{dt} - v = \frac{JK}{\Pi}\,\theta.$$

This is a linear equation in z with a second member, if for a moment we put $z = \log t$. Integrating it, and replacing t, we find, as the complete integral,

$$v = t\left\{\frac{v_0}{t_0} + \frac{JK}{\Pi}\int_{t_0}^{t}\frac{\theta dt}{t^2}\right\}\dots\dots\dots\dots\dots(37).$$

61. *Expansions of different gases, pressure constant, calculated from the Joule and Thomson experiment.* We have from equation (37)

$$\frac{v - v_0}{v_0} = \frac{t - t_0}{t_0}\left\{1 + \frac{JKt_0}{\Pi v_0}\frac{1}{1 - \dfrac{t_0}{t}}\int_{t_0}^{t}\frac{\theta dt}{t^2}\right\}\dots\dots\dots(38).$$

For each of the gases experimented on, except hydrogen, θ was found to vary nearly in the inverse ratio of t^2. Putting then $\theta = A\,(t_0/t)^2$, we find

$$t_0\int_{t_0}^{t}\theta dt/t^2 = \tfrac{1}{3}A\left[1 - \left(\frac{t_0}{t}\right)^3\right].$$

Hence, for these gases at pressures from 0 to 5 or 6 atmos, (38) becomes

$$\frac{v - v_0}{v_0} = \frac{t - t_0}{t_0}\left\{1 + \tfrac{1}{3}A\frac{JK}{\Pi v_0}\left[1 + \frac{t_0}{t} + \left(\frac{t_0}{t}\right)^2\right]\right\}\dots\dots(39).$$

This shows that the "proper mean cooling effect" (M in the table of § 58) is

$$\tfrac{1}{3}\left[1+\frac{t_0}{100+t_0}+\left(\frac{t_0}{100+t_0}\right)^2\right]A,$$

$$=\tfrac{1}{3}\left(1+\frac{1}{1\cdot3663}+\frac{1}{1\cdot3663^2}\right)A=\cdot756A,$$

which differs so little from

$$\tfrac{1}{2}\left(1+\frac{1}{1\cdot3663^2}\right)A=\cdot769A,$$

the arithmetic mean of the cooling effects at 0° and 100° C., that if we had simply taken the arithmetic mean for each of the other gases, as for want of knowing better we took it for hydrogen, the difference in the result would have been barely perceptible.

62. Modifying (39) or (38) to suit any two temperatures, t, t', we have

$$\frac{v-v'}{v'}=\frac{t-t'}{t'}\left(1+\frac{JK}{\Pi v'}\,M\right)\dots\dots\dots\dots(40),$$

M denoting the proper mean cooling effect per atmo in the Joule and Thomson experiment (to be reckoned as negative in the case of hydrogen or any other gas, if there is any other, in which the experiment shows a heating effect). This "proper mean" may be taken as the arithmetic mean of the values for t and t', unless $t-t'$ considerably exceeds 100.

To reduce (40) to numbers, let V_0 be the volume of unit mass of the gas when at the temperature of melting ice, and under one atmo of pressure. Regnault (vol. II. p. 303) finds that the value of K/V_0 is within 1 per cent. the same for oxygen, nitrogen, and hydrogen as for common air. He also (vol. II. pp. 224—226) finds K to be the same for common air at from 1 to 12 atmos; for hydrogen, 1 to 9 atmos; for carbonic acid, 1 to 37 atmos.

No doubt similar constancy would be found for oxygen and nitrogen. Hence, as above (§ 56), for common air we still have $JK/\Pi V_0=\cdot0126$, and thus (40) becomes

$$\frac{v-v'}{v'}=\frac{t-t'}{t'}\left(1+\frac{V_0}{v'}\cdot0126M\right)\begin{cases}\text{for common air, oxygen,}\\\text{hydrogen, and nitrogen}\end{cases}\dots(41).$$

If in this formula we take t and t' for the temperatures of 100°C. and 0° C., $(t-t')/t'$ becomes $\dfrac{100}{273\cdot1}$ or $\cdot3662$; and we therefore find

$$E=\cdot3662\left(1+\frac{V_0}{v_0}\cdot0126M\right)\dots\dots\dots\dots(42),$$

which agrees with (33) above.

63. The values given by Joule and Thomson's experiment for M are $-0.039°$ for hydrogen, $+0.208°$ for air, $+0.253°$ for oxygen, and $+0.249°$ for nitrogen.

From these and from the previous results for carbonic acid (§ 58) we have the following table for calculating the expansions from $0°$ C. to $100°$ C. of the gases named :—

Name of Gas.	Expansion under constant pressure $(=E)$.
Hydrogen	$.3662\ (1 - 0.00049\ V_0/v_0)$
Common air	$.3662\ (1 + 0.0026\ V_0/v_0)$
Oxygen	$.3662\ (1 + 0.0032\ V_0/v_0)$
Nitrogen	$.3662\ (1 + 0.0031\ V_0/v_0)$
Carbonic acid	$.3662\ (1 + 0.0163\ V_0/v_0)$

$$......(43).$$

These formulæ must be exceedingly near the truth for all pressures from 0 to 6 atmos, because within this range the thermal effects in the Joule and Thomson experiment were very approximately in simple proportion to the differences of pressure on the two sides of the plug. The following table of results calculated

Name of Gas.	Ratio of Bulk at 0° C. to Bulk supposing Pressure were 1 atmo at the same temperature. $\dfrac{v_0}{V_0}$.	Ratio of Density at 0° C. to Density supposing Pressure were 1 atmo at the same temperature. $\dfrac{V_0}{v_0}$.	Expansion, pressure constant, from 0° to 100° C.	
			According to theory.	According to direct experiments by Regnault.
Hydrogen	∞	0	$.3662$	...
	1	1	$.3660$	$.36613$
	1/3	3	$.3657$	...
	1/3·35	3·35	$.3656$	$.36616$
	1/6	6	$.3651$	...
Common Air	∞	0	$.3662$	...
	1	1	$.3672$	$.36706$
	1/3	3	$.3691$	...
	1/3·38	3·38	$.3694$	$.36954$
	1/6	6	$.3719$	...
Oxygen	∞	0	$.3662$	...
	1	1	$.3674$	...
	1/3	3	$.3697$	...
	1/6	6	$.3732$	...
Nitrogen	∞	0	$.3662$	...
	1	1	$.3673$	...
	1/3	3	$.3696$	...
	1/6	6	$.3730$	...
Carbonic Acid	∞	0	$.3662$	...
	1	1	$.3721$	$.37099$
	1/3	3	$.3841$	...
	1/3·316	3·316	$.3859$	$.38455$
	1/6	6	$.4019$	...

from (43) for several pressures of from 0 to 6 atmos is interesting, as showing such different expansions for the different cases, determined by thermodynamic theory from Regnault's measurements of specific heats and Joule and Thomson's of their particular thermal effect, with absolutely no direct measurement of expansion except the one for common air at one atmo, shown as the third entry of column 5 in the table.   The other five entries of column 5 show a fair amount of agreement between our theoretical results and the only direct measurements by Regnault.   More of direct measurement, to allow a more extensive comparison, is very desirable.

Practical construction of constant-pressure gas thermometer. 64.   We are now quite prepared to make a practical working thermometer directly adapted to show temperature on the absolute thermodynamic scale through the whole range of temperature, from the lowest attainable by any means to the highest for which glass remains solid.   It is to be remarked that our investigation of § 51 above, and all the deduced formulæ and relative calculations, are absolutely independent of the approximate fulfilment of Boyle's law by the gases to which we have applied them, and are equally applicable without any approach to fulfilment of Boyle's law; also that the only experimental data on which are founded our special numerical conclusions of §§ 59 to 63 above are Regnault's measurements of specific heats under constant pressures, and Joule and Thomson's measurements of the thermal effect of forcing gas through a porous plug.   From these experimental data alone we see by formula (38) of § 61 above how to graduate a constant-pressure gas thermometer so that it shall show temperature on the absolute thermodynamic scale.   Hence, notwithstanding the difficulty (§ 24 above) which Regnault found in the thermometric use of air or other gases on the system of constant pressure, and his practical preference for the constant-volume air thermometer, it becomes of the highest importance to construct a practical constant-pressure gas thermometer.   This we believe may be done by avoiding the objectionable expedient adopted by Pouillet and Regnault of allowing a portion (when high temperatures are to be measured the greater portion) of the whole gas to be pressed into a cool volumetric chamber out of the thermometric chamber proper by the expansion of the portion which remains in; and instead fulfilling the condition, stated, but pronounced practically impossible, by Regnault (*Expériences*, vol. I. pp. 168, 169), that the thermometric gas "shall, like the mercury of a mercury thermometer,

be allowed to expand freely at constant pressure in a cali-
brated reservoir maintained throughout
at one temperature." We have accordingly designed a constant-pressure
gas thermometer to fulfil this condition.
It is represented in the accompanying
drawing (fig. 10), and described in the
following section.

65. The vessel containing the
thermometric fluid, which in this case
is to be either hydrogen or nitrogen*,
consists in the main of a glass bulb and
tube placed vertically with bulb up and
mouth down; but there is to be a
secondary tube of much finer bore
opening into the bulb or into the main
tube near its top, as may be found
most convenient in any particular case.
The main tube which, to distinguish it
from the secondary tube, will be called
the volumetric tube, is to be of large
bore, not less than 2 or 3 centimetres,
and is to be ground internally to a truly
cylindric form. To allow this to be
done it must be made of thick, well-
annealed glass, like that of the French
glass-barrelled air - pumps. The secondary tube, which will be called the

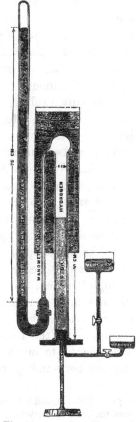

Fig. 10.   Constant-Pressure
Hydrogen Thermometer.

manometric capillary, is to be of round bore, not very fine, say
from half a millimetre to a millimetre diameter.   Its lower end is
to be connected with a mercury manometer to show if the pressure
of the thermometric air is either greater or less than the definite

* Common air is inadmissible because even at ordinary temperatures its oxygen
attacks mercury.   The film of oxide thus formed would be very inconvenient at the
surface of the mercury caulking, round the base of the piston, and on the inner
surface of the glass tube, to which it would adhere.   Besides sooner or later the
whole quantity of oxygen in the air must be diminished to a sensible degree by the
loss of the part of it which combines with the mercury.   So far as we know,
Regnault did not complain of this evil in his use of common air in his normal air
thermometer (see §§ 24, 25 above), nor in his experiments on the expansion of air
(*Expériences*, vol. i.), though probably it has vitiated his results to some sensible

pressure to which it is to be brought every time a thermometric measurement is made by the instrument. The change of volume required to do this for every change of temperature is made and measured by means of a micrometer screw* lifting or lowering a long solid glass piston, fitting easily in the glass tube, and caulked air-tight by mercury between its lower end and an iron sole-plate by which the mouth of the volumetric tube is closed. To perform this mercury caulking, when the piston is raised and lowered, mercury is allowed to flow in and out through a hole in the iron sole-plate by an iron pipe, connected with two mercury cisterns at two different levels by branches each provided with a stopcock. When the piston is being raised the stopcock of the branch leading to the lower cistern is closed, and the other is opened enough to allow the mercury to flow up after the piston and press gently on its lower side, without entering more than infinitesimally into the space between it and the surrounding glass tube (the condition of the upper bounding surface of the mercury in this respect being easily seen by the observer looking at it through the glass tube). When the piston is being lowered the stopcock in the branch leading from the upper cistern is closed, and the one in the branch leading to the lower cistern is opened enough to let the mercury go down before the piston, instead of being forced to any sensible distance into the space between it and the surrounding tube, but not enough to allow it to part

---

degree. But he found it to produce such great irregularities when, instead of common air, he experimented on pure oxygen that from the results he could draw no conclusion as to the expansion of this gas (*Expériences*, vol. I. p. 77). Another reason for the avoidance of air or other gas containing free oxygen is to save the oil or other liquid which is interposed between it and the mercury of the manometer from being thickened or otherwise altered by oxidation.

* This screw is to be so well fitted in the iron sole-plate as to be sufficiently mercury-tight without the aid of any soft material, under such moderate pressure as the greatest it will experience when the pressure chosen for the thermometric gas is not more than a few centimetres above the external atmospheric pressure. When the same plan of apparatus is used for investigation of the expansion of gases under high pressures, a greased leather washer may be used on the upper side of the sole in the screw-hole plate, to prevent mercury from escaping round the screw. It is to be remarked that in no case will a little oozing out of the mercury round the screw while it is being turned introduce any error at all into the thermometric result; because the correctness of the measurement of the volume of the gas depends simply on the mercury being brought up into contact with the bottom of the piston, and not more than just perceptibly up between the piston and volumetric tube surrounding it.

company with the lower surface of the piston. The manometer is simply a mercury barometer of the form commonly called a siphon barometer, with its lower end not open to the air but connected to the lower end of the manometric capillary. This connexion is made below the level of the mercury in the following manner. The lower end of the capillary widens into a small glass bell or stout tube of glass of about 2 centimetres bore and 2 centimetres depth, with its lip ground flat like the receiver of an air-pump. The lip or upper edge of the open cistern of the barometer (that is to say, the cistern which would be open to the atmosphere were it used as an ordinary barometer) is also ground flat, and the two lips are pressed together with a greased leather washer between them to obviate risk of breaking the glass, and to facilitate the making of the joint mercury tight. To keep this joint perennially good, and to make quite sure that no air shall ever leak in, in case of the interior pressure being at any time less than the external barometric pressure or being arranged to be so always, it is preserved and caulked by an external mercury-jacket not shown in the drawing. The mercury in the thus con-stituted lower reservoir of the manometer is above the level of the leather joint; and the space in the upper part of the reservoir over the surface of the mercury, up to a little distance into the capillary above; is occupied by a fixed oil or some other practically vapourless liquid. This oil or other liquid is introduced for the purpose of guarding against error in the reckoning of the whole bulk of the thermometric gas, on account of slight irregular changes in the capillary depression of the border of the mercury surface in the reservoir.

66. In the most accurate use of the instrument, the glass and mercury and oil of the manometer are all kept at one definite temperature according to some convenient and perfectly trust-worthy intrinsic thermoscope (§§ 15 and 16 above), by means of thermal appliances not represented in the drawing but easily imagined. This condition being fulfilled, the one desired pressure of the thermometric gas is attained with exceedingly minute accuracy by working the micrometer screw up or down until the oil is brought precisely to a mark upon the manometric capillary.

In fact, if the glass and mercury and oil are all kept rigorously at one constant temperature, the only access for error is through

irregular variations in the capillary depressions in the borders of
the mercury surfaces.   With so large a diameter as the 2 centimetres chosen in the figured dimensions of the drawing, the error
from this cause can hardly amount to $\frac{1}{100}$ per cent. of the whole
pressure, supposing this to be one atmo or thereabouts.

For ordinary uses of our constant-pressure gas thermometer,
where the most minute accuracy is not needed, the rule will still
be to bring the oil to a fixed mark on the manometric capillary;
and no precaution in respect to temperature will be necessary
except to secure that it is approximately uniform throughout the
mercury and containing glass, from lower to higher level of the
mercury.   The quantity of oil is so small that, whatever its
temperature may be, the bringing of its free surface to a fixed
mark on the capillary secures that the mercury surface below the
oil in the lower reservoir is very nearly at one constant point
relatively to the glass, much more nearly so than it could be made
by direct observation of the mercury surface, at all events without
optical magnifying power.   Now if the mercury surface be at a
constant point of the glass, it is easily proved that the difference
of pressures between the two mercury surfaces will be constant,
notwithstanding considerable variations of the common temperature of the mercury and glass, provided a certain easy condition
is fulfilled, through which the effect of the expansion of the glass
is compensated by the expansion of the mercury.   This condition
is that the whole volume of the mercury shall bear to the volume
in the cylindric vertical tube from the upper surface to the level
of the lower surface the ratio $(\lambda - \frac{1}{3}\sigma)/(\lambda - \sigma)$, where λ denotes the
cubic expansion of the mercury, and σ the cubic expansion of the
solid for the same elevation of temperature, it being supposed for
simplicity of statement that the tube is truly cylindric from the
upper surface to the level of the lower surface, and that the
sectional area of the tube is the same at the two mercury surfaces.
The cubic expansion of mercury is approximately seven times the
cubic expansion of glass.   Hence

$$(\lambda - \tfrac{1}{3}\sigma)/(\lambda - \sigma) = (7 - \tfrac{1}{3})/6 = 1\cdot111.$$

Hence the whole volume of the mercury is to be about $1\cdot111$
times the volume from its upper surface to the level of the lower
surface; that is to say, the volume from the lower surface in
the bend to the same level in the vertical branch is to be $\frac{1}{9}$ of the

volume in the vertical tube above this surface. A special experiment on each tube is easily made to find the quantity of mercury that must be put in to cause the pressure to be absolutely constant when the surface in the lower reservoir is kept at a fixed point relatively to the glass, and when the temperature is varied through such moderate differences of temperature as are to be found in the use of the instrument at different times and seasons.

A sheet-iron can containing water or oil or fusible metal, with external thermal appliances of gas or charcoal furnace, or low-pressure or high-pressure steam heater, and with proper internal stirrer or stirrers, is fitted round the bulb and manometric tube to produce uniformly throughout the mass of the thermometric gas the temperature to be measured. This part of the apparatus, which will be called for brevity the heater, must not extend so far down the manometric tube that when raised to its highest temperature it can warm the caulking mercury to as high a temperature as 40° C., because at somewhat higher temperatures than this the pressure of vapour of mercury begins to be perceptible (see Table V. below), and would vitiate the thermometric use of the pure hydrogen or nitrogen of our thermometer. To secure sufficient coolness of the mercury it will probably be advisable to have an open glass jacket of cold water (not shown in the drawing) round the volumetric tube, 2 or 3 centimetres below the bottom of the heater, and reaching to about half a centimetre above the highest position of the bottom of the piston.

67. It seems probable that the constant-pressure hydrogen Conclu-sion of ther-mometry. or nitrogen gas thermometer which we have now described may give even more accurate thermometry than Regnault's constant-volume air thermometers (§ 24 above), and it seems certain that it will be much more easily used in practice.

We have only to remark here further that, if Boyle's law were rigorously fulfilled, thermometry by the two methods would be identical, provided the scale in each case be graduated or calculated so as to make the numerical reckoning of the temperature agree at two points,—for example, 0° C. and 100° C. The very close agreement which Regnault found among his different gas thermometers and his air thermometers with air of different densities

(§ 25 above), and the close approach to rigorous fulfilment of
Boyle's law which he and other experimenters have ascertained
to be presented by air and other gases used in his thermometers,
through the ranges of density, pressure, and temperature at which
they were used in these thermometers, renders it certain that in
reality the difference between Regnault's normal air thermometry
and thermometry by our hydrogen gas constant-pressure thermometer must be exceedingly small.   It is therefore satisfactory to
know that for all practical purposes absolute temperature is to
be obtained with very great accuracy from Regnault's thermometric system by simply adding 273 to his numbers for temperature on the centigrade scale.   It is probable that at the
temperatures of 250° or 300° C. (or 523 or 573 absolute) the greatest
deviation of temperature thus reckoned from correct absolute
temperature is not more than half a degree.

68.   The thermometric scale being now thoroughly established
in theory and practice (§§ 33—69 and §§ 18—30), we are prepared
to define, without any ambiguity, the expressions thermal capacity
and specific heat with reference to matter at any temperature and
in any physical condition.

"Thermal
capaci-
ties" and
"specific
heats."
Definition 1.—The thermal capacity of a body (whether it be a
portion of matter homogeneous throughout or of homogeneous
substance in two different conditions as liquid and steam, or solid
and vapour of solid, or a piece of apparatus consisting of different
parts as glass and metals, and containing as the case may be
liquids or gases,—subject only to the condition that the whole
matter considered is at one temperature) is the quantity of heat
required to raise its temperature by one degree on the absolute
thermodynamic scale: the external circumstances of the body,
whether geometrical, as regards its bounding surface, or dynamical,
as regards force on its bounding surface, being determinately defined.   When the substance is fluid the circumstantial definition
may be: (1) that the volume be constant, or (2) that the pressure
be constant, or (3) that the pressure vary as a given arbitrary
function of the volume.   Each of the three kinds of the circumstantial definition occurs in practice, and is practically essential for
the completion of our statement.   When the substance is solid
more varied and complex circumstantial definition may occur.
An important case is, that the solid be free to expand in all

directions under any uniform constant pressure, and this condition
is nearly enough approximated to in all ordinary modes of dealing
with solids for determination of specific heats. Other conditions as to external stress, or as to shape or bulk of the solid,
which may occur in practice, are carefully considered in books on
Thermodynamics.

Definition 2.—The specific heat of a substance is the thermal
capacity of a stated quantity of it. This stated quantity is
generally understood to be the unit of mass, unless some other
definite quantity is explicitly designated, as for instance the
quantity of the substance which occupies unit of volume at some
definite pressure and temperature, for instance, one atmo and
temperature 273° absolute. It is of no consequence what unit of
mass is chosen provided it be the same as that which is used in
defining the thermal unit; but, unless the contrary be explicitly
stated, we always understand one gramme as the unit of mass
and the thermal unit as the quantity of heat required to raise
one gramme of water from 273° to 274° absolute (compare § 6
above).

69. There is scarcely any subject upon which more skilled
labour in scientific laboratories, chemical and physical, has been
spent than the measurement of specific heats, whether of solids,
liquids, or gases. An ample and well-arranged table of results
is to be found in Clarke's *Constants of Nature*, a compilation of
numerical results of scientific experiments made in all parts of
the world by various observers and experimenters, a most valuable
aid to scientific knowledge given to the world as No. 255 of the
Smithsonian Miscellaneous Collections. It is most interesting as
showing how very differently different substances behave in respect
to constancy or variation of specific heat with temperature. Thus
it shows that, according to the results of all the experimenters, the
specific heats of all the substances experimented on, whether
simple or compound, are very nearly constant at all events for
ranges between −10° and 200° or 300° C., except the three
elementary substances, boron, carbon, silicon. The specific heats
of these three have been found by F. Weber to vary greatly with
temperature. Thus for diamond he finds the specific heat to be
1 at 0° C. and ·27 at 206° C., or nearly threefold of the amount
at 0°; at −50° C. the specific heat is ·063; and at +985° it is

459 or about seven and a half times the specific heat at − 50°
(a curious practical commentary, we may remark in passing, on
the doctrine of the calorists on specific heat referred to in § 8
above).   The specific heats of carbon in its other forms of graphite
and charcoal through wide ranges of temperature according to the
same observer, F. Weber, are particularly interesting and significant.   The approximate equality of the product of specific heat
into the atomic weight for the simple metals is interesting and
important ; no less so is the utter want of constancy and uniformity
in the corresponding product for other substances, whether simple
or compound.   If we were to define a metal as a substance for
which through the range of temperature from 0° to 250° C. the
product of the specific heat into the atomic weight is not less than
5·86 and not greater than 6·93, we should include every substance
commonly called a metal, and no substance not commonly called a
metal, except phosphorus, and solid sulphur lately fused.

TRANSFERENCE OF HEAT.

Trans-
ference
of heat
by con-
duction;

70.   When two contiguous portions of matter are at different
temperatures, heat is transferred from the warmer to the colder.
This process is called conduction of heat.

by ra-
diation.

Light and
radiant
heat
identical.

When two bodies at different temperatures are separated by a
transparent medium, such as air, or water, or glass, or ice, heat
passes from the warmer to the colder irrespectively of the temperature of the intervening medium, except in so far as its trans-
parency may in some slight degree be affected by the temperature.
Thus the colder of the two bodies becomes actually heated above
the temperature of the intervening medium if the warmer be kept
above this temperature, and if heat is not otherwise drawn off
from the colder body in greater quantity than the heat entering
it from the warmer.   This process of transference from one body
to another body at a distance through an intervening medium
is called radiation of heat.   The condition, which we know
to be a state of wave-motion, of the intervening matter in
virtue of which heat is thus transferred is called light ; and
radiant heat is light if we could but see it with the eye, and
not merely discern with the mind, as we do, that it is perfectly
continuous in quality with the species of radiant heat which we
see with the material eye through its affecting the retina with

the sense of light. Thus a white hot poker in a room perfectly
darkened from all other lights is seen as a brilliant white light
gradually becoming reddish and less bright, until it absolutely
fades from vision in a dull red glow. Long after it has ceased to
be visible to the eye, the fact that heat is being transferred from it
to colder bodies all round it, or above it or below it, is proved by
our sense of heat in a hand or face held near it on any side or
above it or below it. By considering the whole phenomenon of
the white hot mass, without much of experimental investigation,
we judge that there is perfect continuity through the whole
process, in the first part of which the radiant heat is visible, and
in the second part invisible, to the human eye : and thorough
experimental investigation confirms this conclusion. Thus radiant
heat is brought under the undulatory theory of light, which in its
turn becomes annexed to heat as a magnificent outlying province
of the kinetic theory of heat.

71. In this article we confine ourselves to a practical evalua- Thermal
tion of rate of gain or loss of heat across the surface of an isolated emissivity
solid placed in a medium such as air, and enclosed in a solid defined.
surface all at one temperature, as is approximately the case with
the air and the floor, walls, and ceiling of an ordinary room. A
rough approximation to the law of this action, founded on supposing the rate of motion to be in simple proportion to the excess
of the temperature of the isolated solid above the temperature
of the surrounding medium and enclosure was used by Fourier in
those of his solutions in which surface emissivity or, as he called it,
" Conductibilité extérieure," is concerned. Without adopting any
hypothesis, we define thermal emissivity as the quantity of heat
per unit of time, per unit of surface, per degree of excess of temperature, which the isolated body loses in virtue of the combined
effect of radiation and convection by currents of air. This definition does not involve the hypothesis of simple proportionality ;
and the surface emissivity is simply to be determined by experiment for any given temperature of the enclosure, and any
given temperature of the isolated body. Dulong and Petit made
elaborate experiments on this subject, but did not give any results
in absolute measure (App. C, below).

So far as we know the first thoroughly trustworthy experiments
giving emissivities in absolute measure were made in the Laboratory
of the Natural Philosophy class in the University of Glasgow by

Mr D. Macfarlane, in a series of experiments on the cooling of a copper ball. The results are given in Table IX. The ball experimented on was 4 centimetres in diameter, and was suspended in the interior of a double-walled tin-plate vessel. The space between the double walls of this vessel was filled with water at the temperature of the air, and the interior surface was coated with lampblack. Two thermo-electric junctions, one at the centre of the ball the other in contact with the exterior surface of the enclosure, in circuit with a sensitive mirror galvanometer, served to measure the difference of temperatures between the centre of the ball and the exterior surface of the enclosure. By this arrangement the exterior junction was kept very uniformly at a temperature of 14° C., while the other had the varying temperature of the centre of the ball. Two sets of experiments were made. In one the ball had a bright surface, in the other it was coated with soot from the flame of a lamp, and in both the air was kept moist by a saucer of water placed in the interior of the tinplate enclosure. The results are given in terms of the number of units of heat lost per second, per square centimetre of surface of the copper, per degree of difference between the temperatures of the two junctions (App. B, below).

Conduction of heat.

72. Returning to the conduction of heat, we have first to say that the theory of it was discovered by Fourier and given to the world through the French Academy in his *Théorie Analytique de la Chaleur*[1], with solutions of problems naturally arising from it, of which it is difficult to say whether their uniquely original quality, or their transcendently intense mathematical interest, or their perennially important instructiveness for physical science, is most to be praised. Here we can but give the very slightest sketch of the elementary law of conduction in an isotropic substance, the mathematical expression for it in terms of orthogonal plane or curved coordinates, and a few of the elementary solutions in Fourier's theory.

73. Consider a slab of homogeneous solid bounded by two parallel planes. Let the substance be kept at two different temperatures over these parallel planes by suitable sources of heat and cold. For example, let one side be kept cold by a stream of

[1] A translation into English by Freeman has been recently published, in 1 vol. 8vo. by the Cambridge University Press, 1879.

cold water, or by a large quantity of ice and water in contact with it, and the other kept warm by a large quantity of warm water or by steam blown against it. Whatever particular plans of heater and refrigerator be adopted, care must be taken that the temperature be kept uniform over the whole, or over a sufficiently large area of each side of the slab, to render the isothermal surfaces sensibly parallel planes through the whole of the slab intercepted between the two calorimetric areas, and that the temperature at each side is prevented from varying with time. It will be found that heat must continually be applied at one side and removed from the other, to keep the circumstances in the constant condition thus defined. When this constant condition of surface temperature is maintained long enough, the temperature at every point of the slab settles towards a constant limiting value; and when this limiting value has been sensibly reached by every point of the slab, the temperature throughout remains sensibly constant so long as the surface temperatures are kept constant. In this condition of affairs the temperature varies continuously from one side of the slab to the other; and it is constant throughout each interior plane parallel to the sides; in other words, the isothermal surfaces are parallel planes. Let V and V' be the temperatures in two of these isothermals and a the distance between them. The quotient $(V - V')/a$ is the average rate of variation of temperature per unit of length between these two isothermals. Let Q be the quantity of heat taken in per unit of time at a certain area A on one side, and emitted at the corresponding area of the other side of the slab, measured by proper calorimetrical appliances to these areas, which we shall call the calorimetric areas of the apparatus. It will generally be found that the value of the quotient $(V - V')/a$ is not the same for consecutive isothermal surfaces.

74. Circumstances being as described in § 73, the thermal conductivity (k) of the substance between the isothermals v and v' is the value of

$$\frac{Qa}{A(v - v')} \qquad \dots\dots\dots\dots\dots\dots\dots\dots\dots(1).$$

It must be remembered that the temperatures v, v' used in this definition are temperatures of the substance itself. Some experimenters have given largely erroneous results through assuming that the temperatures of the two sides of the slab were equal to those of the calorimetric fluids, such as warm water or steam on one side, and iced water or cold water, with its temperature measured by a thermometer, on the other side. To obtain correct results, the actual temperatures at two points *in* the conducting body itself must be ascertained by aid of suitable thermometers, or thermometers and differential thermoscopes, applied in such a way as not sensibly to disturb the isothermal surfaces. This, so far as we know, has not been done by any experimenter hitherto, in attempting to measure thermal conductivity directly by the method indicated in the definition; and therefore if any results hitherto obtained by this method are trustworthy, it is only in a few cases,—cases in which the substance experimented upon has been of such small conducting power, and the stirring of the calorimetric fluids on its two sides so energetic, that we can feel sure that the observed or assumed temperatures of these fluids, or of the portions of them of which the temperatures have been measured by thermometers, have not differed sensibly from the temperatures of the slab at its surfaces in contact with them.

75. What utter confusion has permeated scientific literature, from experiments on thermal conductivity vitiated through non-fulfilment of this condition, is illustrated by results quoted in Everett's *Units and Physical Constants* (1st ed., London, 1879), among which we find ·19 for the conductivity of copper according to Péclet, and 1·1 according to Ångström (which we now know to be correct). When we look to Péclet's and Ångström's own papers the confusion becomes aggravated. Péclet, in his *Mémoire sur la détermination des coefficients de conductibilité des métaux pour la chaleur*[*], quotes old experiments of Clément, and others more recent of Thomas and Laurent, regarding which he gives certain details. Taking his information no doubt from Péclet's paper, Ångström gives a statement[†] for the conductivity of copper, according to experimenters who had preceded him, which, with the

---

[*] *Annales de Chimie et de Physique*, Paris, 1841.

[†] In Ångström's own statement the unit quantity of heat is that required to raise 1000 grammes of water 1°. The conduction is reckoned per square metre of the copper plate per second of time, and the unit chosen for the rate of variation of temperature across the plate is 1° per millimetre. To reduce his numbers to the C. G. S. system we must therefore multiply by $10^3 \times 10^{-4} \times 10^{-1} = 10^{-2}$.

decimal point shifted two places to the left to reduce to C. G. S., is as follows :—

Clément	00231
Thomas and Laurent	·0122
Péclet	·178

But Angström did not notice that Péclet had stated the thickness of the plate experimented on by Clément to be between 2 and 3 millimetres. Péclet himself in his next sentence seems to have forgotten this when he compares the figure ·23 which he had calculated from Clément's results, without taking account of the thickness of the plate, with 1·22 which he calculates from Thomas and Laurent's experiments on copper, without stating any thickness for the tube of copper on which (instead of a flat plate) they had experimented. Thus we have no data for finding what their results really were in either Péclet's or Ångström's paper; but Péclet seems to show enough regarding it to let us now feel perfectly sure that it is only a question of whether it is tens or hundreds of times too small. Omitting it then from the preceding statement, completing the correction by multiplying the ·0023 by 2½ (assuming the thickness of the plate to be 2½ millimetres, as Péclet says it was between 2 and 3) to give Clément's result, and appending Ångström's result, which we now know to be more nearly correct, we have the following statement for thermal conductivity of copper in C. G. S. units :—

| ·0057, according to Clément. |
| ·178, „ „ Péclet *. |
| 1·1, „ „ Ångström. |

76. The comparison of these results is highly instructive. Clément's result is two hundred times too small, and Péclet's six times too small. Clément experimented by exposing one side of a plate of copper of a square metre surface and about two and a half millimetres thick to steam at 100° C., and the other side to water at 28° C. It was assumed that the difference of temperature between the two sides was 72°. The difference really was about ·36 of a degree, as we know from the quantity of heat actually conducted through it in Clément's experiment, indicated

* This result was published by Péclet in 1853, in a work entitled *Nouveaux documents relatifs au chauffage et à la ventilation*.

by the amount of steam which he found to be condensed into water. In fact, the amount of steam condensed did not differ sensibly from what it would have been if the copper plate had been infinitely thin, or its substance of infinite thermal conductivity. It is important in engineering, and in many of the arts and manufactures involving thermal processes, and particularly in that one of them of greater everyday value to the human race than all the others put together, cookery, to know that for copper or iron boilers, or steam-pipes, or pots or frying-pans, the transmission of heat from radiant burning coal or charcoal, or red or white hot fireclay or other solids, and from hot air in contact with them, on one side, to hot water or steam or oil or melted fat on the other side, or hot liquid or steam on one side and cool air on the other side, is for practical purposes sensibly the same as if the thermal conductivity of the metal were infinite, or its resistance to the transmission of heat nothing. The explanation is obvious to us now with the definite and sure knowledge regarding thermal conductivities of different substances and of matter in different conditions, solid, liquid, and gaseous, gained within the last twenty years. Angström, Forbes, F. Neumann, and Tait have given, each one of them with thoroughly sufficient experimental evidence to leave no room to doubt the substantial accuracy of his results, absolute values for the thermal conductivities of copper and iron. Clausius and Maxwell have given us thermal conductivities of air and other gases, from their splendid development of the kinetic theory, which are undoubtedly trustworthy as somewhat close approximations to the true values, and which it is quite possible are more accurate than we can hope to see obtained from direct measurements of the conduction of heat through gases. J. T. Bottomley has given a trustworthy and somewhat closely accurate direct measurement of the thermal conductivity of water. From the results of these experimenters' work, reduced to uniform C.G.S. reckoning in our tables (XII. and XIII.) of thermal conductivities we see that the thermal conductivity of iron is 80 times, and that of copper 500 times that of water. The thermal conductivity of iron is 3500 times, and that of copper is 20,000 times that of air. Hence, although water or air at the interface of its contact with the metal is essentially at the same temperature as the metal, there must be great differences of temperature in very thin layers of the fluid close to the interface when there is large

flux of heat through the metal, and the temperature of the fluid
as measured by any practicable thermometer, or inferred from
knowledge of the average temperature of the whole fluid, or from
the temperatures of entering and leaving currents of fluid, may
differ by scores of degrees from the actual temperature of the
solid at the interface. It is remarkable that Péclet, while perceiving that Clément's result was largely erroneous on this account,
and improving the mode of experimenting by introducing a
rotating mechanical stirrer to change very rapidly the fluid in
contact with the solid, only multiplied Clément's conductivity by
30 instead of by 200, which would have been necessary to annul
the error. Notwithstanding his failure to obtain accurate results
for metals, we have ventured to include his results for wood, and
solids of lower conductivity than wood, in our table, because we
perceive that he was alive to the necessity for very energetic
stirring of the liquid, and the mechanical means which he adopted
for it, though utterly insufficient for the case of even the least
conductive of the metals, were probably not so for wood and solids
of lower conductivity than wood; and because it is not probable
that the complication of heat generated by the stirring (which
Ångström suggests as an objection to Péclet's method) was in any
case sufficient to produce a sensible influence upon the experimental results.

77. The first correct determinations of thermal conductivities
were given by Forbes in his paper on the temperature of the
earth, in the *Transactions of the Royal Society of Edinburgh*, for
1846 (Vol. XVI. Part II.), as calculated from his observations of
underground temperature at three localities in the neighbourhood
of Edinburgh—the trap rock of Calton Hill, the sand of the Experimental Garden, and the sandstone of Craigleith Quarry—by
an imperfect approximate method indicated by Poisson. A more
complete analytical treatment of the observational results, analysed harmonically and interpreted by application of Fourier's
formula (equation (19) of Art. LXXII. Vol. II. above) to each term
separately*, gave results (quoted in Table XII. below) for the
conductivities, which differed but little from Forbes' approximate
determinations.

* "On the Reduction of Observations of Underground Temperature; with ap-
plication to Prof. Forbes' Edinburgh Observations, and the continued Calton Hill
Series," by Prof. (Sir) W. Thomson. *Trans. R. S. E.*, April, 1860. Vol. XXII. Part II.
[Art. XCIII. below.]

Estimat-
ing and
compar-
ing the
annual
periodic
variations
of tempe-
rature at
the depths
of the ther-
mometers.

78. It has always seemed to us that the best mode of experimenting on the conductivity of metals must, without doubt, be by an artificial imitation in a metallic bar, of the natural periodic variations of underground temperature, produced by periodically varied thermal appliances at one end of the bar. The effect of loss or gain of heat through the sides (or lateral surface) of the bar (ideally annullable by a coating of ideal varnish impermeable to heat) may be practically annulled by making the period of the variation small enough.

Let k be the thermal conductivity of the substance and c its thermal capacity per unit bulk. Let e be the emissivity (§ 71 above) of its surface. Let the bar be circular-cylindric, and r the radius of its cross section. At time t, let v be the mean temperature in a cross section at distance x from the end, and v' the surface-temperature at the circular boundary of this section,—all temperatures being reckoned as differences from the temperature of the surrounding medium, called zero temporarily for brevity. The heat lost from the circumference of the bar between the cross sections $x - \tfrac{1}{2}dx$ and $x + \tfrac{1}{2}dx$ in time dt is $ev'\, 2\pi r dx dt$, and the heat conducted lengthwise, across the cross section x, in the same time, is $\pi r^2 \dfrac{dv}{dx}$. Hence we readily find (see *Fourier, Mathematics,* Art. LXXII. Vol. II. above) as the equation of conduction of heat along the bar, very approximately if v' differs very little from v (that is to say, if the temperature is very nearly uniform throughout each cross section),

$$c\,\frac{dv}{dt} = \frac{d}{dx}\left(k\,\frac{dv}{dx}\right) - \frac{2e}{r}\,v' \quad\ldots\ldots\ldots\ldots\ldots(1).$$

To estimate v', let v'' be the temperature at the centre of the cross section x, and let $\left(\dfrac{-dv}{dr}\right)'$ denote the rate of decrease of temperature from within outwards in the substance of the bar close to its surface. We have clearly

$$v'' - v' < r\left(\frac{-dv}{dr}\right)';$$

and, because the emission is supplied by conduction from within,

$$ev' = k\left(\frac{-dv}{dr}\right)'.$$

Hence $\qquad\qquad v'' - v' < \dfrac{er}{k}\,v'.$

The value of e for a blackened globe hung within a hollow, with blackened bounding surface, is about $\frac{1}{4000}$ according to Macfarlane's experiments* (Table IX. below), and considerably less for surfaces with any degree of polish. We may therefore take $\frac{1}{4000}$ as a maximum value for e. The values of k for copper and iron at ordinary temperatures are, in C. G. S., approximately ·95 and 18. Hence, if $r = 5$ cms. (or the diameter of the bar 10 cms., which is more than it is likely to be in any laboratory experiments), we find

$$v'' - v' \begin{cases} < \dfrac{v'}{880} \text{ for copper,} \\[2mm] < \dfrac{v'}{160} \text{ for iron.} \end{cases}$$

Hence the error will be practically nothing if we take $v' = v$. Thus, and if we suppose k to be independent of temperature, (1) becomes

$$\frac{dv}{dt} = \frac{k}{c} \frac{d^2v}{dx^2} - \frac{2e}{rc} v \dots\dots\dots\dots\dots(2),$$

or

$$\frac{dv}{dt} = \kappa \frac{d^2v}{dx^2} - hv, \dots\dots\dots\dots\dots(3),$$

which is Fourier's equation for the conduction of heat along a bar or the circumference of a thin ring. Its solution to express simple harmonic variations of temperature produced in an infinitely long bar by properly varied thermal appliances at one end is

$$v = R\epsilon^{-gx} \cos{(nt - fx + e)}\dots\dots\dots\dots(4),$$

n, R, e being arbitrary constants,—the "speed," the semi-range, and the epoch for $x = 0$; and f, g constants given by the formulæ

$$\left. \begin{aligned} f &= \sqrt{\frac{1}{2\kappa} \{(n^2 + h^2)^{\frac{1}{2}} - h\}^{\frac{1}{2}}} \\[2mm] g &= \sqrt{\frac{1}{2\kappa} \{(n^2 + h^2)^{\frac{1}{2}} + h\}^{\frac{1}{2}}} \end{aligned} \right\} \dots\dots\dots\dots(5).$$

For iron and copper the values of c are respectively ·95 and ·845. Hence, with the previously used values of k for these metals, and with 1/4000 for e, we find $\kappa = 1\cdot1$ for copper and $\kappa = \cdot2$ for iron; and for either, $h = 1/1700r$ nearly enough.

Suppose, for example, $r = 2$ cm., this makes $h = 1/3400$; and suppose the period to be 32 m. (the greatest of those chosen by

---

* Appendix B to present Article.

Ångström), this makes $n = 2\pi/(60 \times 32)$, or roughly $n = 1/310$ and $h/n = 1/17$. Now when h/n is small, we have approximately

$$f = \sqrt{\frac{n}{2\kappa}} \cdot \left(1 - \tfrac{1}{2}\frac{h}{n}\right),$$

$$g = \sqrt{\frac{n}{2\kappa}} \cdot \left(1 + \tfrac{1}{2}\frac{h}{n}\right);$$

and therefore with the assumed numbers

$$f = \sqrt{\frac{n}{2\kappa}}\left(1 - \frac{1}{34}\right), \text{ and } g = \sqrt{\frac{n}{2\kappa}} \cdot \left(1 + \frac{1}{34}\right),$$

by which we see that the propagation of the variation of temperature is but little affected by the lateral surface emissivity. Little as this effect is, it is very perfectly eliminated by the relation

$$fg = \frac{n}{2\kappa} \quad\quad\quad\dots\dots\dots\dots\dots\dots\dots\dots\dots(6),$$

which we find from (5).

It is convenient to remark that g is the rate of diminution of the Napierian logarithm of the range, and f the rate of retardation of the epoch (reckoned in radians) per centimetre of the bar. Were there no lateral emissivity these would be equal, and the diffusivity (see § 82 below) might be calculated from each separately. This was done in our analysis of the Edinburgh underground temperature observations (see note to § 77 above). But in the propagation of periodic variation of temperature along a bar (as of electric potential along the conductor of a submarine cable) lateral emissivity (or imperfect insulation) augments the rate of diminution of the logarithm of the amplitude, and diminishes the rate of retardation of the phase, leaving the product of the two rates unaffected, and allowing the diffusivity to be calculated from it by equation (6). This was carried out for copper and iron by Ångström in Sweden, and the results communicated to the Royal Swedish Academy in January 1861. German and English editions of his paper have been published in Pogg *Annalen*, Vol. 114, 1861, p. 513, and the *Phil. Mag.* for 1863 (first half year). The details of the apparatus and of the actual experiments, in which Ångström had the assistance of Thalen, are sufficiently described in this paper*, and in a subsequent paper

---

* The first paper is marred unhappily by two or three algebraic and arithmetical errors. One algebraic error is very disturbing to a careful reader, and might even to a hasty judgment seem to throw doubt on the validity of the experimental use

(Pogg. *Annalen*, Vol. 118, 1863, p. 428), to allow us to feel perfect confidence in the very approximate accuracy of the results. Hence we have included them in our Table.

79. The question, Does thermal conductivity vary with temperature? was experimentally investigated by Forbes about thirty years ago; and in a first provisional statement of results communicated to the British Association at Belfast in 1852 it was stated that the thermal conductivity of iron is less at high temperatures than at low. Forbes' investigation was conducted by an elaborate method of experimenting, in which the static temperature of a long bar of metal is observed after the example of the earlier experiments of Despretz, with a most important additional experiment and measurement by means of which the static result is reduced to give conductivity in absolute measure, and not merely as in Despretz's experiments to give comparisons between the conductivities of different metals. In 1861 and 1865 Forbes published in the *Transactions Roy. Soc. Edin.*, Vols. XXIII. and XXIV., results calculated from his experiments, including the first determination of thermal conductivity of a metal (iron) in absolute measure, and a confirmation of his old result that the conductivity of iron diminishes with rise of temperature. Forbes's bars have been inherited and further utilized, and bars of copper, lead, and other metals have been made and experimented upon according to the same method, by his successor in the university of Edinburgh, Professor Tait. The investigation was conducted

---

which is made of the formulæ. There is, however, no real foundation for any such doubt. The following little correction suffices to put the matter right. For the general term as printed in Ångström's paper read

$$\epsilon^{-xg_i\sqrt{i}}\sin\left(\frac{2i\pi t}{T} - xg_i'\sqrt{i} + \beta'\right),$$

with the following values for g_i, g_i':—

$$g_i = \sqrt{\sqrt{\frac{\pi^2}{K^2T^2} + \frac{H^2}{4K^2i^2}} + \frac{H}{2Ki}};$$

$$g_i' = \sqrt{\sqrt{\frac{\pi^2}{K^2T^2} + \frac{H^2}{4K^2i^2}} - \frac{H}{2Ki}};$$

instead of these formulæ without the i, as Ångström gives them. Here we see that $g_i g_i' = \frac{\pi}{KT}$; and it is the product $g_i g_i'$ that Ångström uses in his experimental application, not the separate values of either g_i or g_i'. Hence no error is introduced by his having overlooked that g_i is not equal to g_1 except for $i = 1$.

partly with a view to test whether the electric conductivities and the thermal conductivities of different metals, more or less approximately pure, and of metallic alloys, are in the same order, and, further, if their thermal conductivities are approximately in the same proportion as their electric conductivities. The following results quoted from his paper on "Thermal and Electric Conductivity" (*Transactions R. S. E.*, Vol. XXVIII. 1878) are valuable as an important instalment, but expressly only an instalment, towards the answering of this interesting question :—

"Taking the inferior copper ('Copper C') as unit both for thermal and for electric conductivity, we find the following table of conductivities at ordinary temperatures, with the rough results as to specific gravity and specific heat referred to in § 15 above :—

	Thermal.	Electric.
Copper, Crown	1·41	1·729
„ C.	1·00	1·000
Forbes's iron	0·29	0·264
Lead	0·12	0·149
German silver	0·14	0·117

"The agreement of these numbers is by no means so close as is generally stated; but this is no longer remarkable, for it is well known that the electric conductivity of all pure metals alters very much with the temperature, while we have seen that as regards thermal conductivity there is but slight change with either copper or lead, though there is a large change with iron. This accords with some results of my own on the electric conductivity of iron at high temperatures (*Proc. R. S. E.*, Vol. VIII. 1872–75, p. 32), and with the results of the repetition of these experiments by a party of my laboratory students. *Proc. R. S. E.*, Vol. VIII. 1872–75, p. 629."

80.   The absolute values of Tait's results for the first three metals of the preceding list are given in square centimetres per second in our Table XII. below. As to change of thermal diffusivity with temperature, Tait finds but little difference in the diffusivity of copper through the wide range of temperatures from 0° to 300° C., and that difference an augmentation instead of a diminution at the high temperatures, as shown in the following

results, of which the first factors denote square centimetres per second*:—

Copper, Crown...............$1{\cdot}177 \,(1 + 0{\cdot}0004t)$

,,   C..................... $\cdot836 \,(1 + 0{\cdot}00055t)$

Iron .... $\cdot232 \,(1 - 0{\cdot}00144t)$.

On the other hand, Ångström finds for copper from experiments described in his second paper referred to above, at mean temperatures of from $28°{\cdot}8$ to $71°{\cdot}5$ C., results which reduced to square centimetres per second, are as follows :—

Copper, first specimen... $1{\cdot}216 \,(1 - \cdot00214t)$

,,   second specimen $1{\cdot}163 \,(1 - \cdot001519t)$

Iron .... $\cdot224 \,(1 - \cdot002874t)$.

Comparing these two sets of values it is interesting to see so close an agreement between results obtained by methods of experiment differing so much as that of Forbes and Tait differs from Ångström's. The change of sign from + to − in the change of diffusivity due to increase or decrease of temperature, does not really imply any great difference between the two sets, but it is desirable that Ångström's experiments should be repeated, and especially for copper, through a much wider range of temperatures. This can be done with great ease, from the lowest temperature obtainable by freezing mixtures to temperatures up to the melting point of copper, so excellently plastic is Ångström's method. Our proposed extension of it is to be carried out by proper thermal appliances to the end of the bar, which Ångström left to itself,—appliances by which in one series of experiments it may be kept constant at − 50° or − 60° C., in others left to itself to take nearly the atmospheric temperature, in others kept at high temperatures limited only by the melting temperature of copper, if the experimenter desires to go so far. We would also suggest that the thermo-electric method first introduced by Wiedemann and Franz in their experiments on the static temperature of bars or wires heated at one end and allowed to lose heat by convection and radiation from their sides, (which

---

* Tait gave his results in terms of the foot and minute, with, for the unit of heat, the amount of heat required to raise the temperature of a cubic foot of the substance by 1°. In other words, his results are "diffusivities" (§ 82) in square feet per minute. They are multiplied by 15·48 to reduce to square centimetres per second as given in the text.

was rejected, not, we think, judiciously, by Ångström), might be used with advantage instead of the mercury thermometers inserted in holes in the bar as in Ångström's apparatus; or that, if thermometers are to be used, air thermometers in which the bulb of the thermometer is itself a very small hole in the bar experimented on, and the tube a fine-bore glass tube fitted to this hole, would be much preferable to the mercury thermometers hitherto employed in, we believe, all experiments except those of Wiedemann and Franz, on the conduction of heat along metallic bars.

Diffusion of heat.

81. Fourier's ninth chapter is entitled "De la Diffusion de la Chaleur." The idea embodied in this title is the spreading of heat in a solid tending to ultimate equalization of temperature throughout it, instead of the transference of heat from one body to another by conduction through the solid considered. Though Fourier makes the special subject of his chapter on "Diffusion" the conduction of heat through an infinite solid, we may conveniently regard as coming under the several designations "Diffusion of Heat" every case of thermal conduction in which the heat conducted across any part of the solid has the effect of warming contiguous parts on one side of it, or of leaving contiguous parts on the other side cooler,—in other words, every case in which the temperature of the body through which the conduction of heat takes place is varying with time, as distinguished from what Fourier calls "Uniform Motion of Heat," or the class of cases in which the temperature at every point of

Investigations of thermal conductivity by "uniform motion of heat."

the body is constant. The experiments of Péclet, Despretz, Forbes and Tait, Wiedemann and Franz, were founded on the uniform conduction of heat across slabs or along bars, and their determinations of relative and absolute conductivities were made by comparing or by measuring absolutely quantities of heat that were *conducted out of the body tested*. On the other hand, it is the diffusion of heat through the body that is used in the

By diffusion of heat.

determinations of thermal conductivity in absolute measure by Forbes and W. Thomson* from the periodic variations of underground temperature; in those of Ångström, from his experiments on the spreading of periodic variations of temperature through bars of iron and copper, and a series of valuable experiments a year or two later by F. Neumann, applying the same general

* See Art. xciii. below.

method to bars of brass, zinc, German silver, and iron; in experiments by F. Neumann on substances of lower conductivity (coal, cast sulphur, ice, snow, frozen earth, gritstone) formed into cubes or globes of 5 or 6 inches diameter, and heated uniformly, and then left to cool in an atmosphere of lower temperature, and from time to time during the cooling explored by thermo-electric junctions imbedded in them to show their internal distribution; in similar experiments on the cooling of globes of 14 cm. diameter of porphyritic trachyte by Ayrton and Perry in Japan ; and in Kirchhoff and Hansemann's recent experiments*, to find the thermal conductivity of iron by the not well-chosen method of suddenly cooling one side of a cube of iron of 14 cm., and observing the temperatures by aid of thermo-electric junctions in several points of the line perpendicular to this side through its middle.

82.　When the effect of heat conducted across any part of a body, in heating the substance on one side or leaving the substance on the other side cooler, is to be reckoned, it is convenient to measure the thermal conductivity in terms, not of the ordinary gramme water-unit of heat, but of a special unit, the quantity of heat required to raise the temperature of unit bulk of the substance by 1° C.　In other words, if k be the conductivity in terms of any thermal unit, and c the thermal capacity of unit bulk of the substance, it is k/c, not merely k, that expresses the quality of the substance on which the phenomenon chiefly depends.　We therefore propose to give to k/c the name of *thermal diffusivity* (or simply *diffusivity* when heat is understood to be the subject), while still using the term *thermal conductivity* to denote the conducting power as defined in § 73, without restriction as to the thermal unit employed.　It is interesting and important to remark that "diffusivity" is essentially to be reckoned in units of area per unit of time, and that its "dimensions" are L^2/T.　Its regular C. G. S. reckoning is therefore in square centimetres per second.　Diffusion of electricity through a submarine cable, has been shown† to follow the same law as the "linear" diffusion of heat, which Fourier calls the diffusion of heat when the isothermal surfaces are parallel planes.　The

Thermal diffusivity.

---

* Wiedemann's (late Poggendorff's) *Annalen*, 1880, No. 1.

† *Proc. Roy. Soc.*, May 1855, "On the Theory of the Electric Telegraph" [Art. LXXIII. Vol. II. above].

curves of the following diagram and Tables A, B, and C show
in a practically useful way the result in the course of the times
noted, of from fractions of a second to thousands of millions of
years, of linear diffusion of two different qualities in an infinite
line from an initial condition in which there is sudden transition
from one quality to the other, for the thoroughly practical cases
specified in the accompanying explanations.

No. 1
Curve.

No. 2
Curve.

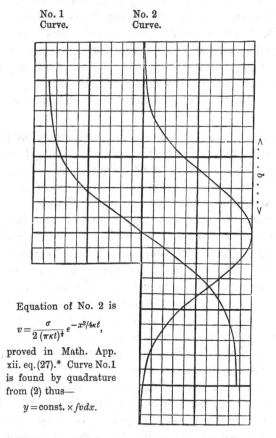

Equation of No. 2 is

$$v = \frac{\sigma}{2\,(\pi\kappa t)^{\frac{1}{2}}}\,\epsilon^{-x^2/4\kappa t},$$

proved in Math. App.
xii. eq. (27).* Curve No.1
is found by quadrature
from (2) thus—

$$y = \text{const.} \times \int v\,dx.$$

Fig. 11.—Diagram of Diffusion.

Curve No. 1 shows temperature; or quantity of substance in
solution; or potential in the conductor of a submarine cable through
which electricity is diffusing. Curve No. 2 shows rate per unit

* See Art. LXXII. above, Vol. II. p. 48.

of distance of variation of the temperature, or of the quantity of substance in solution. Vertical ordinates are actual distances through the medium. Horizontal ordinates represent in No. 1 curve, temperature or quantity of diffusing substances or electric potential; and in No. 2 curve rate of variation of temperature or of diffusing substance or of electric potential.

Diffusions.—Table A.

Substance.	Time in Seconds from the commencement of the Diffusion until the Condition represented by the Curves on the Actual Scale ($b = 2$ Centimetres) is reached.
Carbonic acid through air	6·97 seconds.
Heat through hydrogen	·89 of a second.
,, ,, copper	·93 ,,
,, ,, iron	5·5 seconds.
,, ,, air	6·25 ,,
,, ,, underground strata	100·0 ,,
,, ,, wood	770 ,,
Common salt through water	87150 ,,
Electricity through Suez-Aden cable	$1·087 \times 10^{-16}$ of a second.
,, ,, Aden-Bombay cable	$·739 \times 10^{-16}$,,
,, ,, Persian Gulf cable	$·635 \times 10^{-16}$,,
,, ,, Atlantic cable	$·440 \times 10^{-16}$,,
,, ,, French Atlantic Cable	$·396 \times 10^{-16}$,,
,, ,, Direct U.S. cable	$·340 \times 10^{-16}$,,

Diffusions (Secular).—Table B.

Substance.	Time in Years from the commencement of the Diffusion until the Condition represented by the Curves on the Scale of $b = 20$ Kilometres, or 1,000,000 times the Actual Scale, is reached.
Carbonic acid through air *	220,000 years.
Heat through hydrogen	28,500 ,,
,, ,, copper	29,200 ,,
,, ,, iron	174,000 ,,
,, ,, air	198,000 ,,
,, ,, underground strata †	317,000 ,,
,, ,, wood	24,700,000 ,,
Common salt through water ‡	2,760,000,000 ,,

* Instructive as to the proportion of carbonic acid in air at different heights, proving its approximate uniformity due to convection, not to diffusion.

† Instructive as to geological theories respecting terrestrial temperature.

‡ Instructive as to theories respecting the saltness of the sea.

Diffusions (Electrical).—Table C.

Name of Cable.	Time in Seconds from the commencement of the Diffusion until the Condition represented by the Curves on the Scale of $b = 1000$ Nautical Miles, or 92,615,000 times the Actual Scale, is reached.
Suez-Aden	·932 of a second.
Aden-Bombay	·634 ,,
Persian Gulf	·545 ,,
Atlantic	·377 ,,
French Atlantic	·339 ,,
Direct United States	·292 ,,

83. The following tables (I. to XXIV.) contáin useful information regarding various thermal properties of matter.

TABLE I.—*Linear Expansions of Solids**.

Name.	Mean Expansion per degree C. through Range stated.	Range †.	Authority.
Silver	·00002120	0° to 100°	Muschenbröck.
,, ...	,, 1910	0° ,, 100	Kupffer.
,, ...	,, 1943	0° ,, 100°	Matthiessen.
Thallium	,, 3021	40°	⎫
Sulphur, Sicily	,, 6413	40°	⎬ Fizeau.
Selenium, cast.........	,, 3680	40°	
Tellurium, ,, ...	,, 1675	40°	⎭
Lead	,, 2799	0° to 100°	Matthiessen.
,, ...	,, 2924	40°	Fizeau.
Iron	,, 1156	0° to 100°	Borda.
,, ...	,, 1190	0° ,, 100°	⎰ Calvert, Johnson, ⎱ and Lowe.
Steel annealed.........	,, 1220	0° ,, 100°	Muschenbröck.
,, French cast, tempered ...	,, 1322	40°	⎫
,, French cast, annealed ...	·, 1101	40°	⎬ Fizeau.
,, English cast, annealed ...	,, 1095	40°	⎭
Steel, soft..............	,, 103	...	⎱ Calvert, Johnson.
Cast iron	,, 112	0° to 100°	⎰ and Lowe.
Cobalt, red. by H. compressed.........	,, 1236	40°	⎫
Nickel, red. by H. compressed.........	,, 1279	40°	⎬ Fizeau.
Copper	,, 1883	0° to 300°	Dulong and Petit.
,, ...	,, 1866	0° to 100°	Matthiessen.
,, native L. Superior......	,, 1698	40°	⎫
,, commercial...	,, 1678	40°	⎬ Fizeau.
Ruthenium, semi-fused	,, 0963	40°	
Rhodium, semi-fused	,, 0850	40°	⎭

* Abridged from Clarke's *Constants of Nature.*

† Where only one number is given for the range in this and the following table, the corresponding statement is to be understood as applying through a small range on either side of the number stated.

TABLE I. (*continued*).—*Linear Expansions of Solids.*

Name.	Mean Expansion per degree C. through Range stated.	Range.	Authority.
Palladium	·00001104	0° to 100°	Matthiessen.
„ forged	„ 1176	40°	Fizeau.
Platinum	„ 0918	0° to 300°	Dulong and Petit.
„	„ 0680	0° „ 100°	Calvert, Johnson, and Lowe.
„	„ 0886	0° „ 100°	Matthiessen.
Iridium	„ 0700	40°	Fizeau.
Osmium, semi-fused	„ 0657	40°	
Zinc	„ 2200	0° to 100°	Calvert, Johnson, and Lowe.
„	„ 2976	0° „ 100°	Matthiessen.
„ distilled	„ 2918	40°	Fizeau.
Cadmium	„ 332	0° to 100°	Calvert, Johnson, and Lowe.
„	„ 3159	0° „ 100°	Matthiessen.
„ distilled	„ 3069	40°	
Magnesium, cast	„ 2694	40°	Fizeau.
Indium, cast	„ 4170	40°	
Arsenic, sublimed	„ 0559	40°	
Antimony	„ 0980	0° to 100°	Calvert, Johnson, and Lowe.
„	„ 1056	0° „ 100°	Matthiessen.
„ along axis ⎫ Crystal.	„ 1692	40°	
„ normal to axis ⎬	„ 0882	40°	Fizeau.
„ mean value ⎭	„ 1152	40°	
Bismuth	„ 133	0° to 100°	Calvert, Johnson, and Lowe.
„	„ 1316	0° „ 100°	Matthiessen.
„ along axis ⎫ Crystal.	„ 1621	40°	
„ normal to axis ⎬	„ 1208	40°	Fizeau.
„ mean value ⎭	„ 1346	40°	
Gold annealed	„ 1460	0° to 100°	Muschenbröck.
„ „	„ 138	0° „ 100°	Calvert, Johnson, and Lowe.
„ „	„ 1470	0° „ 100°	Matthiessen.
„ cast	„ 1443	40°	
Carbon, diamond	„ 00000	38°·8	
„ „	„ 00562	0°	
„ „	„ 00852	20°	
„ „	„ 01286	50°	Fizeau.
„ graphite	„ 0786	40°	
„ gas carbon	„ 0540	40°	
„ anthracite	„ 2078	40°	

TABLE I. (*continued*).—*Linear Expansions of Solids.*

Name.	Mean Expansion per degree C. through Range stated.	Range.	Authority.
Tin	+ ·00002330	16° to 99°	Kopp.
,, ...	,, 2234	40°	Fizeau.
,, ...	,, 2296	0° ,, 100°	Matthiessen.
Aluminium, com- ⎱ mercial .... ⎰	,, 222	0° ,, 100°	⎰ Calvert, Johnson, ⎱ and Lowe.
Fluor spar, CaF$_2$......	,, 19504	0° ,, 100°	Pfaff.
Silver iodide, AgI, cylinder, precipitated and compressed—			⎞
Lengthwise	− ·00000166	40°	⎟
Transversely......	− ·00000122	40°	⎟
Mean value	− ·00000137	40°	⎬ Fizeau.
Mercuric iodide HgI$_2$.	+ ·00002387	40°	⎟
Lead iodide, PbI$_2$ ...	,, 3359	40°	⎟
Cadmium iodide, CdI$_2$	,, 2916	40°	⎠
Hæmatite, Fe$_2$O$_3$,⎱ along axis .... ⎰	·00000829	40°	⎱
Hæmatite, Fe$_2$O$_3$,⎱ normal to axis ... ⎰	,, 0836	40°	⎰ Fizeau.
Magnetic oxide of ⎱ Iron, Fe$_3$O$_4$ .... ⎰	,, 09540	0° to 100°	Pfaff.
Copper oxide, CuO ...	− ·000000095	0°	⎞
,, ,,	,, 00000	4°·1	⎟
,, ,,	+ ·000000136	10°	⎬
,, ,,	,, 00597	30°	⎟
,, ,,	,, 01059	50°	⎟
Zinc oxide, ZnO, ⎱ along axis .... ⎰	,, 0316	40°	⎬ Fizeau.
,, normal to ⎱ axis .... ⎰	,, 0539	40°	⎟
Quartz SiO$_2$, along ⎱ axis .... ⎰	,, 0781	40°	⎟
,, ,, normal to axis .... ⎰	,, 1419	40°	⎠
Pyrite, FeS$_2$	,, 10084	0° to 100°	⎞
Galena, PbS............	,, 18594	0° ,, 100°	⎬ Pfaff.
Beryl, longit. axis ...	,, 017214	0° ,, 100°	⎟
,, horiz. axis ...	− ·0000001316	0° ,, 100°	⎠

TABLE I. (*continued*).—*Linear Expansions of Solids.*

Name.	Mean Expansion per degree C. through Range stated.	Range.		Authority.
Emerald, along axis..	− ·00000106	40°		
,, normal to axis	+ ·00000137	40°		Fizeau.
Topaz, lesser horiz. axis	,, 08325	0° to 100°		
,, greater horiz. axis	,, 08362	0° ,, 100°		
,, vertical horiz. axis	,, 04723	0° ,, 100°		
Tourmaline, longit. axis	,, 09369	0° ,, 100°		Pfaff.
,, horiz. axis	,, 077321	0° ,, 100°		
Garnet	,, 08478	0° ,, 100°		
Iceland spar, along the axis	+ ·00002621	40°		
,, ,, normal to the axis	− ·00000540	40°		Fizeau.
Glass tube	+ ·0000083333	0° ,, 100°		Smeaton.
,, ,, ...	,, 08280	0° ,, 100°		Deluc.
,, rod	,, 086130	0° ,, 100°		
,,	,, 091827	100° ,, 200°		Dulong and Petit.
,,	,, 101114	200° ,, 300°		
,, plate	,, 0890890	0° ,, 100°		
,, ,, crown ...	,, 087572	0° ,, 100°		Lavoisier and
,, ,, ,, ...	,, 089760	0° ,, 100°		Laplace.
,, ,, ,, ...	,, 091751	0° ,, 100°		
,, white French...	,, 08510	0° ,, 100°		...
,, tube	,, 08998	...		Hagen.
,, soft Thuringian	,, 1195	...		Weinhold.
Wedgewood ware......	,, 08813	16°·6 to 100°		Daniell:
,, ,, ...	,, 08983	16°·6 ,, 350°		
Bayeux porcelain	,, 165	1000° ,, 1400°		Deville and Troost.
Platin-iridium (one-tenth iridium) ...	,, 0884	40°		Fizeau.
Solder—2 lead, 1 tin .	,, 2508	0° to 100°		Smeaton.
Type metal (lead and antimony) ...	,, 2033	16°·6 ,, 100°		Daniell.
	,, 1952	16°·6 ,, 264°		
Zinc and tin—8 zinc, 1 tin	,, 2692	0° ,, 100°		
Copper and tin—8 copper, 1 tin	,, 1817	0° ,, 100°		Smeaton.
Speculum metal	,, 1933	0° ,, 100°		

TABLE I. (*continued*).—*Linear Expansions of Solids.*

Name.	Mean Expansion per degree C. through Range stated.	Range.		Authority.
Bronze, ¼ tin	+ ·00001844	16°·6 to	100°	⎫
,,　　,,　.........	,,　2116	16°·6 ,,	350°	⎬ Daniell.
,,　　,,　.........	,,　1737	16°·6 ,,	957°	⎭
Brass, cast	,,　1875	0°　,,	100°	⎫ Smeaton.
,,　wire	,,　1930	0°　,,	100°	⎭
,,　,,　............	,,　1783	0°　,,	100°	Borda.
,,　English.........	,,　1893	0°　,,	100°	⎫ Roy.
,,　　,,　.........	,,　1895	0°　,,	100°	⎭
,,　..................	,,　1859	40°		Fizeau.
Pewter	,,　2283	0°　to	100°	Smeaton.
,,　.................	,,　2033	16°·6 to	100°	⎫ Daniell.
,,　.................	,,　1994	16°·6 ,,	206°	⎭
Paraffin, Rangoon ...	·00027854	40°		⎫
Soft coal, Charleroi..	·00002782	40°		⎬ Fizeau.
Ebonite.................	,,　770	16°·7 to	25°·3	⎫ Kohlrausch.
,,　.................	,,　842	25°·3 ,,	35°·4	⎭
Deal wood	,,　0496	0°　,,	100°	Kater.
Brick....................	,,　0550	0°　,,	100°	Adie.
,,　Fire	,,　0493	0°　,,	100°	,,
Granite	,,　0868	0°　,,	100°	Bartlett.
Marble, Black	,,　0445	0°　,,	100°	⎫ Dunn & Sang.
,,　White.........	,,　1072	0°　,,	100°	⎭
Sandstone	,,　1174	0°　,,	100°	Adie.
Slate	,,　1038	0°　,,	100°	,,
Graphite	,,　0786	0°　,,	40°	Fizeau.
Ice	,,　5236	− 27°·5 ,,	1°·25	Schumacher.
Ice, H_2O	·002941	...		Heinrich; (*Baier. Akad. Phys. Abhandl.* 1806).

TABLE II.—*Cubical Expansions of Solids and Liquids**.

Name.	Mean Expansion per degree C.	Range.	Authority.
Bromine	·001016027	− 7°	} Pierre.
,, ...	·001038186	0°	
,, ...	·001318677	63°	
Iodine, solid	·000235	...	} Billet.
,,　upon fusion.........	·1682	..	
,,　liquid	·000856	...	
Silver......................	·00005831	0° to 100°	Matthiessen.
Sulphur, native	·000137	0° ,, 13°·2	}
,, ...	·000223	13°·2 ,, 50°·3	
,, ...	·000259	50°·3 ,, 78°	
,, ...	·000620	78° ,, 96°·5	} Kopp.
,, ...	·003097	96°·5 ,, 109°·9	
,, ...	·05002	{In melting at 115°.	}
Lead	·000089	...	
,, ...	·00008399	0° to 100°	Matthiessen.
Iron	·0000355	0° ,, 100°	} Dulong and Petit.
,, ...	·0000441	0° ,, 300°	
,, ...	·000037	...	Kopp.
Copper	·000055	0° to 100°	Playfair and Joule.
,, ...	·00005649	0° ,, 300°	Dulong and Petit.
,, ...	·000051	...	Kopp.
,, ...	·00004998	0° to 100°	}
Palladium......................	·00003312	0° ,, 100°	} Matthiessen.
Platinum	·00002658	0° ,, 100°	}
Zinc	·000089	...	Kopp.
,, ...	·00008928	0° to 100°	Matthiessen.
Cadmium	·0000940	...	Kopp.
Mercury......................	·00017905	0°	}
,, ...	·00017950	10°	
,, ...	·00018001	20°	
,, ...	·00018051	30°	
,, ...	·00018102	40°	
,, ...	·00018152	50°	
,, ...	·00018203	60°	
,, ...	·00018253	70°	
,, ...	·00018304	80°	} Regnault.
,, ...	·00018354	90°	
,, ...	·00018405	100°	
,, ...	·00018657	150°	
,, ...	·00018909	200°	
,, ...	·00019161	250°	
,, ...	·00019413	300°	}
,, ...	·00019666	350°	
,, ...	·0001812	0° to 100°	Matthiessen.
Phosphorus	·000366	8°·3 ,, 15° 8	} Kopp.
Antimony	·000033	...	
,, ...	·00003167	0° to 100°	Matthiessen.
Bismuth	·0000400	...	Kopp.
Gold	·00004411	0° to 100°	Matthiessen.
Diamond	·00000354	40°	Fizeau.
Tin............................	·0000690	...	Kopp.

* Abridged from Clarke's *Constants of Nature*.

TABLE II. (*continued*).—*Cubical Expansions of Solids and Liquids.*

Name.	Mean Expansion per degree C.	Range.		Authority.
Water, H_2O	- ·000034	0° to	4°	Weidner.
,, ...	+ ·0002500	4° ,,	50°	} Matthiessen.
,, ...	·0004496	4° ,,	100°	
Ice	·0001585	0° ,,	1°	Plücker and Geissler.
Sulphur dioxide	·00202	5° ,,	10°	D'Andreéf.
Carbon dioxide, liquid, CO_2	·00475	-10° ,,	- 5°	
,, ,, ,,	·00492	- 5° ,,	0°	
,, ,, ,,	·00540	0° ,,	5°	
,, ,, ,,	·00629	5° ,,	10°	} D'Andreéf.
,, ,, ,,	·00769	10° ,,	15°	
,, ,, ,,	·00975	15° ,,	20°	
,, ,, ,,	·01277	20° ,,	25°	
,, disulphide, CS_2 ...	·0011016	-50° ,,	0°	
,, ,, ,, ...	·00119625	0° ,,	40°	} Muncke.
,, ,, ,, ...	·0012517	0° ,,	70°	
,, ,, ,, ...	·0012366	0° ,,	40°	
,, ,, ,, ...	·0013260	0° ,,	80°	} Hirn.
,, ,, ,, ...	·0014596	0° ,,	120°	
,, ,, ,, ...	·0016608	0° ,,	180°	
Copper Oxide, CuO	·00000279	40°		Fizeau.
Zinc Oxide, ZnO	·00001394			
Quartz, SiO_2	·000040	...		Kopp.
Emerald	·00000168	40°		Fizeau.
Beryl	·00000105	0° to	100°	
Topaz	·00002137	0° ,,	100°	
Tourmaline	·00002181	0° ,,	100°	} Pfaff.
Garnet	·000025434	0° ,,	100°	
Analcime	·000027783	0° ,,	100°	
Idocrase	·00002700	0° ,,	100°	......
Zircon	·00002835	0° ,,	100°	......
Glass, white tube...........	·00002648	0° ,,	100°	
,, ,, globule	·00002553	0° ,,	100°	
,, green tube...........	·00002299	0° ,,	100°	
,, ,, globule	·00002132	0° ,,	100°	
,, Swedish tube	·00002363	0° ,,	100°	
,, ,, globule ...	·00002426	0° ,,	100°	
,, hard French tube...	·00002142	0° ,,	100°	} Regnault.
,, ,, ,, globule	·00002242	0° ,,	100°	
,, crystal tube	·00002101	0° ,,	100°	
,, ,, globule......	·00002330	0° ,,	100°	
,, globe...............	·00002326	0° ,,	100°	
,, common tube	·00002579	0° ,,	1°	

TABLE III.—Expansions of Liquids.

Pierre, having measured the expansions of many different liquids, embodies the results in the empirical formula $\delta_t = at + bt^2 + ct^3$, where δ_t represents the expansion of unit volume from 0° to t° cent., and a, b, c have the values given in the Table for the different liquids specified. The substances and formulæ in quotation marks are quoted from Dixon's *Heat*, Dublin, 1849.

	Name of Liquid.	Atomic Constitution.	Temperature of Boiling Point.	Observed Pressure of Boiling Point. (mms.)	Value of Coeff. a.	Value of Coeff. b.	Value of Coeff. c.
1	Ethyl (Ether)	$(C_2H_5)_2O$	35°.5	755.8	0·001 513	0·000 002 36	0·000 000 040 0
2	Ethylic alcohol	C_2H_5HO	78°.3	758.0	0·001 049	0·000 001 75	0·000 000 003 3
3	Methylic alcohol	CH_3HO	66°.7	753.0	0·001 186	0·000 001 56	0·000 000 009 1
4	Amylic alcohol	$C_5H_{11}HO$	131°.8	751.3	{ 0·000 890 ; 0·000 899 }	{ 0·000 000 66 ; 0·000 000 69 }	{ 0·000 000 011 8[1] ; 0·000 000 010 1[2] }
5	Chloride of ethyl	C_2H_5Cl	12°.5	760.0	0·001 575	0·000 002 81	0·000 000 015 7
6	Bromide of ethyl	C_2H_5Br	38°.75	760.0	0·001 338	0·000 001 50	0·000 000 016 9
7	Iodide of ethyl	C_2H_5I	70°.0	751.7	0·001 142	0·000 001 96	0·000 000 006 2
8	Bromide of methyl	CH_3Br	13°.0	759.0	0·001 415	0·000 003 32	0·000 000 113 8
9	Iodide of methyl	CH_3I	43°.8	750.2	0·001 200	0·000 002 16	0·000 000 010 1
10	Formiate of ethyl	$C_2H_5CHO_2$	52°.9	752.0	0·001 325	0·000 002 86	0·000 000 006 6
11	Acetate of ethyl	$C_2H_5C_2H_3O_2$	74°.1	760.0	0·001 258	0·000 002 96	0·000 000 001 5
12	Butyrate of ethyl	$C_2H_5C_4H_7O_2$	119°.0	746.5	{ 0·001 203 ; 0·001 296 }	{ 0·000 012 76 ; 0·000 002 91 }	{ 0·000 000 022 6[3] ; 0·000 000 052 1[4] }
13	Acetate of methyl	$CH_3C_2H_3O_2$	56°.5	760.0	0·001 240	0·000 000 63	0·000 000 004 3
14	Butyrate of methyl	$CH_3C_4H_7O_2$	103°.5	760.0	0·001 129	0·000 000 87	0·000 000 013 1
15	Terchloride of phosphorus	PCl_3	75°.95	760.0	0·000 847	0·000 000 44	0·000 017 923 6
16	Terbromide of phosphorus	PBr_3	172°.9	760.0	{ 0·000 824 ; 0·000 979 }	{ 0·000 000 91 ; 0·000 000 97 }	{ 0·000 000 002 5[5] ; 0·000 000 001 8[6] }
17	Terchloride of arsenic	$AsCl_3$	130°.21	760.0	0·001 133	0·000 000 91	0·000 000 007 6
18	Bichloride of tin	$SnCl_4$	113°.9	760.0	0·000 943	0·000 001 35	0·000 000 000 9
19	Tetrachloride of titanium	$TiCl_4$	136°.4	760.0	0·001 294	0·000 002 18	0·000 000 040 9
20	"Terchloride of silicon"	"$SiCl_4$"	57°.74	760.0	0·000 953	0·000 000 76	0·000 000 000 3
21	"Terbromide of silicon"	"$SiBr_4$"	153°.6	762.0	0·001 119	0·000 001 05	0·000 000 010 3
22	Bichloride of ethylene[7]	$C_2H_4Cl_2$	84°.9	761.9	0·000 953	0·000 001 32	0·000 000 001 1[8]
23	Bibromide of ethylene	$C_2H_4Br_2$	131°.45	760.0	0·001 017	0·000 000 10	0·000 000 008 8[9]
24	Bromine	Br	59°.27	760.0	0·001 038	0·000 001 71	0·000 000 005 4
25	Bisulphide of carbon	CS_2	46°.2	760.0	0·001 140	0·000 001 37	0·000 000 019 1

[1] From 15° to 80°. [2] From 80° to 131°.8. [3] From 13° to 90°. [4] From 90° to 118°. [5] From 0° to 100°. [6] From 100° to 175°·3.
[7] For this liquid t is counted from 20°·09, its point of congelation. [8] From 20°·09 to 100°·16. [9] From 100°·16 to 132°·6.

TABLE IV.—*Density of Water*.*

Temp. C°.	Density.	Temp. C°.	Density.	Temp. C°.	Density.
0°	·999884	13°	·999443	35°	·99419
1	·999941	14	·999312	40	·99236
2	·999982	15	·999173	45	·99038
3	1·000004	16	·999015	50	·98821
4	1·000013	17	·998854	55	·98583
5	1·000003	18	·998667	60	·98339
6	·999983	19	·998473	65	·98075
7	·999946	20	·998272	70	·97795
8	·999899	22	·997839	75	·97499
9	·999837	24	·997380	80	·97195
10	·999760	26	·996879	85	·96880
11	·999668	28	·996344	90	·96557
12	·999562	30	·995778	100	·95866

* From Kupffer's observations, as reduced by Prof. W. H. Miller, according to which the absolute density at 4° C. is not 1, but 1·000013.

TABLE V.—*Steam Pressures (in Centimetres of Mercury).*

Water.		Mercury*.		Sulphur.	
Temp.	Pressure.	Temp.	Pressure.	Temp.	Pressure.
− 32°	0·03	0°	·002	390°	27·23
− 30	0·04	10	·003	400	32·90
− 25	0·06	20	·004	410	39·52
− 20	0·09	30	·005	420	47·21
− 15	0·14	40	·008	430	56·10
− 10	0·21	50	·011	440	66·31
− 5	0·31	60	·016	450	77·99
0	0·46	70	·024	460	91·27
5	0·65	80	·035	470	106·32
10	0·92	90	·051	480	123·27
15	1·27	100	·075	490	142·29
20	1·74	110	·107	500	163·53
25	2·35	120	·153	510	187·16
30	3·15	130	·218	520	213·33
35	4·18	140	·306	530	242·20
40	5·49	150	·427	540	273·92
45	7·14	160	·590	550	308·65
50	9·20	170	·809	560	346·53
55	11·75	180	1·100	570	387·71
60	14·88	190	1·48		
65	18·69	200	1·99		
70	23·31	210	2·63		
75	28·85	220	3·47		
80	35·46	230	4·53		
85	43·30	240	5·88		
90	52·54	250	7·57		
95	63·38	260	9·67		
100	76·00	270	12·30		
105	90·64	280	15·52		
110	107·54	290	19·45		
115	126·94	300	24·22		
120	149·13	310	29·97		
125	174·39	320	36·87		
130	203·03	330	45·09		
135	235·37	340	54·83		
140	271·76	350	66·32		
145	312·55	360	79·77		
150	358·12	370	95·46		
155	408·86	380	113·96		
160	465·16	390	134·67		
165	527·45	400	158·80		
170	596·17	410	186·37		
175	671·75	420	217·75		
180	754·64	430	253·30		
185	845·32	440	293·40		
190	944·27	450	338·44		
195	1051·96	460	388·81		
200	1168·90	470	444·94		
205	1295·57	480	507·24		
210	1432·48	490	576·13		
215	1580·13	500	652·03		
220	1739·04	510	725·34		
225	1909·70	520	826·50		
230	2092·64				

* See Table VI., p. 219, below for more correct values of mercury steam-pressure for Temperatures from 0° to 220°.

TABLE VI.

Steam Pressure of Mercury (in Centimetres of Mercury).*

Tempera-ture.	Regnault.	Hagen[1].	Hertz[2].	Ramsay and Young[3].
0°	·002	·0015	·000019	—
10	·00268	·0018	·000050	—
20	·00372	·0021	·00013	—
30	·00530	·0026	·00029	—
40	·00767	·0033	·00063	·0008
50	·01120	·0042	·0013	·0015
60	·01643	·0055	·0026	·0029
70	·02410	·0074	·0050	·0052
80	·03528	·0102	·0093	·0092
90	·05142	·0144	·0165	·0160
100	·07455	·0210	·0285	·0270
120	·15341	—	·0779	·0719
140	·30592	—	·193	·1763
160	·59002	—	·438	·4013
180	1·1000	—	·923	·8535
200	1·990	—	1·825	1·7015
220	3·470	—	3·490	3·1957

[1] Hagen, *Ann. Phys. Chem.*, New Series, Vol. 16, p. 610, July 1, 1882.

[2] Hertz, *Ann. Phys. Chem.*, New Series, Vol. 16, p. 610, Aug. 15, 1882.

[3] Ramsay and Young, *Journal of the Chemical Society*, 1886.

* [Note of 6th June, 1887. The values given in Table V. for the pressure of mercury steam at various temperatures are from Regnault, but as recent observations have shewn that these values are much too large for low temperatures, I have included Table VI. above, taken from a paper "On the Vapour-pressures of Mercury," by William Ramsay, Ph.D., and Sydney Young, D.Sc. (*Jour. of the Chemical Society*, January, 1886). W. T.]

TABLE VII.

Steam Pressures (in Centimetres of Mercury)*.

Chlorobenzene.		Bromobenzene.		Aniline.		Methyl Salicylate.		Bromonaphthalene.	
Temp. C°.	Pressure Cms.	Temp. C°.	Pressure Cms.	Temp. C°.	Pressure Cms.	Temp. C°.	Pressure Cms.	Temp. C°.	Pressure Cms.
70°	9·79	120°	24·79	150°	28·37	175°	21·51	215°	15·89
75	11·94	125	32·08	155	33·17	180	24·94	220	18·18
80	14·48	130	37·27	160	38·60	185	28·78	225	20·74
85	17·43	135	43·08	165	44·71	190	33·09	230	23·60
90	20·84	140	49·58	170	51·56	195	37·89	235	26·79
95	24·77	145	56·84	175	59·21	200	43·24	240	30·34
100	29·28	150	64·91	180	67·72	205	49·17	245	34·28
105	34·42	155	73·86	185	77·15	210	55·75	250	38·64
110	40·26	160	83·75			215	63·02	255	43·45
115	46·85					220	71·00	260	48·74
120	54·28					225	79·81	265	54·54
125	62·62							270	60·88
130	71·90							275	67·79
								280	75·29

* From a paper, "A Method for obtaining Constant Temperatures," by Prof. W. Ramsay and Sydney Young, D.Sc.; Journal of the Chemical Society, Sept., 1885, Vol. XLVII.

TABLE VIII.—*Steam Pressures (in Centimetres of Mercury).*

Temperature.	Essence of Turpentine, $C_{10}H_{16}$	Ethyl Alcohol, C_2H_6O	Benzine, C_6H_6	Methyl Alcohol, CH_4O	Ethyl Iodide, C_2H_5I	Chloroform, $CHCl_3$	Carbon Bisulphide, CS_2	Ethyl Bromide, C_2H_5Br	Ethyl Ether, $C_4H_{10}O$
− 30°	...	...	...	·27	...	...	...	3·22	...
− 25	...	...	...	·41	...	...	..	4·41	...
− 20	...	·33	·58	·63	...	...	4·73	5·92	6·89
− 15	...	·51	·88	·93	...	...	6·16	7·81	8·93
− 10	...	·65	1·29	1·35	...	...	7·94	10·15	11·47
− 5	...	·91	1·83	1·92	...	...	10·13	13·06	14·61
0	·21	1·27	2·53	2·68	4·19	...	12·79	16·56	18·44
+ 5	...	1·76	3·42	3·69	5·41	...	16·00	20·72	23·09
10	·29	2·42	4·52	5·01	6·92	...	19·85	25·74	28·68
15	...	3·30	5·89	6·71	8·76	...	24·41	31·69	35·36
20	·44	4·45	7·56	8·87	11·00	16·05	29·80	38·70	43·28
25	...	5·94	9·59	11·60	13·69	20·02	36·11	46·91	52·59
30	·69	7·85	12·02	15·00	16·91	24·75	43·46	56·45	63·48
35	...	10·29	14·93	19·20	20·71	30·35	51·97	67·49	76·12
40	1·08	13·37	18·36	24·35	25·17	36·93	61·75	80·19	90·70
45	...	17·22	22·41	30·61	30·38	44·60	72·95	94·73	107·42
50	1·70	21·99	27·14	38·17	36·40	53·50	85·71	111·28	126·48
55	...	27·86	32·64	47·22	43·32	63·77	100·16	130·03	148·11
60	2·65	35·02	39·01	57·99	51·22	75·54	116·45	151·19	172·50
65	...	43·69	46·34	70·73		88·97	134·75	174·95	199·89
70	4·06	54·11	54·74	85·71		104·21	155·21	201·51	230·49
75	...	66·55	64·32	103·21		121·42	177·99	231·07	264·54
80	6·13	81·29	75·19	123·85		140·76	203·25	263·86	302·28
85	...	98·64	87·46	147·09		162·41	231·17	300·06	343·95
90	9·06	118·93	101·27	174·17		186·52	261·91	339·89	389·83
95	...	142·51	116·75	205·17		213·28	296·63	383·55	440·18
100	13·11	169·75	134·01	240·51		242·85	332·51	431·23	495·33
105	...	201·04	153·18	280·63		275·40	372·72	483·12	555·62
110	18·6	236·76	174·41	325·96		311·10	416·41	539·40	621·46
115	...	277·34	197·82	376·98		350·10	463·74	600·24	693·33
120	25·7	323·17	223·54	434·18		392·57	514·8	665·80	771·92
125	...	374·69	251·71	498·05		438·66	569·97	736·22	
130	34·9	432·30	282·43	569·13		488·51	629·16	811·65	
135	...	496·42	315·85	647·93		542·25	692·59	892·19	
140	46·4	567·46	352·07	733·71		600·02	760·40	977·96	
145	...	645·81	391·21	830·89		661·92	832·69		
150	60·5	731·84	433·37	936·13		728·06	909·59		
155	68·6	825·92	478·65			798·53			
160	77·5		527·14			873·42			
165	87·13		568·30			952·78			
170	97·54		634·07						

TABLE VIII. (*continued*).—*Steam Pressures (in Centimetres of Mercury).*

Temperature.	Cyanogen Chloride, ClCN	Ethyl Chloride, C_2H_5Cl	Sulphurous Acid, SO_2	Methylic Ether, (C_2H_6O)	Methyl Chloride, CH_3Cl	Ammonia, NH_3	Sulphuretted Hydrogen, H_2S	Carbonic Acid, CO_2	Nitrous Oxide, N_2O
− 30°	6·83	11·02	28·75	57·65	57·90	86·61	...	...	...
− 25	10·34	14·50	37·38	71·61	71·78	110·43	374·93	1300·70	1569·49
− 20	14·82	18·75	47·95	88·20	88·32	139·21	443·85	1514·24	1758·66
− 15	20·36	23·96	60·79	107·77	107·92	173·65	519·65	1758·25	1968·43
− 10	27·05	30·21	76·25	130·66	130·96	214·16	608·46	2034·02	2200·80
− 5	35·02	37·67	94·69	157·25	157·87	262·42	706·60	2344·13	2457·92
0	44·41	46·52	116·51	187·90	189·10	318·33	820·63	2690·66	2742·10
+ 5	55·40	56·93	142·11	222·99	225·11	383·03	949·08	3075·38	3055·86
10	68·19	69·11	171·95	262·90	266·38	457·40	1089·63	3499·86	3401·91
15	83·03	83·26	206·49	307·98	313·41	542·34	1244·79	3964·69	3783·17
20	100·19	99·62	246·20	358·60	366·69	638·78	1415·15	4471·66	4202·79
25	119·98	118·42	291·60	415·10	426·74	747·70	1601·24	5020·73	4664·14
30	142·74	139·90	343·18	477·80	494·05	870·10	1803·53	5611·90	5170·85
35	168·87	164·32	401·48		569·11	1007·02	2022·43	6244·73	5726·81
40	198·80	191·96	467·02			1159·53	2258·25	6918·44	6335·98
45	232·98	223·07	540·35			1328·73	2495·43	7631·46	
50	271·93	257·94	622·00			1515·83	2781·48		
55	316·21	266·84	712·50			1721·98	3069·07		
60	366·42	340·05	812·38			1948·21	3374·02		
65	423·22	387·85	922·14			2196·51	3696·15		
70	457·32	440·50				2467·55	4035·32		
75	559·46	498·27				2763·00			
80		561·41				3084·31			
85		630·16				3433·09			
90		704·75				3810·92			
95		785·39				4219·57			
100		872·28				4660·82			

TABLE IX.—*Emissivity for Heat of Polished ·and Blackened Copper Surfaces**.

Difference of Temperature.	Amount of Heat lost per second per square centimetre of the surface per degree of difference of temperatures.		Ratio.
	Polished Surface.	Blackened Surface.	
5°	·000178	·000252	·707
10	·000186	·000266	·699
15	·000193	·000279	·692
20	·000201	·000289	·695
25	·000207	·000298	·694
30	·000212	·000306	·693
35	·000217	·000313	·693
40	·000220	·000319	·693
45	·000223	·000323	·690
50	·000225	·000326	·690
55	·000226	·000328	·690
60	·000226	·000328	·690

* See Appendix B, below.

TABLE X. *Radiation**.

Quantity of heat, in gramme-water-centigrade units, lost per square centimetre per second from a bright platinum surface, at the temperatures stated, and under a pressure of ·00115 mms. of mercury.

Temperature of surroundings.	Temperature of Platinum wire.	Radiation.
15°	317°	$186·7 \times 10^{-4}$
„	440°	493·5 „
„	628°	1689· „
„	759°	3918· „
„	821°	5163· „
„	858°	6658· „
„	868°	6767· „

* From an investigation now (June, 1887) being made by Mr J. T. Bottomley, the results of which are from time to time being communicated to the Royal Society.

TABLE XI.—*Radiation and Communication to Air**.

Quantity of heat, in gramme-water-centigrade units, lost per square centimetre per second from a bright platinum surface, at a temperature of 412° C. and under the pressures stated.

Temperature of surroundings.	Pressure in millimetres of mercury.	Radiation.
16°	740·	8137×10^{-4}
,,	560·	8004 ,,
,,	440·	7971 ,,
,,	340·	7956 ,,
,,	240·	7941 ,,
,,	140·	7875 ,,
,,	90·	7818 ,,
,,	64·	7686 ,,
,,	52·	7658 ,,
,,	49·	7643 ,,
17·2	34·	7563 ,,
,,	24·	7408 ,,
,,	17·2	7249 ,,
,,	13·2	7104 ,,
,,	5·7	6314 ,,
12·5	2·5	5125 ,,
,,	1·7	4364 ,,
,,	·88	3487 ,,
,,	·444	2683 ,,
15°	·141	1502 ,,
,,	·070	1045 ,,
,,	·053	910·5 ,,
,,	·034	727·3 ,,
,,	·011	539·2 ,,
,,	·0071	459·1 ,,
,,	·0051	436·4 ,,
,,	·00015	392·7 ,,
,,	·00007	378·8 ,,

* From Mr J. T. Bottomley's experiments. See note to Table X. above.

TABLE XII.—*Thermal Conductivities.*

Substances.	Temperatures.	Conductivity. k.	Authority.
Silver	0°	1·0960	H. F. Weber.
Copper (crown).........................	15°	·9996	Tait.
,, C.	,,	·7129	Do.
,, first specimen	,,	·9946	Ångström.
,, second ,, ...	,,	·9599	Do.
,, ...	,,	·7202	Lorenz.
Iron, 1 in. sq. bar.....................	,,	·1491	Forbes.
,, 1¼ in.sq. bar	,,	·1980	Do.
,, Forbes 1¼ in. bar	,,	·2010	Tait.
,, ...	,,	·1896	Ångström.
,, ...	,,	·1660	Lorenz.
Mercury	0°	·01479	H. F. Weber.
,, ...	50°	·01893	Do.

TABLE XII. (continued)—Thermal Conductivities.

Substances.	Tempera-tures.	Conductivity. k.	Authority.
Phosphor bronze	15°	·4152	Kirchhoff and Hansemann.
Zinc	0°	·3056	H. F. Weber.
,, ...	15°	·2545	Kirchhoff and Hansemann.
		0° 100°	
Aluminium..................	...	·3435 ·3619	
Antimony	...	·0442 ·0396	
Bismuth.....................	...	·0177 ·0164	
Brass, red	...	·2460 ·2827	
,, yellow..................	...	·2041 ·2540	
Cadmium	...	·2200 ·2045	} Lorenz.
German Silver	...	·0700 ·0887	
Lead..........................	...	·0836 ·0764	
Magnesium..................	...	... ·3760	
Tin	...	·1528 ·1423	
Air..........................			
Oxygen			Clausius and Maxwell, ac-
Nitrogen	...	·000049	cording to kinetic theory.
Carbonic oxide			
Carbonic acid	...	·000038	Do. do.
Hydrogen	...	·00034	Do. do.
Oxygen	7° to 8°	·0000563	Winkelmann.
Nitrogen.....................	7° to 8°	·0000524	Do.
Carbonic oxide	0°	·0000499	Do.
,, ,, ...	7° to 8°	·0000510	Do.
Carbonic acid	0°	·0000305	Do.
,, ,, ...	100°	·0000466	Do.
Hydrogen	7° to 8°	·0003324	Do.
,, ...	0°	·0003190	Grätz.
,, ...	100°	·0003693	Do.
Air	0°	·0000492	Kundt and Warburg.
,, ...	0°	·0000513	Winkelmann.
,, ...	0°	·00004838	Grätz.
,, ...	100°	·0000653	Winkelmann.
,, ...	100°	·00005734	Grätz.
Stone, porphyritic trachyte	...	·00590	Ayrton and Perry, Phil. Mag., 1878.
Underground strata	...	·005	Forbes and Wm. Thomson.
Sandstone, Craigleith	...	·01068	Do. do.
Traprock of Calton Hill ...	...	·00415	Do. do.
Sand of Exper. Garden ...	...	·00262	Do. do.
Water.........................	...	·002	J. T. Bottomley.
Ice	...	·00568	Neumann.
Fir, across fibres	...	·00026	Péclet, in Everett's Units and Physical Constants.
,, along fibres.............	...	·00047	Do. do.
Walnut, across fibres	...	·00029	Do. do.
,, along fibres.........	...	·00048	Do. do.
Oak, across fibres	...	·00059	Do. do.
Cork	...	·000397	Do. do.
Hempen cloth, new	...	·000144	Do. do.
,, ,, old	...	·000119	Do. do.
Writing paper, white	...	·000119	Do. do.
Grey paper, unsized.........	...	·000094	Do. do.
Calico (new), all densities..	...	·000139	Do. do.
Wool (carded), all densities	...	·000122	Do. do.
Finely carded cotton wool..	...	·000111	Do. do.
Eider down	...	·000108	Do. do.
Glass	...	·0021	Péclet.
,, Ger. and Eng. plated	...	·00198	Profs. Herschel and Le-brun, and J. F. Dunn.
,, ,, ,,	...	·00234	
,, Ger. toughened	...	·00185	

TABLE XIII.—*Diffusivities (Thermal, Laminar-motional, Material, and Electric).*

Substance.	Thermal Conductivity. k.	Thermal Capacity of Unit Bulk. c.	Diffusivity, in square centi-metres per second. k/c.	Authority.
Copper	·99	·845	1·18	Ångström & Tait.
Iron	·20	·886	·225	Ångström & Tait.
Air				
Oxygen				
Nitrogen	·000049	·000307	·16	Clausius and Max-well, according to kinetic theory.
Carbonic oxide				
Carbonic acid	·000038	·000428	·0888	
Hydrogen	·00034	·000307	1·12	
Underground strata (rough average)	·005	·5	·01	Forbes and W. Thomson.
Wood	·0005	·39	·013	
Water	·002	1·00	·002	J. T. Bottomley.
Traprock of Calton Hill	·00415	·5283	·00786	
Sand of Experimen-tal Garden	·00262	·3006	·00872	Forbes and W. Thomson.
Sandstone of Craig-leith Quarry	·01068	·4623	·02311	
Gravel of Greenwich Observatory Hill	...	...	·01249	Everett, *Units and Phys. Constants,* 1886 Ed., p. 110.
Stone, porphyritic trachyte	·00590	·5738	·01028	Ayrton and Perry, *Phil. Mag.,* 1878, first half-year, p. 241.

Laminar-motional Diffusivities.

Laminar motion through air	·053	Stokes & Maxwell.
Laminar motion through water at 10° C.	·0132	Poiseuille, Stokes.

Material Diffusivities.

Common salt, through water	·0000116	Fick.
Cane sugar, through water	·00000365	Voit.
Caramel, through water at 10° C.	·00000054	Prof. Mach of Prague, *Imperial Academy of Sciences of Vienna,* January 13, 1879.
Albumen, through water at 13° C.	·00000073	
Cane sugar, through water at 9° C.	·0000036	
Water and common salt, through water at 5° C.	·00000885	
Do. do. at 9° C.	·0000106	
Hydrochloric acid, through water	·0000202	
Potassium chloride, 10 % solution through water	·0000127	T. Schuhmeister, *Wiener Sitzungs-berichte,* April 3, 1879.
Potassium bromide, ,, ,,	·0000131	
Potassium iodide, ,, ,,	·0000130	
Potassium nitrate, ,, ,,	·00000926	
Potassium carbonate, ,, ,,	·00000694	
Potassium sulphate, ,, ,,	·00000868	
Ammonium chloride, ,, ,,	·0000120	
Sodium bromide, ,, ,,	·00000995	
Sodium iodide, ,, ,,	·00000926	
Sodium chloride, ,, ,,	·00000972	
Sodium nitrate, ,, ,,	·00000694	
Sodium carbonate, ,, ,,	·00000463	

TABLE XIII. *(continued).—Diffusivities (Thermal, Laminar-motional, Material, and Electric).*

Material Diffusivities (continued).	Diffusivity, in square centimetres per second. k/c.	Authority.
Sodium sulphate, 10°/₀ solution through water...	·00000764	
Lithium bromide, „ „ ...	·00000926	
Lithium iodide, „ „ ...	·00000926	
Lithium chloride, „ „ ...	·00000810	
Calcium chloride, „ „ ...	·00000787	T. Schuhmeister.
Cupric chloride, „ „ ...	·00000498	
Copper sulphate, „ „ ...	·00000243	
Zinc sulphate, „ „ ...	·00000231	
Magnesium sulphate, „ „ ..	·00000324	
Carbonic acid and air	·1423	
Carbonic acid and hydrogen	·5556	
Oxygen and hydrogen	·7214	
Carbonic acid and oxygen	·1409	Loschmidt, *Im-*
Carbonic acid and carbonic oxide	·1406	*perial Academy*
Carbonic acid and marsh gas	·1586	*of Sciences of*
Carbonic acid and nitrous oxide....................	·0983	*Vienna,* 10th
Sulphurous acid and hydrogen	·4800	March, 1870.
Oxygen and carbonic oxide	·1802	
Carbonic oxide and hydrogen	·6422	

Electric Cable Diffusivities.

Name of Cable.	Resistance per Nautical Mile in Ohms[1].	Reciprocal of the Resistance per Centimetre of length in electro-magnetic C.G.S. Measure. k.	Electrostatic Capacity per Nautical Mile in Microfarads[2].	Electrostatic Capacity per Centimetre of length in electro-magnetic C.G.S. Measure. c.	Diffusivity, in square centimetres per second. k/c.	
Suez-Aden......	10·42	·0000178	·3580	$1·932 \times 10^{-21}$	$·922 \times 10^{16}$	
Aden-Bombay.	7·02	·0000264	·3610	$1·955 \times 10^{-21}$	$1·356 \times 10^{16}$	
Persian Gulf Cable, 1864...	6·25	·0000296	·3486	$1·882 \times 10^{-21}$	$1·578 \times 10^{16}$	
Atlantic, 1865.	4·27	·0000434	·3535	$1·908 \times 10^{-21}$	$2·274 \times 10^{16}$	
French Atlantic, 1869	3·16	·0000586	·4295	$2·318 \times 10^{-21}$	$2·528 \times 10^{16}$	
Direct United States Cable.	2·88	·0000643	·4095	$2·211 \times 10^{-21}$	$2·909 \times 10^{16}$	

[1] The ohm is equal to 10^9 electro-magnetic C.G.S. units.
[2] One microfarad is one-millionth of the farad, or British Association unit of electrostatic capacity, and is equal to 10^{-15} C.G.S. electro-magnetic units of capacity.

TABLE XIV.—*Diffusivities of Non-magnetic Metals for Electric Current**.

Specific resistance (C. G. S.) is the resistance, in centimetres per second, of one centimetre length of conductor having its cross-sectional area equal to one square centimetre. Specific resistance (C. G. S.) divided by 4π is equal to the diffusivity in square centimetres per second.

Substance.	Diffusivity for electric current, in square centimetres per second.
Copper . . .	128
Lead . . .	1550
German Silver .	1652
Platinoid .	2706

TABLE XV.   *Melting Points.*

	C°.		C°.
Aluminium . .	850°	Lead	326°
Ammonia . .	− 75	Manganese . .	1500
Antimony . .	432	Mercury . . .	− 39
Bismuth . .	268	Nickel . . .	1450
Brass . . .	1015	Palladium . .	1500
Britannia Metal .	250	Phosphorus (yellow) .	44
Bromine . .	− 7	,, ,, (red) .	255
Cadmium . .	321	Platinum . . .	1775
Cobalt . .	1500	Potassium . .	62
Copper . .	1054	Salt, NaCl . .	772
Gold . . .	1045	Selenium, Vitreous .	104
Iodine . .	115	,, Crystalline .	217
Iridium . .	1950	Silver . . .	954
Iron Cast (White) .	1100	Sodium . . .	95
,, ,, (Grey) .	1200	Sulphur, Rhombic .	114
,, Hammered .	1600	,, Prismatic .	120
,, Pure . .	1587	,, Amorphous .	120
,, Wrought .	2000	Tin	233
,, Steel . .	1400	Zinc	433
,, ,, Cast .	1375	Water . . .	0

* See, on this subject, Art. XCVI. at the end of the present volume.

TABLE XVI. *Boiling Points.*

		Boiling Points. C°.	Pressure in mms. of Mercury.
Ammonia....................	H_3N	− 33°·7	749
Bromine	Br	+ 59·27	760
Carbon Monoxide	CO	−193·	760
„ Dioxide	CO_2	− 78·2	760
Chlorine	Cl	− 33·6	760
„ Petroxide......	Cl_2O_4	+ 9·9	731
Hydrochloric acid (abt.)	HCl	−102·0	
Hydrogen..................	H	−200·	
„ Sulphide......	H_2S	− 61·8	760
Iodine	I	+250·	
Mercury	Hg	+357·25	760
Nitrogen	N	−193·1	
„ Monoxide......	N_2O	− 92·	
Oxygen....................	O	−184·	760
Phosphorus (yellow) ...	P	+287·3	762
Potassium	K	+725·	
Selenium	Se	+666·	
Sodium (about)	Na	+900·	
Sulphur....................	S	+448·4	760
Sulphuric acid............	H_2SO_4	+325·	760
Sulphurous Anhydride..	SO_2	− 10·5	744
Sulphur Trioxide.........	SO_3	+ 46·	760
Zinc Chloride	$ZnCl_2$	+680·	

TABLE XVII. *Boiling Points.*

		Boiling Point C°.	Pressure in mms. of Mercury.
Chloride of Ethyl	C_2H_5Cl	12°·5	760
Aldehyd..............................	C_2H_4O	20·8	759
Methyl Formiate	$C_2H_4O_2$	32·3	760
Ether	$(C_2H_5)_2O$	34·97	760
Bromide of Ethyl	C_2H_5Br	38·75	760
Carbon di-sulphide	CS_2	46·2	760
Formic Ether	$C_3H_6O_2$	54·4	760
Methylic Acetate......................	$CH_3C_2H_3O_2$	56·3	760
Acetone	C_3H_6O	56·53	760
Silicic Chloride	$SiCl_4$	57·74	760
Chloroform	$CHCl_3$	61·2	760
Wood spirit (Methyl Alcohol).........	CH_4O	66·7	753
Ethyl Iodide............................	C_2H_5I	72·34	760
Acetic Ether	$C_4H_8O_2$	74·3	760
Trichloride of Phosphorus	PCl_3	75·95	760
Alcohol (Ethyl)	C_2H_6O	78·4	760
Benzol	C_6H_6	80·4	760
Dutch liquid (Ethylene Dichloride).	$C_2H_4Cl_2$	84·5	760
Water..................................	H_2O	100·0	760
Formic Acid............................	CH_2O_2	101·0	
Methylic Butyrate, normal............	$CH_3C_4H_7O_2$	103·5	760
Stannic Chloride	$SnCl_4$	113·9	760
Methyl Iso-valerate.....................	$CH_3C_5H_9O_2$	117·0	764
Acetic Acid	$HC_2H_3O_2$	117·1	749
Butyric Ether, normal	$C_6H_{12}O_2$	121·07	760
Ethylene Dibromide	$C_2H_4Br_2$	131·45	760
Arsenious Chloride	$AsCl_3$	130·21	760
Titanic Chloride	$TiCl_4$	136·41	760
Fusel Oil, normal	$C_5H_{12}O$	137·0	740
Silicic Bromide.........................	$SiBr_4$	153·6	762
Butyric Acid, normal	$HC_4H_7O_2$	162·32	
Phosphorus Bromide	PBr_3	172·9	760
Oxalic Acid.............................	$C_2H_2O_4$	203·0	
Glycerine...............................	$C_3H_8O_3$	290·0	760
Sulphuric Acid..........................	H_2SO_4	325·0	760

TABLE XVIII. *Boiling Points of Common Oils.*

	C°.		C°.
Oil of Rosemary .	168°	Petroline . .	280°
Hemp oil . .	258	Olive oil . .	315
Sage oil . .	260	Anthracene . .	350
Rose oil . .	274		

TABLE XIX. *Latent Heats of Fusion,*

per unit mass of the substance; in terms of the quantity of heat
required to raise unit mass of water from 0° to 1° C.

Bismuth	. .	12·64	Palladium . .	36·3
Bromine	. .	16·18	Phosphorus . .	5·24
Cadmium	. .	13·66	Platinum . .	27·18
Gallium	. .	19·11	Sea Water . .	54·69
Ice	. . .	79·25	Silver . . .	21·07
Iron, Cast, Grey .		23·0	Sulphur . .	9·37
Iron, Cast, White		33·0	Tin . . .	14·25
Lead . . .		5·37	Zinc . . .	28·13
Mercury	. .	2·83		

TABLE XX. *Latent Heats of Evaporation.*

See Table XIX. above.

Acetic Acid..................	$C_2H_4O_2$	121·0
Acetate Ethyl..............	$C_4H_8O_2$	105·80
„ Methyl	$C_3H_6O_2$	110·20
Alcohol, Cityl..............	$C_{16}H_{34}O$	58·48
„ Ethyl	C_2H_6O	208·92
„ Methyl	CH_4O	263·86
Ammonia	NH_3	297·38
Arsenious Chloride........	$AsCl_3$	53·0
Bromine	Br	45·60
Carbon dioxide	CO_2	49·32
„ disulphide	CS_2	86·67
Ethyl Bromide	C_2H_5Br	61·65
„ Chloride	C_2H_5Cl	89·30
„ Oxide (Ether)	$(C_2H_5)_2O$	91·0
Formiate, Ethyl	$C_3H_6O_2$	105·3
„ Methyl	$C_2H_4O_2$	117·1
Formic Acid	CH_2O_2	120·7
Iodine	I	23·95
Mercury	Hg	62·0
Nitric Acid.................	HNO_3	115·08
„ Anhydride	N_2O_5	44·81
Nitrous Oxide.............	N_2O	100·6
Phosphorus Chloride	PCl_3	51·42
Sulphur	S	362·0
„ Chloride	S_2Cl_2	49·37
„ Dioxide...........	SO_2	94·56
Sulphuric Acid	H_2SO_4	122·1
„ Anhydride	SO_3	147·5
Tin Chloride	S_nCl_4	30·53
Water	H_2O	$\begin{cases} 535·9 \\ 536·5 \end{cases}$

TABLE XXI. *Specific Heats of Solids.*

Substances.	Specific Heat.	At. Weight.	Sp. Ht. × At. Wt.
Aluminium	·2122	27·0	5·7
Antimony	·0507	119·6	6·1
Arsenic	·0814	74·9	6·1
Bismuth.............	·0305	210·	6·4
Brass...................	·0939		
Bromine.............	·0843	79·76	6·7
Cadmium	·0548	111·6	6·1
Copper	·0950	63·	6·2
Glass	·198		
Gold..................	·0324	196·2	6·3
Graphite (about) ...	·310		
Ice	·504		
Iodine	·0541	126·54	6·8
Iridium	·0323	196·7	6·3
Iron	·1124	55·9	6·3
Lead..................	·0315	206·4	6·5
Magnesium	·245	23·94	5·9
Marble	·216		
Mercury	·03192	199·8	6·4
Nickel	·1092	58·6	6·4
Palladium	·0592	106·2	6·3
Phosphorus (yellow)	·1699	30·96	5·3
Platinum	·0324	194·3	6·4
Platinum-iridium...	·0410		
Potassium	·1655	39·03	6·4
Quartz	·19		
Silver.................	·0599	107·66	6·4
Sodium...............	·2934	22·99	6·7
Steel..................	·118		
Sulphur	·1844	31·98	5·9
Tin	·0559	117·35	6·6
Zinc	·0935	64·9	6·0

TABLE XXII. *Specific Heats—Liquids.*

Substances.	Specific Heat.	Temp. C°.
Acid, Acetic	·4599	10° to 15°
,, Hydrochloric		
(+ 10H$_2$O)	·749	18°
,, Nitric (,,)	·768	18°
,, Sulphuric	·3315	16° to 20°
,, ,, (+ 5H$_2$O)	·5764	16° to 20°
Alcohol, Amylic	·564	26° to 44°
,, Ethylic..........	·615	30°
,, ,, ...	·505	− 20°
,, Methylic.........	·601	15° to 20°
Ethyl, Oxide of...........	·5113	− 30°
,, ,, ...	·5467	+ 30
,, Chloride of.......	·4276	− 28° and 4°
,, Bromide of........	·2164	5° to 10°
,, Acetate...........	·4960	− 30°
,, ,, ...	·5588	30°
Benzol	·4158	19° to 30°
Bisulphide of Carbon ...	·2401	30°
Bromine	·1071	13° to 45°
Chloroform	·2354	30°
Mercury	·0333	5° to 36°
Nitro Benzol..............	·3478	10° to 15°
Paraffin	·683	
Phosphorus (yellow)	·2045	49° to 98°
Sulphur	·2346	119° to 147°
Sulphur Chloride.........	·2024	10° to 15°
Turpentine	·4537	40°

TABLE XXIII. *Specific Heats of Gases.*

Substances.		Pressure constant.	Volume constant.
Acetone	C_3H_6O	·4125	·378
Air........................		·2375	·1684
Alcohol, Ethyl.........	C_2H_6O	·4534	·410
„ Methyl.......	CH_4O	·4580	·395
Ammonia...............	H_3N	·5356	·391
Arsenious Chloride ...	$AsCl_3$	·1122	·101
Benzol	C_6H_6	·3754	·350
Bromine	Br	·0555	·0429
Carbon Monoxide	CO	·2425	·1736
„ Dioxide	CO_2	·2169	·172
„ Disulphide	CS_2	·1596	·131
Chlorine	Cl_2	·1241	·0928
Chloroform	$CHCl_3$	·1489	·140
Dutch liquid	$C_2H_4Cl_2$	·2293	·209
Ether	$C_4H_{10}O$	·4797	·453
„ Acetic	$C_4H_8O_2$	·4008	·378
Ethyl Chloride.........	C_2H_5Cl	·2738	·243
„ Cyanide	C_3H_5N	·4261	
„ Bromide.........	C_2H_5Br	·1896	·171
„ Sulphide	$C_4H_{10}S$	·4008	·379
Hydrochloric Acid	HCl	·1867	·1304
Hydrogen...............	H	3·4090	2·411
„ Sulphide ...	H_2S	·2451	·184
Marsh Gas..............	CH_4	·5929	·468
Nitrogen	N	·2438	·1727
Nitrous Oxide	N_2O	·2262	·181
Nitric Oxide	NO	·2317	·1652
Olefiant Gas............	C_2H_4	·4040	·359
Oxygen..................	O_2	·2175	·1551
Phosphorus Chloride .	PCl_3	·1347	·120
Silicic Chloride	$SiCl_4$	·1322	·120
Stannic „	$SnCl_4$	·0939	·086
Steam	H_2O	·4805	·370
Sulphur Dioxide	SO_2	·1544	·123
Titanic Chloride	$TiCl_4$	·1290	·119
Turpentine	$C_{10}H_{16}$	·5061	·491

TABLE XXIV.   *Specific Heats—Saline Solutions**.

Substances.	Density.	Specific Heat.	Density into Specific Heat.
Copper Sulphate	1·184	·8354	·9891
	1·142	·8788	1·0036
	1·109	·8998	·9979
	1·0871	·9193	·9994
Lead Acetate	1·216	·8319	1·0116
,, Nitrate	1·1334	·8816	·9992
Iron Sulphate	1·1523	·8468	·9758
	1·146	·8814	1·0101
Potassium Bichromate	1·0577	·9474	1·0021
Sodium Carbonate.....	1·0893	·9222	1·0046
	1·080	·9308	1·0053
Sodium Chloride	1·185	·8390	·9942
Zinc Sulphate	1·327	·7526	·9987
	1·258	·7755	·9756
	1·161	·8606	·9992
	1·075	·9230	·9922

* T. Gray, "On the Specific Heats of Saline Solutions." *Proceedings*, Royal Society of Edinburgh, 21st June, 1880.

APPENDIX A.

On the Alteration of Temperature Accompanying Changes of Pressure in Fluids.

[*Proc. Roy. Soc.*, June, 1857; *Phil. Mag.*, June Suppl., 1858.]

Let a mass of fluid, given at a temperature t and under a pressure p, be subjected to the following cycle of four operations in order.

(1) The fluid being protected against gain or loss of heat, let the pressure on it be increased from p to $p + \varpi$.

(2) Let heat be added, and the pressure of the fluid maintained constant at $p + \varpi$, till its temperature rises by dt.

(3) The fluid being again protected against gain or loss of heat, let its pressure be reduced from $p + \varpi$ to p.

(4) Let heat be abstracted, and the pressure maintained at p, till the temperature sinks to t again.

At the end of this cycle of operations, the fluid is again in the same physical condition as it was at the beginning, but, as is shown by the following considerations, a certain transformation of heat into work or the reverse has been effected by means of it.

In two of these four operations the fluid increases in bulk, and in the other two it contracts to an equal extent. If the pressure were uniform during them all, there would be neither gain nor loss of work; but inasmuch as the pressure is greater by ϖ during operation (2) than during operation (4), and rises during (1) by the same amount as it falls during (3), there will, on the whole, be an amount of work equal to $\varpi \, dv$, done by the fluid in expanding, over and above that which is spent on it by pressure from without while it is contracting, if dv denote a certain augmentation of volume which, when ϖ and dt are infinitely small, is infinitely nearly equal to the expansion of the fluid during operation (2), or its contraction during operation (4). Hence, considering the bulk of the fluid primitively operated on as unity, if we take

$$\frac{dv}{dt} = e,$$

to denote an average coefficient of expansion of the fluid under
constant pressure of from p to $p + \varpi$, or simply its coefficient of
expansion at temperature t and pressure p, when we regard ϖ as
infinitely small, we have an amount of work equal to

$$\varpi \, e \, dt$$

gained from the cycle. The case of a fluid such as water
below 4° Cent., which contracts under constant pressure, with an
elevation of temperature, is of course included by admitting
negative values for e, and making the corresponding changes in
the statement.

Since the fluid is restored to its primitive physical condition at
the end of the cycle, the source from which the work thus gained
is drawn, must be heat, and since the operations are each perfectly
reversible, Carnot's principle must hold; that is to say, if θ denote
the excess of temperature of the body while taking in heat above
its temperature while giving out heat, and if μ denote "Carnot's
function," the work gained, per unit of heat taken in at the higher
temperature, must be equal to

$$\mu \, \theta.$$

But while the fluid is giving out heat, that is to say, during
operation (4), its temperature is sinking from $t + dt$ to t, and may
be regarded as being on the average $t + \frac{1}{2}dt$; and while it is taking
in heat, that is, during operation (2), its temperature is rising from
what it was at the end of operation (1) to a temperature higher
by dt, or on the average exceeds by $\frac{1}{2}dt$, the temperature at the
end of operation (1). The average temperature while heat is
taken in consequently exceeds the average temperature while heat
is given out, by just as much as the body rises in temperature
during operation (1). If, therefore, this be denoted by θ, and if
$K \, dt$ denote the quantity of heat taken in during operation (2),
the gain of work from heat in the whole cycle of operations must
be equal to $\mu \, \theta \, K \, dt$, and hence we have

$$\mu \, \theta . K \, dt = \varpi \, e \, dt.$$

From this we find

$$\theta = \frac{e}{\mu K} \varpi,$$

where, according to the notation that has been introduced, θ is the
elevation of temperature consequent on a sudden augmentation of
pressure from p to $p + \varpi$; e is the coefficient of expansion of the

fluid, and K its capacity for heat, under constant pressure; and μ is Carnot's function, being, according to the absolute thermo-dynamic scale of temperature*, simply the reciprocal of the temperature, multiplied by the mechanical equivalent of the thermal unit. If then t denote the absolute temperature, which we have shown by experiment† agrees sensibly with temperature by the air-thermometer Cent. with 274° added, and if J denote the mechanical equivalent of the thermal unit centigrade, we have

$$\theta = \frac{t\,e}{JK}\varpi.$$

This expression agrees in reality, but is somewhat more convenient in form, than that first given, *Dynamical Theory of Heat*, § 49, *Trans. R.S.E.* 1851 [Art. XLVIII. Vol. I. above].

Thus for water the value of K, the thermal capacity of a cubic centimetre under constant pressure, is unity, and e varies from 0 to about 1/2200, for temperatures rising from that of maximum density to 50° Cent., and the elevation of temperature produced by an augmentation of pressure amounting to n times 1033 grammes per square centimetre (that is to say, to n atmospheres), is

$$\theta = \frac{t\,e \times 1033}{42350}\,n;$$

and for mercury, we have, since $K = (13\cdot6 \times \cdot0338 =)\; \cdot459$

$$\theta = \frac{t\,e \times 1033}{19439}\,n.$$

If, as a rough estimate, we take

$$e = \frac{t - 278}{46} \times \frac{1}{2200},$$

the formula becomes

$$\theta = \frac{t\,(t - 278)}{4,150,000}\,n.$$

If, for instance, the temperature be 300° on the absolute scale (that is, 26° C.), we have

$$\theta = \frac{n}{636}.$$

---

* Proposed by Dr Joule and myself in our paper " On the Thermal Effects of Fluids in Motion," *Trans. Roy. Soc.*, 1853, 1854 and 1860 [Art. XLIX. Vol. I. above].

† See Part II. of our Paper "On the Thermal Effects of Fluids in Motion," *Philosophical Transactions*, 1854. [Art. XLIX. Vol. I. above.]

as the heating effect produced by the sudden compression of water at that temperature: so that ten atmospheres of pressure would give 1/64 of a degree cent., or about five divisions on the scale of the most sensitive of the ether thermometers we have as yet had constructed.

Again, for mercury, if we take 1/5500 as the value of e, the formula becomes

$$\theta = \frac{t}{103600}\, n\,;$$

and at temperature 26° cent., the heating effect of ten atmospheres is found to be 1/34 of a degree cent.

TABLE giving the thermal effects of a pressure of ten atmospheres on water and mercury*.

Temperature.	Increase or decrease of temperature in water.		Increase of temperature in mercury.
0°	·00026	decrease	·026
3°·95			·0264
10°	·00040	increase	·027
20°	·00112	do.	·028
30°	·00188	do.	·029
40°	·00269	do.	·030
50°	·00354	do.	·031
60°	·00445	do.	·032
70°	·00540	do.	·033
80°	·00640	do.	·034
90°	·00736	do.	·035
100°	·00855	do.	·036

APPENDIX B.

EXPERIMENTS MADE TO DETERMINE SURFACE CONDUCTIVITY FOR HEAT IN ABSOLUTE MEASURE. By DONALD MACFARLANE.

[*Proc. Roy. Soc. Jan.* 11, 1872.]

The experiments described in this paper were made in the Physical Laboratory of the University of Glasgow, under the direction of Sir William Thomson, during the summer of 1871. A set of similar experiments were made in 1865†; but being merely

* Added August 1, 1858.

† [Note of Aug. 22, 1887. The only data, in absolute measurement, of surface conductivity which I knew to be in existence previous to this date, are referred to in

preliminary, carried on by different individuals, and embracing only a limited range of temperatures, it is thought unnecessary to allude further to them here*.

A copper ball, 2 centimetres radius, having a thermo-electric junction at its centre, was suspended in the interior of a double-walled tin-plate vessel which had the space between the double sides filled with water at the atmospheric temperature, and the interior coated with lamp-black. The other junction was in metallic contact with the outside of the vessel, and the circuit was completed through the coil of a mirror galvanometer. One junction was thus kept at a nearly constant temperature of about 14° Cent., while the other had the gradually diminishing temperature of the ball.

Having adjusted the galvanometer to the degree of sensitiveness desired, the copper ball was heated in the flame of a spirit-lamp till its temperature was considerably above that required to throw the spot of light off the scale; it was then put into position in the interior of the tin-plate vessel, and as soon as the spot of light came within range, the deflections from the zero position were noted at intervals of one minute exactly till the change of deflection was reduced to about two scale-divisions per minute.

Two series of experiments were made in this way, each consisting of several sets of readings. In the first the ball had a bright surface, and in the second it was coated with a thin covering of soot from the flame of a lamp, and in both the air was kept moist by a saucer containing a quantity of water placed in the interior of the tin-plate vessel.

the following question extracted from No. XVII. of the weekly examination papers for the class of Natural Philosophy in the University of Glasgow, session 1864–5: "According to Péclet a warm body with any ordinary non-metallic surface, loses heat from each square centimetre of surface at the rate of $\frac{1}{1000}$ of a thermal unit per second if its temperature is 1° above that of the air and solids around it. Hence show that if a large flat mass of rock of conductivity ·003, is kept 1° warmer than the air by the conduction of heat from within, it must be warmer within than at its surface at the rate of 1° per 12 centimetres inwards." W. T.]

* These experiments consisted of two series, one with the air moist by a little water placed in the interior of the vessel, the other having the air dried by substituting sulphuric acid for water as in the first series; and the results in the two cases were so nearly alike, that any effect due to the moisture or dryness of the air could not be distinguished from errors of observation. From this circumstance, as well as the limited range of temperatures, these results are not given here.

As the range of differences of temperatures of the junctions extended over 50° Cent., the change in the difference of thermo-electric qualities of the copper and iron wires forming the junctions was very considerable, and it was necessary to make a careful thermometric comparison of the temperatures of the junctions and galvanometer deflections. For this purpose the junctions were tied to the bulbs of two previously compared thermometers, having their stems divided to tenths of a degree Cent.; these were then placed in two vessels of water, one at the temperature of the air, and the other heated by small additions of hot water, and kept well stirred; simultaneous readings of the thermometers and gal-vanometer deflections were then taken at various points of the scale*, from which the formula

$$\theta = 0°{\cdot}0924 + 0°{\cdot}0000227x$$

was obtained, where θ is the value of a scale-division in terms of a degree Centigrade, and x the galvanometer deflection; and the difference of temperature of the junctions is therefore

$$\theta x = 0°{\cdot}0924x + 0°{\cdot}0000227x^2,$$

from which the numbers in col. II. of the following Tables were calculated.

The method adopted in reducing the observations was this:— Each single set of readings was arranged in a vertical column, and the whole series placed side by side with corresponding numbers in the same horizontal line; the means of the horizontal lines were formed into a similar column, and divided into groups, each con-sisting of four consecutive numbers, and the means of these groups form the numbers in col. I. of the Tables.

Col. II. contains the differences of the temperatures of the junctions at intervals of four minutes, corresponding to the mean deflections in col. I.

---

* These readings were plotted, and the curve drawn through the points agreed very closely with a portion of a parabolic curve whose equation is

$$\delta = 2{\cdot}4 + 10{\cdot}6x - {\cdot}019x^2,$$

δ denoting the deflections of the galvanometer, and x the difference of tempera-ture; δ is a maximum when $x = \dfrac{10{\cdot}6}{{\cdot}038} = 279°$, and, the colder junction having been at 16° Cent., we get 295° as the neutral point of the specimens of copper and iron wires used—a very close agreement with former observations, considering the great distance of the neutral point from the temperature of the observations.

Col. III. contains the common logarithms of the numbers in col. II.

Col. IV. contains the differences of the successive numbers in col. III. divided by 4.

Col. V. is formed from col. IV., by multiplying by the Napierian logarithm of 10, and is the rate at which the difference of temperature varies per minute.

Col. VI. shows the quantity of heat emitted from the ball in gramme-water units per square centimetre per second per degree of difference of temperatures, and is formed by multiplying the numbers in col. V. by ·009385*, a constant depending on the surface of the ball and its capacity for heat.

Col. VII. shows the mean differences of temperatures corresponding to the numbers in cols. IV., V. and VI.

Col. VIII. the numbers found in col. VI. and VII. were plotted on squared paper, and a mean curve drawn through the points; and, assuming the quantity of heat emitted to be represented by the formula

$$x = a + bt + ct^2,$$

where t is the difference of temperature, the coordinates of the curve were employed to determine a, b, and c; and col. VIII., calculated by the formula, is added to show the degree of approximation to which the results of the experiment are represented by it.

* The surface of the ball was 50·26 sq. centimetres, and its capacity for heat 28·31 gramme-water units. Let x denote the heat emitted per second, per sq. centimetre per degree of difference of temperature, and C the rate at which the difference of temperature varies per minute; then

$$\frac{x \times 60 \times 50·26}{28·31} = C,$$

and therefore $x = ·009385\,C.$

TABLE I. *First Series.*

Atmosphere moist. Copper Ball polished bright.

Means of nine sets of Observations.

I. Mean deflections of nine sets of observations.	II. Difference of temperature, (D).	III. $\log_{10} D$.	IV. $\dfrac{\log_{10} D' - \log_{10} D''}{4}$.	V. $\dfrac{\log_e D - \log_e D''}{4}$, ($C$).	VI. Emissivities by observation.	VII. Mean difference of temperature, $\frac{1}{2}[D'+D'']$, (t).	VIII. Emissivities calculated by formula, (x).
627·19	66·88	·82527	·01037	·02387	·000223	63·83	·000230
576·28	60·79	·78380	·01045	·02406	·000226	58·00	·000227
528·82	55·21	·74200	·01055	·02429	·000227	52·65	·000226
484·50	50·10	·69981	·01041	·02396	·000225	47·81	·000224
444·14	45·51	·65815	·01033	·02378	·000223	43·45	·000222
410·15	41·38	·61681	·01028	·02366	·000222	39·51	·000220
373·37	37·64	·57570	·01015	·02337	·000219	35·96	·000217
342·26	34·28	·53509	·01020	·02348	·000220	32·75	·000215
313·64	31·21	·49429	·00992	·02284	·000214	29·85	·000213
287·89	28·48	·45459	·00982	·02261	·000213	27·25	·000210
264·42	26·02	·41529	·00943	·02171	·000212	24·94	·000208
243·58	23·85	·37756	·00962	·02215	·000204	22·84	·000205
223·96	21·83	·33907	·00945	·02175	·000208	20·92	·000202
206·15	20·01	·30127	·00915	·02106	·000204	19·20	·000200
190·20	18·39	·26467	·00905	·02083	·000198	17·62	·000198
175·57	16·92	·22845	·00909	·02093	·000195	16·24	·000196
161·99	15·56	·19207	·00901	·02074	·000196	14·94	·000194
149·51	14·32	·15603			·000195		

Formula for calculating column VIII.:—

$$x = ·000168 + ·00000198\,t - ·000000017\,t^{2}.$$

TABLE II. *Second Series.*
Atmosphere moist. Copper Ball blackened.
Ten sets of Observations.

I. Mean deflections of ten sets of observations.	II. Difference of temperature, (D).	III. $\log_{10} D$.	IV. $\dfrac{\log_{10} D' - \log_{10} D''}{4}$	V. $\dfrac{\log_e D' - \log_e D''}{4}$, (C).	VI. Emissivities by observation.	VII. Mean difference of temperature, $\frac{1}{4}[D'+D'']$, (t).	VIII. Emissivities calculated by formula, (x).
631·85	67·44	·82893	·01511	·03478	·000326	63·06	·000328
558·46	58·68	·76849	·01510	·03477	·000326	54·87	·000328
492·92	51·06	·70808	·01488	·03426	·000322	47·79	·000325
435·27	44·52	·64856	·01476	·03399	·000319	41·69	·000321
384·29	38·86	·58950	·01460	·03362	·000315	36·42	·000315
339·49	33·98	·53122	·01433	·03300	·000310	31·88	·000311
300·20	29·78	·47392	·01391	·03202	·000301	27·99	·000303
266·11	26·20	·41830	·01377	·03170	·000297	24·64	·000298
236·12	23·08	·36324	·01358	·03126	·000293	21·72	·000292
209·65	20·87	·30999	·01325	·03050	·000286	19·20	·000287
186·59	18·03	·25600	·01311	·03018	·000283	17·00	·000283
166·14	15·98	·20358	·01290	·02970	·000279	15·09	·000279
148·16	14·19	·15198	·01258	·02896	·000272	13·41	·000274
132·46	12·64	·10165	·01265	·02902	·000272	11·94	·000271
118·29	11·25	·05104	·01230	·02832	·000266	10·70	·000268
105·97	10·04	·00186	·01286	·02846	·000267	9·50	·000265
94·79	8·96	·95240	·01261	·02903	·000272	8·47	·000262
84·59	7·98	·90195					

Formula for calculating column VIII.:—

$$x = ·000238 + ·000005306t - ·00000026t^{3}.$$

The following Table gives the results calculated by the formula for every fifth degree within the limits of the experiments*:—

TABLE III.

Difference of temperature.	Emissivity.		Ratio of emissive power of polished to that of blackened surface.
	Polished surface.	Blackened surface.	
$\overset{\circ}{5}$	·000178	·000252	·707
10	·000186	·000266	·699
15	·000193	·000279	·692
20	·000201	·000289	·695
25	·000207	·000298	·694
30	·000212	·000306	·693
35	·000217	·000313	·693
40	·000220	·000319	·693
45	·000223	·000323	·690
50	·000225	·000326	·690
55	·000226	·000328	·690
60	·000226	·000328	·690

APPENDIX C.

NOTE ON DULONG AND PETIT'S LAW OF COOLING.   By DONALD MACFARLANE.

[*Proc. Roy. Soc.* June 10, 1875.]

THE *Journal de Physique* for December 1873 contains a friendly notice by Professor A. Cornu of experiments made to determine surface-conductivity for heat (or, as we may call it, "thermal emissivity") in absolute measure, an account of which was communicated to the Royal Society, and read January 1872 (see Appendix B above). On the results there given M. Cornu remarks:—

"Ces nombres vérifient la conclusion de Dulong et Petit, à savoir que les vitesses de refroidissement ne dépendent de l'état des surfaces que par une constante de proportionalité.

---

* The results here given in Table III. agree as closely as might be expected with the results obtained by Mr J. P. Nichol from experiments made in the Natural Philosophy Laboratory of the University of Edinburgh, and described by Prof. Tait in "Notes" read before the Royal Society of Edinburgh, June 6, 1870 (*Proc. R. S. E.* Vol. VII. p. 206).

" L'accélération négative du rapport des pouvoirs émissifs n'infirme pas sensiblement cette conclusion; elle est si faible qu'elle peut être attribuée à une petite erreur régulière dans l'évaluation des différences de température; en effet, l'auteur ne paraît tenir aucun compte d'une cause délicate d'erreur qui avait préoccupé Dulong et Petit, à savoir la résistance inégale à la transmission de la chaleur dans les deux cas. Il est évident que, dans le refroidissement le plus rapide, la température est distribuée moins uniformément que dans le cas d'un refroidissement lent ; l'aiguille thermoélectrique indique donc moins bien la température moyenne de la masse que les boules de mercure des physiciens français."

On this it is to be remarked that a rigorous proportionality in the rates of cooling of different surfaces is in itself not probable; and my experiments in fact disprove it, so far as it is not at all likely that the errors of observation could be so great or so consistently regular in the same direction as the truth of the supposed law would require.

As to the variation of temperature from centre to surface occasioned by the rapid cooling of the ball, this was certainly not overlooked in planning the experiments. Sir William Thomson considered the matter carefully, and selected copper, on account of its high conductivity, estimating that in a copper ball of the dimensions used (diameter 4 centimetres) the temperature must be sensibly uniform throughout. A very simple calculation (made in consequence of M. Cornu's criticism, and appended below) from Fourier's celebrated formula for the cooling of a homogeneous solid globe shows, in fact, that, in the case of a copper globe of 2 centimetres radius, the centre is warmer than the surface by only about 1/4000 of the excess of its temperature above that of the surrounding medium. There would be a much greater difference of temperature between surface and centre in a globe of mercury of the same dimensions, because mercury is a much worse conductor of heat than copper, and because a much greater difference of temperatures than that which there is in the copper would be required to produce any considerable convection of heat by currents in the liquid. Moreover the glass envelope containing the mercury in a thermometer-bulb of ordinary dimensions produces a sensible difference of temperature between the outer surface of the glass exposed to the external medium and the surface of the mercury. For let b be the thickness of the glass, E the "emissivity" of its outer surface, and k the conductivity of its substance; let the excess of temperature of the outer surface of the glass above that of the surrounding medium be v, and the excess of temperature of the inner surface of the glass above the outer δv; we have

$$k\,\frac{\delta v}{b} = Ev.$$

Now by the Glasgow experiments it has been found that E is approximately $\frac{1}{4000}$ of a gramme-water thermal unit per square centimetre per second; and by the determinations of conductivities of stones and underground strata in absolute measure by Peclet and Forbes the value of k for glass may be roughly estimated at $\frac{1}{400}$, in terms of centimetre, second, and gramme-water thermal unit. Hence

$$\frac{\delta v}{v} = \frac{1}{10}\,b.$$

Thus, if the thickness of the glass be half a millimetre (i.e. $b = \frac{1}{20}$), we have

$$\delta v = \frac{1}{200}\,v.$$

This is a small difference, but by no means imperceptible in the delicate experiments of Dulong and Petit; and it is twenty times the difference of temperature between the centre and surface of the cooling copper globe of 4 centimetres diameter.

APPENDIX.

Distribution of temperature in a cooling copper globe of 4 centimetres diameter, calculated from Fourier's formula

$$v = \Sigma P_i \frac{\sin \theta_i}{\theta_i}\,\epsilon^{-\rho_i t} \quad\dots\dots\dots\dots\dots\dots(1),$$

where

$$\rho_i = \frac{\varpi_i^2 k}{a^2 c}, \quad \theta_i = \frac{x}{a}\,\varpi_i,$$

ϖ_i roots of the transcendental equation

$$\frac{\varpi}{\tan \varpi} = 1 - \frac{Ea}{k}\dots\dots\dots\dots\dots\dots\dots(2),$$

and P_i coefficients determined to give (according to Fourier's method) any arbitrary function of x from $x = 0$ to $x = a$, for the value of v, when $t = 0$:

v temperature at time t and distance x from centre of globe,

a the radius of the globe,

k the thermal conductivity of its substance,

c the thermal capacity per unit volume of its substance,

E the thermal emissivity of its surface.

Taking the centimetre, second, and gramme-water thermal units for the fundamental units, we have, as stated above,

$$E = \frac{1}{4000} \text{ (rough approximation)};$$

and Ångström's experiments gave for copper

$$k = 1 \text{ approximately.}$$

Therefore

$$\frac{Ea}{k} = \frac{a}{4000};$$

and for the globe of 4 centimetres diameter used in the Glasgow experiments,

$$\frac{Ea}{k} = \frac{1}{2000}.$$

In all cases in which $\dfrac{Ea}{k}$ is small, the smallest root of the transcendental equation (2) is approximately equal to

$$\sqrt{\frac{3Ea}{k}}$$

Calling this ϖ_1, we have

$$\rho_1 = \frac{3E}{ac}$$

and

$$\frac{\sin \theta_1}{\theta_1} = 1 - \frac{1}{6}\theta_1^2 \text{ approximately,}$$

$$= 1 - \frac{1}{2}\frac{Ea}{k}\frac{x^2}{a^2}.$$

Now any chosen term of (1) is a particular solution of the problem; that is to say, it is *the* solution for the case for which the initial distribution of temperature is that which it expresses when $t = 0$. Hence

$$v = \left(1 - \frac{1}{2}\frac{Ea}{k}\frac{x^2}{a^2}\right)\epsilon^{-\frac{3E}{ac}t}$$

expresses the temperature at time t, if when $t = 0$ the temperature is expressed by

$$v = 1 - \frac{1}{2}\frac{Ea}{k}\frac{x^2}{a^2}.$$

Taking, for instance, the copper globe of 4 centimetres diameter, we have

$$v = \left(1 - \frac{1}{4000}\frac{x^2}{a^2}\right)\epsilon^{-\frac{3E}{ac}t} \quad \ldots\ldots\ldots\ldots\ldots(3);$$

and we see that in the Glasgow experiments the difference of temperatures between surface and centre was just $\frac{1}{4000}$ of the excess of either above the temperature of the surrounding medium, when time enough had elapsed to allow the first term of Fourier's series to be the predominating one. *Before* that time the difference of temperatures must have been *less* than 1/4000 of either, if initially the temperature was uniform from surface to centre. The Fourier analysis of the transition from the supposed initial uniform distribution to the state represented by (3) is exceedingly interesting, but unnecessary for the settlement of the present question.

APPENDIX D.

APPROXIMATE PHOTOMETRIC MEASUREMENTS OF MOON, CLOUDY SKY, AND ELECTRIC AND OTHER ARTIFICIAL LIGHTS.

[Read before the Philosophical Society of Glasgow, 29th November, 1882 (*Proc.* Vol. XIV. p. 80).]

SIR WILLIAM THOMSON pointed out that the light and heat perceived in the radiations from hot bodies were but the different modes in which the energy of vibration induced by the heat was conveyed to our consciousness. A hot kettle, red-hot iron, incandescent iron, or platinum, or carbon, the incandescence in the electric arc, all radiate energy in the same manner, and according as it is perceived through the sense of sight, by its organ the eye, or by the sense of heat[*], we speak of it as light or heat. When

---

[*] Sometimes wrongly called the sense of touch. The true list of the senses, first given, I believe, by Dr Thos. Reid (born 1710, and Professor of Moral Philosophy in the University of Glasgow from 1764 to 1780), makes two of what used to be called the sense of touch, so that, instead of the still too common wrong-reckon-

the period of vibration is longer than one four-hundred-million-millionth of a second, the radiation can only be perceived by the sense of heat; when the period of vibration is shorter than one four-hundred-million-millionth of a second, and longer than one eight-hundred-million-millionth of a second, the radiation is perceived as light, by the eye.

Pouillet, from a series of experiments, deduced a value of the energy radiated by the sun, equal, in British units, to about 86 foot-pounds per second per square foot of the earth's surface*, or about 1 horse-power to every 6½ square feet at the earth's surface. We may estimate from this the value of the solar radiation at the surface of the sun. The sun is merely an incandescent molten mass losing heat by radiation, and surrounded by an atmosphere of incandescent vapour, so that the radiant energy really comes

ing of five senses, we have six, as follows :—

Sense of Force.	Sense of Light.
„ Heat.	„ Taste.
„ Sound.	„ Smell.

The sense of force is the department left to the sense of touch, when the sense of heat or of temperature is taken from it. The sense of touch, other than temperature, has sometimes been, not very judiciously or logically, called "Tactile sense," those who have so called it not having noticed that "tactile" is merely the adjective of or belonging to touch. Physiologists have justly objected to the name "muscular sense of touch," by which, I believe, Thomas Reid himself, and certainly some of his successors in Glasgow, who, teaching his philosophy in this matter, have designated the sense of touch other than temperature. The perception of roughness, as distinguished from smoothness (when we touch a piece of sandstone or of loaf sugar, and compare the sensation with what we perceive when we touch a piece of glass), would not be regarded by physiologists as a muscular perception. But it is a sense of force and of places of application of force which, in the case of touching a rough body, are the little areas of greater pressure distributed among area of less pressure or of no pressure. When we perceive resistance to our two hands pressing on solid matter, or holding up weights, we perceive force and places of application of force; the places of application of the force being the surfaces of the two hands. In respect to this case of the perception of force, physiologists would no doubt accept Thomas Reid's name of "muscular sense." But whether the person perceiving the force is conscious of a sensation in the muscles of his arms, or, as in the case of perceiving roughness, is merely conscious of a sensation in the sensitive material of his fingers, or is conscious of roughness in the tips of his fingers, and of muscular stress in the muscles of his fingers, the thing perceived is still force and places of application of force. The simplest proper name of the sense to which these perceptions belong is the sense of "force."

* More recent observers have found larger and larger results. Langley (*Comptes Rendus*, 11th Sept., 1882, and *American Journal of Science*, vol. xxv., March, 1883) finds 1·7 times the value given by Pouillet, and quotes MM. Soret, and Crova and Violle, who found respectively 1·3 and 1·5 times Pouillet's result.

out from any square foot or square mile of the sun's surface, as from a pit of luminous fluid which we cannot distinguish as either gaseous or liquid. Take, however, instead of the sun, an ideal radiating surface of a solid globe of 440,000 miles radius. The distance of the earth being taken as 93 million miles, the radius of the sun is equal to, say in round numbers, one two-hundredth of the earth's distance; hence the area at the earth's distance corresponding to one square foot of the sun's surface, is equal to 40,000 square feet. The radiation on this surface is (40,000 × 86, or) 3,440,000 foot-pounds, which is therefore the amount of radiation from each square foot of the sun's surface. This amounts to about 7,000 horse-power, which, according to our brain-wasting British measure, we must divide by 144, if we wish to know the radiation per square inch of the sun's surface, which we thus find to be about 50 horse-power.

The normal current through a Swan lamp giving a 20-candle light is equal to 1·4 amperes with a potential of 43 to 45 volts. Hence the activity of the electric working in the filament is 61·6 ampere-volts or watts (according to Dr Siemens' happy designation of the name of Watt, to represent the unit of activity constituted by the ampere-volt). To reduce this to horse-power we must divide by 746, and we thus find about 1-12th of a horse-power for the electric activity in a Swan lamp. The filament is $3\frac{1}{2}$ inches long, and ·01 of an inch in diameter of circular section; the area of the surface is thus 1-9th of a square inch, and therefore the activity is at the rate of 3-4ths of a horse-power per square inch. Hence the activity of the sun's radiation is about sixty-seven times greater than that of a Swan lamp per equal area, when incandesced to 240 candles per horse-power.

In this country, the standard light to which photometric measurements are referred, is that obtained from what is known as a "standard candle." Latterly, however, objections have been raised against its accuracy. It has been said that differences of as much as 14 per cent. have been found in the intensity of the light given by different standard candles, and that serious differences have been observed, in the intensity of the light from different parts of the same candle, in the course of its burning. The Carcel lamp, the standard in use in France, has been regarded as the only reliable standard. It is, no doubt, very reliable and accurate in its indications; but it should be remembered that its accuracy is greatly owing to the careful method and the laborious precautions taken to secure accuracy. If something akin to the precautions applied to the Carcel lamp by Regnault and Dumas, were applied to the production and use of the standard candle, there is little doubt but that sufficient accuracy for most practical purposes could also be obtained with it; probably as good results as are already obtained by the use of the Carcel lamp.

At the Conference on Electrical Units which met in Paris lately (Oct. 1882), a suggestion was made to use as a standard for photometric measurements the incandescence of melting platinum, and very interesting results and methods in connection with the proposal were presented to the meeting. According to experiments by M. Violle, which M. Dumas reported to the Conference, a square centimetre of liquid platinum at the melting temperature gives of yellow light seven times, and of violet light twelve times, the quantities of the same colours given by a Carcel lamp. The apparent area of the Swan filament, being one-ninth of a square inch, is ·23 of a square centimetre, and when incandesced to 20 candles must be about as bright as the melting platinum of M. Violle's experiment, as the 7 carcels of yellow and 12 of violet must correspond to something like 10 carcels or 85 candles, in the ordinary estimation of illumination by our eyes. The tint of M. Violle's glowing platinum cannot be very different from that of the ordinary Swan lamp incandesced to its "20 candles." Thus, both as to tint and brightness, it appears that melted platinum at its freezing temperature is nearly the same as a carbon filament in vacuum incandesced to 240 candles per horse-power.

For photometric measurements in which the lights compared are nearly enough points, and are sufficiently bright, to give good shadows, a very convenient method is that of Rumford, by a comparison of the shadows cast by the sources of light on a white surface. The apparatus necessary is only a piece of white paper, a small cylindrical body such as a pencil, and a means of measuring distances. Ordinary healthy eyes are usually quite consistent in estimating the strength of shadows, even when the shadows examined are of different colours, and with a reasonable amount of care photometric measurements by this method may be obtained within 2 or 3 per cent. of accuracy. The difference in the colours of the shadows is, of course, due to each shadow being illuminated by the other light.

An observation on moonlight made in this way showed the moonlight at the time and place of the observation (at York early in September, 1881, about midnight, near the time of full moon) to be equal to that of a candle at a distance of 230 centimetres. The moon's distance $(3.8 \times 10^{10}$ cms.) is 1.65×10^8 times the distance of the candle. Hence, ignoring for a moment the loss of moonlight in transmission through the earth's atmosphere, we find $(1.65 \times 10^8)^2$, or twenty-seven thousand million million as the number of candles that must be spread over the moon's earthward hemisphere painted black, to send us as much light as we receive from her. Probably about one and a half times as many candles, or say forty thousand million million would be required, because the absorption by the earth's atmosphere may have stopped about one-third of the light from reaching the place where the observation was made. The moon's diameter is 3.5×10^8 centimetres, and therefore half the area of her surface is 19×10^{16} square centimetres, which is nearly five times forty-thousand million million. Thus it appears that if the hemisphere of the moon facing the earth were painted black and covered with candles standing packed in square order touching one another (being say one candle to every five square centimetres of surface), all burning normally, the light received at the earth would be about the same in quantity, as estimated by our eyes, as it really is. It would have very much the same tint and general appearance as an ordinary theatrical moon, except that it would be brightest at the rim and continuously less bright from the rim to the centre of the circle where the brightness would be least.

The luminous intensity of a cloudy sky he found, about 10 A.M. one day in York during the meeting of the British Association, to be such that light from it through an aperture of one square inch area was equal to about one candle. The colour of its shadow compared with that from a candle was as deep buff yellow to azure blue—the former shadow being illuminated by the candle alone, the latter by the light coming through the inch hole in the window shutter.

Arago[*] has compared the luminous intensity of the sun with that of a candle, and estimates it as equal to about 15,000 times that of a candle flame. This, as will be seen below, is probably less than 1/3 of the truth, for a bright Paris sun.

[*] *Astronomie Populaire*, livre xiv. chap. xxv.

Since the reading of this paper, a hasty comparison of sunlight with a candle, last Friday (December 8) showed, at one o'clock on that day, the sunlight reaching his house in the University to be of such brilliancy that the amount of it coming through a pin-hole in a piece of paper of ·09 of a centimetre diameter, produced an illumination equal to that of 126 candles. This is 6·3 times the 20-candle Swan light, of which the apparent area of incandescent surface is ·23 of a square centimetre, or 3·8 times the area of the pin-hole. Hence the sun's surface, as seen through the atmosphere at the time and place of observation, was 24 times as bright as the Swan carbon when incandesced to 240 candles per horse-power. By cutting a piece of paper of such shape and size as just to eclipse the flame of the candle, and measuring the area of the piece of paper, he found about 2·7 square centimetres as the corresponding area of the flame. This is 420 times the area of the pin-hole, and therefore the intensity of the light from the sun's disc was equal to (126 × 420) about 53,000 times that of a candle-flame. This is more than three times the value found by Arago for the intensity of the light from the sun's disc as compared with that from a candle-flame—so much for a Glasgow December sun! Yet, as we shall see presently, the loss of light in passage through the atmosphere must have been much more than in the measurement of moonlight at York.

The ·09 cm. diameter of the pin-hole, of the Glasgow observation, subtends, at 230 centimetres distance, an angle of 1/2556 of a radian, which is 23·7 times the sun's diameter (1/107 of a radian). But at 230 cms. distance the sunlight through the pin-hole amounted to 126 times the York moonlight (which was one candle at 230 cms. distance). Hence the Glasgow sunlight was [(23·7)² × 126 times or] 71,000* times the York moonlight. Now the moon's apparent area† is 1/193000 of the whole spherical area‡, and

---

* For thorough and accurate investigations regarding the relative brightnesses of Sun, Moon, Planets and Stars;—see F. Zöllner in the Jubelband (1874) of Poggendorf's *Annalen;* also a paper by Ludwig Seidel, "Untersuchungen über die gegenseitige Helligkeit der Fixsterne erster Grösse und über die Extinction des Lichtes in der Atmosphäre. Nebst einem Anhange über die Helligkeit der Sonne verglichen mit Sternen, und über die Lichte reflectirende Kraft der Planeten": *Abhand. der math. phys. Classe der Königlich Bayerischen Akad. der Wissen.* Band 6, Munich, 1852.

† Which is $\pi/(4.109 \cdot 7^2)$.

‡ Which is 4π.

therefore if she were perfectly white, she would, when shining full on the earth, give only 1/193000 of sunlight, in conditions of equal loss by absorption in the atmosphere.

APPENDIX E.

ON THE CONVECTIVE EQUILIBRIUM OF TEMPERATURE IN THE ATMOSPHERE.

[Read before the Literary and Philosophical Society of Manchester, January 21st, 1862, and published in the *Memoirs*, Vol. II. of 3rd Series, p. 125.]

THE particles composing any fluid mass are subject to various changing influences, in particular of pressure, whenever they are moved from one situation to another. In this way they experience changes of temperature altogether independent of the effects produced by the radiation or conduction of heat. When all the parts of a fluid are freely interchanged and not sensibly influenced by radiation and conduction, the temperature of the fluid is said to be in a state of convective equilibrium. The equations of convective equilibrium in the atmosphere are as follows, Π, T, and W denoting the pressure, temperature, and mass per cubic cm. of the air at the earth's surface, and p, t, and ρ the same qualities of the air at any height x:—

$$\left(\frac{p}{\Pi}\right)^{1-\frac{1}{k}} = \frac{t}{T}* \quad\dots\dots\dots\dots\dots\dots\dots(1),$$

which is the known relation between temperature and pressure;

$$\frac{p}{\Pi} = \left(\frac{\rho}{W}\right)^{k} \quad\dots\dots\dots\dots\dots\dots\dots(2),$$

the deduced relation between pressure and density; and

$$dp = -\rho dx \quad\dots\dots\dots\dots\dots\dots\dots(3),$$

the hydrostatic equation, the variation of gravity at different heights being neglected, and the weight of unit mass (1 gm.) being taken as unit of force. Hence by integration,

$\dfrac{t}{T} = 1 - \dfrac{Wx}{\Pi} \cdot \dfrac{k-1}{k}$, or if, for brevity, we denote $\dfrac{\Pi}{W}$ by H,

$$\frac{t}{T} = 1 - \frac{x}{H} \cdot \frac{k-1}{k} \quad\dots\dots\dots\dots\dots(4).$$

---

* For proof, see foot-note, p. 258, below. See also Part I. of present article § 74, pp. 65—68.

From (4), (1), and (2), it appears that temperature, pressure, and density would all vanish at the very moderate height $\frac{1\cdot41}{\cdot41} \times H$, which is about 27500 metres (or between 27 and 28 kilometres), if convective equilibrium existed and if the gaseous laws had application to so low temperatures and densities. It has always appeared to me to be most improbable that there is any limit to our atmosphere; and no one can suppose that there is a limit at any height nearly so small as 27 or 28 kilometres. It is difficult to make even a plausible conjecture as to the effects of deviations from the gaseous laws in circumstances of which we know so little as those of air at very low temperatures ; but it seems certain that the other hypothesis involved in the preceding equations is violated by actions tending to *heat* the air in the higher regions.   For at moderate elevations above the surface, where we have air following very strictly the gaseous laws, the rate of decrease of temperature would, according to equation (4), be $\frac{\cdot41 \times T}{1\cdot41\, H}$ per metre, that is to say, $\cdot01°$ per metre, since $H = 7988 \times \frac{T}{274}$ or $1°$ Cent. per 100 metres.   Now, the actual decrease, according to Mr Welsh, is $1°$ Cent. in 161 metres, or not much more than half that according to convective equilibrium.

It seems that radiation, instead of partially accounting for the greater warmth of the air below, as commonly supposed, may actually diminish the cooling effect, in going up, which convection produces.   In fact, since direct conduction is certainly insensible, we have only convection and radiation to deal with, except when condensations of moisture, &c., have to be taken into account.   In fair and cloudless weather, then, the lower and lowest air being on the whole warmer (the lowest being of course at the same temperature as the earth's surface), it is perfectly certain that the upper air must gain heat by radiation from the lower—and that the convective difference of temperature must be diminished by the mutual inter-radiation.

There are difficulties connected with the radiation of heat from air and earth out into space, and from the sun to air and earth ; but I think a full consideration of all the circumstances must explain the smallness of the decrease of temperature which observation shows.

Dr Joule having suggested that condensation of vapour in upward currents of air might account, to a considerable extent if not perfectly, for the smallness of the lowering of temperature actually found in going up, I have added the following investigation, in which the effect of condensation is taken into account.

If a quantity of air, dry or moist, is allowed to expand from bulk v to bulk $v + dv$, it will do an amount of work equal to pdv on the surrounding matter. Now, by the principle established approximately by Dr Joule, in his experiments on air in 1844* the change of temperature which the mass will experience will be almost exactly equal to what would be produced by keeping it at constant volume, $v + dv$, and removing a quantity of heat equal to the thermal equivalent of pdv. This is expressed by $\frac{1}{J}pdv$, if we adopt the usual notation, J, for the dynamical equivalent of the thermal unit. Now, if t and $t + dt$ denote the primitive and the cooled temperatures, so that $- dt$ expresses the cooling effect (which is positive, dt being negative), the bulk of the vapour, if at saturation in each case, would tend to be $v\dfrac{s + ds}{s}$; if s denote the volume of a gramme of vapour at saturation at any temperature t, and $s + ds$ its volume at temperature $t + dt$. Hence if, as it will be seen is the case, $v\dfrac{ds}{s}$ is greater than dv, a portion equal in bulk to $v\dfrac{ds}{s} - dv$ of the water primitively in vapour, must become condensed. Hence the abstraction of the heat $\frac{1}{J}pdv$ produces two effects; it cools the mass of air at constant volume from temperature t to temperature $t + dt$, and it condenses a bulk

$$v\frac{ds}{s} - dv$$

of vapour. Hence, if L denote the latent heat of a cubic cm. of vapour of water at temperature t, and N the specific heat of one gramme of air at constant volume, we have

---

* "On the Changes of Temperature produced by the Rarefaction and Condensation of Air," communicated to the Royal Society, June 20, 1844, also published in the *Philosophical Magazine*, 1845, first half year: Joule's *Scientific Papers*, Vol. I., p. 171.

$$\frac{1}{J}pdv = N \times (-dt) + L\left(v\frac{ds}{s} - dv\right)^{*},$$

if we suppose the mass of air considered to weigh one gramme (with or without the vapour, which will make but little difference on the whole weight).   Hence

$$\frac{dv}{-dt} = \frac{JN + JLv\dfrac{d\log s}{-dt}}{p + JL},$$

where, for brevity, $d\log s$ is written in place of $\dfrac{ds}{s}$, $\log s$ denoting the Napierian logarithm of s.

To find L and $\dfrac{d\log s}{-dt}$, it is necessary to know the bulk of a gramme of steam at different temperatures.   Dr Joule and I have demonstrated†, by experiments on air and by dynamical reasoning, that

$$L = \frac{J}{t}\frac{dp}{dt}\left(1 - \frac{\lambda}{\gamma}\right),$$

where p denotes the pressure of vapour at saturation at the temperature t, and $\dfrac{\lambda}{\gamma}$ denotes the ratio of the bulk of liquid to vapour.   Since $\dfrac{\lambda}{\gamma}$ is very small, we have $L = \dfrac{J}{t}\dfrac{dp}{dt}$ approximately.

It was shown also in the same Paper, that the density of saturated vapour was to be obtained more accurately from this equation, and Regnault's experiments on the latent heat of a stated weight of vapour, than from any direct experiments on the density of vapour made up to that time.   This conclusion has been verified

---

* If $L=0$, this equation becomes

$$\frac{1}{J}pdv = N \times (-dt),$$

or, since $JN = \dfrac{pv}{t} \cdot \dfrac{1}{k-1}$ (by an elementary thermodynamic formula for a perfect gas),

$$\frac{dv}{v} = \frac{1}{k-1} \times \frac{-dt}{t},$$

whence, by integration, $\dfrac{T}{t} = \left(\dfrac{V}{v}\right)^{k-1}$.

This expresses the elevation of temperature experienced by a perfect gas when compressed and not allowed to part with heat.

† "On the Thermal Effects of Fluids in Motion," Part II., "Theoretical Deductions," Section II., *Transactions of the Royal Society*, June, 1854, (Art. XLIX., Vol. I., above).

by the recent experiments of Messrs Fairbairn and Tate. With
the assistance of some excellent tables in Rankine's *Steam Engine
and other Prime Movers*, calculated on these principles, I have
obtained the following results :—

Temperature Centigrade, or $t - 273 \cdot 7$.	Volume of 1 grm. of air at pressure 1034 grms. per square cm.	Dynamical value of latent heat of 1 cubic cm. of saturated vapour.	Proportionate diminution of bulk of saturated vapour per 1° Cent. of elevation of temperature.	Augmentation of volume of 1 grm. of moist air required to cool it 1° Cent.	Elevation from earth's surface required to cool moist air by 1° Cent.
$t - 273 \cdot 7$	v	JL	$\dfrac{d \log s}{- dt}$	$\dfrac{dv}{- dt}$	$\dfrac{dx}{- dt}$
°	cubic cms.	gm.-cms.		cubic cms.	metres.
0	771	122	·0698	11·891	152
5	787	170	·0671	13·421	168
10	801	235	·0644	15·193	186
15	815	319	·0617	17·185	207
20	829	430	·0592	19·326	229
25	844	571	·0569	21·567	252
30	858	750	·0546	23·720	274
35	872	975	·0524	25·757	284

The column of this Table headed $\dfrac{dv}{- dt}$ is calculated from the
preceding formula. It expresses the expansion on the bulk of a
cubic cm. required to produce a cooling effect $- dt$ (along with an
infinitesimal lowering of pressure below the standard pressure of
1034 gms. per square cm., denoted by p), when the mass is not
allowed either to absorb or to emit heat

The last column $\left(\text{headed } \dfrac{dx}{- dt}\right)$ is calculated from the column
headed $\dfrac{dv}{- dt}$ by the following formula,

$$dx = pdv + pv \frac{- dt}{t} ,$$

and shows the height, dx, that must be reached to get a lowering
of temperature, $- dt$, when air saturated with moisture ascends.
The pressure, p, is taken as 1034 grms. per square cm.; and the
value of $\dfrac{v}{t}$, which is the same for the same pressure, whatever is

17—2

the temperature, is $\frac{773}{274}$. The results, for temperatures from 0°
to 35° Cent., are exhibited in the last column of the Table. For
the temperatures 0°, 5°, and 10°, they agree very well with the
height for which Mr Welsh found a lowering of temperature of
1° Cent.; and we may conclude that at the times and places of his
observations the lowering of temperature upwards was nearly the
same as that which air saturated with moisture would experience
in ascending.

It is to be remarked that, except when the air is saturated
and when, therefore, an ascending current will always keep
forming cloud, the effect of vapour of water, however near saturation, will be scarcely sensible on the cooling effect of expansion.
Hence the law of convective equilibrium of temperature in upward
or downward currents of cloudless air must agree very closely with
that investigated above, and must give a variation of 1° Cent. in
not much more or less than 100 metres.

It appears, therefore, that the explanation suggested by
Dr Joule is correct; and that the condensation of vapour in
ascending air is the chief cause of the cooling effect being so
much less than that which would be experienced by dry air.

ART. XCIII.—ON THE REDUCTION OF OBSERVATIONS OF UNDER-
    GROUND TEMPERATURE; WITH APPLICATION TO PROFESSOR
    FORBES' EDINBURGH OBSERVATIONS, AND THE CONTINUED
    CALTON HILL SERIES.

[*Transactions Royal Society of Edinburgh* (Read 30th April, 1860), Vol. XXII.,
Part II., p. 405.]

I.—*Analysis of Periodic Variations.*

1. EVERY purely periodical function is, as is well known,
expressible by means of a series of constant coefficients multiplying sines and cosines of the independent variable with a constant
factor and its multiples. This important truth was arrived at by
an admirable piece of mathematical analysis, called for by Daniel
Bernoulli, partially given by Lagrange, and perfected by Fourier.

2. To simplify my references to the mathematical propositions
of this theory, I shall commence by laying down the following
definitions :—

DEF. 1. A simple harmonic function is a function which
varies as the sine or cosine of the independent variable, or. of an
angle varying in simple proportion with the independent variable.
The harmonic curve is the well known name applied to the
graphic representation, on the ordinary Cartesian system, of what
I am now defining as a simple harmonic function. It is the form
of a string vibrating in such a manner as to give the simplest
and smoothest possible character of sound ; and, in this case, the
displacement of each particle of the string is a harmonic function
of the time, besides being a harmonic function of the distance of
its position of equilibrium from either end of the string. The
sound in this case may be called a perfect unison.

DEF. 2. The argument of a simple harmonic function is the
angle to the sine or cosine of which it is proportional.

Cor. The argument of a harmonic function is equal to the independent variable multiplied by a constant factor, with a constant added; that is to say, it may be any linear function of the independent variable.

Def. 3. When time is the independent variable, the epoch is the interval which elapses from the era of reckoning till the function first acquires a maximum value. The augmentation of argument corresponding to that interval will be called "the epoch in angular measure," or simply "the epoch" when no ambiguity can exist as to what is meant.

Def. 4. The period of a simple harmonic function is the augmentation which the independent variable must receive to increase the argument by a circumference.

Cor. If c denote the coefficient of the independent variable in the argument, the period is equal to $\frac{2\pi}{c}$. Thus, if T denote the period, ϵ the epoch in angular measure, and t the independent variable, the argument proper for a cosine is

$$\frac{2\pi t}{T} - \epsilon,$$

and the argument for a sine

$$\frac{2\pi t}{T} - \epsilon + \frac{\pi}{2}.$$

3. *Composition and Resolution of Simple Harmonic Functions of one Period.*

Prop. The sum of any two simple harmonic functions of one period, is equal to one simple harmonic function whose amplitude is the diagonal of a parallelogram described upon lines drawn from one point to lengths equal to the amplitudes of the given functions, at angles measured from a fixed line of reference equal to their epochs, and whose epoch is the inclination of the same diagonal to the same line of reference.

Cor. 1. If A, A' be the amplitudes of two simple harmonic functions of equal period, and ϵ, ϵ' their epochs; that is to say, if $A\cos(mt - \epsilon)$, $A'\cos(mt - \epsilon')$ be two simple harmonic functions; the one simple harmonic function equal to their sum has for its amplitude and its epoch the following values respectively:—

(amplitude) $\qquad \{(A \cos \epsilon + A' \cos \epsilon')^2 + (A \sin \epsilon + A' \sin \epsilon')^2\}^{\frac{1}{2}};$

or $\qquad \{A^2 + 2AA' \cos (\epsilon' - \epsilon) + A'^2\}^{\frac{1}{2}},$

(epoch) $\qquad \tan^{-1} \dfrac{A \sin \epsilon + A' \sin \epsilon'}{A \cos \epsilon + A' \cos \epsilon'}.$

COR. 2. Any number of simple harmonic functions, of equal period, added together, are equivalent to a single harmonic function of which amplitude and epoch are derived from the amplitude and epochs of the given functions, in the same manner as the magnitude and inclination to a fixed line of reference, of the resultant of any number of forces in one plane, are derived from the magnitudes and the inclinations to the same line of reference of the given forces.

COR. 3. The physical principle of the superposition of sounds being admitted, any number of simple unisons of one period co-existing, produce one simple unison of the same period, of which the intensity (measured by the square of the amplitude) and the epoch are determined in the manner just specified.

COR. 4. The sum of any number of simple harmonic functions of one period vanishes for every argument, if it vanishes for any two arguments not differing by a semi-circumference, or by some multiple of a semi-circumference.

COR. 5. The co-existence of perfect unisons may constitute perfect silence.

COR. 6. A simple harmonic function of any epoch may be resolved into the sum of two whose epochs are respectively zero and a quarter period, and whose amplitudes are respectively equal to the value of the given function for the arguments zero and a quarter period respectively.

4. *Complex Harmonic Functions.*—Harmonic functions of different periods added can never produce a simple harmonic function. If their periods are commensurable their sum may be called a complex harmonic function.

COR. A complex harmonic function is the proper expression for a perfect harmony in music.

5. *Expressibility of Arbitrary Functions by Trigonometrical series.*

Prop. A complex harmonic function, with a constant term added, is the proper expression, in mathematical language, for any arbitrary periodic function.

6. *Investigation of the Trigonometrical Series expressing an Arbitrary Function.*—Any arbitrary periodic function whatever being given, the amplitudes and epochs of the terms of a complex harmonic function, which shall be equal to it for every value of the independent variable, may be investigated by the "method of indeterminate coefficients," applied to determine an infinite number of coefficients from an infinite number of equations of condition, by the assistance of the integral calculus, as follows:—

Let $F(t)$ denote the function, and T its period. We must suppose the value of $F(t)$ known for every value of t, from $t=o$ to $t=T$. Let M_0 denote the constant term, and let M_1, M_2, M_3, &c., denote the amplitudes, and $\epsilon_1, \epsilon_2, \epsilon_3$, &c., the epochs of the successive terms of the complex harmonic functions by which it is to be expressed; that is to say, let these constants be such that

$$F(t) = M_0 + M_1 \cos\left(\frac{2\pi t}{T} - \epsilon_1\right) + M_2 \cos\left(\frac{4\pi t}{T} - \epsilon_2\right)$$
$$+ M_3 \cos\left(\frac{6\pi t}{T} - \epsilon_3\right) + \&c.$$

Then, expanding each cosine by the ordinary formula, and assuming

$$M_1 \cos \epsilon_1 = A_1, \quad M_2 \cos \epsilon_2 = A_2, \&c.$$
$$M_1 \sin \epsilon_1 = B_1, \quad M_2 \sin \epsilon_2 = B_2, \&c.$$

we have

$$F(t) = A_0 + A_1 \cos \frac{2\pi t}{T} + A_2 \cos \frac{4\pi t}{T} + A_3 \cos \frac{6\pi t}{T} + \&c.$$
$$+ B_1 \sin \frac{2\pi t}{T} + B_2 \sin \frac{4\pi t}{T} + B_3 \sin \frac{6\pi t}{T} + \&c.$$

Multiplying each member by $\cos \frac{2i\pi t}{T} dt$ where i denotes o or any integer, and integrating from $t=o$ to $t=T$, we have,—

$$\int_0^T F(t) \cos \frac{2i\pi t}{T} dt = A_i \int_0^T \left(\cos \frac{2i\pi t}{T}\right)^2 dt;$$
$$= A_i \times \tfrac{1}{2}T, \text{ when } i \text{ is any integer;}$$

or $\qquad\qquad = A_0 \times T, \text{ when } i=0.$

Hence
$$A_0 = \frac{1}{T} \int_0^T F(t)\, dt$$

$$A_i = \frac{2}{T} \int_0^T F(t) \cos \frac{2i\pi t}{T}\, dt;$$

and similarly we find

$$B_i = \frac{2}{T} \int_0^T F(t) \sin \frac{2i\pi t}{T}\, dt:$$

equations by which the coefficients in the double series of sines and cosines are expressed in terms of the values of the function supposed known from $t = o$ to $t = T$. The amplitudes and epochs of the single harmonic terms of the chief period and its sub-multiples are calculated from them, according to the following formula:

$$\tan \epsilon_i = \frac{B_i}{A_i}; \quad M_i = (A_i^2 + B_i^2)^{\frac{1}{2}}$$

(or for logarithmic calculation, $M_i = A_i \sec \epsilon_i$).

The preceding investigation is sufficient as a solution of the problem,—to find a complex harmonic function expressing a given arbitrary periodic function, when once we are assured that the problem is possible; and when we have this assurance, it proves that the resolution is *determinate*; that is to say, that no other complex harmonic function than the one we have found can satisfy the conditions. For a thorough and most interesting analysis of the subject, supplying all that is wanting to complete the investigation, and giving admirable views of the problem from all sides, the reader is referred to Fourier's delightful treatise. A concise and perfect synthetical investigation of the harmonic expression of an arbitrary periodic function is to be found in Poisson's *Théorie Mathématique de la Chaleur*, Chap. VII.

II.—*Periodic Variations of Terrestrial Temperature.*

7. If the whole surface of the earth were at each instant of uniform temperature, and if this temperature were made to vary as a perfectly periodic function of the time, the temperature at any internal point must ultimately come to vary also as a periodic function of the time, with the same period, whatever may have been the initial distribution of temperature throughout the whole. Fourier's principles show how the periodic variation of internal temperature is to be conceived as following, with diminished

amplitude and retarded phase, from the varying temperature at the surface supposed given : and by his formulæ the precise law according to which the amplitude would diminish and the phase would be retarded, for points more and more remote from the surface, if the figure were truly spherical and the substance homogeneous, is determined.

8.   The largest application of this theory to the earth as a whole is to the analysis of imaginable secular changes of temperature, with at least thousands of millions of years for a period. In such an application, it would be necessary to take into account the spherical figure of the earth as a whole.   Periodic variations at the surface with any period less than a million* of years will, at points below the surface, give rise to variations of temperature not appreciably influenced by the general curvature, and sensibly agreeing with what would be produced if the surface were an infinite plane, except in so far as they are modified by superficial irregularities.   Hence Fourier's formulæ for an infinite solid, bounded on one side by an infinite plane, of which the temperature is made to vary arbitrarily, contain the proper analysis for diurnal or annual variations of terrestrial temperature, unless a theory of the effect of inequalities of surface (upon which no investigator has yet ventured) is aimed at.

9.   The effect of diurnal variations of temperature becomes insensible at so small a distance below the surface, that in most localities irregularities of soil and drainage must prevent any very satisfactory theoretical treatment of their inward progression and extinction from being carried out.   At depths exceeding three feet below the surface, all periodic effects of daily variations of

---

* A periodic variation of external temperature of one million years' period would give variations of temperature within the earth sensible to one thousand times greater depths than a similar variation of one year's period.   Now the ordinary annual variation is reduced to $\frac{1}{20}$ of its superficial amount at a depth of 25 French feet, and is scarcely sensible at a depth of 50 French feet (being there reduced, in such rock as that of Calton Hill, to $\frac{1}{400}$).   Hence, at a depth of 50,000 French feet, or about ten English miles, a variation having one million years for its period would be reduced to $\frac{1}{400}$.   If the period were ten thousand million years, the variation would similarly be reduced to $\frac{1}{400}$ at 1000 miles' depth, and would be to some appreciable extent affected by the spherical figure of the whole earth, although to only a very small extent, since there would be comparatively but very little change of temperature (less than $\frac{1}{20}$ of the superficial amount) beyond the first layer of 500 miles' thickness.

temperature become insensible in most soils, and the observable
changes are those due to a daily average, varying from day
to day. If now the annual variation of temperature were
truly periodic, a complex harmonic function could be determined
to represent for all time the temperature at three feet or any
greater depth. But in reality the annual variation is very far
from recurring in a perfectly periodic manner, since there are
both great differences in the annual average temperatures, and
never-ceasing irregularities in the progress of the variation within
each year. A full theory of the consequent variations of temperature propagated downwards, must include the consideration
of non-periodic changes; but the most convenient first step is
that which I propose to take in the present communication, in
which the average annual variations for groups of years will be
discussed according to the laws to which periodic variations are
subject.

10. The method which Fourier has given for treating this
and other similar problems is founded on the principle of the
independent superposition of thermal conductions. This principle
holds rigorously in nature, except in so far as the conductivity or
the specific heat of the conducting substance may vary with the
changes of temperature to which it is subjected; and it may be
accepted with very great confidence in the case with which we
are now concerned, as it is not at all probable that either the
conductivity or the specific heat of the rock or soil can vary at
all sensibly under the influence of the greatest changes of temperature experienced in their natural circumstances; and, indeed,
the only cause we can conceive as giving rise to sensible change
in these physical qualities is the unequal percolation of water,
which we may safely assume to be confined in ordinary localities
to depths of less than three feet below the surface. The particular
mode of treatment which I propose to apply to the present subject
consists in expressing the temperature at any depth as a complex
harmonic function of the time, and considering each term of this
function separately, according to Fourier's formulæ for the case
of a simple harmonic variation of temperature, propagated inwards
from the surface. The laws expressed by these formulæ may be
stated in general terms as follows.

11. *Fourier's Solution stated.*—If the temperature at any
point of an infinite plane, in a solid extending infinitely in all

directions, be subjected to a simple harmonic variation, the temperature throughout the solid on each side of this plane will follow everywhere according to the simple harmonic law, with epochs retarded equally, and with amplitudes diminished in a constant proportion for equal augmentations of distance. The retardation of epoch expressed in circular measure (arc divided by radius) is equal to the diminution of the Napierian logarithm of the amplitude; and the amount of each per unit of distance is equal to $\sqrt{\dfrac{\pi c}{Tk}}$, if c denote the capacity for heat of a unit bulk of the substance, and k its conductivity*.

12. Hence, if the complex harmonic functions expressing the varying temperature at two different depths be determined, and each term of the first be compared with the corresponding term of the second, the value of $\sqrt{\dfrac{\pi c}{Tk}}$ may be determined either by dividing the difference of the Napierian logarithms of the amplitudes or the difference of the epochs by the distance between the points. The comparison of each term in the one series with the corresponding term in the other series gives us, therefore, two determinations of the value of $\sqrt{\dfrac{\pi c}{k}}$, which should agree perfectly, if (1) the data were perfectly accurate, if (2) the isothermal surfaces throughout were parallel planes, and if (3) the specific heat and conductivity of the soil were everywhere and always constant.

As these conditions are not strictly fulfilled in any natural application, the first thing to be done in working out the theory is to test how far the different determinations agree, and to judge accordingly of the applicability of the theory in the circumstances. If the test thus afforded prove satisfactory, the value of the conductivity in absolute measure may be deduced from the result with the aid of a separate experimental determination of the specific heat.

13. The method thus described differs from that followed by Professor Forbes in substituting the separate consideration of

---

* That is to say, the quantity of heat conducted per unit of time across a unit area of a plate of unit thickness, with its two surfaces permanently maintained at temperatures differing by unity.

separate terms of the complex harmonic function for the examination of the whole variation unanalysed, which he conducted
according to the plan laid down by Poisson.

This plan consists in using the formulæ for a simple harmonic
variation, as approximately applicable to the actual variation. At
great depths the amplitudes of the second and higher terms of
the complex harmonic function become so much reduced as not
sensibly to influence the variation, which is consequently there
expressed with sufficient accuracy by a single harmonic term of
yearly period; but at even the greatest depths for which continuous observations have actually been made, the second (or
semi-annual) term has a very sensible influence, and the third
and fourth terms are by no means without effect on the variations
at three feet and six feet from the surface. A close agreement
with theory is therefore not to be expected, until the method of
analysis which I now propose is applied. It may be added, that
in the theoretical reductions hitherto made, either by Professor
Forbes or others, the amplitudes of the variations for the different
depths have alone been compared, and the very interesting conclusion of theory, as to the relation between the absolute amount
of retardation of phase and the diminution of amplitude for any
increase of depth, has remained untested.

14. In Professor Forbes' paper*, the very difficult operations
which he had performed for effecting the construction and the
sinking of the thermometers, and the determination of the corrections to be applied to obtain the true temperatures of the
earth at the different depths from the readings of the scales
graduated on their stems protruding above the surface, are fully
described. The results of five years' observations—1837 to 1842
—are given, along with most interesting graphical representations
and illustrations. A process of graphic interpolation, for estimating the temperatures at times intermediate between those of
observations, is applied for the purpose of obtaining data from
which the complex harmonic functions expressing the temperatures
actually observed for the different depths are determined. I am
thus indebted to Professor Forbes for the mode of procedure
(described below) which I have myself followed in expressing

* Account of Some Experiments on the Temperature of the Earth at Different
Depths and in Different Soils near Edinburgh; *Transactions R.S.E.*, Vol. xvi.
Part ii. Edinburgh, 1846.

the variations of temperature during the succeeding thirteen years for the Calton Hill station (where alone the observations were continued). The only variation from his process which I have made is, that instead of taking twelve points of division for the yearly period I have taken thirty-two, with a view to obtaining a more perfect representation of all the features of the observed variations, and a more exact average for the principal terms, especially the annual and the semi-annual terms of the complex harmonic function expressing them.

15. *Application of the General Theory to Five Years' Observations—1837 to 1842—at* Professor Forbes's *three Thermometric Stations.*—The first application which I made of the analytical theory explained above, was to the harmonic terms which Professor Forbes had found for expressing the average annual progressions of temperature during the five years' term of observations at the three stations. These terms (which I have recalculated to get their values true to a greater number of significant figures), with alterations of notation which I have found convenient for the analytical expressions, are as follows :—

Three Feet below Surface.

Observatory . . . $45 \cdot 49 + 7 \cdot 39 \cos 2\pi \, (t - \cdot 63) + 0 \cdot 362 \cos 2\pi \, (2t - \cdot 669)$

Experimental Garden . $46 \cdot 13 + 9 \cdot 00 \cos 2\pi \, (t - \cdot 616) + 0 \cdot 737 \cos 2\pi \, (2t - \cdot 183)$

Craigleith . . . $45 \cdot 88 + 8 \cdot 16 \cos 2\pi \, (t - \cdot 617) + 0 \cdot 284 \cos 2\pi \, (2t - \cdot 154)$

Six Feet below Surface.

Observatory . . . $45 \cdot 86 + 5 \cdot 06 \cos 2\pi \, (t - \cdot 686) + 0 \cdot 433 \cos 2\pi \, (2t - \cdot 731)$

Experimental Garden . $46 \cdot 42 + 6 \cdot 66 \cos 2\pi \, (t - \cdot 665) + 0 \cdot 501 \cos 2\pi \, (2t - \cdot 182)$

Craigleith . . . $45 \cdot 92 + 6 \cdot 16 \cos 2\pi \, (t - \cdot 649) + 0 \cdot 368 \cos 2\pi \, (2t - \cdot 305)$

Twelve Feet below Surface.

Observatory . . . $46 \cdot 36 + 2 \cdot 44 \cos 2\pi \, (t - \cdot 799) + 0 \cdot 075 \cos 2\pi \, (2t - \cdot 833)$

Experimental Garden . $46 \cdot 76 + 3 \cdot 38 \cos 2\pi \, (t - \cdot 782) + 0 \cdot 230 \cos 2\pi \, (2t - \cdot 390)$

Craigleith . . . $45 \cdot 92 + 4 \cdot 22 \cos 2\pi \, (t - \cdot 713) + 0 \cdot 067 \cos 2\pi \, (2t - \cdot 819)$

Twenty-four Feet below Surface.

Observatory . . . $46 \cdot 87 + 0 \cdot 655 \cos 2\pi \, (t - 1 \cdot 013)$

Experimental Garden . $47 \cdot 09 + 0 \cdot 920 \cos 2\pi \, (t - \cdot 986)$

Craigleith . . . $46 \cdot 07 + 1 \cdot 940 \cos 2\pi \, (t - \cdot 849)$

The semi-annual terms in these equations present so great irregularities (those for the Calton Hill station, for instance, showing a greater amplitude at 6 feet deep than at 3 feet), that no satisfactory result can be obtained by including them in the theoretical discussion on which we are now about to enter. We shall see later, however, that when an average for the whole

XCIII.] OBSERVATIONS OF UNDERGROUND TEMPERATURE. 271

period of eighteen years for the Calton Hill station is taken, the semi-annual terms are, for the 3 feet and 6 feet depths, in fair agreement with theory; and for the two greater depths are as small as is necessary for the verification of the theory, and so small as to be much influenced by errors of observation and of reduction, or of "corrections" for temperature of the thermometer tubes. For the present, we attend exclusively to the annual terms. The amplitudes and epochs of these terms, extracted from the preceding equations, are shown in the following table :—

TABLE I.—ANNUAL HARMONIC VARIATIONS OF TEMPERATURE.

Depths below surface in French feet.	CALTON HILL.			EXPERIMENTAL GARDEN.			CRAIGLEITH QUARRY.		
	Amplitudes in degrees Fahr.	Epochs of Maximum.		Amplitudes in Degrees Fahr.	Epochs of Maximum.		Amplitudes in degrees Fahr.	Epochs of Maximum.	
		In Degs. and Mins.	In Months and Days.		In Degs. and Mins.	In Months and Days.		In Degs. and Mins.	In Months and Days.
Feet. 3	7·386	226° 52'	Aug. 19	9·063	221° 40'	Aug. 13	8·069	222° 0'	Aug. 14
6	5·063	247 5'	Sept. 8	6·661	239 20'	„ 31	6·148	233 43'	„ 26
12	2·455	287 30'	Oct. 19	3·408	281 27'	Oct. 13	4·216	256 42'	Sept. 17
24	0·655	365 6'	Jan. 6	0·920	355 0'	Dec. 27	1·836	305 46'	Nov. 7

By taking the differences of the Napierian logarithms of the amplitudes, and the differences of epochs reduced to circular measure (arc divided by radius), thus shown for the different depths, and dividing each by the corresponding difference of depths, we find the following numbers.

TABLE II.—RATES OF LOGARITHMIC DIMINUTION IN AMPLITUDE, AND OF RETARDATION IN EPOCH, OF ANNUAL HARMONIC VARIATIONS DOWNWARDS.

Depths below surface in French feet.	CALTON HILL.		EXPERIMENTAL GARDEN.		CRAIGLEITH QUARRY.	
	Rate of Diminution of Napierian Logarithm of Amplitude per foot of Descent.	Rate of Retardation of Epoch in Circular Measure, per foot of Descent.	Rate of Diminution of Napierian Logarithm of Amplitude per foot of Descent.	Rate of Retardation of Epoch in Circular Measure, per foot of Descent.	Rate of Diminution of Napierian Logarithm of Amplitude per foot of Descent.	Rate of Retardation of Epoch in Circular Measure, per foot of Descent.
3 to 6 feet.	·1259	·1176	·1004	·1163	·09372	·06399
6 to 12	·1206	·1176	·1130	·1193	·06304	·06690
12 to 24	·1101	·1129	·1084	·1062	·06476	·06690
3 to 24	·1154	·1149	·1082	·1114	·06841	·06648

16. All the numbers here shown for each station would be equal, if the conditions of uniformity supposed in the theoretical

solution were fulfilled. The discrepancies are, with the exception of one of the numbers for Craigleith Quarry, on the whole small —smaller, indeed, than might be expected, when the very notable deviations of the true circumstances from the theoretical conditions are considered. The mean results over the 21 feet, shown in the last line, present very remarkable agreements: the numbers derived from amplitudes being identical with that derived from epochs for the Calton Hill station; while the differences between the corresponding numbers for the two other stations are in each case only about 3 per cent. Taking that one number for the first station, and the mean of the slightly differing numbers derived from amplitudes and from epochs respectively, for the second and third, we have undoubtedly very accurate determinations of the value of $\sqrt{\dfrac{\pi c}{k}}$ for the three stations, which are as follows :—

Calton Hill Trap Rock.	Experimental Garden Sand.	Craigleith Quarry Sandstone.
$\sqrt{\dfrac{\pi c}{k}} = \cdot 1154.$	$\sqrt{\dfrac{\pi c}{k}} = \cdot 1098.$	$\sqrt{\dfrac{\pi c}{k}} = \cdot 06744.$

A continuation of the observations at Calton Hill not only leads, as we shall see, to almost identical results, both by diminution of amplitude and by retardation, on the whole 21 feet, but also reproduces some of the features of discrepance presented by the progress of the variation through the intermediate depths; and therefore confirms the general accuracy of the preceding results, for all the stations, so far as it might be questioned because of only five years' observations having been available. Further consideration of these results, and deduction of the conductivities of the different portions of the earth's crust involved, is deferred until after we have taken into account the farther data for Calton Hill, to the reduction of which we now proceed.

17. *Application to Thirteen Years' Observations* (1842–1854) *at the Thermometric Station, Calton Hill.*—The observations on thermometers fixed by Professor Forbes at the different depths in the rock of Calton Hill, have been regularly continued weekly till the present time by the staff of the Royal Edinburgh Observatory, and regularly corrected to reduce to true temperatures of the bulbs, on the same system as before. Tables of these

TABLE III.

Year	Feet	A_1	B_1	A_2	B_2	A_3	B_3	A_4	B_4
1842	3	−6·19	−5·00	+·01	+·25	+·60	+·06	+·23	−·71
	6	−2·85	−4·80	−·15	+·03	+·10	+·10	+·12	−·26
	12	+·34	−2·73	−·12	−·13	−·08	−·04	+·01	−·04
	24	+·68	−·14	+·00	−·07	−·02	−·04	−·01	−·02
1843	3	−4·75	−5·11	+·17	+·91	+1·23	+·30	+·79	−·17
	6	−1·63	−4·38	−·20	+·61	+·45	+·42	+·32	+·30
	12	+·83	−2·04	−·18	−·08	−·05	+·17	−·03	+·10
	24	+·62	+·12	−·00	−·02	−·01	−·01	−·00	+·00
1844	3	−5·29	−4·53	−·05	+·70	+·74	+·71	+·08	+·49
	6	−2·11	−4·09	+·22	+·50	+·20	+·50	−·06	+·20
	12	+·52	−2·15	+·18	+·05	+·11	+·13	−·05	−·01
	24	+·59	−·02	−·03	−·02	+·00	−·03	−·01	−·02
1845	3	−5·17	−5·01	−·17	+·56	+·67	+·29	−·28	+·02
	6	−2·02	−4·38	+·07	+·30	·00	+·18	−·04	−·08
	12	+·63	−2·15	+·12	+·06	−·01	−·03	−·00	+·02
	24	+·65	+·13	+·04	+·00	−·01	+·02	+·00	+·02
1846	3	−5·65	−5·17	+·03	+1·05	+·86	+·64	+·11	−·49
	6	−2·37	−4·64	−·38	+·44	−·63	+·39	+·00	−·22
	12	+·47	−2·70	−·30	−·17	−·14	−·45	−·03	−·07
	24	+·64	−·22	−·02	−·17	+·03	−·11	−·02	−·06
1847	3	−5·36	−5·31	+·69	+·24	−·18	−·81	−·03	−·14
	6	−2·08	−4·58	+·18	+·32	−·11	−·39	−·05	−·04
	12	+·70	−2·37	−·03	+·17	+·12	+·14	+·03	−·02
	24	+·66	+·16	−·01	+·04	+·01	+·03	−·01	+·03

TABLE III. *continued.*

Year.	Feet.	A_1	B_1	A_2	B_2	A_3	B_3	A_4	B_4
1848	3	− 5·83	− 4·46	+ ·33	+ ·27	+ ·29	+ ·35	+ ·45	+ ·30
	6	− 2·32	− 4·16	+ ·13	+ ·27	+ ·02	+ ·23	+ ·28	+ ·09
	12	+ ·56	− 2·15	+ ·04	+ ·16	+ ·01	+ ·09	− ·04	+ ·11
	24	+ ·66	− ·10	− ·01	+ ·03	− ·00	+ ·02	− ·01	+ ·01
1849	3	− 4·56	− 4·44	+ ·05	− 1·14	− ·66	− ·10	− ·48	− ·69
	6	− 1·85	− 3·97	+ ·20	+ ·45	− ·28	− ·15	− ·01	− ·25
	12	+ ·49	− 2·06	− ·23	− ·04	− ·04	− ·06	+ ·09	− ·05
	24	+ ·57	+ ·03	− ·00	− ·02	− ·01	− ·02	+ ·00	− ·01
1850	3	− 5·40	− 4·50	− ·12	+ ·70	+ ·01	+ ·82	− ·15	+ ·42
	6	− 2·43	− 4·15	− ·22	+ ·31	+ ·54	+ ·47	− ·11	+ ·17
	12	+ ·17	− 2·27	− ·15	− ·04	+ ·03	+ ·05	− ·04	− ·01
	24	+ ·61	− ·04	− ·01	− ·03	+ ·10	− ·00	− ·01	− ·01
1851	3	− 4·18	− 4·53	− ·12	+ ·96	− ·01	− ·31	− ·22	+ ·18
	6	− 1·65	− 3·92	− ·19	+ ·53	+ ·09	− ·07	− ·03	− ·14
	12	+ ·61	− 1·99	− ·22	− ·01	+ ·18	− ·06	− ·05	− ·02
	24	+ ·56	+ ·02	− ·01	− ·05	− ·04	− ·01	− ·14	− ·01
1852	3	− 4·92	− 4·80	− ·20	+ 1·32	− ·00	− ·24	− ·46	− ·31
	6	− 1·87	− 4·25	− ·23	+ ·71	+ ·64	− ·10	− ·31	− ·02
	12	+ ·54	− 2·24	− ·26	+ ·05	+ ·15	− ·09	− ·01	− ·07
	24	+ ·61	− ·03	− ·12	− ·07	− ·01	− ·04	− ·00	− ·02
1853	3	− 5·08	− 5·43	+ ·83	+ ·30	− ·01	− ·27	− ·18	− ·19
	6	− 1·92	− 4·57	+ ·38	+ ·41	+ ·11	− ·17	+ ·06	− ·13
	12	+ ·76	− 3·15	− ·01	+ ·21	+ ·05	+ ·00	− ·01	+ ·03
	24	+ ·62	− ·18	− ·39	+ ·03	− ·01	− ·10	− ·01	+ ·08
1854	3	− 5·69	− 4·56	− ·61	+ ·53	− ·00	− ·15	− ·15	− ·20
	6	− 2·48	− 4·27	− ·50	− ·01	− ·00	− ·13	− ·08	− ·03
	12	+ ·43	− 2·31	− ·12	− ·21	+ ·02	− ·03	− ·02	− ·01
	24	+ ·63	− ·03	+ ·02	− ·02	+ ·00	− ·01	− ·01	− ·01
Average for 13 years 1842 to '54	3	− 5·236	− 4·835	+ ·114	+ ·687	+ ·150	+ ·0778	+ ·05462	+ ·14846
	6	− 2·122	− 4·320	− ·0838	+ ·375	− ·00615	+ ·0185	+ ·02923	+ ·01615
	12	+ ·5415	− 2·332	− ·0385	+ ·00923	− ·01846	− ·00778	− ·006154	− ·003078
	24	+ ·6231	− ·0200	− ·0385	− ·0285	− ·00231	− ·00462	− ·01462	− ·003846

corrected observations, for the thirteen years 1842 to 1854 inclusive, having been supplied to me through the kindness of Professor
Piazzi Smyth, I have had the first five terms of the harmonic
expression for each year determined in the following manner*:—
In the first place, the observations were laid down graphically, and
an interpolating curve drawn through the points, according to the
method of Professor Forbes. The four curves thus obtained represent the history of the varying temperature at the four different
depths respectively, as completely and accurately as it can be
inferred from the weekly observations. The space corresponding
to each year was then divided into 32 equal parts (the first point
of division being taken at the beginning of the year), and the
corresponding temperatures were taken from the curve. The coefficients of the double harmonic series (cosines and sines) for
each year were calculated from these data, with the aid of the
forms given by Mr Archibald Smith, and published by the Board
of Admiralty, for deducing the harmonic expression of the error
of a ship's compass from observations on the 32 points. The
general form of the harmonic expression being written thus—

$$V = A_0 + A_1 \cos 2\pi t + B_1 \sin 2\pi t + A_2 \cos 4\pi t + B_2 \sin 4\pi t + \&c.$$

where V denotes the varying temperature to be expressed, and
t the time, in terms of a year as unit. Table III. shows the
results which were obtained, with the exception of the values
of A_0:—

The values which were found for A_0 should represent the
annual mean temperatures. They differ slightly from the annual
means shown in the Royal Observatory Report, which, derived as
they are from a direct summation of all the weekly observations,
must be more accurate. The variations, and the final average
values of these annual means, present topics for investigation of
the highest interest and importance, as I have remarked elsewhere
(see British Association's Report, section A, Glasgow, 1855 [Art.
LXXXVII. Vol. II. above]); but as they do not belong to the special
subject of the present paper, their consideration must be deferred
to a future occasion.

* The operations here described, involving, as may be conceived, no small
amount of labour, were performed by Mr D. M‘Farlane, my laboratory assistant,
and Mr J. D. Everett, now [1887] Professor of Natural Philosophy in Queen's
College, Belfast.

18. *Theoretical Discussion.*—The mean value of the coefficients in the last line of Table III., being obtained from so considerable a number of years, can be but very little influenced by irregularities from year to year, and must therefore correspond to harmonic functions for the different depths, which would express truly periodic variations of internal temperature consequent upon a continued periodical variation of temperature at the surface.

19. According to the principle of the superposition of thermal conductions, the difference between this continuous harmonic function of five terms for any one of the depths, and the actual temperature there at the corresponding time of each year, would be the real temperature consequent upon a certain real variation of superficial temperature. Hence the coefficients shown in the preceding table afford the data, first by their mean values, to test the theory explained above for simple harmonic variations, and to estimate the conductivity of the soil or rock, as I propose now to do; and secondly, as I may attempt on a future occasion, to express analytically the residual variations which depend on the inequalities of climate from year to year, and to apply the mathematical theory of conduction to the nonperiodic variations of internal temperature so expressed.

20. Let us, accordingly, now consider the complex harmonic functions corresponding to the mean coefficients of Table III. above, and, in the first place, let us reduce the double harmonic series in each case, to series in each of which a single term represents the resultant simple harmonic variation of the period to which it corresponds, in the manner shown by the proposition and formulæ of § 3 above.

21. On looking to the annual and semi-annual terms of the series so found, we see that their amplitudes diminish, and their epochs of maximum augment, with considerable regularity, from the less to the greater depths. The following Table IV. shows, for the annual terms, the logarithmic rate of diminution of the amplitudes, and the rate of retardation of the epoch between the points of observation in order of depth:—

TABLE IV.—AVERAGE OF THIRTEEN YEARS, 1842 TO 1854;
TRAP ROCK OF CALTON HILL.

Depths below surface, in French feet.	Rate of diminution of Napierian Logarithm of Amplitude per French foot of descent.	Rate of retardation of Epoch in Circular Measure, per French foot of descent.
3 to 6 feet	·1310	·1233
6 to 12 ,,	·1163	·1140
12 to 24 ,,	·1121	·1145
3 to 24 feet	·1160	·1156

22. The numbers here shown would all be the same, if the conditions of uniformity supposed in the theoretical solution were fulfilled. Although, as in the previous comparisons, the agreement is on the whole better than might have been expected, there are certainly greater differences than can be attributed to errors of observation. Thus, the means of the numbers in the two columns are for the three different intervals of depth in order as follows :—

	Mean deductions from amplitude and epoch.
3 to 6 feet......................	·127
6 to 12 ,, ...	·115
12 to 24 ,, ...	·113

—numbers which seem to indicate an essential tendency to diminish at the greater depths. This tendency is shown very decidedly in each column separately; and it is also shown in each of the corresponding columns, in Table II. (p. 272) above, of results derived from Professor Forbes' own series of a period of five years.

23. There can be no doubt but that this discrepance is not attributable to errors of observation, and it must therefore be owing to deviation in the natural circumstances from those assumed for the foundation of the mathematical formulæ. In reality, none of the conditions assumed in Fourier's solution is rigorously fulfilled in the natural problem; and it becomes a most interesting subject for investigation to discover to what particular violation or violations of these conditions, the remarkable and systematic difference discovered between the deductions from the formula and the results of observation is due. In the first

place, the formula is strictly applicable only to periodic variations, and the natural variations of temperature are very far from being precisely periodic; but if we take the average annual variation through a sufficiently great number of years, it may be fairly presumed that irregularities from year to year will be eliminated; and that the discrepance we have now to explain does not depend on residual inequalities of this kind seems certain, from the fact that it exists in the average of Professor Forbes' first five years' series no less decidedly than in that of the period of thirteen years following.

24. For the true explanation we must therefore look either to inequalities (formal or physical) in the surface at the locality, or to inequalities of physical character of the rock below. It may be remarked, in the first place, that if the rates of diminution of logarithmic amplitude and of retardation of epoch, while less, as they both are, at the greater depths, remained exactly equal to one another, the conductivity must obviously be greater, and the specific heat less in the same proportion inversely, at the greater depths. For in that case, all that would be necessary to reconcile the results of observations with Fourier's formula, would be to alter the scale of measurement of depths so as to give a nominally constant rate of diminution of the logarithmic amplitude and of the retardation of epoch; and the physical explanation would be, that thicker strata at the greater depths, and thinner strata at the less depths (all of equal horizontal area), have all equal conducting powers and equal thermal capacities *.

25. Now, in reality, a portion, but only a portion, of the discrepance may be done away with in this manner; for while the logarithmic amplitudes and the epochs each experience a some-

---

* The "conducting power" of a solid plate is an expression of great convenience, which I define as the quantity of heat which it conducts per unit of time, when its two surfaces are permanently maintained at temperatures differing by unity. In terms of this definition, the specific conductivity of a substance may be defined as the conducting power per unit area of a plate of unit thickness. The conducting power of a plate is calculated by multiplying the number which measures the specific conductivity of its substance by its area, and dividing by its thickness.

The *thermal capacity of a body* may be defined as the quantity of heat required to raise its mass by a unit (or one degree) of temperature. The *specific heat* of a substance is the thermal capacity of a unit quantity of it, which may be either a unit of weight or a unit of bulk.

what diminished rate of variation per French foot of descent at
the greater depths, this diminution is much greater for the former
than for the latter; so that although the mean rates per foot on the
whole 21 feet are as nearly as possible equal for the two, (being
·1160 for the logarithmic amplitudes, and ·1156 for the epoch),
the rate of variation of the logarithmic amplitude exceeds that
of the epoch by about 6 per cent., on the average of the stratum
3 to 6 feet; and falls short of it by somewhat more than 2 per
cent., in the lower stratum, 12 to 24 feet.   To find how much of
the discrepance is to be explained by the variation of conductivity
and specific heat in inverse proportion to one another at the
different depths, we may take the mean of the rates of variation
of logarithmic amplitude and of epoch at each depth, and alter
the scale of longitudinal reckoning downwards, so as to reduce
the numerical measures of these rates to equality.   This, however,
we shall not do in either the five years' or the thirteen years'
term, which we have hitherto considered separately, but for a
harmonic annual variation representing the average of the whole
eighteen years 1837 to 1854.

26.   By taking, for each depth, the coefficients A_1, B_1 (not
explicitly shown above), derived from the first five years' average
and multiplying by 5; taking similarly the coefficients A_1, B_1, for
the succeeding thirteen years' average, and multiplying by 13;
adding each of the former products to the corresponding one of
the latter, and dividing by 18; we obtain, as the proper average
for the whole eighteen years, the values shown in the following
Table (V.), in the columns headed A_1, B_1.   The amplitudes and
epochs shown in the next columns are deduced from these by the
formulæ $\sqrt{(A_1^2 + B_1^2)}$ and $\tan^{-1}\dfrac{B_1}{A_1}$ respectively,—

TABLE V.—ANNUAL HARMONIC VARIATION OF TEMPERATURE IN
CALTON HILL, FROM 1837 TO 1854 INCLUSIVE.

Depths.	A_1 In degrees Fahr.	B_1 In degrees Fahr.	Amplitudes in degrees Fahr.	Epochs in degrees and minutes.
3 feet	− 5°·184	− 4°·989	7°·1949	223°54′
6 feet	− 2 ·080	− 4 ·416	4 ·8812	244 47
12 feet	+ 0 ·5961	− 2 ·3345	2 ·4094	284 19
24 feet	+ 0 ·6311	+ 0 ·0306	0 ·6319	362 47

From these, as before, for the terms of five years and of thirteen years separately, we deduce the following :—

TABLE VI.—AVERAGE OF EIGHTEEN YEARS, 1837 TO 1854;
TRAP ROCK OF CALTON HILL.

Depths below surface, in French feet.	Rate of Diminution of Logarithmic Amplitude per French foot of Descent.	Rate of Retardation of Epoch in Circular Measure, per French foot of Descent.
3 to 6 feet	·1286	·1215
6 to 12 ,,	·1177	·1150
12 to 24 ,,	·1115	·1141
3 to 24 feet	·1157	·1154

27. Hence, we have as final means, of effects on logarithmic amplitudes and on epochs, for the average annual variation on the whole period of eighteen years,—

1. From depth 3 feet to 6 feet, ... ·1250
2.   ,,   6 ,,   12 ,, ... ·1163
3.   ,,   12 ,,   24 ,, ... ·1128

If now, in accordance with the proposed plan, we measure depths, not in constant units of length, but in terms of thicknesses corresponding to equal conducting powers and thermal capacities, and if we continue to designate the thickness of the first stratum by its number 3 of French feet, our reckoning for the positions of the different thermometers will stand as follows :—

TABLE VII.

Thermometers numbered downwards.	Depths in true French feet, below No. 1.	Depths in Terms of Conductive Equivalents.
I.	0	0
II.	3	3
III.	9	$3 + \dfrac{\cdot1163}{\cdot1250} \times 6 = 8 \cdot 58$
IV.	21	$8 \cdot 58 + \dfrac{\cdot1128}{\cdot1250} \times 12 = 19 \cdot 41$

According to this way of reckoning depths, we have the following rates of variation of the logarithmic amplitudes, and of the epochs separately, reduced from the previously stated means for the whole period of eighteen years :—

TABLE VIII.

Portions of Rock.	Rates of Diminution of Logarithmic Amplitude per French foot, and Conductive Equivalents.	Rate of Retardation of Epoch per French foot, and Conductive Equivalents.
Between Thermometers Nos. I. and II.	·1286	·1215
,, ,, II. and III.	·1265	·1236
,, ,, III. and IV.	·1236	·1264
Between Thermometers Nos. I. and IV.	·1252	·1248

28. Comparing this Table (VIII.) with Table VI. above we see that the discrepancies are very much diminished; and we cannot doubt but that the conductive power of the rock is less in the lower parts of the rock, and that the amount of the variation is approximately represented by Table VII. We have, however, in Table VIII. still too great discrepancies to allow us to consider variation in the value of c/k, as the only appreciable deviation from Fourier's conditions of uniformity.

29. In endeavouring to find whether these residual discrepancies are owing to variations of k and c not in inverse proportion one to the other, I have taken Fourier's equation

$$c \frac{dv}{dt} = k \frac{d^2v}{dx^2} + \frac{dk}{dx} \frac{dv}{dx},$$

where v denotes the temperature at time t, and at a distance x from an isothermal plane of reference (a horizontal plane through thermometer No. I., for instance); k the conductivity, varying with x; and c the capacity for heat of a unit of volume, which may also vary with x. In this equation I have taken

$$v = a\epsilon^{-P} \cos\left(\frac{2\pi t}{T} - Q\right),$$

where P and Q are functions of x, assumed so as to express as nearly as may be the logarithmic amplitudes, and the epochs, deduced from observation. I have thus obtained two equations of condition, from which I have determined k and c, as functions of x. The problem of finding what must be the conductivity and the specific heat at different depths below the surface, in order that, with all the other conditions of uniformity perfectly fulfilled, the annual harmonic variation may be exactly that which we have found on the average of the eighteen years' term at Calton Hill, is thus solved. The result is, however, far from satisfactory. The small variations in the values of P and Q which we have found in the representation of the observed temperatures, require very large and seemingly unnatural variations in the values of k and c.

30. I can only infer that the residual discrepancies from Fourier's formula shown in Table VIII. are not with any probability attributable to variations of conductivity and specific heat in the rock, and conclude that they are to be explained by irregularities, physical and formal, in the surface. It is possible, indeed, that thermometric errors may have considerable influence, since there is necessarily some uncertainty in the corrections estimated for the temperatures of the different portions of the columns of liquid above the bulbs; and before putting much confidence in the discrepancies we have found, as true expressions of the deviations in the natural circumstances from Fourier's conditions, a careful estimate of the probable or possible amount of error in the observed temperatures should be made. That even with perfect *data* of observation, as great discrepancies should still be found in final reductions such as we have made, need not be unexpected when we consider the nature of the locality, which is described by Professor Forbes in the following terms:—

The position chosen for placing the thermometer was below the surface "in the Observatory enclosure on the Calton Hill, at a height of 350 feet above the sea. The rock is a porphyritic trap, with a somewhat earthy basis, dull and tough fracture. *The exact position is a few yards east of the little transit house.* There are *also other buildings in the neighbourhood.* The ground rises slightly to the east, and *falls abruptly to the west at a distance of fifteen yards.* The immediate surface is flat, *partly covered with grass, partly with gravel*"*.

I have marked by italics those passages which describe

---

* Professor Forbes on the Temperature of the Earth, *Trans. R.S.E.*, 1846, p. 194.

circumstances such as it appears to me might account for the discrepancies in question.

31. *Application to Semi-annual Harmonic Terms.*—The harmonic expressions given above (§ 15) for the average periodic variations for the three stations of Professor Forbes' original series of five years observations, contain semi-annual terms, which are obviously not in accordance with theory. The retardations of epochs and the diminutions of amplitudes are, on the whole, too irregular to be reconcileable by any supposition as to the conductivities and specific heat of the soils and rocks involved, or as to the possible effects of irregularity of surface; and in two of the three stations, the amplitude of the semi-annual term is actually greater as found for the six feet deep than for the three feet deep thermometer, which is clearly an impossible result. The careful manner in which the observations have been made and corrected, seems to preclude the supposition that these discrepancies, especially for the three feet and six feet thermometers, for which the amplitudes of the semi-annual terms are from $°{\cdot}28$ to $°{\cdot}74$ (corresponding to variations of double those amounts, or from $°{\cdot}56$ to $1°{\cdot}48$), can be attributed to errors in the *data*. It must be concluded, therefore, that the semi-annual terms of those expressions do not represent any truly periodic elements of variation, and that they rather depend on irregularities of temperature in the individual years of the term of observation. Hence, until methods for investigating the conduction inwards of non-periodic variations of temperature are applied, we cannot consider that the special features of the progress of temperature during the five years' period at the three stations, from which our apparent semi-annual terms have been derived, have been theoretically analysed. But, as we have seen, every irregularity depending on individual years is perfectly eliminated when the average annual variation over a sufficiently great number of years is taken. Hence it becomes interesting to examine particularly the semi-annual terms for the eighteen years' average of the Calton Hill thermometers, which we now proceed to do.

32. Calculating as above (§ 26), for the coefficients A_1, B_1, the average values of A_2 and B_2, from Professor Forbes' results for his first five years' term, and from the averages for the next thirteen years shown in Table III. above, we find the values of A_2

and B_2 shown in the following table. The amplitudes and epochs are deduced as usual by the formulæ $\sqrt{(A_2{}^2 + B_2{}^2)}$ and $\tan^{-1}\dfrac{B_2}{A_2}$. These reductions I only make for the three feet deep and the six feet deep thermometers, since, for the two others, as may be judged by looking at the thirteen years' average, shown in the former table, the amounts of the semi-annual variation do not exceed the probable errors in the data of observation sufficiently to allow us to draw any reliable conclusions from their apparent values.

TABLE IX.—AVERAGE SEMI-ANNUAL HARMONIC TERM, FROM EIGHTEEN YEARS' OBSERVATIONS AT CALTON HILL.

Depths below surface, in French feet.	A_2 In degrees Fahr.	B_2 In degrees Fahr.	Amplitudes in degrees Fahr.	Epochs in degrees and minutes.
3 feet.	°·1518	°·5842	°·604	75°26′
6 feet.	·0461	·3911	·394	96 43

The ratio of diminution of the amplitude here is $\dfrac{·604}{·394}$ or 1·53, of which the Napierian logarithm is ·426. Dividing this by 3, we find ·142 as the rate of diminution of the logarithmic amplitude per French foot of descent.

The retardation of epoch shown is 21° 17′; and therefore the retardation per French foot of descent is 7° 6′, or, in circular measure, ·1239. If the data were perfect for a periodical variation, and the conditions of uniformity supposed in Fourier's solution were fulfilled, these two numbers would agree, and each would be equal to $\sqrt{\dfrac{2\pi c}{k}}$. Hence, dividing them each by $\sqrt{2}$, we find

Apparent values of $\sqrt{\dfrac{\pi c}{k}}$

·100   (by amplitudes)

·0877   (by epochs).

The true value of $\sqrt{\dfrac{\pi c}{k}}$ must, as we have seen, be ·116, to a very close degree of approximation.

33. When we consider the character of the reduction we have made, and remember that the data were such as to give no

semblance of a theoretical agreement when the first five years' term of observations was taken separately, we may be well satisfied with the approach to agreement presented by these results, depending as they do on only eighteen years in all, and we may expect that, when the average is of a still larger term of observation, the discrepancies will be much diminished. In the mean time, we may regard the semi-annual term we have found for the three feet deep thermometer as representing a true feature of the yearly vicissitude; and it will be surely interesting to find whether it is a constant feature for the locality of Edinburgh, to be reproduced on averages of subsequent terms of observation.

34. It may be remarked, that the nearer to the equator is the locality, the greater relatively will be the semi-annual term; that within the tropics the semi-annual term may predominate, except at the great depths; and that at the equator the tendency is for the annual term to disappear altogether, and to leave a semi-annual term as the first in a harmonic expression of the yearly vicissitude of temperature. The facilities which underground observation affords for the analysis of periodic variations of temperature, when the method of reduction which I have adopted is followed, will, it is to be hoped, induce those who have made similar observations in other localities to apply the same kind of analysis to their results; and it is much to be desired, that the system of observing temperatures at two if not more depths below the surface may be generally adopted at all meteorological stations, as it will be a most valuable means for investigating the harmonic composition of the annual vicissitudes.

III.—Deduction of Conductivities.

35. Notwithstanding the difficulty we have seen must attend any attempt to investigate all the circumstances which must be understood, in order to reconcile perfectly the observed results with theory, the general agreement which we have found is quite sufficient to allow us to form a very close estimate of the ratio of the conductivity of the rock to its specific heat per unit of bulk. Thus, according to the means deduced from the whole period of eighteen years' observation, the average rate of variation of the logarithmic amplitude of the annual term through the whole space of twenty-one feet is ·1157, and of the epoch of the same

term, ·1154. The mean of these, or ·1156, can differ but very little from the true average value of $\sqrt{\dfrac{\pi c}{k}}$ for the portion of rock between the extreme thermometers.

36. Multiplying π by the square of the reciprocal of this number, we find 235·1 as the value of $\dfrac{k}{c}$, or, as we may call it, the conductivity of the rock in terms of the thermal capacity of a cubic foot of its own substance*. In other words, we infer that all the heat conducted in a year (the unit of time) across each square foot of a plate one French foot thick, with its two sides maintained constantly at temperatures differing by 1°, would, if applied to raise the temperature of portions of the rock itself, produce a rise of 1° in 235 cubic feet. As it is difficult (although by no means impossible) to imagine circumstances in which the heat, regularly conducted through a stratum maintained, with its two sides, at perfectly constant temperatures, could be applied to *raise* the temperatures of other portions of the same substance, we may vary the statement of the preceding result, and obtain the following completely realisable illustration.

37. Let a large plate of the rock, everywhere one French foot thick, have every part of one of its sides (which, to avoid circumlocution, we shall call its lower side) maintained at one constant temperature, and let portions of homogeneous substance, at a temperature 1° lower, be continually placed in contact with the upper surface, and removed to be replaced by other homogeneous portions at the same lower temperature, as soon as the temperature of the matter actually thus applied rises in temperature by $\frac{1}{1000}$ of a degree. If this process is continued for a year, the whole quantity of the refrigerating matter thus used to carry away the heat conducted through the stratum must amount to 235,000 cubic feet for each square foot of area, which will be at the rate of ·00745 of a cubic foot per second. We may therefore imagine the process as effected by applying an extra stratum 00745 of a foot thick every second of time. This extra stratum, after lying in contact for one second, will have risen in temperature by $\frac{1}{1000}$ of a degree. By means of the information contained in this apparently unpractical statement, many interesting

* In respect to this definition see § 82 of Art. XCII. Part II. above.

problems may be practically solved, as I hope to show in a subsequent communication.

38. The value of $\sqrt{\dfrac{\pi c}{k}}$, derived from the whole eighteen-years'
period of observation (·1156), differs so little from that (·1154)
found previously (§ 16) from Professor Forbes' observations and
reductions of the first five of the years, that we may feel much
confidence in the accuracy of the values ·1098 and 06744, which,
from his five years' data alone, we found (§ 16) for the corresponding
constant with reference to the sand at the Experimental Garden
and the sandstone of Craigleith Quarry.   From them, calculating
as above (§ 36), we find 260·5 and 690·7 as the values of $\dfrac{k}{c}$ for
the terrestrial substances of these localities respectively; results
of which the meaning is illustrated by the statements of §§ 36
and 37.

39. To deduce the conductivities of the strata, in terms of
uniform thermal units, Professor Forbes had the "specific heats"
of the substances determined experimentally by M. Regnault.
The results, multiplied by the specific gravities, gave for the
thermal capacities of portions of the three substances, in terms
of that of an equal bulk of water, the values ·5283, 3006, and
·4623 respectively.   Now, these must be the values of c, if the
thermal unit in which k is measured is the thermal capacity of a
French cubic foot of water.   Multiplying the values of $\dfrac{k}{c}$ found
above by these values of c, we find for k the following values :—

Trap-rock of Calton Hill.   Sand of Experimental Gardens.   Sandstone of Craigleith.

124·2,   78·31,   319·3,

The values found by Professor Forbes were—

111·2,   82·6,   298·3.

Although many comparisons have been made between the
conducting powers of different substances, scarcely any data as
to thermal conductivity in absolute measure have been hitherto
published, except these of Professor Forbes, and probably none
approaching to their accuracy.   The slightly different numbers
to which we have been led by the preceding investigation are no
doubt still more accurate.

40. To reduce these results to any other scale of linear measurement, we must clearly alter them in the inverse ratio of the square of the absolute lengths chosen for the units*. The length of a French foot being 1·06575 of the British standard foot, we must therefore multiply the preceding numbers by 1·13581, to reduce them to convenient terms.

41. We may, lastly, express them in terms of the most common unit, which is the quantity of heat required to raise the temperature of a grain of water by 1°; and to do this we have only to multiply each of them by 7000 × 62·447, being the weight of a cubic foot in grains.

42. The following table contains a summary of our results as to conductivity expressed in several different ways, one or other of which will generally be found convenient :—

TABLE X. — THERMAL CONDUCTIVITIES OF EDINBURGH STRATA, IN BRITISH ABSOLUTE UNITS [UNIT OF LENGTH, THE ENGLISH FOOT].

Description of Terrestrial Substance.	Conductivities in Terms of Thermal Capacity of Unit Bulk of Substance (k/c).			Conductivities in Terms of Thermal Capacity of Unit Bulk of Water k.			Conductivities in Terms of Thermal Capacity of One Grain of Water.
	Per Ann.	Per 24ʰ.	Per Second.	Per Ann.	Per 24ʰ.	Per Second.	Per Second.
Trap-rock of Calton Hill	267·0	·7310	·000008461	141·1	·3863	·000004471	1·9544
Sand of Experimental Garden	295·9	·8100	·000009375	88·9	·2435	·000002818	1·2319
Sandstone of Craigleith Quarry	784·5	2·1478	·00002486	362·7	·9929	·00001149	5·0225

* Because the absolute amount of heat flowing through the plate across equal areas will be inversely as the thickness of the plate ; and the effect of equal quantities of heat in raising the temperature of equal areas of the water will be inversely as the depth of the water. The same thing may be perhaps more easily seen by referring to the elementary definition of thermal conductivity (footnote to § 11, above). The absolute quantity of heat conducted across unit area of a plate of unit thickness, with its two sides maintained at temperatures differing by always the same amount, will be directly as the areas, and inversely as the thickness, and therefore simply as the absolute length chosen for unity. But the thermal unit in which these quantities are measured, being the capacity of a unit bulk of water, is directly as the cube of the unit length, and therefore the numbers expressing the quantities of heat compared will be inversely as the cubes of the lengths chosen for unity, and directly as these simple lengths : that is to say, finally, they will be inversely as the squares of these lengths.

43. The statements (§§ 36 and 37) by which the signification of k/c has been defined and illustrated, require only to have *cubic feet of water* substituted for *cubic feet of rock*, in their calorimetric specifications, to be applicable similarly to define and illustrate the meaning of the conductivity denoted by k. The fluidity of the water allows a modified and somewhat simpler explanation, equivalent to that of § 36, to be now given, as follows:—

44. If a long rectangular plate of rock, one foot thick, in a position slightly inclined to the horizontal, have water one foot deep flowing over it in a direction parallel to its length, and if the lower surface of the plate be everywhere kept 1° higher in temperature than the upper, the water must flow at the rate of k times the length of the plate per unit of time, in order that the heat conducted through the plate may raise it just 1° in temperature in its flow over the whole length. [It must be understood here, that the plate becomes warmer, on the whole, under the lower parts of the stream of water, its upper surface being everywhere at the same temperature as the water in contact with it, while its lower surface is, by hypothesis, at a temperature 1° higher.] If, for instance, the plate be of Calton Hill trap-rock, the water must, according to the result we have found, flow at the rate of 141·1 times its length in a year, or of ·3863 of its length in twenty-four hours, to be raised just 1° Fah. in temperature in flowing over it. Thus water, one French foot deep, flowing over a plane bed of such rock at the rate of ·3863 of a mile in twenty-four hours, will, in flowing one mile, have its temperature raised 1° Fah. by heat conducted through the plate. The rates required to fulfil similar conditions for the sand of the Experimental Garden and the sandstone of Craigleith Quarry are similarly found to be ·2435 of the length and ·9929 of the length, in twenty-four hours.

APPENDIX.

On the Reduction of Periodical Variations of Underground Temperature, with applications to the Edinburgh Observations.

[Brit. Assoc. Report for 1859, pp. 54—56.]

The principle followed in the reductions which form the subject of this communication may be briefly stated thus :—

The varying temperature during a year, shown by any one of the underground thermometers on an average for a series of years, is expressed by the ordinary method in a trigonometrical series of terms representing simple harmonic variations*,—the first having a year for its period, the second a half-year, the third a third part of a year, and so on. The yearly term of the series is dealt with separately for the thermometers at the different depths, the half-yearly term also separately, and so on, each term being treated as if the simple periodic variation which it represents were the sole variation experienced. The elements into which the whole variation is thus analysed are examined so as to test their agreement with the elementary formulæ by which Fourier expressed the periodic variations of temperature in a bar protected from lateral conduction, and experiencing a simple harmonic variation of temperature at one end, or in an infinite solid experiencing at every point of an infinite plane through it a variation of temperature according to the same elementary law. In any locality in which the surface of the earth is sensibly plane and uniform all round to distances amounting at least to considerable multiples of the depth of the lowest thermometer, and in which the conducting power of the soil or rock below the surface is perfectly uniform to like distances round and below

* By a simple harmonic variation is meant a variation in proportion to the height of a point which moves uniformly in a vertical circle.

the thermometers, this theory must necessarily be found in excessively close agreement with the observed results. The comparison which is made in the investigations now brought forward must be regarded, therefore, not as a test of the correctness of a theory which has mathematical certainty, but as a means of finding how much the law of propagation of heat into the soil is affected by the very notable deviations from the assumed conditions of uniformity as to surface, or by possible inequalities of underground conductivity existing in the localities of observation. When those conditions of uniformity are perfectly fulfilled both by the surface and by the substance below it, the law of variation in the interior produced by a simple harmonic variation of temperature at the surface, as investigated by Fourier, may be stated in general terms in the three following propositions :—(1) The temperature at every interior point varies according to the simple harmonic law, in a period retarded by an equal interval of time, and with an amplitude diminished in one and the same proportion, for all equal additions of depth. (2) The absolute measure, in ratio of arc to radius, for the retardation of phase, is equal to the diminution of the Neperian logarithm of the amplitude ; and each of these, reckoned per unit of length as to augmentation of distance from the surface, is equal to the square root of the quotient obtained by dividing the product of the ratio of the circumference of a circle to its diameter, into the thermal capacity of a unit of bulk of the solid, by the thermal conductivity of the same estimated for the period of the variation as unity of time. (3) For different periods, the retardations of phase, measured each in terms of a whole period, and the diminutions of the logarithm of the amplitude, all reckoned per unit of depth, are inversely proportional to the square roots of the periods.

The first series of observations examined by the method thus described were those instituted by Professor Forbes, and conducted under his superintendence during five years, in three localities of Edinburgh and the immediate neighbourhood : (1) the trap-rock of Calton Hill ; (2) the sand below the soil of the Experimental Garden ; and (3) the sandstone of Craigleith Quarry. In each place there were, besides a surface thermometer, four thermometers at the depths of 3, 6, 12, and 24 French feet respectively. The diminution in the amplitude, and the retardation of phase in going downwards, have been determined for the annual, for the half-yearly, third-yearly, and the quarterly term, on the average for these five years for each locality. The same has been determined for the average of twelve years of observation, continued on Calton Hill by the staff of the Royal Edinburgh Observatory.

The following results with reference to the annual harmonic term are selected for example:—

Average of five years, 1837 *to* 1842.

	Retardation of phase in days, per French foot of descent.	Retardation of phase in circular measure, per French foot of descent.	Diminution of Napierian logarithm of amplitude, per French foot of descent.
Calton Hill.			
3 feet to 6 feet.	......	·11635	·12625
6 ,, 12 ,,	......	·11344	·12156
12 ,, 24 ,,	......	·11490	·10959
Mean	13⅓ days.	·1149	·11914
Experimental Gardens.			
3 feet to 6 feet.	......	·11635	·10037
6 ,, 12 ,,	......	·11929	·11304
12 ,, 24 ,,	......	·10617	·10844
Mean	13⅙ days.	·11314	·10728
Craigleith Quarry.			
3 feet to 6 feet.	......	·063995	·09372
6 ,, 12 ,,	......	·066903	·06304
12 ,, 24 ,,	......	·066903	·06476
Mean	7½ days.	·065934	·07384

If Fourier's conditions of uniformity, stated above, were fulfilled strictly, the numbers shown in the second column would be all equal among one another, and equal to those in the third column. The differences between the actual numbers are surprisingly small, but are so consistent that they cannot be attributed to errors of observation. It is possible they may be due to a want of perfect agreement in the values of a degree on the different thermometric scales; but it seems more probable that they represent true discrepancies from theory, and are therefore excessively interesting, and possibly of high importance with a view to estimating the effects of inequalities of surface and of the interior conductivity. The final means of the numbers in the second and third columns are, for

> Calton Hill.................. ·11702
> Experimental Gardens..... ·11061
> Craigleith Quarry........... ·06988

The thermal capacities of specimens of the trap-rock, the sand, and the sandstone of the three localities were, at the request of Professor Forbes, measured by Regnault, and found to be respectively

> ·5283, ·3006, and ·4623.

Hence, according to proposition (3), stated above, the thermal conductivities are as follows:—

> Trap-rock of Calton Hill............ 121·2
> Sand of Experimental Gardens ... 77·19
> Sandstone of Craigleith Quarry ... 273·6

These numbers do not differ much from those given by Professor Forbes, who for the first time derived determinations of thermal conductivity in absolute measure from observations of terrestrial temperature. In consequence of the peculiar mode of reduction followed in the present investigation, it may be assumed that the estimates of conductivity now given are closer approximations to the truth. To reduce to the English foot as unit of length, we must multiply by the square of 1·06575; to reduce, further, to the quantity of heat required to raise 1 lb. of water by 1° Fah. as unit of heat, we must multiply by 66·447; and lastly, to reduce to a day as unit of time, we must divide by 365¼. We thus find the following results:—

> Trap-rock of Calton Hill............ 23·5
> Sand of Experimental Gardens.... 15·0
> Sandstone of Craigleith Quarry.... 53·5

These numbers show the quantities of heat per square foot conducted in a day through a layer of the material 1 foot thick, kept with its two surfaces at a difference of temperature of 1° Fah., —the unit of heat being, for instance, the quantity required to raise 1000 lbs. of water by $\frac{1}{1000}$th of a degree in temperature.

Art. XCIV. On the Secular Cooling of the Earth.

[*Transactions of the Royal Society of Edinburgh*, Vol. XXIII.
Read April 28, 1862.]

1. For eighteen years it has pressed on my mind, that essential principles of Thermo-dynamics have been overlooked by those geologists who uncompromisingly oppose all paroxysmal hypotheses, and maintain not only that we have examples now before us, on the earth, of all the different actions by which its crust has been modified in geological history, but that these actions have never, or have not on the whole, been more violent in past time than they are at present.

2. It is quite certain the solar system cannot have gone on even as at present, for a few hundred thousand or a few million years, without the irrevocable loss (by dissipation, not by *annihilation*) of a very considerable proportion of the entire energy initially in store for sun heat, and for Plutonic action. It is quite certain that the whole store of energy in the solar system has been greater in all past time, than at present; but it is conceivable that the rate at which it has been drawn upon and dissipated, whether by solar radiation, or by volcanic action in the earth or other dark bodies of the system, may have been nearly equable, or may even have been less rapid, in certain periods of the past. But it is far more probable that the secular rate of dissipation has been in some direct proportion to the total amount of energy in store, at any time after the commencement of the present order of things, and has been therefore very slowly diminishing from age to age.

3. I have endeavoured to prove this for the sun's heat, in an

article recently published in *Macmillan's Magazine**, where I have shown that most probably the sun was sensibly hotter a million years ago than he is now. Hence, geological speculations assuming somewhat greater extremes of heat, more violent storms and floods, more luxuriant vegetation, and hardier and coarser-grained plants and animals, in remote antiquity, are more probable than those of the extreme quietist, or "uniformitarian," school. A "middle path," not generally safest in scientific speculation, seems to be so in this case. It is probable that hypotheses of grand catastrophes destroying all life from the earth, and ruining its whole surface at once, are greatly in error; it is impossible that hypotheses assuming an equability of sun and storms for 1,000,000 years, can be wholly true.

4. Fourier's mathematical theory of the conduction of heat is a beautiful working out of a particular case belonging to the general doctrine of the "Dissipation of Energy†." A characteristic of the practical solutions it presents is, that in each case a distribution of temperature, becoming gradually equalised through an unlimited future, is expressed as a function of the time, which is infinitely divergent for all times longer past than a definite determinable epoch. The distribution of heat at such an epoch is essentially *initial*—that is to say, it cannot result from any previous condition of matter by natural processes. It is, then, well called an "*arbitrary* initial distribution of heat," in Fourier's great mathematical poem, because that which is rigorously expressed by the mathematical formula could only be realised by action of a power able to modify the laws of dead matter. In an article published about nineteen years ago in the *Cambridge Mathematical Journal* ‡," I gave the mathematical criterion for an essentially initial distribution; and in an inaugural essay, *De Motu Caloris per Terræ Corpus*, read before the Faculty of the University of Glasgow in 1846, I suggested, as an application of these principles, that a perfectly complete geothermic survey

---

* "On the Age of the Sun's Heat," March, 1862: [also *Popular Lectures and Addresses*, Macmillan, 1889].

† *Proceedings Royal Soc. Edin.* Feb. 1852, "On a Universal Tendency in Nature to the Dissipation of Mechanical Energy." Also, "On the Restoration of Energy in an Unequally Heated Space," *Phil. Mag.*, 1853, first half year. [Articles LIX. and LXII. Vol. I. above.]

‡ February, 1844.—"Note on Certain Points in the Theory of Heat:" [Art. X. Vol. I. above.]

would give us data for determining an initial epoch in the problem of terrestrial conduction.   At the meeting of the British Association in Glasgow in 1855, I urged that special geothermic surveys should be made for the purpose of estimating absolute dates in geology, and I pointed out some cases, especially that of the salt-spring borings at Creuznach, in Rhenish Prussia, in which eruptions of basaltic rock seem to leave traces of their igneous origin in residual heat*. I hope this suggestion may yet be taken up, and may prove to some extent useful; but the disturbing influences affecting underground temperature, as Professor Phillips has well shown in a recent inaugural address to the Geological Society, are too great to allow us to expect any very precise or satisfactory results.

5. The chief object of the present communication is to estimate from the known general increase of temperature in the earth downwards, the date of the first establishment of that *consistentior status*, which, according to Leibnitz's theory, is the initial date of all geological history.

6. In all parts of the world in which the earth's crust has been examined, at sufficiently great depths to escape large influence of the irregular and of the annual variations of the superficial temperature, a gradually increasing temperature has been found in going deeper. The rate of augmentation (estimated at only $\frac{1}{110}$th of a degree, Fahr., in some localities, and as much as $\frac{1}{15}$th of a degree in others, per foot of descent) has not been observed in a sufficient number of places to establish any fair average estimate for the upper crust of the whole earth.   But $\frac{1}{50}$th is commonly accepted as a rough mean; or, in other words, it is assumed as a result of observation, that there is, on the whole, about 1° Fahr. of elevation of temperature per 50 British feet of descent.

7. The fact that the temperature increases with the depth implies a continual loss of heat from the interior, by conduction outwards through or into the upper crust.   Hence, since the upper crust does not become hotter from year to year, there must be a secular loss of heat from the whole earth. It is possible that no cooling may result from this loss of heat, but only an exhaustion of potential energy, which in this case could scarcely

---

* See *British Association Report of* 1855 (Glasgow) Meeting, [Art. LXXXVII. Vol. II. above].

be other than chemical affinity between substances forming part of the earth's mass.   But it is certain that either the earth is becoming on the whole cooler from age to age, or the heat conducted out is generated in the interior by temporary dynamical (that is, in this case, chemical) action *   To suppose, as Lyell, adopting the chemical hypothesis, has done †, that the substances, combining together, may be again separated electrolytically by thermo-electric currents, due to the heat generated by their combination, and thus the chemical action and its heat continued in an endless cycle, violates the principles of natural philosophy in exactly the same manner, and to the same degree, as to believe that a clock constructed with a self-winding movement may fulfil the expectations of its ingenious inventor by going for ever.

8.   It must indeed be admitted that many geological writers of the Uniformitarian school, who in other respects have taken a profoundly philosophical view of their subject, have argued in a most fallacious manner against hypotheses of violent action in past ages.   If they had contented themselves with showing that many existing appearances, although suggestive of extreme violence and sudden change, may have been brought about by long-continued action, or by paroxysms not more intense than some of which we have experience within the periods of human history, their position might have been unassailable; and certainly could not have been assailed except by a detailed discussion of their facts.   It would be a very wonderful, but not an absolutely in-credible result, that volcanic action has never been more violent on the whole than during the last two or three centuries; but it is as certain that there is now less volcanic energy in the whole earth than there was a thousand years ago, as it is that there is less gunpowder in a "Monitor" after she has been seen to dis-charge shot and shell, whether at a nearly equable rate or not, for five hours without receiving fresh supplies, than there was at the beginning of the action.   Yet this truth has been ignored or

---

* Another kind of dynamical action, capable of generating heat in the interior of the earth, is the friction which would impede tidal oscillations if the earth were partially or wholly constituted of viscous matter.   See a paper by Prof. G. H. Darwin, "On problems connected with the tides of a viscous spheroid," *Phil. Proc. Roy.* 1879, Part II.   W. T.   July, 1883.

† *Principles of Geology*, chap. xxxi. ed. 1853.

denied by many of the leading geologists of the present day*, because they believe that the facts within their province do not demonstrate greater violence in ancient changes of the earth's surface, or do demonstrate a nearly equable action in all periods.

9. The chemical hypothesis to account for underground heat might be regarded as not improbable, if it was only in isolated localities that the temperature was found to increase with the depth; and, indeed, it can scarcely be doubted that chemical action exercises an appreciable influence (possibly negative, however) on the action of volcanoes; but that there is slow uniform "combustion," *eremacausis*, or chemical combination of any kind going on, at some great unknown depth under the surface everywhere, and creeping inwards gradually as the chemical affinities in layer after layer are successively saturated, seems extremely improbable, although it cannot be pronounced to be absolutely impossible, or contrary to all analogies in nature. The less hypothetical view, however, that the earth is merely a warm chemically inert body cooling, is clearly to be preferred in the present state of science.

10. Poisson's celebrated hypothesis, that the present underground heat is due to a passage, at some former period, of the solar system through hotter stellar regions, cannot provide the circumstances required for a paleontology continuous through that epoch of external heat. For from a mean of values of the conductivity, in terms of the thermal capacity of unit volume, of the earth's crust, in three different localities near Edinburgh, which I have deduced from the observations on underground temperature instituted by Principal Forbes there, I find that if the supposed transit through a hotter region of space took place between 1250 and 5000 years ago, the temperature of that supposed region must have been from 25° to 50° Fahr. above the present mean temperature of the earth's surface, to account for the present general rate of underground increase of temperature, taken as 1° Fahr. in 50 feet downwards. Human history negatives this supposition. Again, geologists and astronomers will, I presume, admit that the earth cannot, 20,000 years ago, have been

* It must be borne in mind that this was written in 1862. The opposite statement concerning the beliefs of geologists would probably be now [1889] nearer the truth. W. T.

in a region of space 100° Fahr. warmer than its present surface. But if the transition from a hot region to a cool region supposed by Poisson took place more than 20,000 years ago, the excess of temperature must have been more than 100° Fahr., and must therefore have destroyed animal and vegetable life. Hence, the farther back and the hotter we can suppose Poisson's hot region, the better for the geologists who require the longest periods; but the best for their view is Leibnitz's theory, which simply supposes the earth to have been at one time an incandescent liquid, without explaining how it got into that state. If we suppose the temperature of melting rock to be about 10,000° Fahr. (an extremely high estimate), the consolidation may have taken place 200,000,000 years ago. Or, if we suppose the temperature of melting rock to be 7000° Fahr. (which is more nearly what it is generally assumed to be), we may suppose the consolidation to have taken place 98,000,000 years ago.

11. These estimates are founded on the Fourier solution demonstrated below. The greatest variation we have to make on them, to take into account the differences in the ratios of conductivities to specific heats of the three Edinburgh rocks, is to reduce them to nearly half, or to increase them by rather more than half. A reduction of the Greenwich underground observations recently communicated to me by Professor Everett of Windsor, Nova Scotia [now, 1889, of Queen's College, Belfast], gives for the Greenwich rocks a quality intermediate between those of the Edinburgh rocks. But we are very ignorant as to the effects of high temperatures in altering the conductivities and specific heats of rocks, and as to their latent heat of fusion. We must, therefore, allow very wide limits in such an estimate as I have attempted to make; but I think we may with much probability say that the consolidation cannot have taken place less than 20,000,000 years ago, or we should have more underground heat than we actually have, nor more than 400,000,000 years ago, or we should not have so much as the least observed underground increment of temperature. That is to say, I conclude that Leibnitz's epoch of emergence of the *consistentior status* was probably between those dates.

12. The mathematical theory on which these estimates are founded is very simple, being in fact merely an application of

one of Fourier's elementary solutions to the problem of finding
at any time the rate of variation of temperature from point to
point, and the actual temperature at any point, in a solid extending
to infinity in all directions, on the supposition that at an initial
epoch the temperature has had two different constant values on
the two sides of a certain infinite plane.   The solution for the
two required elements is as follows :—

$$\frac{dv}{dx} = \frac{V}{\sqrt{\pi \kappa t}} \epsilon^{-\frac{x^2}{4\kappa t}},$$

$$v = v_0 + \frac{2V}{\sqrt{\pi}} \int_0^{\frac{x}{2\sqrt{\kappa t}}} dz \epsilon^{-z^2},$$

where κ denotes the conductivity of the solid, measured in terms
of the thermal capacity of the unit of bulk;

V, half the difference of the two initial temperatures;

v_0, their arithmetical mean;

t, the time;

x, the distance of any point from the middle plane;

v, the temperature of the point x at time t;

and, consequently (according to the notation of the differential
calculus), dv/dx the rate of variation of the temperature per unit
of length perpendicular to the isothermal planes.

13.   To demonstrate this solution, it is sufficient to verify—

(1) That the expression for v satisfies Fourier's equation for
the linear conduction of heat, viz.;

$$\frac{dv}{dt} = \kappa \frac{d^2v}{dx^2};$$

(2) That when $t = 0$, the expression for v becomes $v_0 + V$ for
all positive, and $v_0 - V$ for all negative, values of x;

and (3), That the expression for dv/dx is the differential coefficient of the expression for v with reference to x.

The propositions (1) and (3) are proved directly by differentiation.   To prove (2), we have, when $t = 0$, and x positive,

$$v = v_0 + \frac{2V}{\pi} \int_0^\infty dz \epsilon^{-z^2},$$

or according to the known value, $\frac{1}{2}\sqrt{\pi}$, of the definite integral
$\int_0^\infty dz \epsilon^{-z^2}$, $v = v_0 + V$;

and for all values of t, the second term has equal positive and negative values for equal positive and negative values of x, so that when $t = 0$ and x negative,

$$v = v_0 - V.$$

The admirable analysis by which Fourier arrived at solutions including this, forms a most interesting and important mathematical study. It is to be found in his *Théorie Analytique de la Chaleur.* Paris, 1822.

14. The accompanying diagram (page 303) represents, by two curves, the preceding expressions for dv/dx, and v respectively.

15. The solution thus expressed and illustrated applies, for a certain time, without sensible error, to the case of a solid sphere, primitively heated to a uniform temperature, and suddenly exposed to any superficial action, which for ever after keeps the surface at some other constant temperature. If, for instance, the case considered is that of a globe, 8000 miles in diameter, of solid rock, the solution will apply with scarcely sensible error for more than 1000 millions of years. For, if the rock be of a certain average quality as to conductivity and specific heat, the value of κ, as I have shown in a previous communication to the Royal Society[*], will be 400, to unit of length a British foot and unit of time a year; and the equation expressing the solution becomes

$$\frac{dv}{dx} = \frac{V}{35 \cdot 4} \cdot \frac{1}{\sqrt{t}} \cdot \epsilon^{-x^2/1600t};$$

and if we give t the value 1,000,000,000, or anything less, the exponential factor becomes less than $\epsilon^{-5 \cdot 6}$ (which being equal to about 1/270, may be regarded as insensible), when x exceeds 3,000,000 feet, or 568 miles. That is to say, during the first 1000 million years the variation of temperature does not become sensible at depths exceeding 568 miles, and is therefore confined to so thin a crust, that the influence of curvature may be neglected.

16. If, now, we suppose the time to be 100 million years from the commencement of the variation, the equation becomes

$$\frac{dv}{dx} = \frac{1}{3 \cdot 54 \times 10^5} V \epsilon^{-x^2/1600 \times 10^8}$$

[*] "On the Reduction of Observations of Underground Temperature." *Trans. Roy. Soc. Edin.*, March, 1860 [Art. xciii. above].

INCREASE OF TEMPERATURE DOWNWARDS IN THE EARTH.

$ON = x.$

$NP' = be^{-x^2/a^2} = y'.$

$NP = \text{area } ONP'A \div a = \dfrac{1}{a} \displaystyle\int_0^x y' dx\, {}^*.$

$a = 2\sqrt{\kappa t}.$

$\dfrac{dv}{dx} = \dfrac{V}{a} \cdot \dfrac{NP}{b\frac{1}{2}\sqrt{\pi}}.$

$v - v_0 = V \cdot \dfrac{NP}{b \cdot \frac{1}{2}\sqrt{\pi}}.$

The curve OPQ shows excess of temperature above that of the surface.

The curve $AP'R$ shows rate of augmentation of temperature downwards.

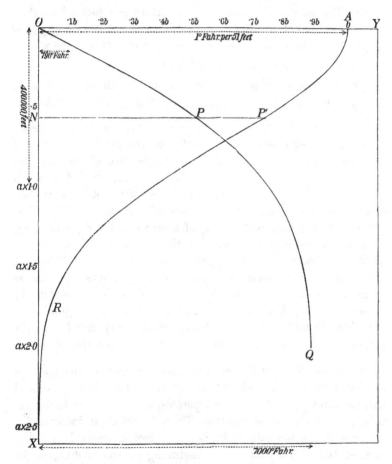

* A table of the values of this integral, sometimes now called the "Error Function," is to be found in Table III. of De Morgan's article on "The Theory of Probabilities," *Encyclopaedia Metropolitana*, Edition 1845, Vol. II. W. T. March 27, 1889.

The diagram, therefore, shows the variation of temperature which would now exist in the earth, if its whole mass being first solid and at one temperature 100 million years ago, the temperature of its surface had been everywhere suddenly lowered by V degrees, and kept permanently at this lower temperature: the scales used being as follows :—

(1) For depth below the surface,—scale along OX, length a, represents 400,000 feet.

(2) For rate of increase of temperature per foot of depth,—scale of ordinates parallel to OY, length b, represents $\frac{1}{354000}$ of V per foot. If, for example, $V = 7000°$ Fahr., this scale will be such that b represents $\frac{1}{50}$ of a degree per foot.

(3) For excess of temperature,—scale of ordinates parallel to OY, length b, represents $V/\frac{1}{2}\sqrt{\pi}$, or $7900°$, if $V = 7000°$ Fahr.

Thus the rate of increase of temperature from the surface downwards would be sensibly $\frac{1}{51}$ of a degree per foot for the first 100,000 feet or so. Below that depth the rate of increase per foot would begin to diminish sensibly. At 400,000 feet it would have diminished to about $\frac{1}{141}$ of a degree per foot. At 800,000 feet it would have diminished to less than $\frac{1}{50}$ of its initial value, —that is to say, to less than $\frac{1}{2550}$ of a degree per foot; and so on, rapidly diminishing, as shown in the curve. Such is, on the whole, the most probable representation of the earth's present temperature, at depths of from 100 feet, where the annual variations cease to be sensible, to 100 miles; below which the whole mass, or all, except a nucleus cool from the beginning, is (whether liquid or solid) probably at, or very nearly at, the proper melting temperature for the pressure at each depth.

17.   The theory indicated above throws light on the question so often discussed as to whether terrestrial heat can have influenced climate through long geological periods, and allows us to answer it very decidedly in the negative. There would be an increment of temperature at the rate of 2° Fahr. per foot downwards near the surface, 10,000 years after the beginning of the cooling, in the case we have supposed. The radiation from earth and atmosphere into space (of which we have yet no satisfactory absolute measurement) would almost certainly be so rapid in the earth's actual circumstances, as not to allow a rate of increase of 2° Fahr. per

foot underground to augment the temperature of the surface by much more than about 1°; and hence I infer that the general climate cannot be sensibly affected by conducted heat, at any time more than 10,000 years after the commencement of superficial solidification. No doubt, however, in particular places there might be an elevation of temperature by thermal springs, or by eruptions of melted lava, and everywhere vegetation would, for the first three or four million years, if it existed so soon after the epoch of consolidation, be influenced by the sensibly higher temperature met with by roots extending a foot or more below the surface.

18. Whatever the amount of such effects is at any one time, it would go on diminishing according to the inverse proportion of the square roots of the times from the initial epoch. Thus, if at 10,000 years we have 2° per foot of increment below ground,

At   40,000 years we should have   1° per foot.
„   160,000   „   „   „ $\frac{1}{2}$°   „
„   4,000,000   „   „   „ $\frac{1}{10}$°   „
„   100,000,000   „   „   „ $\frac{1}{50}$°   „

It is, therefore, probable that for the last 96,000,000 years the rate of increase of temperature underground has gradually diminished from about $\frac{1}{10}$th to about $\frac{1}{50}$th of a degree Fahrenheit per foot, and that the thickness of the crust through which any stated degree of cooling has been experienced has in that period gradually increased up to its present thickness from $\frac{1}{5}$th of that thickness. Is not this, on the whole, in harmony with geological evidence, rightly interpreted? Do not the vast masses of basalt, the general appearances of mountain ranges, the violent distortions and fractures of strata, *the great prevalence of metamorphic action* (which must have taken place at depths of not many miles if so much), all agree in demonstrating that the rate of increase of temperature downwards must have been much more rapid, and in rendering it probable that volcanic energy, earthquake shocks, and every kind of so-called Plutonic action, have been, on the whole, more abundantly and violently operative in geological antiquity than in the present age?

19. But it may be objected to this application of mathematical theory—(1), That the earth was once all melted, or at least melted all round its surface, and cannot possibly, or rather cannot with any probability, be supposed to have been ever a uni-

formly heated solid, 7000° Fahr. warmer than our present surface
temperature, as assumed in the mathematical problem; and (2),
No natural action could possibly produce at one instant, and
maintain for ever after, a seven thousand degrees' lowering of
the surface temperature. Taking the second objection first, I
answer it by saying, what I think cannot be denied, that a large
mass of melted rock, exposed freely to our air and sky, will, after
it once becomes crusted over, present in a few hours, or a few
days, or at the most a few weeks, a surface so cool that it can be
walked over with impunity. Hence, after 10,000 years, or, indeed,
I may say after a single year, its condition will be sensibly the
same as if the actual lowering of temperature experienced by the
surface had been produced in an instant and maintained constant
ever after. I answer the first objection by saying, that if experimenters will find the latent heat of fusion, and the variations of
conductivity and specific heat of the earth's crust up to its melting
point, it will be easy to modify the solution given above, so as
to make it applicable to the case of a liquid globe gradually
solidifying from without inwards, in consequence of heat conducted
through the solid crust to a cold external medium. In the meantime, we can see that this modification will not make any considerable change in the resulting temperature of any point in
the crust, unless the latent heat parted with on solidification
proves, contrary to what we may expect from analogy, to be considerable in comparison with the heat that an equal mass of the
solid yields in cooling from the temperature of solidification to the
superficial temperature. But, what is more to the purpose, it is
to be remarked that the objection, plausible as it appears, is
altogether fallacious, and that the problem solved above corresponds much more closely, in all probability, with the actual
history of the earth, than does the modified problem suggested by
the objection. The earth, although once all melted, or melted
all round its surface, did, in all probability, really become a solid
at its melting temperature all through, or all through the outer
layer, which had been melted; and not until the solidification was
thus complete, or nearly so, did the surface begin to cool. That
this is the true view can scarcely be doubted, when the following
arguments are considered.

20. In the first place, we shall assume that at one time the
earth consisted of a solid nucleus, covered all round with a very

deep ocean of melted rocks, and left to cool by radiation into space. This is the condition that would supervene, on a cold body much smaller than the present earth meeting a great number of cool bodies still smaller than itself, and is therefore in accordance with what we may regard as a probable hypothesis regarding the earth's antecedents. It includes, as a particular case, the commoner supposition, that the earth was once melted throughout, a condition which might result from the collision of two nearly equal masses. But the evidence which has convinced most geologists that the earth had a fiery beginning, goes but a very small depth below the surface, and affords us absolutely no means of distinguishing between the actual phenomena, and those which would have resulted from either an entire globe of liquid rock, or a cool solid nucleus covered with liquid to any depth exceeding 50 or 100 miles. Hence, irrespectively of any hypothesis as to antecedents from which the earth's initial fiery condition may have followed by natural causes, and simply assuming, as rendered probable by geological evidence, that there was at one time melted rock all over the surface, we need not assume the depth of this lava ocean to have been more than 50 or 100 miles; although we need not exclude the supposition of any greater depth, or of an entire globe of liquid.

21. In the process of refrigeration, the fluid must (as I have remarked regarding the sun, in the recent article in *Macmillan's Magazine** already referred to, and regarding the earth's atmosphere, in a communication to the Literary and Philosophical Society of Manchester†) be brought by convection, to fulfil a definite law of distribution of temperature which I have called "convective equilibrium of temperature." That is to say, the temperatures at different parts in the interior must, in any great fluid mass which is kept well stirred, differ according to the different pressures by the difference of temperatures which any one portion of the liquid would present, if given at the temperature and pressure of any one part, and then subjected to variation of pressure, while prevented from losing or gaining heat. The

* "On the Age of the Sun's Heat," March, 1862: also *Popular Lectures and Addresses*, Vol. I. Macmillan, 1889.

† "On the Convective Equilibrium of Temperature in the Atmosphere," read Jan. 21, 1862: published in the *Memoirs*, Vol. II. of 3rd Series; [Art. XCII. above, Appendix E.]

reason for this is the extreme slowness of true thermal conduction; and the consequently preponderating influence of great
currents throughout a continuous fluid mass, in determining the
distribution of temperature through the whole.

22. The thermo-dynamic law connecting temperature and
pressure in a fluid mass, not allowed to lose or gain heat, investigated theoretically, and experimentally verified in the cases of air
and water, by Dr Joule and myself*, shows, therefore, that the
temperature in the liquid will increase from the surface downwards, if, as is most probably the case, the liquid contracts in
cooling. On the other hand, if the liquid, like water near its
freezing point, expanded in cooling, the temperature, according to the convective and thermo-dynamic laws just stated
(§§ 21, 22), would actually be lower at great depths than near
the surface, even although the liquid is cooling from the surface;
but there would be a very thin superficial layer of lighter and
cooler liquid, losing heat by true conduction, until solidification at
the surface would commence.

23. Again, according to the thermo-dynamic law of freezing,
investigated by my brother, Professor James Thomson †, and
verified by myself experimentally for water‡, the temperature
of solidification will, at great depths, because of the great pressure
there, be higher there than at the surface if the fluid contracts, or
lower than at the surface if it expands, in becoming solid.

24. How the temperature of solidification, for any pressure,
may be related to the corresponding temperature of fluid con-

---

* Joule, "On the Changes of Temperature produced by the Rarefaction and
Condensation of Air," *Philosophical Magazine*, May, 1845; or Joule's *Scientific
Papers*, Vol. I. London, 1884. Thomson, "On a Method for Discovering Experimentally the Mechanical Work spent, and the Heat produced by the Compression
of a Gaseous Fluid," *Trans. Roy. Soc. Edin.*, Vol. xx. Part II., read April 21, 1851;
Philosophical Magazine, December, 1852 [Art. XLVIII. Part IV. §§ 61–80, Vol. I.
above]. Joule and Thomson, "On the Thermal Effects of Fluids in Motion,"
Trans. Roy. Soc., read June 16, 1853, and June 15, 1854 [Art. XLIX. Vol. I.
above]. Thomson, "On the Alterations of Temperature accompanying Changes
of Pressure in Fluids," *Proc. Roy. Soc.*, Vol. VIII., read June 15, 1857 [Art. XCII.
App. A., above].

† "Theoretical Considerations Regarding the Effect of Pressure on the Freezing
Point of Water," *Trans. R. S. E.*, Jan. 1849. [Appendix to Art. XLI. Vol. I. above.]

‡ *Proceedings R. S. E.*, Jan. 1850; *Philosophical Magazine*, 1850, Vol. XXXVII.
&c. [Art. XLII. Vol. I. above].

vective equilibrium, it is impossible to say, without knowledge which we do not yet possess, regarding the expansion with heat, and the specific heat of the fluid, and the change of volume and the latent heat developed in the transition from fluid to solid.

25. For instance, supposing, as is most probably true, both that the liquid contracts in cooling towards its freezing-point, and that it contracts in freezing; we cannot tell, without definite numerical data regarding those elements, whether the elevation of the temperature of solidification, or of the actual temperature of a portion of the fluid given just above its freezing point, produced by a given application of pressure, is the greater. If the former is greater than the latter, solidification would commence at the bottom, or at the centre if there is no solid nucleus to begin with, and would proceed outwards; and there could be no complete permanent incrustation all round the surface till the whole globe is solid, with, possibly, the exception of irregular, comparatively small spaces of liquid.

26. If, on the contrary, the elevation of temperature, produced by an application of pressure to a given portion of the fluid, is greater than the elevation of the freezing temperature produced by the same amount of pressure, the superficial layer of the fluid would be the first to reach its freezing point, and the first actually to freeze.

27. But if, according to the second supposition of § 22 above, the liquid expanded in cooling near its freezing-point, the solid would probably likewise be of less specific gravity than the liquid at its freezing-point. Hence the surface would crust over permanently with a crust of solid, constantly increasing inwards by the freezing of the interior fluid in consequence of heat conducted out through the crust. The condition most commonly assumed by geologists would thus be produced.

28. But Bischof's experiments, upon the validity of which, so far as I am aware, no doubt has ever been thrown, show that melted granite, slate, and trachyte, all contract by something about 20 per cent. in freezing. We ought, indeed, to have more experiments on this most important point, both to verify Bischof's results on rocks, and to learn how the case is with iron and other unoxydised metals. In the meantime we must assume it as probable

that the melted substance of the earth did really contract by a very considerable amount in becoming solid.

29. Hence, if according to any relations whatever among the complicated physical circumstances concerned, freezing did really commence at the surface, either all round or in any part, before the whole globe had become solid, the solidified superficial layer must have broken up and sunk to the bottom, or to the centre, before it could have attained a sufficient thickness to rest stably on the lighter liquid below. It is quite clear, indeed, that if at any time the earth were in the condition of a thin solid shell of, let us suppose 50 feet or 100 feet thick of granite, enclosing a continuous melted mass of 20 per cent. less specific gravity in its upper parts, where the pressure is small, this condition cannot have lasted many minutes. The rigidity of a solid shell of superficial extent, so vast in comparison with its thickness, must be as nothing, and the slightest disturbance must cause some part to bend down, crack, and allow the liquid to run out over the whole solid. The crust itself must in consequence become shattered into fragments, which must all sink to the bottom, or to meet in the centre and form a nucleus there if there is none to begin with.

30. It is, however, scarcely possible, that any such continuous crust can ever have formed all over the melted surface at one time, and afterwards have fallen in. The mode of solidification conjectured in § 25 above, seems on the whole the most consistent with what we know of the physical properties of the matter concerned. So far as regards the result, it agrees, I believe, with the view adopted as the most probable by Mr Hopkins*. But whether from the condition being rather that described in § 26 above, which seems also possible, for the whole or for some parts of the heterogeneous substance of the earth, or from the viscidity as of mortar, which necessarily supervenes in a melted fluid, composed of ingredients becoming, as the whole cools, separated by crystallising at different temperatures, before the solidification is perfect, and which we actually see in lava from modern volcanoes; it is probable that when the whole globe, or some very thick superficial layer of it, still liquid or viscid, has cooled down to near its temperature of perfect solidification, incrustation at the surface must commence.

* See his Report on "Earthquakes and Volcanic Action," *British Association Report for* 1847.

31.   It is probable that crust may thus form over wide extents of surface, and may be temporarily buoyed up by the vesicular character it may have retained from the ebullition of the liquid in some places; or, at all events, it may be held up by the viscidity of the liquid, until it has acquired some considerable thickness sufficient to allow gravity to manifest its claim, and sink the heavier solid below the lighter liquid.   This process must go on until the sunk portions of crust build up from the bottom a sufficiently close ribbed solid skeleton or frame, to allow fresh incrustations to remain bridging across the now small areas of lava pools or lakes.

32.   In the honey-combed solid and liquid mass thus formed, there must be a continual tendency for the liquid, in consequence of its less specific gravity, to work its way up; whether by masses of solid falling from the roofs of vesicles or tunnels, and causing earthquake shocks, or by the roof breaking quite through when very thin, so as to cause two such hollows to unite, or the liquid of any of them to flow out freely over the outer surface of the earth; or by gradual subsidence of the solid, owing to the thermo-dynamic melting, which portions of it, under intense stress, must experience, according to views recently published by my brother, Professor James Thomson*.   The results which must follow from this tendency seem sufficiently great and various to acount for all that we see at present, and all that we learn from geological in-vestigation, of earthquakes, of upheavals and subsidences of solid, and of eruptions of melted rock.

33.   These conclusions, drawn solely from a consideration of the necessary order of cooling and consolidation, according to Bischof's result as to the relative specific gravities of solid and of melted rock, are in perfect accordance with what I have recently demonstrated† regarding the present condition of the earth's in-terior,—that it is not, as commonly supposed, all liquid within a thin solid crust of from 30 to 100 miles thick, but that it is on the whole more rigid certainly than a continuous solid globe of glass of the same diameter, and probably than one of steel.

* "On Crystallization and Liquefaction as influenced by Stresses tending to Change of Form in Crystals," *Proceedings of the Royal Society*, Vol. xi., read Dec. 5, 1861.

† In a paper "On the Rigidity of the Earth," communicated to the Royal Society a few days ago; April, 1862 [Art. xcv. below].

Art. XCV. On the Rigidity of the Earth; Shiftings of the Earth's Instantaneous Axis of Rotation; and Irregularities of the Earth as a Timekeeper.

[*Transactions Roy. Soc.* 1863; read May 15, 1862: and incorporating, instead of §§ 21–32 of the original paper, the Opening Address to Sec. A. of the British Association, Glasgow meeting, 1876 ; *Brit. Association Report*, 1876.]

1. THAT the earth cannot, as many geologists suppose, be a liquid mass enclosed in only a thin 'shell of solidified matter, is demonstrated by the phenomena of precession and nutation. Mr Hopkins*, to whom is due the grand idea of thus learning the physical condition of the interior from phenomena of rotatory motion presented by the surface, applied mathematical analysis to investigate the rotation of rigid ellipsoidal shells enclosing liquids, and arrived at the conclusion that the solid crust of the earth must be not less than 800 or 1000 miles thick. Although the mathematical part of the investigation might be objected to, I have not been able to perceive any force in the arguments by which this conclusion has been controverted, and I am happy to find my opinion in this respect confirmed by so eminent an authority as Archdeacon Pratt†.

2. It has always appeared to me, indeed, that Mr Hopkins might have pressed his argument further, and have concluded that no continuous liquid vesicle at all approaching to the dimensions of a spheroid 6000 miles in diameter can possibly exist in the earth's interior without rendering the phenomena of precession and nutation very sensibly different from what they are.

3. Considerations regarding the velocities of long waves in deep sea, of tidal waves and of earthquake waves, and the harmonic vibrations of a liquid globe, having recently led me to think

* *Transactions Roy. Soc.*, years 1839, 1840, 1842.
† *Figure of the Earth*, edit. 1860, § 85.

of the relative values of gravitation and elasticity in giving rigidity
to the earth's figure, I was surprised to find that the former would
have a larger share in this effect than the latter, unless the average
substance of the earth had a very high degree of rigidity. For
instance, I found that a homogeneous incompressible liquid globe
of the same density as the mean density of the earth, if changed
to a spheroidal form and then left free, influenced only by mutual
gravitation of its parts, would perform simple harmonic vibrations
in $47^m 12^s$ half-period*. A steel globe of the same dimensions,
without mutual gravitation of its parts, could scarcely oscillate so
rapidly, since the velocity of plane waves of distortion in steel is
only about 10,140 feet per second, at which rate a space equal to
the earth's diameter would not be travelled in less than $1^h 8^m 40^s$.

4. Hence it is obvious that, unless the average substance of
the earth is more rigid than steel, its figure must yield to the distorting forces of the moon and sun, not incomparably less than it
would if it were fluid. To illustrate this conclusion, I have investigated the deformation experienced by a homogeneous elastic
spheroid under the influence of any arbitrarily given disturbing
forces†. I thus find that if $2h'$ denote the difference between the
longest and shortest diameters of the tidal spheroid, calculated on
the supposition that the substance is of homogeneous (and therefore incompressible) fluid, and $2h$ the difference between the longest
and shortest diameters of the spheroid into which the same mass,
if of homogeneous incompressible solid matter, would be deformed
from a naturally spherical figure when exposed to the same lunar
or solid disturbing influence, we have (see § 40, Appendix to this
paper)

$$h = \frac{h'}{1 + \frac{19}{2}\frac{n}{gwr}},$$

where w denotes the mass of unit volume,

  n, the "rigidity" of the substance (see § 71 of Art. XCVI.
    below);

  g, the force of gravity on a unit of mass at the surface ;

and r, the radius of the globe.

* This is demonstrated in §§ 55—58 of "Dynamical Problems regarding Elastic
Spheroidal Shells, &c." *Transactions Roy. Soc.* 1863, read Nov. 27, 1862 [Art. XCVI.
below].

† The solution of this problem will be found in §§ 47, 48 of the paper referred
to above.

5.   The density of iron or steel (7·8 times that of water) does not differ very much from the mean density of the earth (5·5 times that of water according to Cavendish's experiment). The rigidity of iron, according to experiments of my brother, Professor James Thomson*, is 10,800,000 lbs. per square inch. Since the weight of 1 lb. at Glasgow, where the experiment was made, is 32·2 British absolute units of force, we must multiply by 32·2 to reduce to kinetic measure as to force; and we must multiply by 144 to make the unit of area a square foot instead of a square inch. We thus find, in consistent absolute measure,

$$n = 501 \times 10^8,$$

—the unit of mass being 1 lb., the unit of space 1 foot, and the unit of force that force which, acting on one pound of matter during a mean solar second of time, generates a velocity of 1 foot per second. In terms of the same units we have $r = 20,887,700$; $g = 32\cdot14$, being about the average over all the earth; and for iron or steel $w = 487$. Hence

$$h = \frac{h'}{1 + \dfrac{19}{2} \cdot \dfrac{501 \times 10^8}{3308 \times 10^8}} = \frac{h'}{2\cdot44} = \cdot41 h'.$$

Of glass, the rigidity is, according to Wertheim, about one-fifth of the value we have just used as that of iron; and therefore if the earth were homogeneous of its actual mean density, and had throughout the same rigidity as that of glass, the result would be

$$h = \cdot78 h'$$

6.   Hence it appears that if the rigidity of the earth, on the whole, were only as much as that of steel or iron, the earth as a whole would yield about two-fifths as much to the tide-producing influences of the sun and moon as it would if it had no rigidity at all; and it would yield by more than three-fourths of the fluid yielding, if its rigidity were no more than that of glass.

7.   Such a deformation as this would be quite undiscoverable by any direct geodetical or astronomical observations; but if it existed, it would largely influence the actual phenomena of the tides and of precision and nutation.

---

* "On the Elasticity and Strength of Spiral Springs, and of Bars subjected to Torsion," *Cambridge and Dublin Mathematical Journal*, 1848.

§§ 8—20.   *Effect of the Earth's Elastic Yielding on the Tides.*

8.   To find the effect of the earth's elastic yielding on the tides, let $2H$ denote the difference between the greatest and least diameters of the spheroidal surface perpendicular to the resultant of the lunar or solar disturbing force*, and let terrestrial gravitation be supposed perfectly symmetrical about the centre, then H/r will be the ellipticity of that spheroid; and we shall call it the *ellipticity of level produced by the lunar or solar influence on a rigid earth.*   It may be remarked that H is the height of high water above low water in the "equilibrium tide" of an ocean of infinitely small density covering a rigid earth.

9.   Let H' denote the height of the equilibrium tide for an ocean of density $1/N$ of the earth's mean density, the earth being still supposed *perfectly* rigid and covered by the ocean.   Then the terrestrial gravitation level will be disturbed (as is proved in the theory of the attraction of ellipsoids) from the spherical surface to the spheroidal surface of ellipticity $\frac{3}{5} \cdot \frac{1}{N} \cdot \frac{H'}{r}$, by the attraction of the ocean in its altered figure.   The ellipticity of level induced by lunar or solar influence must be added to this to give the ellipticity of actual level, which is of course the ellipticity of the free equilibrium surface of the ocean, or according to our notation H'/r.   Hence

$$\frac{H'}{r} = \frac{H}{r} + \frac{3}{5} \cdot \frac{1}{N} \cdot \frac{H'}{r},$$

by which we find

$$H' = \frac{H}{1 - \frac{3}{5} \cdot \frac{1}{N}}$$

For sea-water the value of N is about 5·5; and therefore

$$H' = \frac{9 \cdot 2}{8 \cdot 2} H = 1 \cdot 12 H,$$

or only 12 per cent. more than for an ocean of infinitely small density.

---

* This "disturbing force" is of course the resultant of the actual attraction of either body on a unit of mass in any position, and a force equal and opposite to its attraction on a unit of mass at the earth's centre.

10. What we have denoted in § 4 above by h', is the value of H', for $N = 1$; and therefore

$$h' = \frac{5}{2} H,$$

and
$$h = \frac{5}{2} \cdot \frac{H}{1 + \frac{19}{2} \cdot \frac{n}{gwr}}.$$

11. Now, according to a proposition regarding the attraction of ellipsoids already used, we have $\frac{3}{5} \cdot \frac{h}{r}$ for the ellipticity in the terrestrial gravitation level, produced by the ellipticity of deformation h/r experienced in consequence of want of perfect rigidity. Hence the ellipticity of the terrestrial gravitation level, as disturbed by lunar or solar influence, is $\frac{3}{5} \cdot \frac{h}{r} + \frac{H}{r}$. This will be the absolute tidal equilibrium ellipticity of an ocean of infinitely small density covering the elastic globe; but since there is a tidal ellipticity h/r induced in the solid itself, the height from low tide to high tide of fluid relatively to solid (that is to say, the difference of depth between high water and low water) will be

$$\left(\frac{3}{5} h + H\right) - h = H - \frac{2}{5} h;$$

or, according to the value of h just found (§ 10),

$$\frac{\frac{19}{2} \cdot \frac{n}{gwr} H}{1 + \frac{19}{2} \cdot \frac{n}{gwr}}.$$

12. This result expresses strictly the height of the equilibrium tide of a liquid of infinitely small density covering an elastic solid globe. It may be regarded as a better expression of the true tidal tendency on the actual ocean, than the slightly different result calculated with allowance for the effect of the attraction of the altered watery figure constituting the equilibrium spheroid, and its influence on the figure of the elastic solid; since the impediments of land and the influence of the sea-bottom, render the actual ocean surface altogether different from that of the equilibrium spheroid.

13. Hence the actual tidal tendency, which would be H if the earth were perfectly rigid, is in reality

$$\frac{\frac{19}{2} \cdot \frac{n}{gwr} \cdot H}{1 + \frac{19}{2} \cdot \frac{n}{gwr}},$$

where n denotes what we may call the *earth's tidal effective rigidity*, being the "rigidity" of a homogeneous incompressible solid globe of equal mass which, with an ocean equal and similar to the earth's, would exhibit the same tides.

14. If, for example, we give n the value for iron or steel above indicated, the formula becomes $\cdot59 \times H$. The comparison between theory and observation, owing to the extreme complexity of the circumstances, has been hitherto so imperfect that we cannot say it disproves this result; and therefore, from tidal phenomena hitherto observed, we cannot infer that the earth is more effectively rigid than steel.

15. The value of n for glass, according to Wertheim, is $2160000 \times 144 \times 32\cdot2$, in British absolute units; and it reduces the formula to $\cdot22\,H$. Now, imperfect as the comparison between theory and observation as to the absolute height of the tides has been hitherto, it is scarcely possible to believe that the height is in reality only two-ninths of what it would be if, as has hitherto been universally assumed in tidal investigations, the earth were perfectly rigid. It seems therefore nearly certain, with no other evidence than is afforded by the tides, that the tidal effective rigidity of the earth must be greater than that of glass.

16. Any approach to a close testing of the absolute amount of the tidal influence can scarcely be expected of either of the two great Kinetic* theories—the Oceanic theory of Laplace, or the

---

* Dynamics meaning properly the science of force, and there being precedents of the very highest kind, for instance, in Delaunay's *Mécanique Rationelle*, of 1861, and Robison's *Mechanical Philosophy* of 1804, in favour of using the term according to its proper meaning—and the modern corrupt usage, which has confined it to the branch of dynamical science in which relative motion is considered, being excessively inconvenient and vexatious,—it has been proposed to introduce the term "kinetics" to express this branch; so that dynamics may be defined simply as the "science of force," and divided into the two branches, Statics and Kinetics. The introduction of this new term, derived from κίνησις, *motion*, or act of moving, does

Channel theory of Airy,—as applied to diurnal or semidiurnal
tides; but notwithstanding the strong contempt which has been
expressed by the last-mentioned naturalist* (no doubt justly as
regards false applications of it) for the equilibrium theory†, we
may look to it confidently for good information when it is applied
to test the difference between mean fortnightly variations of sea-
level at two well-chosen stations, one in a low latitude and the
other in a high latitude. (See Note (§ 39) at the end of this
paper.)

17. The fortnightly tide ‡ at each pole gives high water
when the moon's declination (Δ), whether north or south, is
greatest, and low water when she crosses the equator; and the
whole difference in level produced by it would be

$$H' \sin^2 \Delta$$

if the earth were all covered with water. The mean daily level at
the equator, on the same supposition, would vary by half that
amount, being low water when the moon is furthest from the
equator, and high water when she crosses the equator. But
owing to the actual distribution of land and water, either of those
variations may be diminished by an amount which it is impossible
to estimate theoretically; but then the other must be increased
by nearly the same amount. And if a denote the mean height of
the sea-level above a fixed mark at the earth's north pole, about
the times when the moon's declination is greatest, b the corresponding mean of observations about times when she is crossing
the equator, a' and b' corresponding means derived from observation at an equatorial station, and H something intermediate
between H and H', we must have

$$a - b + b' - a' = \tfrac{3}{2} H \sin^2 \Delta,$$

whatever be the distribution of land and water over the earth,

---

not interfere with Ampère's term, now universally accepted, "kinematics" (from
κίνημα), the *science of movements*.

    * "Naturalist. A person well versed in Natural Philosophy."— Johnson's
Dictionary. Armed with this authority, chemists, electricians, astronomers, and
mathematicians may surely claim to be admitted along with merely descriptive
investigators of nature to the honourable and convenient title of Naturalist, and
refuse to accept so un-English, unpleasing, and meaningless a variation from old
usage as "physicist."

    † *Encyclopædia Metropolitana*, Airy's "Tides and Waves," §§ 64, 539, &c.

    ‡ Airy's *Tides and Waves*, § 45.

only provided the fortnightly tide follows sensibly the equilibrium law, which, for moderately well-chosen stations, we may suppose it must do.

18.   If, instead of being at a pole and at the equator, the stations are in latitudes respectively l and l', we should have

$$a - b + b' - a' = \tfrac{3}{2}\,_{,}H\sin^2\Delta\,(\sin^2 l - \sin^2 l').$$

Now if we suppose the moon's mass to be $\frac{1}{75}$ of the earth's, we have $H = 1\cdot92$ foot.   As $H' = 1\cdot12\,H$, and as there is more area of water than of land over the earth, we cannot be far wrong in taking $_{,}H = 1\cdot08\,H = 2\cdot04$ feet.

The greatest value of Δ is $28°\,37'$; and hence, in the most favourable lunations,

$$a - b + b' - a' = \cdot713 \text{ foot} \times (\sin^2 l - \sin^2 l').$$

19.   Iceland and Teneriffe, in nearly the same longitude, and in latitudes $63°\,20'$ and $28°\,30'$, would probably be very favourable stations.   For them $\sin^2 l - \sin^2 l' = \cdot571$; and therefore

$$a - b + b' - a' = 0\cdot407 \text{ foot},$$

or about 4·9 inches.

It is probable that carefully made and reduced observations, with proper allowance for barometric disturbances, at two such stations, would not only detect this tide, but would give a tolerably accurate determination of its amount.

20.   It would be, for Iceland and Teneriffe, as found above, 4·9 inches if the earth were perfectly rigid; or 3 inches if the tidal effective rigidity is only that of steel; or about an inch if the tidal effective rigidity is only that of glass.

There seems no more hopeful way to ascertain how rigid the earth really is, than to make careful observations with a view to determining the fortnightly tide with all possible accuracy.   It is possible also that very accurate observations on the semi-diurnal tides in a deep inland lake of great extent, or at distant points of the Mediterranean sea-board with only deep water intervening, might help to solve this question.

§§ 21—38.   *Effects of Elastic Yielding on Precession and Nutation.*

[Of date September, 1876: replacing §§ 21—32 of the original paper.]

21.   Whatever may be its age, we may be quite sure the earth is solid in its interior; not, I admit, throughout its whole volume, for there certainly are spaces in volcanic regions occupied by liquid lava; but whatever portion of the whole mass is liquid, whether the waters of the ocean or melted matter in the interior, these portions are small in comparison with the whole; and we must utterly reject any geological hypothesis which, whether for explaining underground heat or ancient upheavals and subsidences of the solid crust, or earthquakes, or existing volcanoes, assumes the solid earth to be a shell of 30, or 100, or 500, or 1000 kilometres thickness, resting on an interior liquid mass.

22.   This conclusion was first arrived at by Hopkins, who may therefore properly be called the discoverer of the earth's solidity. He was led to it by a consideration of the phenomena of precession and nutation, and gave it as shown to be highly probable, if not absolutely demonstrated, by his confessedly imperfect and tentative investigation.   But a rigorous application of the perfect hydrodynamical equations leads still more decidedly to the same conclusion.

23.   I am able to say this now in consequence of a conversation which I had with Professor Newcomb, one evening (June 1876) in Prof. Henry's drawing-room in the Smithsonian Institution, Washington.   Admitting fully my evidence for the rigidity of the earth from the tides, he doubted the argument from precession and nutation.   Trying to recollect what I had written on it fourteen years ago in a paper on the "Rigidity of the Earth," published in the *Transactions of the Royal Society*, my conscience smote me, and I could only stammer out that I had convinced myself that so-and-so and so-and-so, at which I had arrived by a non-mathematical short cut, were true.   He hinted that viscosity might suffice to render precession and nutation the same as if the earth were rigid, and so vitiate the argument for rigidity.   This I could not for a moment admit, any more than when it was first put forward by Delaunay.   But doubt entered my mind regarding the so-and-so and so-and-so; and I had not completed the night

journey to Philadelphia which hurried me away from our un-finished discussion before I had convinced myself that they were grievously wrong. So now I must request as a favour that each one of you on going home will instantly turn up his or her copies of the *Transactions of the Royal Society* for 1863 and of the first edition (1867) of Thomson and Tait's *Natural Philosophy*, Vol. I., and draw the pen through §§ 21—32 of my paper on the "Rigidity of the Earth" in the former, and through every thing in §§ 847, 848, 849 of the latter which refers to the effect on precession and nutation of an elastic yielding of the earth's surface*.

24.   When those passages were written I knew little or nothing of vortex motion ; and until my attention was recalled to them by Professor Newcomb I had never once thought of this subject in the light thrown upon it by the theory of the quasi-rigidity induced in a liquid by vortex motion, which has of late occupied me so much. With this fresh light a little consideration sufficed to show me that (although the old obvious conclusion is of course true, that, if the inner boundary of the imagined rigid shell of the earth were rigorously spherical, the interior liquid could experience no pre-cessional or nutational influence from the pressure on its bounding surface, and therefore if homogeneous could have no precession or nutation at all, or if heterogeneous only as much precession and nutation as would be produced by attraction from without in virtue of non-sphericity of its surfaces of equal density, and therefore the shell would have enormously more rapid precession and nutation than it actually has—forty times as much, for instance, if the thickness of the shell is 60 kilometres) a very slight deviation of the inner surface of the shell from perfect sphericity would suffice, in virtue of the quasi-rigidity due to vortex motion, to hold back the shell from taking sensibly more precession than it would give to the liquid, and to cause the liquid (homogeneous or hetero-

---

* In a paper by Brevet-Major General J. G. Barnard, College of Engineers, U. S. A., entitled "Problems of Rotatory Motion presented by the Gyroscope, the Precession of the Equinoxes, and the Pendulum" of date, October, 1871, and constituting No. 240 of the *Smithsonian Contributions to Knowledge*, I find an anticipation of this correction expressed in the following terms :—"I do not concur with Sir William Thomson in the opinions quoted in note, page 38, from Thomson and Tait, and expressed in his letter to Mr G. Poulett Scrope (*Nature*, Feb. 1, 1872) ; so far as regards fluidity, or imperfect rigidity, within an infinitely rigid envelope, I do not think the rate of precession would be affected."

geneous) and the shell to have sensibly the same precessional motion as if the whole constituted one rigid body. But it is only because of the very long period (26,000 years) of precession, in comparison with the period of rotation (one day), that a very slight deviation from sphericity would suffice to cause the whole to move as if it were a rigid body. A little further consideration showed me :—

(1) That an ellipticity of inner surface equal to $\dfrac{1}{26,000 \times 365}$ would be too small, but that an ellipticity of one or two hundred times this amount would not be too small to compel approximate equality of precession throughout liquid and shell.

(2) That with an ellipticity of interior surface equal to 1/300, if the precession-generating influence were 26,000 times as rapid as it is, the motion of the liquid would be very different from that of a rigid mass rigidly connected with the shell.

(3) That with the actual forces and the supposed interior ellipticity of 1/300, the lunar nineteen-yearly nutation might be affected to about five per cent. of its amount by interior liquidity.

(4) Lastly, that the lunar semiannual nutation must be largely, and the lunar fortnightly nutation enormously, affected by interior liquidity.

25. But although so much could be foreseen readily enough, I found it impossible to discover without thorough mathematical investigation what might be the characters and amounts of the deviations from a rigid body's motion which the several cases of precession and nutation contemplated would present. The investigation, limited to the case of a homogeneous liquid enclosed in an ellipsoidal shell, has brought out results which I confess have greatly surprised me. When the interior ellipticity of the shell is just too small, or the periodic speed of the disturbance just too great, to allow the motion of the whole to be sensibly that of a rigid body, the deviation first sensible renders the precessional or nutational motion of the shell smaller than if the whole were rigid, instead of greater, as I expected. The amount of this difference bears the same proportion to the whole precession or nutation, as the reciprocal of the ellipticity bears to the number of days in the period of the precession or nutation. It is remarkable that this result is independent of the thickness of the shell, assumed how-

ever to be small in proportion to the earth's radius. Thus in the case of precession the effect of interior liquidity would be to diminish the periodic speed of the precession in the proportion stated; in other words, it would add to the precessional period a number of days equal to the number whose reciprocal measures the ellipticity. In the actual case of the earth, if we still take 1/300 as the ellipticity of the inner boundary of the supposed rigid shell, the effect would be to add 300 days to the precessional period of 26,000 years, or to diminish by about 1/600 of a second the annual precession of about 51″, an effect which I need not say would be wholly insensible. But on the lunar nutation of 18·6 years period, the effect of interior liquidity would be quite sensible; 18·6 years being twenty-three times 300 days, the effect would be to diminish the axes of the ellipse which the earth's pole describes in this period, each by 1/23 of its own amount. The semiaxes of this ellipse, calculated on the theory of perfect rigidity from the very accurately known amount of precession, and the fairly accurate knowledge which we have of the ratio of the lunar to the solar part of the precessional motion, are 9″·22 and 6″·86, with an uncertainty not amounting to one half per cent. on account of want of perfect accuracy in the latter part of data. If the true values were less, each by 1/23 of its own amount, the discrepance might have escaped detection, or might *not* have escaped detection; but certainly could be found if looked for. So far nothing can be considered as absolutely proved with reference to the interior solidity of the earth from precession and nutation.

26. Now think of the solar semiannual and the lunar fortnightly nutations. The period of each of these is less than 300 days. The hydrodynamical theory shows that, irrespectively of the thickness of the shell, the nutation of the crust would be zero if the period of the nutational disturbance were 300 times the period of rotation (the ellipticity being 1/300); if the nutational period were any thing between this and a certain smaller critical value depending on the thickness of the crust, the nutation would be negative; if the period were equal to this second critical value, the nutation would be infinite; and if the period were still less, the nutation would be again positive. The 183 days period of the solar nutation falls so little short of the critical 300 days that the amount of the nutation is not sensibly influenced by the thickness

of the crust: it is negative and is equal in absolute value to
183/(300 − 183) times what the amount would be were the earth
solid throughout. Now this amount, as calculated in the *Nautical
Almanac*, gives 0″·55 and 0″·51 for the semiaxes of the ellipse
traced by the earth's axis round its mean position; hence on the
supposition of a solid crust with internal ellipticity 1/300 the semiaxes of the ellipse would be − 0″·86 and − 0″·80. Now I think
we may safely say that if the true nutation placed the earth's axis
on the opposite side of an ellipse having 0″·86 and 0″·80 for its
semiaxes, the discrepance could not possibly have escaped detection.
But, lastly, think of the lunar fortnightly nutation. Its period is
1/20 of 300 days, and its amount, calculated in the *Nautical
Almanac* on the theory of complete solidity, is such that the
greater semiaxis of the approximately circular ellipse described by
the pole is 0″·0325. Were the crust infinitely thin this nutation
would be negative, but its amount nineteen times that corresponding to solidity. This would make the greater semiaxis of the
approximately circular ellipse described by the pole amount to
19 × 0″·0885, which is 1″·7. It would be negative and of some
amount between 1″·7 and infinity, if the thickness of the crust were
any thing from zero to 120 kilometres. This conclusion is absolutely decisive against the geological hypothesis of a thin rigid
shell full of liquid*.

27. But interesting in a dynamical point of view as Hopkins'
problem is, it cannot afford a decisive argument against the earth's
interior liquidity. It assumes the crust to be perfectly stiff and
unyielding in its figure. This, of course, it cannot be, because no
material is infinitely rigid; but, composed of rock and possibly
of continuous metal in the great depths, may the crust not, as a
whole, be stiff enough to practically fulfil the condition of un-
yieldingness? No, decidedly it could not; on the contrary, were
it of continuous steel and 500 kilometres thick, it would yield very
nearly as much as if it were india-rubber to the deforming influences
of centrifugal force and of the sun's and moon's attractions. Now
although the full problem of precession and nutation, and, what is
now necessarily included in it, tides, in a continuous revolving
liquid spheroid, whether homogeneous or heterogeneous, has not

---

* The mathematical investigation which led to the conclusions stated in
§§ 24—26 above, has not yet been published. W. T., Feb. 28, 1889.

yet been coherently worked out, I think I see far enough towards a complete solution to say that precession and nutation will be practically the same in it as in a solid globe, and that the tides will be practically the same as those of the equilibrium theory. From this it follows that precession and nutation of the solid crust, with the practically perfect flexibility which it would have even though it were 100 kilometres thick and as stiff as steel, would be sensibly the same as if the whole earth from surface to centre were solid and perfectly stiff. Hence precession and nutation yield nothing to be said against such hypotheses as that of Darwin*, that the earth as a whole takes approximately the figure due to gravity and centrifugal force, because of the fluidity of the interior and the flexibility of the crust. But, alas for this "attractive sensational idea" (as Poulett Scrope called it)? "that a molten interior to the globe underlies a superficial crust, its surface agitated by tidal waves, and flowing freely towards any issue that may here and there be opened for its outward escape" the solid crust would yield so freely to the deforming influence of sun and moon that it would simply carry the waters of the ocean up and down with it, and there would be no sensible tidal rise and fall of water relatively to land.

28.   The state of the case is shortly this :—The hypothesis of a perfectly rigid crust containing liquid, violates physics by assuming preternaturally rigid matter, and violates dynamical astronomy in the solar semiannual and lunar fortnightly nutations ; but tidal theory has nothing to say against it. On the other hand, the tides decide against any crust flexible enough to perform the nutations correctly with a liquid interior, or as flexible as the crust must be unless of preternaturally rigid matter.

29.   But now thrice to slay the slain : suppose the earth this moment to be a thin crust of rock or metal resting on liquid matter ; its equilibrium would be unstable ! And what of the upheavals and subsidences ? They would be strikingly analogous to those of a ship which has been rammed—one portion of crust up and another down, and then all down. Now whatever may be the relative densities of rock, solid and melted, at or about the

---

* "Observations on the Parallel Roads of Glen Roy and other parts of Lochaber in Scotland, with an attempt to prove that they are of marine origin," *Transactions of the Royal Society* for Feb. 1839, p. 81.

temperature of liquefaction, it is, I think, quite certain that cold solid rock is denser than hot melted rock; and no possible degree of rigidity in the crust could prevent it from breaking in pieces and sinking wholly below the liquid lava. Something like this may have gone on, and probably did go on, for thousands of years after solidification commenced—surface-portions of the melted material losing heat, freezing, sinking immediately, or growing to thicknesses of a few metres, when the surface would be cool and the whole solid dense enough to sink. "This process must go on until the sunk portions of crust build up from the bottom a sufficiently close-ribbed skeleton or frame to allow fresh incrustations to remain, bridging across the now small areas of lava pools or lakes.

30.   "In the honeycombed solid and liquid mass thus formed there must be a continual tendency for the liquid, in consequence of its less specific gravity, to work its way up; whether by masses of solid falling from the roofs of vesicles or tunnels and causing earthquake-shocks, or by the roof breaking quite through when very thin, so as to cause two such hollows to unite or the liquid of any of them to flow out freely over the outer surface of the earth, or by gradual subsidence of the solid owing to the thermodynamic melting which portions of it under intense stress must experience, according to views recently published by my brother, Prof. James Thomson*. The results which must follow from this tendency seem sufficiently great and various to account for all that we learn from geological evidence of earthquakes, of upheavals and subsidences of solid, and of eruptions of melted rock†."

31.   Leaving altogether now the hypothesis of a hollow shell filled with liquid, we must still face the question, How much does the earth, solid throughout, except small cavities or vesicles filled with liquid, yield to the deforming (or tide-generating) influences of sun and moon? This question can only be answered by observation. A single infinitely accurate spirit-level or plummet, far enough away from the sea to be not sensibly affected by the attraction of the

* "On Crystallization and Liquefaction as influenced by Stresses tending to Change of Form in Crystals," *Proc. Roy. Soc.* Vol. XI., read Dec. 5, 1861.

† "Secular Cooling of the Earth," *Transactions of the Royal Society of Edinburgh*, Vol. XXIII. 1862, and Thomson and Tait's *Natural Philosophy*, Edition 1883, Part II., Appendix C; [Art. XCIV. §§ 31, 32 above].

rising and falling water, would enable us to find the answer. Observe by level or plummet the changes of direction of apparent gravity relatively to an object rigidly connected with the earth, and compare these changes with what they would be were the earth perfectly rigid, according to the known masses and distances of sun and moon. The discrepance, if any is found, would show distortion of the earth, and would afford data for determining the dimensions of the elliptic spheroid into which a non-rotating globular mass of the same dimensions and elasticity as the earth would be distorted by centrifugal force if set in rotation, or by tide-generating influences of sun or moon. The effect on the plumb-line of the lunar tide-generating influence is to deflect it towards or from the point of the horizon nearest to the moon, according as the moon is above or below the horizon. The effect is zero when the moon is on the horizon or overhead, and is greatest in either direction when the moon is 45° above or below the horizon. When this greatest value is reached, the plummet is drawn from its mean position through a space equal to 1/12,000,000 of the length of the thread. No ordinary plummet or spirit-level could give any perceptible indication whatever of this effect; and to measure its amount it would be necessary to be able to observe angles as small as 1/120,000,000 of the radian, or about 1/600 of a second. A submerged water-pipe of considerable length, say 12 kilometres, with its two ends turned up and open, might answer. Suppose, for example, the tube to lie north and south, and its two ends to open into two small cisterns, one of them, the southern for example, of half a decimetre diameter (to escape disturbance from capillary attraction), and the other of two or three decimetres diameter (so as to throw nearly the whole rise and fall into the smaller cistern). For simplicity, suppose the time of observation to be when the moon's declination is zero. The water in the smaller or southern cistern will rise from its lowest position to its highest position while the moon is rising to maximum altitude, and fall again after the moon crosses the meridian till she sets; and it will rise and fall again through the same range from moonset to moonrise. If the earth were perfectly rigid, and if the locality is in latitude 45°, the rise and fall would be half a millimetre on each side of the mean level, or a little short of half a millimetre if the place is within 10° north or south of latitude 45°. If the air were so absolutely quiescent during the observations as to give no varying differential pressure on the two water-surfaces to the amount of 1/100 millimetre of water, or 1/1400 of mercury, the observation would be satisfactorily practicable, as it would not be difficult by aid of a microscope to observe the rise and fall of the water in the smaller cistern to 1/100 of a millimetre; but no such quiescence of the atmosphere could be expected at any time; and it is probable that the variations of the water-level due to difference of the barometric pressure at the two ends would, in all ordinary weather, quite overpower the small effect of the lunar tide-generating influence.   If, however, the two cisterns, instead of being open to the atmosphere, were connected air-tightly by a water-pipe with no water in it, it is probable that the observation might be successfully made: but some other apparatus on a small scale would probably be preferable to any elaborate method of obtaining the result by aid of very long pipes laid in the ground; and I have only called your attention to such an ideal method as leading up to the natural phenomenon of tides.

32.   Tides in an open canal or lake of 12 kilometres length would be of just the amount which we have estimated for the cisterns connected by submerged pipe; but would be enormously more disturbed by wind and variations of atmospheric pressure. A canal or lake of 240 kilometres length in a proper direction and in a suitable locality would give but 10 millimetres rise and fall at each end, an effect which might probably be analyzed out of the much greater disturbance produced by wind and differences of barometric pressure; but no open liquid level short of the *ingens œquor*, the ocean, will probably be found so well adapted as it for measuring the absolute value of the disturbance produced on terrestrial gravity by the lunar and solar tide-generating influence. But observations of the diurnal and semidiurnal tides in the ocean do not (as they would on smaller and quicker levels) suffice for this purpose, because their amounts differ enormously from the equilibrium-values on account of the smallness of their periods, in comparison with the periods of any of the grave enough modes of free vibration of the ocean as a whole.   On the other hand, the lunar fortnightly declinational and the lunar monthly elliptic tides, and the solar semiannual and annual elliptic tides, have their periods so long that their amounts must certainly be very approximately equal to the equilibrium-values.   But there are large

annual and semiannual changes of sea-level, probably both differential, on account of wind and differences of barometric pressure and
differences of temperature of the water, and absolute, depending
on rainfall and the melting away of snow and evaporation, which
altogether swamp the small semiannual and annual tides due to
the sun's attraction.   Happily, however, for our object, there is no
meteorological or other disturbing cause which produces periodic
changes of sea-level in either the fortnightly declinational or the
monthly elliptic period; and the lunar gravitational tides in these
periods are therefore to be carefully investigated in order that we
may obtain the answer to the interesting question, How much does
the earth as an elastic spheroid yield to the tide-generating
influence of sun or moon?   Hitherto in the British Association
Committee's reductions of Tidal Observations we have not succeeded in obtaining any trustworthy indications of either of these
tides.   The St-George's pier landing-stage pontoon, unhappily
chosen for the Liverpool tide-gauge, cannot be trusted for such a
delicate investigation: the available funds for calculation were
expended before the long-period tides for Hilbre Island could be
attacked: and three years of Kurrachee gave our only approach to
a result.   Comparisons of this with an indication of a result from
calculations on West Hartlepool tides, conducted with the assistance
of a grant from the Royal Society, seem to show possibly no
sensible yielding, or perhaps more probably some degree of yielding,
of the earth's figure.   The absence from all the results of any
indication of a 18·6-yearly tide (according to the same law as the
other long-period tides) is not easily explained without assuming
or admitting a considerable degree of yielding.

33.   On the whole we may fairly conclude that, whilst there is
some evidence of a tidal yielding of the earth's mass, that yielding
is certainly small, and the effective rigidity is at least as great as
that of steel.

34.   Closely connected with the question of the earth's rigidity,
and of as great scientific interest and of even greater practical
moment, is the question, How nearly accurate is the earth as a
timekeeper? and another of, at all events, equal scientific interest,
How about the permanence of the earth's axis of rotation?

35.   Peters and Maxwell, about 35 and 25 years ago, respectively,
raised the question, How much does the earth's axis of rotation

deviate from being a principal axis of inertia? and pointed out that an answer to this question is to be obtained by looking for a variation in latitude of any or every place on the earth's surface in a period of 306 days. The model before you illustrates the travelling round of the instantaneous axis, relatively to the earth in an approxi-

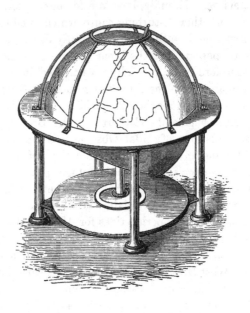

mately circular cone whose axis is the principal axis of inertia, and relatively to space in a cone round a fixed axis. In the model the former of these cones, fixed relatively to the earth, rolls internally on the latter, supposed to be fixed in space. Peters gave a minute investigation of observations at Pulkova in the years 1841—42, which seem to indicate at that time a deviation, amounting to about 3/40 of a second, of the axis of rotation from the principal axis. Maxwell, from Greenwich observations of the years 1851—54, found seeming indications of a very slight deviation, something less than half a second, but differing altogether in phase from that which the deviation indicated by Peters, if real and permanent, would have produced at Maxwell's later time. On my begging Professor Newcomb to take up the subject, he kindly did so at once, and undertook to analyze a series of observations suitable for the purpose which had been made in the United States Naval Observatory, Washington. A few weeks later I received from him

a letter, referring me to a paper* by Dr Nyrén, of Pulkova Observatory, in which a similar negative conclusion as to constancy
of magnitude or direction in the deviation sought for is arrived
at from several series of the Pulkova observations between the
years 1842 and 1872, and containing the following statement of his
conclusions :—

"The investigation of the ten-month period of latitude from
the Washington prime vertical observations from 1862 to 1867 is
completed, indicating a coefficient too small to be measured with
certainty. The declinations with this instrument are subject to an
annual period which made it necessary to discuss those of each
month separately. As the series extended through a full five years,
each month thus fell on five nearly equidistant points of the period.
If x and y represent the coordinates of the axis of instantaneous
rotation on June 30, 1864, then the observations of the separate
months give the following values of x and y :—

	x	y
January	$-0\overset{''}{\cdot}35$	$+0\overset{''}{\cdot}32$
February	$-0\cdot03$	$+0\cdot09$
March	$+0\cdot17$	$+0\cdot16$
April	$+0\cdot44$	$+0\cdot05$
May	$+0\cdot08$	$+0\cdot02$
June	$-0\cdot01$	$-0\cdot01$
July	$-0\cdot05$	$0\cdot00$
August	$-0\cdot24$	$+0\cdot29$
September	$+0\cdot18$	$+0\cdot21$
October	$+0\cdot13$	$-0\cdot01$
November	$+0\cdot08$	$-0\cdot20$
December	$-0\cdot08$	$-0\cdot08$
Mean	$0\cdot01 \pm 0\cdot03$	$+0\cdot05 \pm 0\cdot03$

"Accepting these results as real, they would indicate a radius
of rotation of the instantaneous axis amounting, at the earth's
surface, to 5 feet; and a longitude of the point in which this axis
intersects the earth's surface near the North Pole, such that on July
11, 1864, it was 180° from Washington, or 103° east of Greenwich.

* "Bestimmung der Nutation der Erdachse," of date 16 November, 1871;
published in the *Mémoires l'Académie Impériale des Sciences de St Petersbourg*,
VIIe série, Tome XIX., 1873.

The excess of the coefficient over its probable error is so slight that this result cannot be accepted as any thing more than a consequence of the unavoidable errors of observation."

36. From the discordant character of these results we must not, however, infer that the deviations indicated by Peters, Maxwell, and Newcomb are unreal. On the contrary, any which are outside the limits of probable error of the observations ought properly to be regarded as real. There is, in fact, a *vera causa* in the temporary changes of sea-level due to meteorological causes, chiefly winds, and to meltings of ice in the polar regions and return evaporations, which seems amply sufficient to account for irregular deviations, of from 1/2 to 1/20 of a second, of the earth's instantaneous axis from the axis of maximum moment of inertia, or, as I ought rather to say, of the axis of maximum moment of inertia from the instantaneous axis.

37. As for geological upheavals and subsidences, if on a very large scale of area, they must produce, on the period and axis of the earth's rotation, effects comparable with those produced by changes of sea-level equal to them in vertical amount. For simplicity, calculating as if the earth were of equal density throughout, I find that an upheaval of all the earth's surface in north latitude and east longitude and south latitude and west longitude with equal depression in the other two quarters, amounting at greatest to ten centimetres, and graduating regularly from the points of maximum elevation to the points of maximum depression in the middles of the four quarters, would shift the earth's axis of maximum moment of inertia through 1″ on the north side towards the meridian of 90° W. longitude, and on the south side towards the meridian of 90° E. longitude. If such a change were to take place suddenly, the earth's instantaneous axis would experience a sudden shifting of but 1/300 of a second (which we may neglect), and then, relatively to the earth, would commence travelling, in a period of 306 days, round the fresh axis of maximum moment of inertia. The sea would be set into vibration, one ocean up and another down through a few centimetres, like water in a bath set a-swing. The period of these vibrations would be from 12 to 24 hours, or at most a day or two; their subsidence would probably be so rapid that after at most a few months they would become insensible. Then a regular 306-days period tide of 11 centimetres

from lowest to highest would be to be observed, with gradually diminishing amount from century to century, as, through the dissipation of energy produced by this tide, the instantaneous axis of the earth is gradually brought into coincidence with the fresh axis of maximum moment of inertia. If we multiply these figures by 3600, we find what would be the result of a similar sudden upheaval and subsidence of the earth to the extent of 360 metres above and below previous levels. It is not impossible, although we may regard it as exceedingly improbable, that in the very early ages of geological history such an action as this, and the consequent 400-metres tide producing a succession of deluges every 306 days for many years, may have taken place; but it seems more probable that even in the most ancient times of geological history the great world-wide changes, such as the upheavals of the continents and subsidences of the ocean-beds from the general level of their supposed molten origin, took place gradually through the thermo-dynamic melting of solids and the squeezing out of liquid lava from the interior, to which I have already referred. A slow distortion of the earth as a whole would never produce any great angular separation between the instantaneous axis and the axis of maximum moment of inertia for the time being. Considering, then, the great facts of the Himalayas and the Andes, and Africa and the depths of the Atlantic, and America and the depths of the Pacific, and Australia, and considering further the ellipticity of the equatorial section of the sea-level, estimated by Capt. Clarke at about 1/10 of the mean ellipticity of meridional sections of the sea-level, we need no brush from the comet's tail (a wholly chimerical cause which can never have been put forward seriously except in ignorance of elementary dynamical principles) to account for a change in the earth's axis; we need no violent convulsion producing a sudden distortion on a great scale, with change of the axis of maximum moment of inertia followed by gigantic deluges; and we may not merely admit, but assert as highly probable, that the axis of maximum moment of inertia and the instantaneous axis of rotation, always very near one another, may have been in ancient times very far from their present geographical position, and may have gradually shifted through 10, 20, 30, 40, or more degrees without at any time any perceptible sudden disturbance of either land or water.

38. Lastly, as to variations in the earth's rotational period.

We all know how, in 1853, Adams discovered a correction
to be needed in the theoretical calculation with which Laplace
followed up his brilliant discovery of the dynamical explanation of
an apparent acceleration of the moon's mean motion shown by
records of ancient eclipses; and how he found that when his
correction was applied the dynamical theory of the moon's motion
accounted for only about half of the observed apparent acceleration;
and how Delaunay in 1866 verified Adams's result and suggested
that the explanation may be a retardation of the earth's rotation
by tidal friction. The conclusion is that, since the 19th of March,
721 B.C., a day on which an eclipse of the moon was seen in Babylon,
commencing "when one hour after her rising was fully passed," the
earth has lost rather more than 1/3,000,000 of her rotational
velocity, or, as a timekeeper, is going slower by $11\frac{1}{2}$ seconds per
annum now than then. According to this rate of retardation, if
uniform, the earth at the end of a century would, as a timekeeper,
be found 22 seconds behind a perfect clock, rated and set to agree
with her at the beginning of the century. Newcomb's subsequent
investigations in the lunar theory have on the whole tended to
confirm this result; but they have also brought to light some
remarkable apparent irregularities in the moon's motion, which, if
real, refused to be accounted for by the gravitational theory, without
the influence of some unseen body or bodies passing near enough
to the moon to influence her mean motion. This hypothesis
Newcomb considers not so probable as that the apparent irregularities of the moon are not real, and are to be accounted for by
irregularities in the earth's rotational velocity. If this is the true
explanation, it seems that the earth was going slow from 1850 to
1862, so much as to have got behind by seven seconds in these
twelve years, and then to have begun going faster again so as to
gain eight seconds from 1862 and 1872. So great an irregularity
as this would require somewhat greater changes of sea-level, but
not many times greater than the British Association Committee's
reductions of tidal observations for several places in different parts
of the world, allow us to admit to have possibly taken place. The
assumption of a fluid interior, which Newcomb suggests, and the
flow of a large mass of the fluid "from equatorial regions to a
position nearer the axis," is not, from what I have said to you,
admissible as a probable explanation of the remarkable acceleration of rotational velocity which seems to have taken place about

1862; but happily it is not necessary. A settlement of 14 centimetres in the equatorial regions, with corresponding rise of 28 centimetres at the poles (which is so slight as to be absolutely undiscoverable in astronomical observatories, and which would involve no change of sea-level absolutely disproved by reductions of tidal observations hitherto made), would suffice. Such settlements must occur from time to time; and a settlement of the amount suggested might result from the diminution of centrifugal force due to 150 or 200 centuries tidal retardation of the earth's rotational speed.

Note on the Fortnightly Tide.

39. In water 10,000 feet deep (which is considerably less than the general depth of the Atlantic, as demonstrated by the many soundings taken within the last few years, especially those along the whole line of the Atlantic telegraph cable, from Valencia to Newfoundland) the velocity of long free waves is 567 feet per second*. At this rate the time of advancing through 57° (or a distance equal to the earth's radius) would be only ten hours. Hence it may be presumed that, at least at all islands of the Atlantic, the fortnightly tide should follow sensibly the equilibrium law.

"In the *Philosophical Transactions*, 1839, p. 157, Mr Whewell shows that the observations of high and low water at Plymouth give a mean height of water increasing as the moon's declination increases, and amounting to three inches when the moon's declination is 25°. This is the same direction as that corresponding in the expression above to a high latitude. The effect of the sun's declination is not investigated from the observations. In the *Philosophical Transactions*, 1840, p. 163, Mr Whewell has given the observations of some most extraordinary tides at Petropaulofsk in Kamschatka, and at Novo-Arkangelsk in the island of Sitkhi on the west coast of North America. From the curves in the *Philosophical Transactions*, as well as from the remaining curves relating to the same places (which, by Mr Whewell's kindness, we have inspected), there appears to be no doubt that the mean level of the water at Petropaulofsk and Novo-Arkangelsk

---

* Airy's *Tides and Waves*, § 170.

rises as the moon's declination increases. We have no further information on this point."—Airy's "*Tides and Waves*," § 533.

APPENDIX A, added January 2, 1864.

40.   Let the difference of longest and shortest radii, which
would be produced by lunar and solar influence in the two cases—
of the earth supposed a homogeneous incompressible fluid tending
to the spherical shape by gravitation alone, and supposed a homogeneous incompressible elastic solid without mutual gravitation
but tending in virtue of its elasticity to the spherical figure—be
denoted by h' and h'' respectively; and let h be the difference
of greatest and least radii when both gravity and elasticity act
jointly to maintain the spherical figure.   We shall have obviously

$$\frac{1}{h} = \frac{1}{h'} + \frac{1}{h''}.$$

For the distorting force, being balanced by elasticity and by
gravity jointly, may be divided into two parts, one h/h'' of the
whole, balanced by elasticity alone, and the other h/h', balanced
by gravity alone; and therefore $\dfrac{h}{h''} + \dfrac{h}{h'} = 1$.

But, by § 53 of the mathematical investigation regarding
elastic spheroids, given in the next Paper [Art. XCVI. below], we
have $h'' = \dfrac{3}{2} \dfrac{m}{c^3} \dfrac{5w}{19n} r^3$, where m denotes the mass of the disturbing
body, and c its distance from the earth's centre.   With the same
notation we have, by the aid of § 51 of the same paper, $H = \dfrac{3}{2} \dfrac{m}{c^3} \dfrac{r^2}{g}$,
where H has the meaning defined above in § 8 of the present
paper; and therefore, § 10, $h' = \dfrac{5}{2} \cdot \dfrac{3}{2} \dfrac{m}{c^3} \dfrac{r^2}{g}$.   From this and the value
above for h'', we have $\dfrac{h'}{h''} = \dfrac{19n}{2gwr}$, and, as we have just seen that
$h = \dfrac{h'}{1 + \dfrac{h'}{h''}}$, we have the result used in § 4 above.

APPENDIX B.—*On the Observations and Calculations required to find the Tidal Retardation of the Earth's Rotation**.

[*Philosophical Magazine.* June (Supplement) 1866.]

THE first *publication* of any definite estimate of the possible amount of the diminution of rotatory velocity experienced by the earth through tidal friction is due, I believe, to Mr William Ferrel, and is to be found in the Number for December 8, 1853, of the *Astronomical Journal* of Cambridge, United States. It is founded on calculating the moment round the earth's centre of the attraction of the moon, on a regular spheroidal shell of water symmetrical about its longest axis, this being (through the influence of fluid friction) kept in a position inclined backwards at an acute angle to the line from the earth's centre to the moon. One of the simplest ways of seeing the result is this:—First, by the known conclusions as to the attractions of ellipsoids, or still more easily by the consideration of the proper "spherical harmonic"† (or Laplace's coefficient) of the second degree, we see that an equipotential surface lying close to the bounding surface of a nearly spherical homogeneous solid ellipsoid, is approximately an ellipsoid with axes differing from one another by three-fifths of the amounts of the differences of the corresponding axes of the ellipsoidal boundary. From this it follows‡ that a homogeneous prolate spheroid of revolution attracts points outside it approximately as if its mass were collected in a uniform bar having its ends in the foci of the equipotential spheroid. If, for example, a globe of water of 21,000,000 feet radius (this being nearly enough the earth's radius) be altered into a prolate spheroid with longest radii exceeding the shortest radii by two feet, the equipotential spheroid will have longest and shortest radii differing by 6/5 of a foot. The foci of this latter will be at 7100 feet on each side of the centre; and there-

---

* From the Rede Lecture, Cambridge, May 23, 1866, "On the Dissipation of Energy."

† Thomson and Tait's *Natural Philosophy*, § 536 (4).

‡ Ibid. § 501 and § 480 (*e*).

fore the resultant of gravitation between the supposed spheroid of
water and external bodies will be the same as if its whole mass
were collected in a uniform bar of 14,200 feet length. But by a
well-known proposition *, a uniform line FF'' (a diagram is unnecessary) attracts a point M in the line MK bisecting the angle FMF''.
Let CQ be a perpendicular from C, the middle point of $F'F$, to this
bisecting line MK. If CM be $60 \times 21 \times 10^6$ (the moon's distance),
and if the angle FCM be $45°$, we find, by elementary geometry,
$CQ = \cdot 02$ of a foot (about $\frac{1}{4}$ inch). The mass of a globe of water
equal in bulk to the earth is $1 \cdot 1 \times 10^{21}$ tons†. And, the moon's
mass being about $1/75$ of the earth's, the attraction of the moon on
a ton at the earth's distance is $(1/75) \times (1/60^2)$, or $1/270,000$ of a ton
force, if, for brevity, we call a ton force the ordinary terrestrial
weight of a ton—that is to say, the amount of the earth's attraction
on a ton at its surface. Hence the whole force of the moon on the
earth is $1 \cdot 1 \times 10^{21}/270,000$, or $4 \cdot 1 \times 10^{15}$ tons force. If, then, the
tidal disturbance were exactly what we have supposed, or if it were
(however irregular) such as to have the same resultant effect, the
retarding influence of the moon's attraction would be that of
$4 \cdot 1 \times 10^{15}$ tons force acting in the plane of the equator and in a
line passing the centre at $1/50$ of a foot distance. Or it would be
the same as a simple frictional resistance (as of a friction-brake)
consisting of $4 \cdot 1 \times 10^{15}$ tons force acting tangentially against the
motion of a pivot or axle of about $1/2$ inch diameter. To estimate
the retardation produced by this, we shall suppose the square of
the earth's radius of gyration, instead of being $2/5$, as it would be
if the mass were homogeneous, to be $1/3$, of the square of the
radius of figure, as it is made to be, by Laplace's probable law
of the increasing density inwards, and by the amount of precession
calculated on the supposition that the earth is quite rigid. Hence
(if we take $g = 32 \cdot 2$ feet per second generated per second, and the
earth's mass as $6 \cdot 1 \times 10^{21}$ tons) the loss of angular velocity per
second, on the other suppositions we have made, will be

$$\frac{32 \cdot 2 \times 4 \cdot 1 \times 10^{15} \times \cdot 02}{6 \cdot 1 \times 10^{21} \times \frac{1}{3}\,(21 \times 10^6)^2}, \text{ or } 2 \cdot 94 \times 10^{-21}.$$

---

* Thomson and Tait's *Natural Philosophy*, § 481 (b) and (a).

† In stating large masses, if English measures are used at all, the ton is convenient, because it is 1000 kilogrammes nearly enough for many practical purposes
and rough estimates. It is 1016·047 kilogrammes; so that a ton diminished by
about 1·6 per cent. would be just 1000 kilogrammes.

The loss of angular velocity in a century would be $31\frac{1}{2} \times 10^8$ times this, or $\cdot 93 \times 10^{-11}$, which is as much as $1\cdot 28/10^7$ of $2\pi/86400$, the present angular velocity. Thus in a century the earth would be rotating so much slower that, regarded as a time-keeper, it would lose about one second and a quarter in ten million, or four seconds in a year. And the accumulation of effect of uniform retardation at that rate would throw the earth as a time-keeper behind a perfect chronometer (set to agree with it in rate and absolute indication at any time) by 200 seconds at the end of a century, 800 seconds at the end of two centuries, and so on. In the present very imperfect state of clock-making (which scarcely produces an astronomical clock two or three times more accurate than a marine chronometer or good pocket-watch), the only chronometer by which we can check the earth is one which goes much worse—the moon. The marvellous skill and vast labour devoted to the lunar theory by the great physical astronomers Adams and Delaunay, seem to have settled that the earth has really lost in a century about eleven seconds of time on the moon corrected for previously thought of perturbations. M. Delaunay has suggested that the true cause may be tidal friction, which he has proved to be probably sufficient by some such estimate as the preceding[*]. This, by the *forward* tangential force on the moon in her orbit which it implies, causes a gradual diminution of the mean angular velocity of her radius vector round the earth: which, when taken into account by Prof. Adams[†], on the hypothesis of tidal influence of moon and sun being the sole cause of the unexplained part of the apparent acceleration of the moon's motion, raises from the seeming eleven seconds, to twenty-two seconds, the time which must be actually lost by the earth in a century, on an absolute time-keeper set to keep time according to sidereal seconds as shown by the earth at the beginning of the century.

But the many disturbing influences to which the earth is exposed render it a very untrustworthy time-keeper. For instance,

---

[*] It seems hopeless, without waiting for some centuries, to arrive at any approach to an exact determination of the amount of the actual retardation of the earth's rotation by tidal friction, except by extensive and accurate observation of the amounts and times of the tides on the shores of continents and islands in all seas, and much assistance from *true* dynamical theory to estimate these elements all over the sea. But supposing them known for every part of the sea, the retardation of the earth's rotation is easily calculated by quadratures.

[†] See Thomson and Tait's *Natural Philosophy*, § 830.

let us suppose ice to melt from the polar regions (20° round each pole, we may say) to the extent of something more than a foot thick, enough to give 1·1 foot of water over those areas, or ·066 of a foot of water if spread over the whole globe, which would in reality raise the sea-level by only some such almost undiscoverable difference as 3/4 of an inch, or an inch. This, or the reverse, which we may believe might happen any year, and could certainly not be detected without far more accurate observations and calculations for the mean sea-level than any hitherto made, would slacken or quicken the earth's rate as a time-keeper by one-tenth of a second per year*.

Again, an excellent suggestion, supported by calculations which show it to be not improbable, has been made to the French Academy by M. Dufour, that the retardation of the earth's rotation indicated by M. Delaunay, or some considerable part of it, may be due to an increase of its moment of inertia by the incorporation of meteors falling on its surface. If we suppose the previous average moment of momentum of the meteors round the earth's axis to be zero, their influence will be calculated just as I have calculated that of the supposed melting of ice. Thus meteors falling on the earth in fine powder (as is in all probability the lot of the greater number that enter the earth's atmosphere and do not escape into external space again) enough to form a layer about 1/20 of a foot thick in 100 years, if of twice the density of water, would produce the seeming retardation of 11^s on the time shown by the earth's rotation. But this would also accelerate the moon's mean motion by half the same proportional amount; and therefore a layer of meteor-dust accumulating at the rate of 1/30 of

---

* The calculation is simply this. Let E be the earth's whole mass, a its radius, k its radius of gyration before, and k' after, the supposed melting of the ice, and W the mass of ice melted. Then, since $\frac{2}{3}a^2$ is the square of the radius of gyration of the thin shell of water supposed spread uniformly over the whole surface, and that of either ice-cap is very approximately $\frac{1}{2}a^2(\sin 20°)^2$, we have

$$Ek'^2 = Ek^2 + Wa^2[\tfrac{2}{3} - \tfrac{1}{2}(\sin 20°)^2].$$

And, by the principle of the conservation of moments of momentum, the rotatory velocity of the earth will vary inversely as the square of its radius of gyration. To put this into numbers, we take, as above, $k^2 = \frac{1}{3}a^2$ and $a = 21 \times 10^6$. And as the mean density of the earth is about $5\frac{1}{2}$ times that of water, and the bulk of a globe is the area of its surface into 1/3 of its radius,

$$E : W :: \frac{5 \cdot 5a}{3} : \cdot 066.$$

a foot per century, or 1 foot in 3000 years, would suffice to explain Messrs. Adams and Delaunay's result. I see no other way of directly testing the probable truth of M. Dufour's very interesting hypothesis than to chemically analyze quantities of natural dust taken from any suitable localities (such dust, for instance, as has accumulated in two or three thousand years to depths of many feet over Egyptian, Greek, and Roman monuments). Should a considerable amount of iron with a large proportion of nickel be found or not found, strong evidence for or against the meteoric origin of a sensible part of the dust would be afforded.

Another source of error in the earth as a time-keeper, which has often been discussed, is its shrinking by cooling. But I find by the estimates I have given elsewhere* of the present state of deep underground temperatures, and by taking 1/100000 as the vertical contraction per degree Centigrade of cooling in the earth's crust, that the gain of time by the earth, regarded as a clock, would not in a century amount to more than 1/30 of a second, or 1/6000 of the amount estimated above as conceivably due to tidal friction.

APPENDIX C.—*On the Thermodynamic Acceleration of the Earth's Rotation.*

[*Proc. Royal Society of Edinburgh*, Vol. XI.; read Jan. 16, 1882.]

IT has long been known, having been first, I believe, pointed out by Kant, and more recently brought very near to a practical conclusion by Delaunay, that the earth's rotational velocity is diminished by tidal agency, in virtue of the imperfect fluidity of the ocean. An integral effect of all the consumption of energy by fluid friction (or more properly speaking by continued defor-

---

* "Secular Cooling of the Earth," *Transactions of the Royal Society of Edinburgh*, 1862; and *Philosophical Magazine*, January 1863 [Art. XCIV. above].

mation of fluid matter) in the tidal motions, is to cause the time
of high water on an average for the whole earth to be not exactly

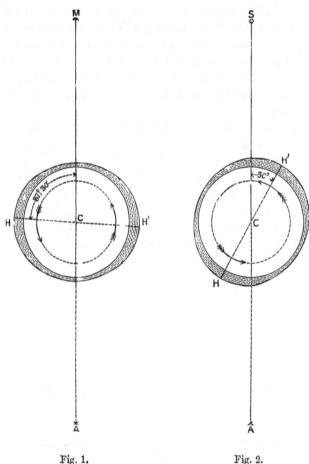

Fig. 1.   Fig. 2.

either transit, or 6 o'clock, as it would be were the ocean a perfect
fluid, but to be some time after transit, and before 6 o'clock*.

* For brevity, I use the word "transit" to denote a time of transit of the tide-
generating body (whether sun or moon), or a time of transit of the point of the
heavens opposite to the tide-generating body, across the meridian of the place;
and "6 o'clock," to denote the middle instant of the interval of time between
consecutive transits. If, to fix the ideas, we first think of the lunar tide alone
as if there were no solar tide, 6 o'clock will mean 6 lunar hours before or
after a lunar transit.

Thus we may imagine the average lunar tide for the whole earth to consist of a displacement of the water, presenting protuberances, not exactly towards moon and anti-moon, but in a line inclined at an angle to the line joining moon and anti-moon, in the direction indicated by the drawing (fig. 1), in which M, A represent the directions of moon and anti-moon, and H, H' the crowns of the ideal spheroid, representing the average water-level for the whole earth. The angle HCM is made $87° 30'$, which would be actually the case if $5^h 50^m$ lunar time were the average time of high water for the whole earth. It is obvious that the resultant force of the moon, on the whole mass of the solid and liquid constituting the earth, is not a single force, exerted in the line MC, but that, after the manner of Poinsot, it may be represented by a single force in this line, and a couple in a direction opposite to that of the arrows indicating in the diagram the direction of the earth's rotation. Thus the lunar attraction produces, as it were, the action of a friction brake resisting the earth's rotation. The same is no doubt also the case in respect to the sun and the water of the ocean.

If HH' were inclined to the line of the attracting body, on the other side from that shown in the first diagram, the effect of the attraction would be to accelerate the earth's rotation. Now this, which is represented in the second diagram (fig. 2), is found by observation to be actually the case in respect to the sun and (not the waters of the ocean, but) the earth's atmosphere. The accompanying table and formula show the result of the Fourier Harmonic Analysis applied for the diurnal period by Mr G. H. Simmonds to barometric observations collected from all parts of the world. In the formula, E denotes the excess of the barometric pressure above its mean value for the day, at the time θ reckoned in degrees from midnight, at the rate of $15°$ per mean solar hour: $R_1 c_1$, $R_2 c_2$, $R_3 c_3$ denote the ranges and angles, corresponding to the times of maximum height, for the first three terms of the Fourier expression which the formula exhibits. The table shows the values of $R_1 c_1$, $R_2 c_2$, $R_3 c_3$, calculated for the different places, from observations at the times stated in column 5.

It is a very remarkable result of this analysis that the amplitude R_2 of the semidiurnal term is for most places, especially those within $40°$ of the equator, considerably greater than the R_1 of the diurnal term. The cause of the semidiurnal variation of barometric pressure cannot be the gravitational tide-generating influence of the sun, because, if it were, there would be a much larger lunar influence of the same kind, while in reality the lunar barometric tide is insensible or nearly so.   It seems therefore certain that the semidiurnal variation of the barometer is due to temperature.   Now the *diurnal* term, in the Harmonic Analysis of the variation of *temperature*, is undoubtedly much larger in all, or nearly all, places than the *semidiurnal*.   It is then very re-markable that the *semidiurnal term of the barometric effect of* the variation of temperature should be greater, and so much greater as it is, than the diurnal.   The explanation probably is to be found by considering the oscillations of the atmosphere, as a whole, in the light of the very formulas which Laplace gave in his *Mécanique Céleste* for the ocean, and which he showed to be also applicable to the atmosphere.   When thermal influence is substituted for gravitational, in the tide-generating force reckoned for, and when the modes of oscillation corresponding respectively to the diurnal and semidiurnal terms of the thermal influence are investi-gated, it will probably be found that the period of free oscillation of the former agrees much less nearly with 24 hours than does that of the latter with 12 hours; and that therefore, with com-paratively small magnitudes of the tide-generating force, the resulting tide is greater in the semidiurnal term than in the diurnal.   Now, if we look to the values of c_2 in the table, we see that, with one exception (Sitka, a place far north, where R_2 is very small), they are all positive acute angles: and we find 61°·3 as the mean of all the 30.   If we assign weights to the different values of c_2, according to the corresponding values of R_2, we should find a somewhat larger number for the true mean value of c_2.   It is enough for our present purpose to say that the mean is 60° or a little more.   Looking now to the formula, we see that the meaning of this is that the times of maximum of the semi-diurnal variation R_2 are a little before 10 o'clock in the morning and a little before 10 o'clock at night (exactly at 10 o'clock if c_2 were exactly 60°).   Without more of observation, or of observation and theory, than has yet been brought to bear on the subject, we cannot tell the law of variation of R_2 with the latitude.   The observations in the table seem to show, what Laplace's Tidal Theory prepares us to expect, that it diminishes more in the Polar regions than it would if it followed the elliptic spheroidal

Extracted from the *Quarterly Journal of the Meteorological Society* for January 1880. "The Diurnal Range of Atmospheric Pressure," by Robert Strachan, F.M.S.

Harmonic Constituents of the Diurnal Variation of Atmospheric Pressure, calculated by G. H. Simmonds, F.M.S.

Name of Place.	Latitude.	Longitude.	Height.	Time of Observation.		Diurnal Constituents.		Semi-diurnal Constituents.		Ter-diurnal Constituents.	
	° ′	° ′	Feet.	Years		R_1 Inches.	c_1 °	R_2 Inches.	c_2 °	R_3 Inches.	c_3 °
Singapore,	1 27 N	103 49	Small No.	5	from 1841 to 1845,	·0210	280·3	·0387	66·0	·0015	333·3
Trevandrum,	8 31 N	77 0	195	5	June 1837 to May 1842,	·0154	290·3	·0424	68·2	·0013	293·1
Madras,	13 4 N	80 14	22	7	1844 to 1850,	·0234	268·9	·0432	67·6	·0007	270·0
Bombay,	18 53 N	72 48	38	17	1846 to 1862,	·0198	242·9	·0382	66·4	·0014	286·3
Calcutta,	22 31 N	88 21	18	15	1855 to 1869,	·0270	250·9	·0394	61·6	·0012	258·5
Simla,	31 6 N	77 12	6953	5·5	June 1841 to December 1846,	·0100	185·7	·0210	48·7	·0015	248·3
Lisbon,	38 43 N	9 9	335	6·9	January 1864 to November 1870,	·0053	246·6	·0176	62·1	·0022	272·4
Pekin,	39 57 N	116 29	101	6	1850 to 1855,	·0295	270·8	·0217	54·8	·0026	256·0
Washington,	38 54 N	77 3	103	9	1861 to 1869,	·0168	265·6	·0201	73·8	·0024	172·9
Girard College,	39 58 N	75 11	112	5·1	June 1840 to June 1845,	·0183	266·6	·0179	75·8	·0018	282·1
Toronto,	43 40 N	79 21	342	7·3	1841 to 1847,	·0140	242·4	·0127	81·5	·0020	291·7
Tiflis (Awlaba),	41 42 N	44 50	1501	7·3	January 1855 to April 1862,	·0246	288·6	·0142	67·5	·0021	242·3
„ (Kuki),	41 43 N		1343	9·7	May 1862 to December 1871,	·0266	294·3	·0162	70·9	·0018	267·1
Vienna,	48 13 N	16 23	650	7·9	1849 to 1856 (less one month of April),	·0068	262·4	·0113	59·5	·0012	272·2
Cracow,	50 4 N	19 58	712	7	1850 to 1856,	·0050	287·5	·0066	42·5	·0016	256·4
Prague,	50 5 N	14 25	351	27	1842 to 1868,	·0100	271·5	·0086	55·1	·0011	281·8
Brussels,	50 51 N	4 22	190	28	1842 to 1869,	·0019	268·8	·0095	56·1	·0012	278·7
Greenwich,	51 29 N	·	159	7	1841 to 1847,	·0011	133·9	·0104	57·9	·0004	326·6
Oxford,	51 46 N	1 15	212	16	1855 to 1870,	·0046	312·7	·0096	66·1	·0013	273·3
Nertchinsk,	51 18 N	119 30	2230	19	1842 to 1845, 1848 to 1855, and 1856 to 1862,	·0126	283·1	·0098	71·1	·0016	236·0
Barnaoul,	53 20 N	83 37	400	17	1842 to 1845, 1850 to 1855, and 1856 to 1862,	·0046	189·2	·0044	72·0	·0011	262·3
Catherinenburg,	56 50 N	60 34	813	18	1842 to 1845, 1849 to 1855, and 1856 to 1862,	·0036	325·4	·0035	62·8	·0003	212·7
Sitka,	57 9 N	135 18	15	10	1843 to 1845, 1848, 1850 to 1854, and 1856,	·0028	135·9	·0037	- 6·3	·0003	147·3
St Petersburg,	59 57 N	30 28	15	22	1841 to 1862,	·0014	152·7	·0035	6·2	·0006	206·9
Batavia,	6 11 S	106 50	24	7	1866 to 1872,	·0239	293·6	·0369	67·3	·0016	281·8
Ascension,	7 55 S	0 58	53	2	September 1863 to August 1865,	·0106	287·4	·0279	66·5	·0004	206·6
St Helena,	15 57 S	5 41	1764	6	1841 to 1846,	·0071	234·1	·0293	63·4	·0014	348·0
Santiago de Chile,	33 26 S	70 38	1790	2·9	November 1849 to September 1852,	·0065	253·8	·0157	77·2	·0014	105·0
Cape of Good Hope,	33 56 S	18 29	Small No.	5·2	April 1841 to June 1846,	·0047	257·8	·0195	72·0	·0014	281·8
Hobarton,	42 52 S	147 27	105	7	1841 to 1847,	·0123	317·5	·0197	84·1	·0018	287·6

law of proportionality to the square of the cosine of the latitude. We may, however, take by inspection from the table

$$R_2 = \cos^2 \text{lat} \times \cdot 032 \text{ inch}$$

as a rough estimate of a barometric variation distributed over the whole earth in the form of an elliptic spheroid, which would give the same resisting couple in the calculation of the solar gravitational influence on the disturbed atmosphere; or (getting quit of the intolerable British inch),

$$R_2 = \cos^2 \text{lat} \times \cdot 08 \text{ cm}.$$

Now the height of the barometer corresponds always to the mass of the air over a given horizontal area of the locality, independently of the temperature of the air; and, in averages for the different places, no doubt independently of the wind also*. Thus for every centimetre of higher or lower mercury in the barometer, there is more or less mass of air over the locality to the extent of 13·596, or say 14 grms. over every square centimetre of horizontal surface. Thus the second diagram with its angle of 30° (corresponding to $c_2 = 60°$) represents the state of things, as regards the quantity of air over different parts in the circle of any parallel of latitude, or at all events of any circle farther from the pole than 60° north or south latitude. It represents the state of things for every parallel of latitude in the imagined elliptic spheroid, constituting the terms we have to deal with in the spherical harmonic expression of the actual effect: and definitively, if we suppose half the excess of the greatest above the least radius of the elliptic spheroid in the diagram to be equal to the square of the sine of the latitude multiplied into ·08 cm., the diagram shows the distribution of a mass of matter of the same density

---

* In strong winds the barometer may stand sensibly above or below the proper value for the weight of the atmosphere over the place, according as the room containing the barometer is more exposed by openings on the windward or on the leeward side of the house in which it is placed. The error due to this cause may be sensible in the diurnal averages for one particular barometer, because of the daily periodic variations in the direction of the wind; but it is not probably large for any well-placed barometer, and, such as it is, it must be fairly well eliminated in the averages for different barometers in variously arranged buildings and in different parts of the world. In passing, it may be remarked, that it is probably not a matter of no importance that the barometer-room of a well-appointed meteorological observatory should be as nearly as may be symmetrically arranged in respect to openings to the external air in different directions, and in respect to shelter against wind from other parts of the building.

as mercury, over the whole surface of the earth, which would experience the same resultant couple from the sun as does the earth's atmosphere in reality. To evaluate this couple we may use the known formula (Thomson and Tait's *Natural Philosophy*, 2nd ed. Vol. I. Part II., § 539) relative to the mutual attraction between a mass M, not concentrated in a point, and a portion of matter m, concentrated in a very distant point—

$$L = 3m \frac{(B-C)yz}{(x^2+y^2+z^2)^{\frac{5}{2}}} \dots\dots\dots\dots\dots\dots (1),$$

where x, y, z denote coordinates of m relatively to rectangular lines OX, OY, OZ coincident with the principal axes of inertia of M through its centre of inertia; B and C the moments of inertia of M round OY and OZ; and L the component round OX, of the couple obtained by transposing, after the manner of Poinsot, the resultant attraction of m, from its actual line through x, y, z, to a parallel line through o, the centre of inertia of M. Suppose now M to be a homogeneous ellipsoid of revolution, having for semi-axes a, b, c, we have

$$B - C = \tfrac{1}{5} M (c^2 - b^2)$$
$$= \tfrac{1}{5} M (c + b)(c - b).$$

Hence for a prolate spheroid of the dimensions stated above, we have

$$B - C = \tfrac{1}{5} Mr\, 0 \cdot 32 \text{ cm} \dots\dots\dots\dots\dots (2),$$

where r denotes the earth's radius in centimetres. To fit the formula (1) to the case represented by the diagram in fig. 2, we have

$$yz = D^2 \sin 30° \cos 30° \dots\dots\dots\dots\dots (3),$$

where D denotes the sun's distance from the earth. With this and (2), (1) becomes

$$L = \frac{3}{5} \frac{mMr\, 0 \cdot 32^{\text{cm}} \sin 30° \cos 30°}{D^3} \dots\dots\dots\dots\dots (4),$$

where M denotes the mass of a quantity of mercury equal in bulk to the earth, so that if E denotes the earth's mass $M = 2 \cdot 5 E$. Now mE/D^2 is the attraction of the earth on the sun: hence if we call this force F,

$$L = \frac{3}{5} 2 \cdot 5 \frac{r}{D} F . 0 \cdot 32^{\text{cm}} . \sin 30° \cos 30°$$

$$= \frac{r}{D} F . 0 \cdot 21^{\text{cm}}.$$

Now if S denote the number of grammes in the sun's mass, we have

$$F = \frac{r^2}{D^2} S . 980 \text{ dynes,}$$

since the earth's attraction on a gramme of matter at its surface is about 980 dynes; and so we find

$$L = \frac{r^3}{D^3} S . 980 . 0\cdot21 = \frac{r^3}{D^3} S . 207 \dots\dots\dots (5).$$

Now if $\dot{\omega}$ denote the acceleration of the earth's angular velocity produced by this couple, we have

$$\dot{\omega} = \frac{L}{I} \dots\dots\dots\dots\dots\dots (6),$$

where I denotes the earth's moment of inertia; and, allowing for the increase of the earth's density from the surface inwards, according to Laplace's probable law, we have, approximately,

$$I = \tfrac{1}{3} r^2 E$$

(instead of $I = \tfrac{2}{5} r^2 E$, as it would be if the mass were homogeneous) E denoting the earth's mass. Hence

$$\dot{\omega} = 3 \frac{r^3}{D^3} \frac{S}{E} \frac{207}{r^2}.$$

Now $D^3/r^3 = 12\cdot3 . 10^{12}$, $S/E = 31\cdot9 . 10^4$, $r = 6\cdot370 . 10^8$ centimetres, which gives $r^2 = 40\cdot6 . 10^{16}$. Hence

$$\dot{\omega} = 3 \frac{31\cdot9 . 10^4}{12\cdot3 . 10^{12}} \frac{207}{40\cdot6 . 10^{16}} = 4\cdot0 . 10^{-23}.$$

This is the rate per second of gain of angular velocity. The earth's angular velocity at present is $2\pi/86400$, or approximately $1/13700$. Calling this ω, we have

$$\frac{\dot{\omega}}{\omega} = 5\cdot5 . 10^{-1}$$

for the proportionate gain per second. There are 31·5 million seconds in a year, and 3150 in a century. Hence the ratio to the earth's present angular velocity, of the gain per second, amounts to $1\cdot73 \times 10^{-9}$.

To interpret the result, suppose two chronometers, A and B, to be kept going for a century, according to the following conditions:—

Chronometer A to be an absolutely perfect timekeeper, and to be regulated to sidereal time at the beginning of the century, in the usual manner, by astronomical observation.

Chronometer B to be kept constantly regulated to sidereal time by astronomical observations from day to day, and from year to year, during the century.

At the end of the century B will be found to be gaining on A to the amount of $1\cdot73 \times 10^{-9}$ of a second per second. This rate of gain has been uniformly acquired; and, therefore, on the average of the century, B has been going faster than A, at the rate of $\cdot86 \times 10^{-9}$ of a second per second. Hence, in the whole century (or $3\cdot15 \times 10^9$ sidereal seconds), B has gained on A to the extent of $2\cdot6$ seconds.

In reality a tenfold greater difference, in the opposite direction, would be observed between the two chronometers. Adams, from his correction of Laplace's dynamical investigation of the acceleration of the moon's mean motion, produced by the sun's attraction, found that our supposed chronometer B, regulated to sidereal time, would be 22 seconds* behind the perfect chronometer A at the end of a century. (See Thomson and Tait's *Natural Philosophy*, 1st ed., § 830; or 2nd ed., vol. i. part 1, § 405.) The retardation of the earth's rotation thus definitively specified, which may be regarded as a well-established result of observation and theory, received from Delaunay what we cannot doubt to be its true explanation, —retardation by tidal friction. The preceding formulas, with the proper change of data, may be readily modified to show the tidal retardation instead of the thermodynamic acceleration. Thus if we go back to fig. 1, and suppose the spheroidal layer to be water, instead of the earth's atmosphere, and take 100 cms. as the excess of the greatest above the least semi-diameter, we have what we may fairly assume to be a not improbable estimate of the equivalent, over the whole earth's surface, to the true tidal deformation of the water of the oceans. If the obliquity HCS were the same in the two cases, and if the sun were the external attracting body in each case, the value of L would be $(50/\cdot08 . 13\cdot596 =) 45\cdot9$ times greater in the second case (fig. 1) than in the first case (fig. 2). Suppose now the moon, instead of the sun, to be the influencing body in the second case (fig. 1), other things being the same, the couple will be $91\cdot8$ times as great in the second case (fig. 1) as in

* See Appendix B, page 339 above.

the first case (fig. 2). (Because the moon's mass, divided by the cube of her distance from the earth, is about double the sun's mass divided by the cube of his distance from the earth.) Now, we must make the couple to be only 10 times as great in the second case (fig. 1) as in the first case (fig. 2) to bring out Adams' result, according to Delaunay's explanation of it. Hence we must suppose, in fig. 1, $\sin HCM \cos HCM$ to be $\frac{1}{10}$ of $\sin 30° \cos 30°$; and we may fulfil this condition by taking $HCM = 87° 30'$.

Thus with the approximate results of observation used above in respect to the earth's atmosphere, and the assumptions we have now made regarding the lunar tide, we have a state of things in which our supposed chronometer B gains on A 2·5 seconds in the course of the century through the thermodynamic acceleration, and loses 25 seconds through the tidal retardation; that is, loses in all 22·5 seconds, or say 22 seconds, which is Adams' result.

ART. XCVI. Dynamical Problems regarding Elastic Spheroidal Shells and Spheroids of Incompressible Liquid.

1. The theory of elastic solids in equilibrium presents the following general problem:—

A solid of any shape being given, and displacements being arbitrarily produced or forces arbitrarily applied over its whole bounding surface, it is required to find the displacement of every point of its substance. The chief object of the present communication is to show the solution of this problem for the case of a shell consisting of isotropic elastic material, and bounded by two concentric spherical surfaces, with the natural restriction that the whole alteration of figure is very small.

2. Let the centre of the spherical surfaces be taken as origin, and let x, y, z be the rectangular co-ordinates of any particle of the solid, in its undisturbed position, and $x+\alpha$, $y+\beta$, $z+\gamma$ the co-ordinates of the same particle when the whole is in equilibrium under the given superficial disturbing action. Then, by the known equations of equilibrium of elastic solids, we have

$$\left.\begin{aligned}
n\left(\frac{d^2\alpha}{dx^2}+\frac{d^2\alpha}{dy^2}+\frac{d^2\alpha}{dz^2}\right)+m\,\frac{d}{dx}\left(\frac{d\alpha}{dx}+\frac{d\beta}{dy}+\frac{d\gamma}{dz}\right)&=0,\\
n\left(\frac{d^2\beta}{dx^2}+\frac{d^2\beta}{dy^2}+\frac{d^2\beta}{dz^2}\right)+m\,\frac{d}{dy}\left(\frac{d\alpha}{dx}+\frac{d\beta}{dy}+\frac{d\gamma}{dz}\right)&=0,\\
n\left(\frac{d^2\gamma}{dx^2}+\frac{d^2\gamma}{dy^2}+\frac{d^2\gamma}{dz^2}\right)+m\,\frac{d}{dz}\left(\frac{d\alpha}{dx}+\frac{d\beta}{dy}+\frac{d\gamma}{dz}\right)&=0
\end{aligned}\right\}\ \dots(1),$$

$m-\tfrac{1}{3}n$ and n denoting the two coefficients of elasticity, which may be called respectively the *elasticity of volume*, and the *rigidity*. A demonstration of these equations, with definitions of the coefficients, will be found in § 71 of an Appendix to the present communication.

3. For brevity let

$$\delta = \frac{d\alpha}{dx}+\frac{d\beta}{dy}+\frac{d\gamma}{dz} \dots\dots\dots\dots\dots\dots (2),$$

so that δ shall denote the cubic dilatation at the point (x, y, z) of the solid. Also, for brevity, let the operation $\dfrac{d^2}{dx^2} + \dfrac{d^2}{dy^2} + \dfrac{d^2}{dz^2}$ be denoted by ∇^2. Then the preceding equations become

$$\left. \begin{aligned} n\nabla^2\alpha + m\frac{d\delta}{dx} &= 0, \\ n\nabla^2\beta + m\frac{d\delta}{dy} &= 0, \\ n\nabla^2\gamma + m\frac{d\delta}{dz} &= 0 \end{aligned} \right\} \dots\dots\dots\dots\dots (3).$$

4. In certain cases, especially the ideal one of an incompressible elastic solid, the following notation is more convenient:—

p the mean normal pressure per unit of area on all sides of any small portion of the solid, round the point x, y, z. Then (below, § 21)

$$p = -\left(m - \tfrac{1}{3}n\right)\left(\frac{d\alpha}{dx} + \frac{d\beta}{dy} + \frac{d\gamma}{dz}\right) \dots\dots\dots (4);$$

and the equations of equilibrium become

$$\left. \begin{aligned} n\nabla^2\alpha - \frac{m}{m - \tfrac{1}{3}n}\frac{dp}{dx} &= 0, \\ n\nabla^2\beta - \frac{m}{m - \tfrac{1}{3}n}\frac{dp}{dy} &= 0, \\ n\nabla^2\gamma - \frac{m}{m - \tfrac{1}{3}n}\frac{dp}{dz} &= 0 \end{aligned} \right\} \dots\dots\dots\dots (5).$$

5. If the solid were incompressible, we should have $m = \infty$ and
$$\frac{d\alpha}{dx} + \frac{d\beta}{dy} + \frac{d\gamma}{dz} = 0,$$

which must be taken instead of (4), and, along with (5), would constitute the four differential equations required for the four unknown functions α, β, γ, p*.

6. To solve the general equations (3) or (5), take d/dx of the first, d/dy of the second, and d/dz of the third; and add. We have thus
$$(n + m)\,\nabla^2\delta = 0 \dots\dots\dots\dots\dots\dots (6),$$
or, which is in general sufficient,
$$\nabla^2\delta = 0 \dots\dots\dots\dots\dots\dots\dots (7).$$

* See Professor Stokes's paper "On the Friction of Fluids in Motion, and the Equilibrium and Motion of Elastic Solids," *Cambridge Philosophical Society's Transactions*, April, 1845.

If, now, an appropriate solution of this equation for δ is found, the three equations (3) may be solved by known methods, the first of them for α, the second for β, and the third for γ,—the arbitrary part of the solution in each case being merely a solution of the equation $\nabla^4 u = 0$. These arbitrary parts must be determined so as to fulfil equation (2) and the prescribed surface conditions.

The complete particular determination of δ cannot, however, in most cases be effected without regard to α, β, γ; and the order of procedure which has been indicated is only convenient for determining the proper forms for general solutions of the equations.

7. First, then, to solve the equation in δ generally, we may use a theorem belonging to the foundation of Laplace's remarkable analysis of the attraction of spheroids, which may be enunciated as follows.

If the equation $\nabla^4\delta = 0$ is satisfied for every point between two concentric spheres of radii a (greater) and a' (less), the value of δ for any point of this space, at distance r from the centre, may be expressed by the double series

$$V_0 \quad + V_1 \quad + V_2 \quad + \&c.$$
$$+ V'_0 r^{-1} + V'_1 r^{-3} + V'_2 r^{-5} + \&c.,$$

of which the first part converges at least as rapidly as the geometrical progression

$$\frac{r}{a}, \quad \left(\frac{r}{a}\right)^2, \quad \left(\frac{r}{a}\right)^3, \dots$$

and the second at least as rapidly as

$$\frac{a'}{r}, \quad \left(\frac{a'}{r}\right)^2, \quad \left(\frac{a'}{r}\right)^3, \dots$$

—if V_i, V'_i denote homogeneous functions of x, y, z of the order i, each satisfying, continuously, for all values of x, y, z, the equation

$$\nabla V = 0.$$

A proof of this proposition is given in Thomson and Tait's *Natural Philosophy*, Vol. I. Part I. chap. i. Appendix B. It is also there shown, what I believe has been hitherto overlooked, that V_i, V'_i, as above defined, cannot but be rational and integral, if i is any positive integer.

8.   To avoid circumlocution, we shall call any homogeneous function of (x, y, z) which satisfies the equation

$$\nabla^2 V = 0$$

a "spherical harmonic function," or, more shortly, a "spherical harmonic." Thus V_i and V_i', as defined in § 7, are spherical harmonics of degree or order i; and $V_i' r^{-2i-1}$, being also a solution of $\nabla^2 V = 0$, is a spherical harmonic of degree $-(i+1)$. We shall sometimes call the latter a spherical harmonic of inverse order i. Thus u_i being any spherical harmonic of integral degree i, and therefore necessarily a rational integral function of this degree, $u_i r^{-2i-1}$ is a spherical harmonic of degree $-(i+1)$, or of inverse order i.

If we put $-(i+1)=j$, and denote this last function by ϕ_j, then we have

$$\phi_j r^{-2j-1} = u_i ;$$

and thus it appears that the relation between a spherical harmonic of positive degree i and of negative degree j is reciprocal. The general (well known) proposition on which this depends is that if V_i is any homogeneous function of (x, y, z) of degree i, positive or negative, integral or fractional, $V_i r^{-2i-1}$ is also a solution of the equation $\nabla^2 V = 0$ (see Thomson and Tait's *Natural Philosophy*, chap. i. Appendix B).

A spherical harmonic of integral, whether positive or negative, degree, satisfying the differential equation continuously for all values of the variables, will be called an "entire spherical harmonic," because such functions are suited for the solution of acoustical and other physical problems regarding entire spheres or entire spherical shells.

A spherical harmonic function of (x, y, z) will be called a "spherical surface-harmonic" when the point (x, y, z) lies anywhere on a spherical surface having its centre at the origin of coordinates. A *spherical surface-harmonic* is therefore a function of two variables, angular coordinates of a point on a spherical surface. If Y_i denote such a function of order i, positive and integral, then $Y_i r^i$ and $Y_i r^{-i-1}$ are what we now call simply *spherical harmonics;* but sometimes we shall call them, by way of distinction, "spherical solid harmonics." Functions Y_i, or spherical surface-harmonics of integral orders, have been generally called "Laplace's coefficients" by English writers.

9. From the theorem enunciated in § 7, we see that the general solution of our problem, so far as δ is concerned, is this:—

$$\delta = \Sigma_{i=0}^{i=\infty} \left(V_i + V'_i r^{-2i-1} \right) \dots\dots\dots\dots(8).$$

10. Now because the equation $\nabla^2 u = 0$ is linear, it follows that differential coefficients of any solution, with reference to x, y, z, or linear functions of such differential coefficients, are also solutions. Hence the terms V_i and $V'_i r^{-2i-1}$, of δ, give harmonics of the degrees $i-1$ and $-(i+2)$, in $\dfrac{d\delta}{dx}$, $\dfrac{d\delta}{dy}$, $\dfrac{d\delta}{dz}$. To solve equations (3) we have therefore only to solve

$$\nabla^2 u = \phi_n,$$

where ϕ_n denotes an entire spherical harmonic of any positive or negative degree, n. Trying

$$u = A r^2 \phi_n,$$

which is obviously the right form, we have

$$\nabla^2 u = A \left\{ r^2 \nabla^2 \phi_n + 4 \left(x \frac{d}{dx} + y \frac{d}{dy} + z \frac{d}{dz} \right) \phi_n + \phi_n \nabla^2 (r^2) \right\}.$$

But, because ϕ_n is a homogeneous function of x, y, z of degree n,

$$\left(x \frac{d}{dx} + y \frac{d}{dy} + z \frac{d}{dz} \right) \phi_n = n \phi_n;$$

and because it is a spherical harmonic,

$$\nabla^2 \phi_n = 0.$$

We have also

$$\nabla^2 (r^2) = 6,$$

by differentiation. Hence

$$\nabla^2 u = A \,.\, 2 \,(2n + 3) \,\phi_n,$$

and therefore the complete solution of the equation

$$\nabla^2 u = \phi_n$$

is

$$u = V + \frac{r^2}{2(2n + 3)} \phi_n,$$

where V denotes any solution of the equation

$$\nabla^2 V = 0.$$

23—2

11. Hence, by taking for ϕ_n the terms of $\dfrac{d\delta}{dx}$, $\dfrac{d\delta}{dy}$, $\dfrac{d\delta}{dz}$ referred to (§ 10) above, and giving n its proper value, $i-1$, or $-(i+2)$, for each term as the case may be, we find, for the complete solution of (3), the following :—

$$\left. \begin{aligned} \alpha &= \Sigma\left\{ u_i + u'_i r^{-2i-1} - \frac{mr^2}{n \cdot 2\,(2i+1)}\frac{d}{dx}(V_i - V'_i r^{-2i-1}) \right\}, \\ \beta &= \Sigma\left\{ v_i + v'_i r^{-2i-1} - \frac{mr^2}{n \cdot 2\,(2i+1)}\frac{d}{dy}(V_i - V'_i r^{-2i-1}) \right\}, \\ \gamma &= \Sigma\left\{ w_i + w'_i r^{-2i-1} - \frac{mr^2}{n \cdot 2\,(2i+1)}\frac{d}{dz}(V_i - V'_i r^{-2i-1}) \right\} \end{aligned} \right\} \dots (9),$$

where $u_i,\ u'_i,\ v_i,\ v'_i,\ w_i,\ w'_i$ denote six harmonics, each of degree i.

12. But in order that these formulæ may express the solution of the original equations (1), the functions u, v, &c. must be related to the functions V so as to satisfy equations (2) and (3). Now, taking account of the following formula,

$$\frac{d}{dx}\left(r^2\frac{d\phi_i}{dx}\right) + \frac{d}{dy}\left(r^2\frac{d\phi_i}{dy}\right) + \frac{d}{dz}\left(r^2\frac{d\phi_i}{dz}\right) = 2\left(x\frac{d}{dx} + y\frac{d}{dy} + z\frac{d}{dz}\right)\phi_i + r^2\nabla^2\phi_i,$$

which becomes simply $2i\phi_i$,

if ϕ_i is a spherical harmonic of any degree i (whether positive or negative, integral or fractional), we derive from (9) by differentiation, and selection of terms of order i, and of order inverse i (or degree $-i-1$),

$$\frac{d\alpha}{dx} + \frac{d\beta}{dy} + \frac{d\gamma}{dz} = \Sigma\left\{ \psi_i + \psi'_i r^{-2i-1} - \frac{m}{n\,(2i+1)}[iV_i + (i+1)\,V'_i r^{-2i-1}] \right\},$$

where, for brevity, we put

$$\left. \begin{aligned} \psi_i &= \frac{du_{i+1}}{dx} + \frac{dv_{i+1}}{dy} + \frac{dw_{i+1}}{dz}, \\ \psi'_i r^{-2i-1} &= \frac{d(u'_{i-1}r^{-2i+1})}{dx} + \frac{d(v'_{i-1}r^{-2i+1})}{dy} + \frac{d(w'_{i-1}r^{-2i+1})}{dz} \end{aligned} \right\} \dots (10).$$

and

Hence, to satisfy (2) and (8),

$$V_i = \psi_i - \frac{mi}{n\,(2i+1)}\,V_i,$$

and

$$V'_i = \psi'_i - \frac{m\,(i+1)}{n\,(2i+1)}\,V'_i,$$

from which we find

$$
\left.
\begin{aligned}
V_i &= \frac{n\,(2i+1)}{(2n+m)\,i+n}\,\psi_i, \\
V'_i &= \frac{n\,(2i+1)}{(2n+m)\,i+n+m}\,\psi'_i
\end{aligned}
\right\} \quad \dots\dots\dots\dots (11).
$$

13.   Using these in (9), we conclude

$$
\left.
\begin{aligned}
\alpha &= \Sigma_{i=0}^{i=\infty}* \left\{ u_{i+1}+u'_{i-1}r^{-2i+1} - \frac{mr^2}{2}\frac{d}{dx}\left[\frac{\psi_i}{(2n+m)i+n} - \frac{\psi'_i r^{-2i-1}}{(2n+m)i+n+m} \right]\right\}, \\
\beta &= \Sigma_{i=0}^{i=\infty} \left\{ v_{i+1}+v'_{i-1}r^{-2i+1} - \frac{mr^2}{2}\frac{d}{dy}\left[\frac{\psi_i}{(2n+m)i+n} - \frac{\psi'_i r^{-2i-1}}{(2n+m)i+n+m} \right]\right\}, \\
\gamma &= \Sigma_{i=0}^{i=\infty} \left\{ w_{i+1}+w'_{i-1}r^{-2i+1} - \frac{mr^2}{2}\frac{d}{dz}\left[\frac{\psi_i}{(2n+m)i+n} - \frac{\psi'_i r^{-2i-1}}{(2n+m)i+n+m} \right]\right\}
\end{aligned}
\right\}
$$

$$\dots\dots\dots\dots\dots\dots\dots\dots\dots\dots\dots\dots\dots\dots\dots\dots\dots\dots\dots(12)$$

for a complete solution of the general equations (1), the equations of equilibrium of an isotropic elastic solid. The circumstances for which this solution is appropriate will be understood when the general proposition of § 7 is duly considered.

14.   It remains to show how the harmonics u_i, v_i, w_i, u'_i, v'_i, w'_i are to be determined so as to satisfy the superficial conditions. Let us first suppose these to be that the displacement of every point of the bounding surface is given arbitrarily. Let ΣA_i, ΣB_i, ΣC_i be the harmonic series†, expressing the three components of the displacement at any point of the outer surface of the shell, and $\Sigma A'_i$, $\Sigma B'_i$, $\Sigma C'_i$ the corresponding expressions for the given condition of the inner surface. Thus the surface-equations of condition to be fulfilled are

$$
\left.
\begin{aligned}
r=a \quad &\begin{cases} \alpha = \Sigma A_i, \\ \beta = \Sigma B_i, \\ \gamma = \Sigma C_i, \end{cases} \\
r=a' \quad &\begin{cases} \alpha = \Sigma A'_i, \\ \beta = \Sigma B'_i, \\ \gamma = \Sigma C'_i, \end{cases}
\end{aligned}
\right\} \dots\dots\dots\dots(13)
$$

* For the case $i=0$, the terms u'_{-1}, v'_{-1}, w'_{-1} may be omitted; but their full interpretation would be to express a displacement without deformation. Thus u'_{-1}, being of degree -1, cannot but be A/r, where A is a constant; and therefore $u'_{i-1}r^{-2i+1}$ becomes A when $i=0$.

† That is, series of terms each of which is a spherical surface-harmonic of

where a and a' denote the radii of the outer and inner surfaces respectively, and A_i, B_i, C_i, A'_i, B'_i, C'_i spherical surface-harmonics of the order i.

15.   Now collecting from the series (12) of § 13, which constitute the general expressions for α, β, γ, those terms which, being either solid spherical harmonics of degrees i and $-i-1$, or such functions multiplied by r^2, give, at the boundary, surface-harmonics of the order i, and equating the terms of this order on the two sides of equations (13), we have

$$
\begin{aligned}
u_i+u_i' r^{-2i-1}-\frac{mr^2}{2}\frac{d}{dx}\left[\frac{\psi_{i+1}}{(2n+m)i+3n+m}-\frac{\psi'_{i-1}r^{-2i+1}}{(2n+m)i-n}\right]\left\{\begin{aligned}&=A_i \text{ when } r=a,\\&=A'_i \text{when } r=a',\end{aligned}\right. \\
v_i+v_i' r^{-2i-1}-\frac{mr^2}{2}\frac{d}{dy}\left[\frac{\psi_{i+1}}{(2n+m)i+3n+m}-\frac{\psi'_{i-1}r^{-2i+1}}{(2n+m)i-n}\right]\left\{\begin{aligned}&=B_i \text{ when } r=a,\\&=B'_i \text{when } r=a',\end{aligned}\right. \\
w_i+w_i' r^{-2i-1}-\frac{mr^2}{2}\frac{d}{dz}\left[\frac{\psi_{i+1}}{(2n+m)i+3n+m}-\frac{\psi'_{i-1}r^{-2i+1}}{(2n+m)i-n}\right]\left\{\begin{aligned}&=C_i \text{ when } r=a,\\&=C'_i \text{when } r=a'\end{aligned}\right.
\end{aligned}
\right\}\;\ldots(15).
$$

16.   These six equations would suffice to determine the six harmonics $u_i, v_i, w_i, u_i', v_i', w_i'$, if ψ_{i+1} and ψ'_{i-1} were known. For, since each of those six functions is a homogeneous function of x, y, z of order i, each of them divided by r^i is a function of angular coordinates relative to the centre, and independent of r; and therefore if, for instance, we denote u_i by $r^i\varpi$ and u_i' by $r^i\varpi'$, we have two unknown quantities ϖ and ϖ' to be determined by the two equations of condition relative to α for the outer and the inner surface.   These equations may be written as follows, if we further denote $\dfrac{d\psi_{i+1}}{dx}$ by $r^i\vartheta$, and $\dfrac{d(\psi'_{i-1}r^{-2i+1})}{dx}$ by $r^{-i-1}\vartheta'$, because these are homogeneous functions of the orders i and $-i-1$ respectively:

$$\varpi a^{2i+1}+\varpi'=A_i a^{i+1}+\frac{ma^{2i+3}}{2\left[(2n+m)\,i+3n+m\right]}\,\vartheta-\frac{ma^2}{2\left[(2n+m)i-n\right]}\,\vartheta',$$

$$\varpi a'^{2i+1}+\varpi'=A_i' a'^{i+1}+\frac{ma'^{2i+3}}{2\left[(2n+m)\,i+3n+m\right]}\,\vartheta-\frac{ma'^2}{2\left[(2n+m)i-n\right]}\,\vartheta'.$$

Resolving these equations for ϖ and ϖ', and returning to the original notation instead of ϖ, ϖ', ϑ, ϑ',

integral order i.   That any function, arbitrarily given over an entire spherical surface, may be so expressed, is a well-known theorem.   A demonstration of it is given in Thomson and Tait's *Natural Philosophy*, Vol. I. Part I. chap. i. Appendix B, § s.

$$u_i = \frac{(a^{i+1}A_i - a'^{i+1}A_i')\,r^i + (a^{2i+3} - a'^{2i+3})\,M_{i+2}\dfrac{d\psi_{i+1}}{dx} - (a^2 - a'^2)\,M'_{i-2}\dfrac{d(\psi'_{i-1}r^{-2i+1})}{dx}\,r^{2i+1}}{a^{2i+1} - a'^{2i+1}},$$

$$u_i' = \frac{(aa')^{i+1}(a^iA_i' - a'^iA_i)\,r^i - (aa')^{2i+1}(a^2 - a'^2)M_{i+2}\dfrac{d\psi_{i+1}}{dx} - (aa')^2(a^{2i-1} - a'^{2i-1})M'_{i-2}\dfrac{d(\psi'_{i-1}r^{-2i+1})}{dx}\,r^{2i+1}}{a^{2i+1} - a'^{2i+1}},$$

where, for brevity,

$$\left.\begin{aligned}
M_{i+2} &= \frac{m}{2\left[(2n+m)\,i + 3n + m\right]},\\
M'_{i-2} &= \frac{m}{2\left[(2n+m)\,i - n\right]}
\end{aligned}\right\}\dotfill(16).$$

Introducing, also for brevity, the following notation,

$$\left.\begin{aligned}
\mathfrak{A}_i &= \frac{a^{i+1}A_i - a'^{i+1}A_i'}{a^{2i+1} - a'^{2i+1}},\\
\mathfrak{A}'_i &= \frac{(aa')^{i+1}(a^iA_i' - a'^iA_i)}{a^{2i+1} - a'^{2i+1}}
\end{aligned}\right\}\dotfill(17),$$

$$\left.\begin{aligned}
\mathfrak{M}_{i+2} &= \frac{a^{2i+1} - a'^{2i+1}}{a^{2i+3} - a'^{2i+3}}\,M_{i+2}, & \mathfrak{M}'_{i-2} &= \frac{a^2 - a'^2}{a^{2i+1} - a'^{2i+1}}\,M'_{i-2},\\
\mathfrak{P}_{i+2} &= \frac{(aa'^{2i+1})(a^2 - a'^2)}{a^{2i+1} - a'^{2i+1}}M_{i+2}, & \mathfrak{P}'_{i-2} &= \frac{(aa')^2(a^{2i-1} - a'^{2i-1})}{a^{2i+1} - a'^{2i+1}}\,M'_{i-2}
\end{aligned}\right\}(18),$$

we have the expressions for u_i and u_i' given below. Dealing with the equations of condition relative to β and γ, and introducing an abbreviated notation $\mathfrak{B}_i$, $\mathfrak{B}_i'$, $\mathfrak{C}_i$, $\mathfrak{C}_i'$, corresponding to (17), we find similar expressions for v_i, v_i', w_i, w_i', as follows:—

$$\left.\begin{aligned}
u_i &= \mathfrak{A}_i r^i + \mathfrak{M}_{i+2}\frac{d\psi_{i+1}}{dx} - \mathfrak{M}'_{i-2}\frac{d(\psi'_{i-1}r^{-2i+1})}{dx}\,r^{2i+1},\\
v_i &= \mathfrak{B}_i r^i + \mathfrak{M}_{i+2}\frac{d\psi_{i+1}}{dy} - \mathfrak{M}'_{i-2}\frac{d(\psi'_{i-1}r^{-2i+1})}{dy}\,r^{2i+1},\\
w_i &= \mathfrak{C}_i r^i + \mathfrak{M}_{i+2}\frac{d\psi_{i+1}}{dz} - \mathfrak{M}'_{i-2}\frac{d(\psi'_{i-1}r^{-2i+1})}{dz}\,r^{2i+1}
\end{aligned}\right\}\dots(19),$$

$$\left.\begin{aligned}
u_i' &= \mathfrak{A}_i' r^i - \mathfrak{P}_{i+2}\frac{d\psi_{i+1}}{dx} - \mathfrak{P}'_{i-2}\frac{d(\psi'_{i-1}r^{-2i+1})}{dx}\,r^{2i+1},\\
v_i' &= \mathfrak{B}_i' r^i - \mathfrak{P}_{i+2}\frac{d\psi_{i+1}}{dy} - \mathfrak{P}'_{i-2}\frac{d(\psi'_{i-1}r^{-2i+1})}{dy}\,r^{2i+1},\\
w_i' &= \mathfrak{C}_i' r^i - \mathfrak{P}_{i+2}\frac{d\psi_{i+1}}{dz} - \mathfrak{P}'_{i-2}\frac{d(\psi'_{i-1}r^{-2i+1})}{dz}\,r^{2i+1}
\end{aligned}\right\}\dots(20).$$

17. It only remains to determine the functions ψ and ψ', which we can do by combining these last equations with (10) of

§ 12. Thus, changing i into $i+1$ in (17) and into $i-1$ in (18), applying equations (10) of § 12, and taking advantage of the following properties,

$$\nabla^2 \psi_{i+2} = 0, \quad \nabla^2 (\psi_i' r^{-2i-1}) = 0, \quad \&c.,$$

$$x \frac{d (\psi_i' r^{-2i-1})}{dx} + y \frac{d (\psi_i' r^{-2i-1})}{dy} + z \frac{d (\psi_i' r^{-2i-1})}{dz} = - (i+1) \psi_i',$$

and

$$x \frac{d\psi_i}{dx} + y \frac{d\psi_i}{dy} + z \frac{d\psi_i}{dz} = i\psi_i,$$

we find

$$\left. \begin{aligned} \psi_i &= \frac{d(\mathfrak{A}_{i+1} r^{i+1})}{dx} + \frac{d(\mathfrak{B}_{i+1} r^{i+1})}{dy} + \frac{d(\mathfrak{C}_{i+1} r^{i+1})}{dz} + (2i+3)(i+1)\mathfrak{M}'_{i-1}\psi_i', \\ \psi_i' &= \left\{ \frac{d(\mathfrak{A}'_{i-1} r^{-i})}{dx} + \frac{d(\mathfrak{B}'_{i-1} r^{-i})}{dy} + \frac{d(\mathfrak{C}'_{i-1} r^{-i})}{dz} \right\} r^{2i+1} + (2i-1)i\mathfrak{P}_{i+1}\psi_i \end{aligned} \right\} (21).$$

These equations, used to determine the two unknown functions ψ_i and ψ_i', give

$$\left. \begin{aligned} \psi_i &= \frac{\Theta_i + (2i+3)(i+1)\mathfrak{M}'_{i+1}\Theta_i'}{1 - (2i+3)(2i-1)(i+1)i\mathfrak{M}'_{i+1}\mathfrak{P}_{i+1}}, \\ \psi_i' &= \frac{(2i-1)i\mathfrak{P}_{i+1}\Theta_i + \Theta_i'}{1 - (2i+3)(2i-1)(i+1)i\mathfrak{M}'_{i+1}\mathfrak{P}_{i+1}} \end{aligned} \right\} \ldots\ldots(22),$$

where, for brevity,

$$\left. \begin{aligned} \Theta_i &= \frac{d(\mathfrak{A}_{i+1} r^{i+1})}{dx} + \frac{d(\mathfrak{B}_{i+1} r^{i+1})}{dy} + \frac{d(\mathfrak{C}_{i+1} r^{i+1})}{dz}, \\ \Theta_i' &= \left\{ \frac{d(\mathfrak{A}'_{i-1} r^{-i})}{dx} + \frac{d(\mathfrak{B}'_{i-1} r^{-i})}{dy} + \frac{d(\mathfrak{C}'_{i-1} r^{-i})}{dz} \right\} r^{2i+1} \end{aligned} \right\} \ldots(23).$$

18. The functions ψ and ψ' being expressed in terms of the data of the problem by equations (22) (23), (17), (18), (16), we have only to use (19) and (20) in (12) to find the following expression of the complete solution :—

$$\left. \begin{aligned} \alpha &= \Sigma \left\{ \mathfrak{A}_i r^i + \mathfrak{A}_i' r^{-i-1} + (\mathfrak{M}_i - \mathfrak{P}_i r^{-2i+3} - M_i r^2) \frac{d\psi_{i-1}}{dx} \right. \\ &\qquad \left. - (\mathfrak{M}_i' r^{2i+5} + \mathfrak{P}_i' - M_i' r^2) \frac{d(\psi'_{i+1} r^{-2i-3})}{dx} \right\}, \\ \beta &= \Sigma \left\{ \mathfrak{B}_i r^i + \mathfrak{B}_i' r^{-i-1} + (\mathfrak{M}_i - \mathfrak{P}_i r^{-2i+3} - M_i r^2) \frac{d\psi_{i-1}}{dy} \right. \\ &\qquad \left. - (\mathfrak{M}_i' r^{2i+5} + \mathfrak{P}_i' - M_i' r^2) \frac{d(\psi'_{i+1} r^{-2i-3})}{dy} \right\}, \\ \gamma &= \Sigma \left\{ \mathfrak{C}_i r^i + \mathfrak{C}_i' r^{-i-1} + (\mathfrak{M}_i - \mathfrak{P}_i r^{-2i+3} - M_i r^2) \frac{d\psi_{i-1}}{dz} \right. \\ &\qquad \left. - (\mathfrak{M}_i' r^{2i+5} + \mathfrak{P}_i' - M_i' r^2) \frac{d(\psi'_{i+1} r^{-2i-3})}{dz} \right\} \end{aligned} \right\} \ldots(24).$$

19. This solution leads immediately, through an extreme case of its application, to the solution of the general problem for a plate of elastic substance between two infinite parallel planes :—Given the displacement of every point of its surface, required the displacement of any interior point. For if we give infinite values to a and a', and keep $a - a'$ finite, the spherical shell becomes an infinite plane plate.

20. It is, however, less easy to deduce the result in this way from the solution for the spherical shell, than to apply directly the general method of § 6 to the case of the infinite plane plate. We shall return to this subject (§ 31, below), when the details of the investigation will be sufficiently indicated.

21. A very important part of the general problem proposed in § 1 remains to be considered,—that in which not the displacement, but the arbitrarily applied force, is given all over the surface. To express the surface-equations of condition for such data, we must use the formulæ expressing the stress (or force of elasticity) in any part of an elastic solid in terms of the strain (or deformation) of the substance. These are

$$\left.\begin{aligned}
P &= (m+n)\frac{d\alpha}{dx} + (m-n)\left(\frac{d\beta}{dy} + \frac{d\gamma}{dz}\right); \\
Q &= (m+n)\frac{d\beta}{dy} + (m-n)\left(\frac{d\gamma}{dz} + \frac{d\alpha}{dx}\right); \\
R &= (m+n)\frac{d\gamma}{dz} + (m-n)\left(\frac{d\alpha}{dx} + \frac{d\beta}{dy}\right); \\
S &= n\left(\frac{d\beta}{dz} + \frac{d\gamma}{dy}\right); \; T = n\left(\frac{d\gamma}{dx} + \frac{d\alpha}{dz}\right); \; U = n\left(\frac{d\alpha}{dy} + \frac{d\beta}{dx}\right);
\end{aligned}\right\} \dots(25)$$

where P, Q, R are the normal tractions (which when negative are pressures) on the faces of a unit cube respectively perpendicular to the lines of reference OX, OY, OZ; and S, T, U the tangential forces along the faces respectively parallel, and in the directions in these planes respectively perpendicular, to OX, OY, OZ (see Appendix, § 70).

22. In terms of these we have the following expressions for the components F, G, H of the force on a unit area perpendicular to any line whose direction-cosines are f, g, h :—

$$F = Pf + Ug + Th,$$
$$G = Uf + Qg + Sh,$$
$$H = Tf + Sg + Rh$$
$$\left.\rule{0pt}{3.5em}\right\} \quad \dots\dots\dots\dots\dots(26),$$

[see Art. XCII. Part I. above; "Mathematical Theory of Elasticity," Chap. III. pp. 86–7].

23. Using the expressions (25) in (26), we find

$$F = (m-n)\left(\frac{d\alpha}{dx} + \frac{d\beta}{dy} + \frac{d\gamma}{dz}\right)f + n\left(\frac{d\alpha}{dx}f + \frac{d\alpha}{dy}\,g + \frac{d\alpha}{dz}\,h\right)$$
$$+ n\left(\frac{d\alpha}{dx}f + \frac{d\beta}{dx}\,g + \frac{d\gamma}{dx}\,h\right)\dots(27),$$

and symmetrical expressions for G and H.

24. If now we suppose f, g, h to denote the direction-cosines of the normal at any point x, y, z of the surface of an elastic solid, the surface condition, when force, not displacement, is given, will be expressed by equating F, G, H respectively to three functions of the coordinates of a point in the surface, quite arbitrary except in so far as they must balance one another in order that equilibrium in the body may be possible; and therefore they must fulfil the following integral equations:—

$$\iint F d\Omega = 0, \quad \iint G d\Omega = 0, \quad \iint H d\Omega = 0\dots\dots\dots\dots\dots\dots(28),$$

$$\iint (Hy - Gz)\,d\Omega = 0, \ \iint (Fz - Hx)\,d\Omega = 0, \ \iint (Gx - Fy)\,d\Omega = 0 \ \dots(29),$$

where $d\Omega$ denotes an element of the surface at the point (x, y, z), and the double integrals include the whole surface of application of the forces F, G, H.

25. For our case of the spherical shell, with origin of co-ordinates at its centre, we have

$$f = \frac{x}{r}, \quad g = \frac{y}{r}, \quad h = \frac{z}{r} \ \dots\dots\dots\dots\dots\dots(30);$$

and the last triple term in the expression (27) for F may be conveniently written thus:—

$$\frac{n}{r}\frac{d\,(\alpha x + \beta y + \gamma z)}{dx} - \frac{n\alpha}{r} \ \dots\dots\dots\dots\dots(31);$$

Then, for brevity, putting

$$\alpha x + \beta y + \gamma z = \zeta \dots\dots\dots\dots\dots\dots(32),$$

and
$$x\frac{d}{dx} + y\frac{d}{dy} + z\frac{d}{dz} = r\frac{d}{dr} \dots\dots\dots\dots(33),$$

where $\frac{d}{dr}$ prefixed to any function of x, y, z will denote its rate of variation per unit of length in the radial direction; and using (2) of § 3, we have, by (30) and the symmetrical equations for G and H,

$$\left.\begin{aligned}
Fr &= (m-n)\,\delta\,.\,x + n\left\{\left(r\frac{d}{dr}-1\right)\alpha + \frac{d\zeta}{dx}\right\}, \\
Gr &= (m-n)\,\delta\,.\,y + n\left\{\left(r\frac{d}{dr}-1\right)\beta + \frac{d\zeta}{dy}\right\}, \\
Hr &= (m-n)\,\delta\,.\,z + n\left\{\left(r\frac{d}{dr}-1\right)\gamma + \frac{d\zeta}{dz}\right\}
\end{aligned}\right\}\dots\dots(34).$$

26. It is to be remarked that these equations express such functions of (x, y, z), the coordinates of any point P of the solid, that $F\,.\,\omega$, $G\,.\,\omega$, $H\,.\,\omega$ are the three components of the force transmitted across an infinitely small area ω perpendicular to OP, while, for any point of either the outer or the inner bounding spherical surface, $F\omega$, $G\omega$, $H\omega$ are the three components of the force applied to an infinitely small element ω of this surface.

27. To reduce the surface-equations of condition derived from these expressions to harmonic equations, let us consider homogeneous terms of degree i of the complete solution, which we shall denote by α_i, β_i, γ_i, and let δ_{i-1}*, ζ_{i+1} denote the corresponding terms of the other functions. Thus we have

$$\left.\begin{aligned}
Fr &= \Sigma\left\{(m-n)\,\delta_{i-1}x + n\,(i-1)\,\alpha_i + n\frac{d\zeta_{i+1}}{dx}\right\}, \\
Gr &= \Sigma\left\{(m-n)\,\delta_{i-1}y + n\,(i-1)\,\beta_i + n\frac{d\zeta_{i+1}}{dy}\right\}, \\
Hr &= \Sigma\left\{(m-n)\,\delta_{i-1}z + n\,(i-1)\,\gamma_i + n\frac{d\zeta_{i+1}}{dz}\right\}
\end{aligned}\right\}\dots(35).$$

28. The second of the three terms of order i in these equations, when the general solution of § 13 is used, become at the boundary each explicitly the sum of two surface harmonics of orders i and

---

* When $i-1$ is positive, δ_{i-1} will express the same function as V_{i-1} of § 9 above. The suffixes now introduced have reference solely to the algebraic degree, positive or negative, of the functions, whether harmonic or not, to the symbols for which they are applied.

$i - 2$ respectively. To bring the other parts of the expressions to similar forms, it is convenient that we should first express ζ_{i+1} in terms of the general solution (12) of § 13, by selecting the terms of algebraic degree i. Thus we have

$$a_i = u_i - \frac{mr^2}{2\left[(2n + m)\,i - n - m\right]} \frac{d\psi_{i-1}}{dx} \quad \dots\dots\dots (36),$$

and symmetrical expressions for β_i and γ_i, from which we find

$$a_i x + \beta_i y + \gamma_i z = \zeta_{i+1} = u_i x + v_i y + w_i z - \frac{(i-1)\,mr^2\psi_{i-1}}{2\left[(2n+m)\,i - n - m\right]}.$$

Hence, by the proper formulæ [see (42) below] for reduction to harmonics,

$$\zeta_{i+1} = - \frac{1}{2i+1} \left\{ \frac{(2i-1)\left[m\,(i-1) - 2n\right]}{2\left[(2n+m)\,i - n - m\right]} r^2\psi_{i-1} + \phi_{i+1} \right\} \dots (37),$$

where

$$\phi_{i+1} = r^{2i+3} \left\{ \frac{d\,(u_i r^{-2i-1})}{dx} + \frac{d\,(v_i r^{-2i-1})}{dy} + \frac{d\,(w_i r^{-2i-1})}{dz} \right\} \quad \dots\dots (38),$$

and (as before assumed in § 12)

$$\psi_{i-1} = \frac{du_i}{dx} + \frac{dv_i}{dy} + \frac{dw_i}{dz} \quad \dots\dots\dots\dots\dots (39).$$

Also, by (11) of § 12, or directly from (36) by differentiation, we have

$$\delta_{i-1} = \frac{n\,(2i-1)}{(2n+m)\,i - n - m} \cdot \psi_{i-1} \quad \dots\dots\dots\dots (40).$$

Substituting these expressions for δ_{i-1}, a_i, and ζ_{i+1} in (35), we find

$$Fr = \Sigma \left\{ n(i-1)u_i + \frac{n(2i-1)\left[(m-2n)i + 2m + n\right]}{(2i+1)\left[(m+2n)i - m - n\right]} x\psi_{i+1} \right.$$
$$\left. - \frac{n\left[2i(i-1)\,m - (2i-1)\,n\right]}{(2i+1)\left[(m+2n)i - m - n\right]} r^2 \frac{d\psi_{i-1}}{dx} - \frac{n}{2i+1} \frac{d\phi_{i+1}}{dx} \right\} \dots (41).$$

This is reduced to the required harmonic form by the obviously proper formula

$$x\psi_{i-1} = \frac{1}{2i-1} \left\{ r^2 \frac{d\psi_{i-1}}{dx} - r^{2i+1} \frac{d\,(\psi_{i-1} r^{-2i+1})}{dx} \right\} \quad \dots\dots\dots (42).$$

Thus, and dealing similarly with the expressions for Gr and Hr, we have, finally,

$$Fr = n\Sigma\left\{(i-1)u_i - 2(i-2)M_i r^2\frac{d\psi_{i-1}}{dx} - E_i r^{2i+1}\frac{d(\psi_{i-1}r^{-2i+1})}{dx} - \frac{1}{2i+1}\frac{d\phi_{i+1}}{dx}\right\},$$

$$Gr = n\Sigma\left\{(i-1)v_i - 2(i-2)M_i r^2\frac{d\psi_{i-1}}{dy} - E_i r^{2i+1}\frac{d(\psi_{i-1}r^{-2i+1})}{dx} - \frac{1}{2i+1}\frac{d\phi_{i+1}}{dy}\right\},$$

$$Hr = n\Sigma\left\{(i-1)w_i - 2(i-2)M_i r^2\frac{d\psi_{i-1}}{dz} - E_i r^{2i+1}\frac{d(\psi_{i-1}r^{-2i+1})}{dx} - \frac{1}{2i+1}\frac{d\phi_{i+1}}{dz}\right\}$$

$$\qquad\qquad\qquad\qquad\qquad\qquad\qquad\qquad\qquad\qquad\qquad\qquad\qquad..(43),$$

where [as above, (16) of § 16]

$$M_i = \frac{1}{2}\frac{m}{(m+2n)i - m - n},$$

and now further

$$E_i = \frac{(m-2n)i + 2m + n}{(2i+1)\left[(m+2n)i - m - n\right]} \qquad\qquad \dots\dots\dots(44).$$

29. To express the surface conditions by harmonic equations, let us suppose the superficial values of F, G, H to be given as follows:

$$\left.\begin{array}{l}F = \Sigma A_i,\\ G = \Sigma B_i,\\ H = \Sigma C_i,\end{array}\right\} \text{ when } r = a,$$

and

$$\left.\begin{array}{l}F = \Sigma A_i',\\ G = \Sigma B_i',\\ H = \Sigma C_i',\end{array}\right\} \text{ when } r = a' \qquad\qquad \dots\dots\dots\dots(45),$$

where $A_i, B_i, C_i, A_i', B_i', C_i'$ denote surface harmonics of order i. Now the terms of algebraic degree i, exhibited in the preceding expressions (43) for Fr, Gr, Hr, become, at either of the concentric spherical surfaces, sums of surface harmonics of orders i and $i-2$, when i is positive, and of orders $-i-1$ and $-i-3$ when i is negative. Hence, selecting all the terms which lead to surface harmonics of order i, and equating to the proper terms of the data (45), we have

$$\frac{n}{r}\left\{\begin{array}{l}(i-1)u_i - (i+2)u_{-i-1} - 2iM_{i+2}r^2\frac{d\psi_{i+1}}{dx} + 2(i+1)M_{-i+1}r^2\frac{d\psi_{-i}}{dx}\\[2mm] -E_i r^{2i+1}\frac{d(\psi_{i-1}r^{-2i+1})}{dx} - E_{-i-1}r^{-2i-1}\frac{d(\psi_{-i-2}r^{2i+3})}{dx} - \frac{1}{2i+1}\left(\frac{d\phi_{i+1}}{dx} - \frac{d\phi_{-i}}{dx}\right)\end{array}\right\}$$

$$= \left\{\begin{array}{l}A_i \text{ when } r = a,\\ A_i' \text{ when } r = a'\end{array}\right\}\dots(46),$$

and symmetrical equations relative to y and z.

30.   These equations might be dealt with exactly as formerly with the equations (15) of § 15.   But the following order of proceeding is more convenient.   Commencing with the first of the surface equations (46), multiplying it by $\left(\dfrac{r}{a}\right)^i$, attending to the degree of each term, and taking advantage of the principle that, if ψ be any homogeneous function of x, y, z, of degree t, the function of angular coordinates, or of the ratios $x : y : z$, which it becomes at the spherical surface $r = a$, is the same as $\left(\dfrac{a}{r}\right)^t \psi$ for any value of r, we have

$$\frac{n}{a}\left\{\begin{array}{l} (i-1)u_i-(i+2)\left(\dfrac{r}{a}\right)^{2i+1}u_{-i-1}-2iM_{i+2}a^2\dfrac{d\psi_{i+1}}{dx} \\[2ex] \qquad\qquad +2(i+1)M_{-i+1}a^2\left(\dfrac{r}{a}\right)^{2i+1}\dfrac{d\psi_{-i}}{dx} \\[2ex] -E_i r^{2i+1}\dfrac{d(\psi_{i-1}r^{-2i+1})}{dx}-E_{-i-1}a^{-2i-1}\dfrac{d(\psi_{-i-2}r^{-2i+3})}{dx} \\[2ex] \qquad\qquad -\dfrac{1}{2i+1}\left[\dfrac{d\phi_{i+1}}{dx}-\left(\dfrac{r}{a}\right)^{2i+1}\dfrac{d\phi_{-i}}{dx}\right] \end{array}\right\}$$
$$= A_i\left(\frac{r}{a}\right)\ ...(47),$$

where the second member, and each term of the first member, is now a homogeneous function of degree i, of x, y, z (being in fact a solid spherical harmonic of degree and order i).   Taking $\dfrac{d}{dx}$ of this, and $\dfrac{d}{dy}$ and $\dfrac{d}{dz}$ of the two symmetrical equations, adding, taking into account equations (38) and (39), and taking advantage of the equation $\nabla^2 V = 0$ for the solid harmonic functions concerned, we have

$$\frac{n}{a}\left\{[i-1+(2i+1)iE_i]\psi_{i-1}-2(i+1)a^{-2i-1}r^{2i-1}\phi_{-i}\right.$$
$$\left.-2(i+1)(2i+1)iM_{-i+1}\left(\frac{r}{a}\right)^{2i+1}\psi_{-i}\right\}\ ...(48).$$
$$= \frac{1}{a^i}\left\{\frac{d(A_i r^i)}{dx}+\frac{d(B_i r^i)}{dy}+\frac{d(C_i r^i)}{dz}\right\}$$

Again, multiplying (47) by $a^{-2}r^{-2i-1}$, and taking $r^{2i+3}\dfrac{d}{dx}$ of the result, dealing similarly with the two symmetrical equations, and adding, we have

$$
\frac{n}{a}\left\{2ia^{-2}\phi_{i+1}-[i+2-(2i+1)(i+1)E_{-i-1}]\left(\frac{r}{a}\right)^{2i+3}\psi_{-i-2}\right.
$$
$$
\left.+2i(i+1)(2i+1)M_{i+2}\psi_{i+1}\right\} \quad\ldots(49).
$$
$$
=\frac{r^{2i+3}}{a^{i+2}}\left\{\frac{d(A_i r^{-i-1})}{dx}+\frac{d(B_i r^{-i-1})}{dy}+\frac{d(C_i r^{-i-1})}{dz}\right\}
$$

Changing i into $i-2$ in this equation, we have

$$
\frac{n}{a}\left\{2(i-2)a^{-2}\phi_{i-1}-[i-(2i-3)(i-1)E_{-i+1}]\left(\frac{r}{a}\right)^{2i-1}\psi_{-i}\right.
$$
$$
\left.+2(i-2)(i-1)(2i-3)M_i\psi_{i-1}\right\} \quad\ldots(50).
$$
$$
=\frac{r^{2i-1}}{a^{i}}\left\{\frac{d(A_{i-2}r^{-i+1})}{dx}+\frac{d(B_{i-2}r^{-i+1})}{dy}+\frac{d(C_{i-2}r^{-i+1})}{dz}\right\}
$$

Precisely similar equations, derived from the inner surface condition of the shell, are obtained by changing a, A, B, C into a', A', B', C'. We thus have (48), (50), and the two corresponding equations for the inner surface, in all four equations, to determine the four unknown functions ψ_{i-1}, ψ_{-i}, ϕ_{i-1}, ϕ_{-i}, in terms of the data which appear in the second members. The equations being simple algebraic equations, we may regard these four functions as explicitly determined. In other words, we may suppose ϕ_i and ψ_i known for every positive or negative integral value of i. Then equation (47), the two equations symmetrical with it, and the others got by changing A, a, &c. into A', a', &c., give u_i, v_i, w_i explicitly in terms of known functions, and the expressions (36) for α_i, β_i, γ_i complete the solution of the problem.

31. The solution for the infinite plane plate is of course included in the general solution for the spherical shell, as remarked above for the case in which surface displacements, not surface forces, were given; but, as in that case, it will be simpler and practically easier to work out the problem *ab initio*, taking advantage of the appropriate Fourier forms. The relative ease of the independent investigation is indeed still greater in the case in which the surface forces are given than in the other case, since the general expressions for the surface forces assume simple forms when the surface is plane, and require no such transformation as

that which we have found necessary, and which has constituted the special difficulty of the problem, when the surface was spherical. The problem of the plane plate presents many questions of remarkable 'interest and practical importance; and although the object and limits of the present paper preclude any detailed investigation of special cases, we may make a short digression to work out the general solution.

32.   Let the origin of coordinates be taken in one side of the plate and the axis OX perpendicular to it. Then, according to the general expressions (25) of § 21, the three components of the force per unit of area, in or parallel to either side of the plate, are respectively

$$
\left.
\begin{aligned}
\text{parallel to } OX, \; & P = (m+n)\frac{d\alpha}{dx} + (m-n)\left(\frac{d\beta}{dy} + \frac{d\gamma}{dz}\right), \\
\text{parallel to } OY, \; & U = n\left(\frac{d\alpha}{dy} + \frac{d\beta}{dx}\right), \\
\text{parallel to } OZ, \; & T = n\left(\frac{d\alpha}{dz} + \frac{d\gamma}{dx}\right)
\end{aligned}
\right\} \ldots (51).
$$

The surface condition to be fulfilled is that each of these functions shall have an arbitrarily given value at every point of each infinite plane side of the plate.

33.   From the indications of § 6 above, it is easily seen that the following assumptions are correct for a general solution of the equations of internal equilibrium, and convenient for the application at present proposed,

$$
\alpha = u + x\,\frac{d\phi}{dx},
$$

$$
\beta = v + x\,\frac{d\phi}{dy},
$$

$$
\gamma = w + x\,\frac{d\phi}{dz},
$$

where u, v, w, and ϕ denote functions of (x, y, z) which each fulfil the equation $\nabla^2 V = 0$. From these, by differentiation, and by taking $\nabla^2 \phi = 0$ into account, we have

$$
\frac{d\alpha}{dx} + \frac{d\beta}{dy} + \frac{d\gamma}{dz} = \frac{du}{dx} + \frac{dv}{dy} + \frac{dw}{dz} + \frac{d\phi}{dx},
$$

or

$$\delta = \psi + \frac{d\phi}{dx}$$

if

$$\psi = \frac{du}{dx} + \frac{dv}{dy} + \frac{dw}{dz}$$

$$\dotsb (52),$$

and δ be used with the same signification as above (§ 2). Also, by differentiation and application of the equations

$$\nabla^2 u = 0, \quad \nabla^2 \frac{d\phi}{dx} = 0,$$

we find

$$\nabla^2 \alpha = 2\frac{d^2\phi}{dx^2}, \quad \nabla^2 \beta = 2\frac{d^2\phi}{dxdy}, \quad \nabla^2 \gamma = 2\frac{d^2\phi}{dxdz}.$$

Hence, to satisfy the general equations of internal equilibrium (3) of § 3, we must have

$$\frac{d\phi}{dx} = -\frac{m}{m+2n}\psi.$$

Hence the general solution becomes

$$\alpha = u - \frac{mx}{m+2n}\psi,$$
$$\beta = v - \frac{mx}{m+2n}\frac{d\int\psi dx}{dy},$$
$$\gamma = w - \frac{mx}{m+2n}\frac{d\int\psi dx}{dz}$$

$$\dotsb (53),$$

where u, v, w are any functions whatever which satisfy the general equation $\nabla^2 V = 0$, and ψ is given by (52); and where, further, it must be understood that $\int\psi dx$ must be so assigned as to satisfy the equation $\nabla^2 V = 0$, which ψ itself satisfies by virtue of (52).

34. The general form of the solution of $\nabla^2 V = 0$, convenient for the present application, is clearly

$$\epsilon^{\pm px}\,{}^{\sin}_{\cos}(sy)\,{}^{\sin}_{\cos}(tz),$$

where p, s, t are three constants subject to the equation

$$p^2 = s^2 + t^2.$$

If now we suppose, as a particular case, the surface condition to be that

$$P = A \sin (sy) \sin (tz),$$
$$U = B \cos (sy) \sin (tz),$$
$$T = C \sin (sy) \cos (tz),$$
when $x = 0,$

and

$$P = A' \sin (sy) \sin (tz),$$
$$U = B' \cos (sy) \sin (tz),$$
$$T = C' \sin (sy) \cos (tz),$$
when $x = a$

.........(54),

where A, B, C, A', B', C' are six given constants, we must clearly have

$$u = (f\epsilon^{-px} + f'\epsilon^{px}) \sin (sy) \sin (tz),$$
$$v = (g\epsilon^{-px} + g'\epsilon^{px}) \cos (sy) \sin (tz),$$
$$w = (h\epsilon^{-px} + h'\epsilon^{px}) \sin (sy) \cos (tz)$$

............(55),

where f, g, h, f', g', h' are six constants to be determined by six linear equations obtained directly from (54), (51), (53), (52), (55). But, by proper interchanges of sines and cosines, we have in (54) a representation of the general terms of the series or of the definite integrals, representing, according to Fourier's principles, the six arbitrary functions, whether periodic or non-periodic, by which P, U, T are given over each of the two infinite plane sides. Hence the solution thus indicated is complete.

35. To complete the theory of the equilibrium of an elastic spheroidal shell, we must now suppose every point of the solid substance to be urged by a given force. The problem thus presented will be reduced to that already solved, by the following simple investigation.

36. Let X, Y, Z be the components of the force per unit of volume on the substance at any point x, y, z. (That is to say, let qX, qY, qZ be the three components of the actual force on a volume q, infinitely small in all its dimensions, enclosing the point x, y, z). Not to unnecessarily limit the problem, we must suppose X, Y, Z to be each an absolutely arbitrary function of x, y, z.

37. When we remember that x, y, z are the coordinates of the undisturbed position of any point of the substance, and differ by the infinitely small quantities α, β, γ from the actual coordinates of the same point of the substance in the body disturbed by the applied forces, we perceive that $Xdx + Ydy + Zdz$ need not be the differential of a function of three independent variables.

It actually will not be a complete differential if the case be that of the interior kinetic equilibrium of a rigid body starting from rest under the influence of given constant forces applied to its surface, and having for their resultant a couple in a plane perpendicular to a principal axis. Nor will $Xdx + Ydy + Zdz$ be a complete differential in the interior of a steel bar-magnet held at rest under the influence of an electric current directed through one half of its length, as we perceive when we consider Faraday's beautiful experiment showing rotation to supervene in this case when the magnet is freed from all mechanical constraint.

38. The equations of elastic equilibrium are of course now

$$n\nabla^2\alpha + m\frac{d\delta}{dx} = -X,$$

$$n\nabla^2\beta + m\frac{d\delta}{dy} = -Y, \left.\right\} \quad \dots\dots\dots\dots\dots(56).$$

$$n\nabla^2\gamma + m\frac{d\delta}{dz} = -Z$$

Let ϖ, ρ, σ denote some three particular solutions of the equations

$$\nabla^2\varpi = -X,$$

$$\nabla^2\rho = -Y, \left.\right\} \quad \dots\dots\dots\dots\dots\dots(57).$$

$$\nabla^2\sigma = -Z$$

These, ϖ, ρ, σ, we may regard as known functions, being derivable from X, Y, Z by known methods (Thomson and Tait's *Natural Philosophy*, Part II., chap. VI.). Then, if we assume

$$\alpha - \frac{\varpi}{n} = \alpha_{,}$$

$$\beta - \frac{\rho}{n} = \beta_{,} \left.\right\} \quad \dots\dots\dots\dots\dots(58),$$

$$\gamma - \frac{\sigma}{n} = \gamma_{,}$$

and

$$\frac{d\alpha_{,}}{dx} + \frac{d\beta_{,}}{dy} + \frac{d\gamma_{,}}{dz} = \delta_{,} \quad \dots\dots\dots\dots\dots(59),$$

the equations (56) of interior equilibrium become

$$n\nabla^2\alpha_{,} + m\frac{d\delta_{,}}{dx} = -\frac{m}{n}\frac{d\xi}{dx},$$

$$n\nabla^2\beta_{,} + m\frac{d\delta_{,}}{dy} = -\frac{m}{n}\frac{d\xi}{dy}, \left.\right\} \quad \dots\dots\dots\dots(60),$$

$$n\nabla^2\gamma_{,} + m\frac{d\delta_{,}}{dz} = -\frac{m}{n}\frac{d\xi}{dz}$$

where ξ is a known function given by the equation

$$\xi = \frac{d\varpi}{dx} + \frac{d\rho}{dy} + \frac{d\sigma}{dz} \quad\ldots\ldots\ldots\ldots\ldots\ldots(61).$$

Now, as we verify in a moment by differentiation, equations (59) and (60) are satisfied by

$$\left.\begin{aligned}
\alpha_{,} &= \frac{-m}{n\,(m+n)} \frac{d\vartheta}{dx}, \\[1mm]
\beta_{,} &= \frac{-m}{n\,(m+n)} \frac{d\vartheta}{dy}, \\[1mm]
\gamma_{,} &= \frac{-m}{n\,(m+n)} \frac{d\vartheta}{dz}
\end{aligned}\right\} \quad\ldots\ldots\ldots\ldots(62),$$

if ϑ is some particular solution of

$$\nabla^2\vartheta = \xi \quad\ldots\ldots\ldots\ldots\ldots\ldots\ldots(63).$$

Hence (58), (57), (62), (63), (61) express a particular solution of (56).

39.   We conclude that the general solution of (56) may be expressed thus :—

$$\left.\begin{aligned}
\alpha &= \frac{1}{n}\left(\varpi - \frac{m}{m+n}\frac{d\vartheta}{dx}\right) + {'\alpha}, \\[1mm]
\beta &= \frac{1}{n}\left(\rho - \frac{m}{m+n}\frac{d\vartheta}{dy}\right) + {'\beta}, \\[1mm]
\gamma &= \frac{1}{n}\left(\sigma - \frac{m}{m+n}\frac{d\vartheta}{dz}\right) + {'\gamma}
\end{aligned}\right\} \quad\ldots\ldots\ldots\ldots(64),$$

where

$$\left.\begin{aligned}
\varpi &= \nabla^{-2}X, \\
\rho &= \nabla^{-2}Y, \\
\sigma &= \nabla^{-2}Z, \\
\vartheta &= \nabla^{-2}\left(\frac{d\varpi}{dx} + \frac{d\rho}{dy} + \frac{d\sigma}{dz}\right)
\end{aligned}\right\} \ldots\ldots\ldots\ldots\ldots(65),$$

according to an abbreviated notation, which explains itself sufficiently ; and ${'\alpha}, {'\beta}, {'\gamma}$ denote a general solution of the equations

$$\left.\begin{aligned}
n\nabla^{2\prime}\alpha + m\,\frac{d}{dx}\left(\frac{d'\alpha}{dx} + \frac{d'\beta}{dy} + \frac{d'\gamma}{dz}\right) &= 0, \\[1mm]
n\nabla^{2\prime}\beta + m\,\frac{d}{dy}\left(\frac{d'\alpha}{dx} + \frac{d'\beta}{dy} + \frac{d'\gamma}{dz}\right) &= 0, \\[1mm]
n\nabla^{2\prime}\gamma + m\,\frac{d}{dz}\left(\frac{d'\alpha}{dx} + \frac{d'\beta}{dy} + \frac{d'\gamma}{dz}\right) &= 0
\end{aligned}\right\} \quad\ldots\ldots\ldots(66).$$

40. This solution is applicable of course to an elastic body of any shape. It enables us to determine the displacement of every point of it when any given force is applied to every point of its interior, and either displacements or forces are given over the whole surface, if we can solve the general problem for the same shape of body with arbitrary superficial data, but no force on the interior parts. For $'\alpha$, $'\beta$, $'\gamma$ are determined by the solution of ·this problem, to be worked out with the given arbitrary superficial functions modified by the subtraction from them of terms due to the parts of α, β, γ which are explicitly shown in terms of data by equations (64) and (65).

41. Hence the problem of § 35 is completely solved,—whether we have *displacements* given over each of the two concentric spherical bounding surfaces, when the solution of §§ 14—18 determines $'\alpha$, $'\beta$, $'\gamma$; or *forces* given over the boundary, when the solution of §§ 26—30 is available. In the former case the superficial values of the functions

$$\frac{1}{n}\left(\varpi - \frac{m}{m+n}\frac{d\vartheta}{dx}\right),$$

$$\frac{1}{n}\left(\rho - \frac{m}{m+n}\frac{d\vartheta}{dy}\right),$$

$$\frac{1}{n}\left(\sigma - \frac{m}{m+n}\frac{d\vartheta}{dz}\right),$$

known from equations (65), must be subtracted from the arbitrary functions given as the superficial values of α, β, γ, and the residues, expressed in surface-harmonic series by the known method, will be the harmonic expressions for the superficial values of $'\alpha$, $'\beta$, $'\gamma$. In the latter case, we must first substitute those known functions $\frac{1}{n}\left(\varpi - \frac{m}{m+n}\frac{d\vartheta}{dx}\right)$, &c., instead of α, β, γ respectively in (34), and the values of Fr, Gr, Hr thus found must be subtracted from the given arbitrary functions representing the true superficial values of Fr, Gr, Hr. The remainders, which we may denote by $'Fr$, $'Gr$, $'Hr$, must then be reduced to harmonic series, as in (45), and used according to the investigation of § 30, to determine $'\alpha$, $'\beta$, $'\gamma$.

42. The general solution (64) and the expression just indicated for the terms to be subtracted from the data so as to find $'Fr$, $'Gr$, $'Hr$ becomes much simplified when, as in some of the

most important practical applications, $X dx + Y dy + Z dz$ is a complete differential.   Thus let

$$-X = \frac{dW}{dx}, \quad -Y = \frac{dW}{dy}, \quad -Z = \frac{dW}{dz} \quad \dots\dots\dots (67),$$

W denoting any function of x, y, z.   Then, assuming, as we may do according to (65),

$$\varpi = \frac{d}{dx} \nabla^{-2} W,$$

$$\rho = \frac{d}{dy} \nabla^{-2} W,$$

$$\sigma = \frac{d}{dz} \nabla^{-2} W,$$

we have by differentiating, &c.,

$$\frac{d\varpi}{dx} + \frac{d\rho}{dy} + \frac{d\sigma}{dz} = W,$$

and therefore

$$\vartheta = \nabla^{-2} W \dots\dots\dots\dots\dots\dots\dots (68).$$

Hence the solution (64) becomes

$$\left.\begin{aligned}
\alpha &= \frac{1}{m+n} \frac{d\vartheta}{dx} + {}'\alpha, \\
\beta &= \frac{1}{m+n} \frac{d\vartheta}{dy} + {}'\beta, \\
\gamma &= \frac{1}{m+n} \frac{d\vartheta}{dz} + {}'\gamma
\end{aligned}\right\} \dots\dots\dots\dots\dots (69).$$

From this we find

$$\left.\begin{aligned}
\delta &= \frac{1}{m+n} W + {}'\delta, \\
\zeta &= \frac{1}{m+n} r \frac{d\vartheta}{dr} + {}'\zeta, \\
{}'\delta &= \frac{d\,'\alpha}{dx} + \frac{d\,'\beta}{dy} + \frac{d\,'\gamma}{dz}, \\
{}'\zeta &= {}'\alpha x + {}'\beta y + {}'\gamma z
\end{aligned}\right\} \dots\dots\dots\dots (70).$$

and (§ 25)

if

and

Hence, by (34),

$$Fr = \frac{1}{m+n} \left\{ (m-n) Wx + n \left[\left(r \frac{d}{dr} - 1 \right) \frac{d}{dx} + \frac{d}{dx} r \frac{d}{dr} \right] \vartheta \right\} + Fr.$$

But

$$\frac{d}{dx} r \frac{d}{dr} = \left(r \frac{d}{dr} + 1\right) \frac{d}{dx}.$$

Thus for Fr, and the symmetrical expressions, we have

$$\left.\begin{aligned}
Fr &= \frac{1}{m+n} \left\{ (m-n)\, Wx + 2nr \frac{d}{dr} \frac{d\vartheta}{dx} \right\} + {}'Fr, \\
Gr &= \frac{1}{m+n} \left\{ (m-n)\, Wy + 2nr \frac{d}{dr} \frac{d\vartheta}{dy} \right\} + {}'Gr, \\
Hr &= \frac{1}{m+n} \left\{ (m-n)\, Wz + 2nr \frac{d}{dr} \frac{d\vartheta}{dz} \right\} + {}'Hr
\end{aligned}\right\} \quad\dots\dots(71).$$

43. These expressions become further simplified if W is a homogeneous function of any positive or negative integral or fractional order $i+1$, in which case we shall denote it by W_{i+1}. For ϑ will be a homogeneous function of order $i+3$, and $\dfrac{d\vartheta}{dx}$ of order $i+2$. Hence

$$r \frac{d}{dr} \frac{d}{dx} \vartheta = (i+2) \frac{d\vartheta}{dx}.$$

Hence the preceding become

$$\left.\begin{aligned}
Fr &= \frac{1}{m+n} \left\{ (m-n)\, W_{i+1}\, x + 2n\,(i+2) \frac{d\vartheta}{dx} \right\} + {}'Fr, \\
Gr &= \frac{1}{m+n} \left\{ (m-n)\, W_{i+1}\, y + 2n\,(i+2) \frac{d\vartheta}{dy} \right\} + {}'Gr, \\
Hr &= \frac{1}{m+n} \left\{ (m-n)\, W_{i+1}\, z + 2n\,(i+2) \frac{d\vartheta}{dz} \right\} + {}'Hr
\end{aligned}\right\} \quad\dots(72).$$

44. These expressions are the more readily reduced to the harmonic forms proper for working out the solution, if the interior force potential, W_{i+1}, is itself a harmonic function. We then have (§ 10)

$$\left.\begin{aligned}
\vartheta &= \frac{1}{2\,(2i+5)} r^2 W_{i+1}, \quad \frac{d\vartheta}{dx} = \frac{1}{2i+5}\left(x W_{i+1} + \frac{1}{2} r^2 \frac{d W_{i+1}}{dx}\right), \\
\text{and} \quad W_{i+1}x &= \frac{1}{2i+3} \left\{ r^2 \frac{d W_{i+1}}{dx} - r^{2i+5} \frac{d\,(W_{i+1} r^{-2i-3})}{dx} \right\}
\end{aligned}\right\} \quad\dots(73);$$

which give

$$Fr = \frac{1}{m+n} \left\{ \frac{m+(i+1)n}{2i+3} r^2 \frac{d W_{i+1}}{dx} - \frac{m(2i+5)-n}{(2i+3)(2i+5)} r^{2i+5} \frac{d\,(W_{i+1} r^{-2i-3})}{dx} \right\}$$
$$+ {}'Fr\dots\dots(74),$$

and symmetrical expressions for Gr and Hr. Here the terms to be subtracted from the arbitrary functions given to represent the superficial values of Fr, Gr, and Hr are each explicitly expressed in sums of two surface harmonics of orders i or $-i-1$, and $i+2$ or $-i-3$ respectively, viz., in each case, that one of the two numbers which is not negative.

45.   When the shell is in equilibrium under the influence of the forces acting on it through its interior, without any application of force to its surface, we must have

$$\left.\begin{array}{l} Fr = 0, \\ Gr = 0, \\ Hr = 0, \end{array}\right\} \text{ when } r = a \text{ and when } r = a' \text{ ......(75).}$$

Hence, for the case in which W is a spherical harmonic, the preceding equations give the proper harmonic expressions for $'Fr$, $'Gr$, $'Hr$ at the outer and inner bounding surfaces, for determining $'\alpha$, $'\beta$, $'\gamma$ by the method of §§ 28—30. Thus, using all the same notations, with the exception of $'\alpha$, $'\beta$, $'\gamma$, $'F$, $'G$, $'H$, instead of α, β, γ, F, G, H, and, for the present, supposing $i+1$ to be positive*, we have the complete harmonic expressions of $'F$, $'G$, $'H$, each in two terms, of orders i and $i+2$ respectively. Hence the A, A', &c. of (45) are given by the following equations :—

$$\left.\begin{array}{l} \dfrac{A_i}{a^{i+1}} = \dfrac{A_i{}'}{a'^{i+1}} = - \dfrac{m + (i+1)\,n}{(2i+3)\,(m+n)}\, r^{-i}\, \dfrac{dW_{i+1}}{dx}, \\[2.5ex] \dfrac{B_i}{a^{i+1}} = \dfrac{B_i{}'}{a'^{i+1}} = - \dfrac{m + (i+1)\,n}{(2i+3)\,(m+n)}\, r^{-i}\, \dfrac{dW_{i+1}}{dy}, \\[2.5ex] \dfrac{C_i}{a^{i+1}} = \dfrac{C_i{}'}{a'^{i+1}} = - \dfrac{m + (i+1)\,n}{(2i+3)\,(m+n)}\, r^{-i}\, \dfrac{dW_{i+1}}{dz}, \\[2.5ex] \dfrac{A_{i+2}}{a^{i+1}} = \dfrac{A'_{i+2}}{a'^{i+1}} = \dfrac{(2i+5)\,m - n}{(2i+3)\,(2i+5)\,(m+n)}\, r^{i+3}\, \dfrac{d(W_{i+1}r^{-2i-3})}{dx}, \\[2.5ex] \dfrac{B_{i+2}}{a^{i+1}} = \dfrac{B'_{i+2}}{a'^{i+1}} = \dfrac{(2i+5)\,m - n}{(2i+3)\,(2i+5)\,(m+n)}\, r^{i+3}\, \dfrac{d(W_{i+1}r^{-2i-3})}{dy}, \\[2.5ex] \dfrac{C_{i+2}}{a^{i+1}} = \dfrac{C'_{i+2}}{a'^{i+1}} = \dfrac{(2i+5)\,m - n}{(2i+3)\,(2i+5)\,(m+n)}\, r^{i+3}\, \dfrac{d(W_{i+1}r^{-2i-3})}{dz} \end{array}\right\} ...(76).$$

* As we shall not in the present paper consider particularly any case of a shell influenced by centres of force in the hollow space within it, which alone could give a potential W_{i+1} of negative degree, we need not write any of the expressions in forms convenient for making $i+1$ negative.

46. The functions derived from A_i, B_i, C_i, &c., which are required for formulæ (48) and (49), are therefore as follows:—

$$\frac{d\,(A_i r^i)}{dx} \quad + \frac{d\,(B_i r^i)}{dy} \quad + \frac{d\,(C_i r^i)}{dz} \quad = 0,$$

$$\frac{d\,(A_i r^{-i-1})}{dx} \quad + \frac{d\,(B_i r^{-i-1})}{dy} \quad + \frac{d\,(C_i r^{-i-1})}{dz}$$

$$= \frac{(i+1)\,(2i+1)\,[m+(i+1)\,n]}{(2i+3)\,(m+n)}\,\frac{a^{i+1}}{r^{2i+3}}\,W_{i+1},$$

$$\frac{d\,(A_{i+2} r^{i+2})}{dx} + \frac{d\,(B_{i+2} r^{i+2})}{dy} + \frac{d\,(C_{i+2} r^{i+2})}{dz}$$

$$= - \frac{(i+2)\,[(2i+5)\,m-n]}{(2i+3)\,(m+n)}\,a^{i+1} W_{i+1},$$

$$\frac{d\,(A_{i+2} r^{-i-3})}{dx} + \frac{d\,(B_{i+2} r^{-i-3})}{dy} + \frac{d\,(C_{i+2} r^{-i-3})}{dz} = 0$$

...(77),

with the corresponding expressions relative to A_i', B_i', C_i', &c., obtained simply by changing a into a'.

Hence by (48) and (50), and the two corresponding equations for the inner surface, we infer that each of the four functions ψ_{i-1}, ψ_{-i}, ϕ_{i-1}, ϕ_{-i} vanishes. By the same equations, with i changed into $i+2$, we obtain expressions, all of one harmonic form, direct or reciprocal, as follows, for the four functions of order $i+1$:—

$$\begin{aligned}
\psi_{i+1} &= K_{i+1} W_{i+1}, \\
\psi_{-i-2} &= K'_{i+1} r^{-2i-3} W_{i+1}, \\
\phi_{i+1} &= L_{i+1} W_{i+1}, \\
\phi_{-i-2} &= L'_{i+1} r^{-2i-3} W_{i+1}
\end{aligned}$$ ....(78),

K_{i+1}, K'_{i+1}, L_{i+1}, L'_{i+1}, which need not be here explicitly expressed, being four constants obtained from the solutions of four simple algebraic equations. Lastly, by the four equations with $(i+4)$ instead of i, we find that ψ_{i+3}, ψ_{-i-4}, ϕ_{i+3}, ϕ_{-i-4} all vanish. Using these results for ψ and ϕ in (47), we see that each of the functions u must be a harmonic congruent with either $\dfrac{dW_{i+1}}{dx}$ or $\dfrac{d\,(W_{i+1} r^{-2i-3})}{dx}$ Hence, by using (78) in (38) and (39) we find

$$
\left.\begin{aligned}
u_i &= \frac{-L_{i+1}}{(2i+1)\,(i+1)}\frac{dW_{i+1}}{dx}, \\[2mm]
u_{i+2} &= \frac{-K_{i+1}}{(2i+5)\,(i+2)}r^{2i+5}\frac{d\,(W_{i+1}r^{-2i-3})}{dx}, \\[2mm]
u_{-i-1} &= \frac{-K'_{i+1}}{(2i+1)\,(i+1)}r^{-2i-1}\frac{dW_{i+1}}{dx}, \\[2mm]
u_{-i+3} &= \frac{-L'_{i+1}}{(2i+5)\,(i+2)}\frac{d\,(W_{i+1}r^{-2i-3})}{dx}
\end{aligned}\right\} \quad \dots\dots\dots\dots(79);
$$

and symmetrical expressions for v and w. Finally, using these expressions, (79) and (78), in (36), and the result in (69) with (73), we arrive at an explicit solution of the problem in the following remarkably simple form :—

$$
\left.\begin{aligned}
\alpha &= \mathfrak{C}_{i+1}\frac{dW_{i+1}}{dx} + \mathfrak{C}'_{i+1}\frac{d\,(W_{i+1}r^{-2i-3})}{dx}, \\[2mm]
\beta &= \mathfrak{C}_{i+1}\frac{dW_{i+1}}{dy} + \mathfrak{C}'_{i+1}\frac{d\,(W_{i+1}r^{-2i-3})}{dy}, \\[2mm]
\gamma &= \mathfrak{C}_{i+1}\frac{dW_{i+1}}{dz} + \mathfrak{C}'_{i+1}\frac{d\,(W_{i+1}r^{-2i-3})}{dz}
\end{aligned}\right\} \quad \dots\dots\dots(80),
$$

where

$$
\left.\begin{aligned}
\mathfrak{C}_{i+1} &= -\frac{L_{i+1} + K'_{i+1}r^{-2i-1}}{(i+1)\,(2i+1)} + \left\{\frac{1}{2\,(2i+3)\,(m+n)}\right. \\[2mm]
&\qquad\qquad \left. -\tfrac{1}{2}\,m\,\frac{K_{i+1}}{(m+2n)\,i+m+3n}\right\}r^2, \\[2mm]
\mathfrak{C}'_{i+1} &= -\frac{K_{i+1}r^{2i+5} + L'_{i+1}}{(i+2)\,(2i+5)} + \left\{\frac{-r^{2i+5}}{(2i+3)\,(2i+5)\,(m+n)}\right. \\[2mm]
&\qquad\qquad \left. +\frac{mr^2}{2}\,\frac{K'_{i+1}}{(m+2n)\,i+2m+3n}\right\}
\end{aligned}\right\} \quad \dots(81).
$$

47. In conclusion, let us consider the case of a solid sphere. For this we have

$$
\psi_{-i-2} = 0, \text{ and } \phi_{-i-2} = 0,
$$

as we see at once from the character of the problem, or as we find by putting $a' = 0$ in the four equations by which in § 46 we have seen that K_{i+1}, K'_{i+1}, L_{i+1}, L'_{i+1} are to be determined. Then, by (48), with i changed into $i+2$, and by (49), we find

$$
\left.\begin{aligned}
\psi_{i+1} &= -\frac{(i+2)\,[(m+2n)i+m+3n]\,[m\,(2i+5)-n]}{(2i+3)\,n\,\{m\,[2\,(i+2)^2+1]-n\,(2i+3)\}\,(m+n)}W_{i+1}, \\[2mm]
\phi_{i+1} &= a^2\frac{(i+1)^2\,(2i+1)\,[m\,(i+3)-n]}{2n\,\{m\,[2\,(i+2)^2+1]-n\,(2i+3)\}}W_{i+1}
\end{aligned}\right\} \quad \dots(82).
$$

The coefficients of W_{i+1} in these expressions are the values which we must take for K_{i+1} and L_{i+1} respectively in (81); and therefore, after reductions which show $(m+n)$ as a factor of the numerator of each fraction in which it appears at first as a factor of the denominator, we have

$$\left. \begin{aligned} \mathfrak{C}_{i+1} = &- \frac{(i+1)\left[m(i+3)-n\right]a^2}{2n\{m\left[2(i+2)^2+1\right]-n(2i+3)\}} \\ &+ \frac{\left[(i+2)(2i+5)m-(2i+3)n\right]r^2}{2n(2i+3)\{m\left[2(i+2)^2+1\right]-n(2i+3)\}} \cdot \\ \mathfrak{C}'_{i+1} = &\frac{(i+1)\,mr^{2i+5}}{n(2i+3)\{m\left[2(i+2)^2+1\right]-n(2i+3)\}} \end{aligned} \right\} \ \ldots\ldots (83).$$

These, substituted in (80), give expressions for α, β, γ which constitute a complete and explicit solution of the problem.

It is easy to verify this result, by testing that (56) (with $-X = \dfrac{dW_{i+1}}{dx}$, &c.) is satisfied for every point of the solid, and that equations (34) give $F=0$, $G=0$, $H=0$ at the bounding surface, $r = a$.

48. The case of $i=1$ is, as we shall immediately see, of high importance. For it the preceding expressions, (83) and (80), become

$$\left. \begin{aligned} \mathfrak{C}_2 &= \frac{-10\,(4m-n)\,a^2 + (21m-5n)\,r^2}{10n\,(19m-5n)}, \\ \mathfrak{C}_2' &= \frac{4mr^7}{10n\,(19m-5n)}, \\ \alpha &= \mathfrak{C}_2 \frac{dW_2}{dx} + \mathfrak{C}_2' \frac{d\,(W_2 r^{-5})}{dx}, \\ \beta &= \mathfrak{C}_2 \frac{dW_2}{dy} + \mathfrak{C}_2' \frac{d\,(W_2 r^{-5})}{dy}, \\ \gamma &= \mathfrak{C}_2 \frac{dW_2}{dz} + \mathfrak{C}_2' \frac{(dW_2 r^{-5})}{dz} \end{aligned} \right\} \ \ldots\ldots (84).$$

49. As an example of the application of §§ 45—48, let us suppose a spherical shell or solid sphere to be equilibrated under the influence of masses collected in two fixed external points*, and each attracting according to the inverse square of its distance.

* If our limits permitted, a highly interesting example might be made of the case of a shell under the influence of a single attracting point in the hollow space

Let the two masses M, M' be in the axis OX; and, P being the point whose coordinates are x, y, z, let $PM = D$, $PM' = D'$. Let

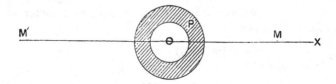

also $OM = c$, $OM' = c'$. Then, if m, m' denote the two masses, for equilibrium we must have

$$\frac{\text{m}}{c^2} = \frac{\text{m}'}{c'^2}.$$

The potential at P, due to the two masses, will be $\dfrac{\text{m}}{D} + \dfrac{\text{m}'}{D'}$, or, according to the notation of § 42, with, besides, w taken to denote the mass of unit volume of the elastic solid,

$$- W = w \left(\frac{\text{m}}{D} + \frac{\text{m}'}{D'} \right).$$

The known forms in the elementary theory of spherical harmonics give immediately the development of this in a converging infinite series of solid harmonic terms. We have only then to apply the solution of §§ 45, 46 to each term, to obtain a series expressing the required solution.

50. We may work out this result explicitly for the case in which both masses are very distant; and for simplicity we shall suppose one of them infinitely more distant than the other; that is to say, we shall suppose it to exercise merely a constant balancing force on the substance of the shell. We shall then have precisely the same bodily disturbing force as that which the earth experiences from the moon alone, or from the sun alone.

51. Referring to the diagram and notation of § 49, we have

$$\frac{1}{D} = \frac{1}{c} \left\{ 1 + \frac{x}{c} + \frac{x^2 - \frac{1}{2}(y^2 + z^2)}{c^2} \right\}$$

within it. The effect will clearly be to keep the whole shell sensibly in equilibrium even if the attracting point is excentric; and under stress even if the attracting point is in the centre.

if we neglect higher powers of x/c, y/c, z/c than the square; and

$$\frac{1}{D'} = \frac{1}{c'}\left(1 - \frac{x}{c}\right)$$

neglecting all higher powers of x/c, y/c, z/c. Hence, taking account of the relation $m/c^2 = m'/c'^2$ required for equilibrium, we have, for the disturbance potential,

$$-W = \frac{m}{c^3}\left(x^2 - \tfrac{1}{2}y^2 - \tfrac{1}{2}z^2\right)w,$$

an irrelevant constant being omitted from the expression which § 49 would give. This being a harmonic of the second degree, we may use it for W_{i+1}, putting $i = 1$ in the formulæ of § 47, and thus solve the problem of finding the deformation of a homogeneous spherical shell under the influence of a distant attracting mass and a uniform balancing force. I hope, in a future communication to the Royal Society, to show the application of this result to the case of the lunar and solar influence on a body such as the earth is assumed to be by many geologists—that is to say, a solid crust, constituting a spheroidal shell, of some thickness less than 100 miles, with its interior filled with liquid. The untenability of this hypothesis is, however, sufficiently demonstrated by the considerations adduced in a previous communication [Art. XCV. above "On the Rigidity of the Earth,"] read May 8, 1862, in which the following explicit solution of the problem for a homogeneous solid sphere only is used.

52.   Using the expression of § 51 for W_2, we have

$$\left.\begin{aligned}
\frac{dW_2}{dx} &= -2\frac{m}{c^3}xw, \\[4pt]
\frac{dW_2}{dy} &= +\frac{m}{c^3}yw, \\[4pt]
\frac{dW_2}{dz} &= +\frac{m}{c^3}zw, \\[4pt]
\frac{d(W_2 r^{-5})}{dx} &= +3\frac{m}{c^3}\frac{(x^2 - \tfrac{3}{2}y^2 - \tfrac{3}{2}z^2)x}{r^{-7}}w, \\[4pt]
\frac{d(W_2 r^{-5})}{dy} &= +3\frac{m}{c^3}\frac{(2x^2 - \tfrac{1}{2}y^2 - \tfrac{1}{2}z^2)y}{r^{-7}}w, \\[4pt]
\frac{d(W_2 r^{-5})}{dz} &= +3\frac{m}{c^3}\frac{(2x^2 - \tfrac{1}{2}y^2 - \tfrac{1}{2}z^2)z}{r^{-7}}w,
\end{aligned}\right\} \quad \dots (85),$$

These formulæ being substituted for the differential coefficients which appear in (84), we have algebraic expressions for the displacement of any point of the solid.

The condition of the body being symmetrical about the axis of x, we may conveniently assume

$$y = \mathsf{y} \cos \phi, \ z = \mathsf{y} \sin \phi,$$

$$\beta^2 + \gamma^2 = \mu^2;$$

so that we shall have (as we see by the preceding expressions)

$$\beta = \mu \cos \phi,$$

$$\gamma = \mu \sin \phi,$$

and μ will denote the component displacement perpendicular to OX. If, further, we assume

$$x = r \cos \theta,$$

$$\mathsf{y} = r \sin \theta,$$

the expressions (84) for the component displacements, with (85) used in them, give

$$\left.\begin{array}{l} \alpha = w \dfrac{m}{c^3} \left\{ -2r\mathfrak{C}_2 + \dfrac{3}{2} \dfrac{\mathfrak{C}'_2}{r^4} (5 \cos^2 \theta - 3) \right\} \cos \theta, \\[3mm] \mu = w \dfrac{m}{c^3} \left\{ \quad r\mathfrak{C}_2 + \dfrac{3}{2} \dfrac{\mathfrak{C}_2'}{r^4} (5 \cos^2 \theta - 1) \right\} \sin \theta \end{array}\right\} \ \ \dots (86).$$

The values given in (84) for $\mathfrak{C}_2$ and $\mathfrak{C}_2'$ are to be used for any internal point, at a distance r from the centre, in these equations (86), and thus we have the simplest possible expression for the required displacement of any point of the solid.

53.   If we resolve the displacement along and perpendicular to the radius, and consider only the radial component, we see that the series of concentric spherical surfaces of the undisturbed globe become spheroids of revolution in the distorted body. The elongation of the axial radius, obtained by putting $\theta = 0$ and taking the value of α, is double the shortening of the equatorial radius, obtained by putting $\theta = \frac{1}{2}\pi$ and taking the value of μ; which we might have inferred from the fact, shown by the general equations (80) above, that there can be no alteration of volume

on the whole within any one of these surfaces. The expression
for the excess of the axial above the equatorial radius is

$$3\,w\,\frac{\mathrm{m}}{c^3}\left(-r\mathfrak{C}_2 + \frac{9}{2}\frac{\mathfrak{C}_2{}'}{r^4}\right),$$

which, if we substitute for $\mathfrak{C}_2$ and $\mathfrak{C}_2{}'$ their values by (84), becomes

$$3w\,\frac{\mathrm{m}}{c^3}\,\frac{2\,(4m-n)\,a^2-(3m-n)\,r^2}{2n\,(19m-5n)}\,r.$$

If in this we take $r = a$, and $m = \infty$, it becomes $3w\,\dfrac{\mathrm{m}}{c^3}\cdot\dfrac{5}{2\,.\,19n}\,.\,r^3$,
which is the result used in § 34 of the paper " On the Rigidity of
the Earth," preceding the present [Art. XCV. above].

54.   In the case of $a' = 0$, the result of § 18 takes the extremely simple form

$$\left.\begin{aligned}
\alpha &= \Sigma\left\{A_i\left(\frac{r}{a}\right)^i + \frac{m\,(a^2-r^2)}{2a^i\,[n\,(2i-1)+m\,(i-1)]}\,\frac{d\Theta_{i-1}}{dx}\right\}, \\[2mm]
\beta &= \Sigma\left\{B_i\left(\frac{r}{a}\right)^i + \frac{m\,(a^2-r^2)}{2a^i\,[n\,(2i-1)+m\,(i-1)]}\,\frac{d\Theta_{i-1}}{dy}\right\}, \\[2mm]
\gamma &= \Sigma\left\{C_i\left(\frac{r}{a}\right)^i + \frac{m\,(a^2-r^2)}{2a^i\,[n\,(2i-1)+m\,(i-1)]}\,\frac{d\Theta_{i-1}}{dz}\right\},
\end{aligned}\right\}\ \dots(87).$$

where

$$\Theta_{i+1} = \frac{d\,(A_i r^i)}{dx} + \frac{d\,(B_i r^i)}{dy} + \frac{d\,(C_i r^i)}{dz}$$

This expresses the displacement at any point within a solid sphere
of radius a, when its surface is displaced in a given manner (ΣA_i,
ΣB_i, ΣC_i). And merely by making i negative we have, in the
same formula, the solution of the same problem for an infinite
solid with a hollow spherical space every point of the surface of
which is displaced to a given distance in a given direction. These
solutions are obtained directly, with great ease, by the method of
§§ 6—15, or are easily proved by direct verification, without any
of the intricacy of analysis inevitable when, as in the general
investigations with which we commenced, a shell bounded by two
concentric spherical surfaces is the subject.

[Added since the reading of the Paper.]

§§ 55 to 58.   *Oscillations of a Liquid Sphere.*

55.   Let V be the gravitation potential at any point $P\,(x, y, z)$, and h the height of the surface (or radial component of its displacement) from the mean spherical surface at a point E in the radius through P.   Then, if

$$h = S_1 + S_2 + \ldots\ldots\ldots\ldots\ldots\ldots\ldots(88)$$

be the expression for h in terms of spherical surface harmonic functions of the position of E, and if μ be the attraction on the unit of mass exercised by a particle equal in mass to the unit bulk of the liquid, we have, by the known methods for finding the attractions of bodies infinitely nearly spherical (Thomson and Tait's *Natural Philosophy*, Part II., chap. VI.),

$$\left.\begin{aligned}
V &= 4\pi a\mu \left\{ \tfrac{1}{2}a - \tfrac{1}{6}\frac{r^2}{a} + \Sigma \left(\frac{r}{a}\right)^i \frac{S_i}{2i+1} \right\} \text{ when } r < a, \\
V &= 4\pi a\mu \left\{ \tfrac{1}{3}\frac{a^2}{r} + \frac{a}{r}\Sigma \left(\frac{a}{r}\right)^i \frac{S_i}{2i+1} \right\} \quad \text{,, } \quad r > a, \\
\text{and} \\
V &= 4\pi a\mu \left\{ \tfrac{1}{3}a + \Sigma \frac{S_i}{2i+1} \right\} \quad\quad \text{,, } \quad r = a
\end{aligned}\right\} \ldots (89).$$

In these $$4\pi\mu a = 3g \ldots\ldots\ldots\ldots\ldots\ldots\ldots(90),$$

if g denote the force of gravity at the surface, due to the mean sphere of radius a.

56.   Now for infinitely small motions the ordinary kinetic equations give

$$-\frac{dp}{dx} = \rho\left(\frac{du}{dt} - \frac{dV}{dx}\right);\ -\frac{dp}{dy} = \rho\left(\frac{dv}{dt} - \frac{dV}{dy}\right);\ -\frac{dp}{dz} = \rho\left(\frac{dw}{dt} - \frac{dV}{dz}\right)$$
$$\ldots\ldots(91);$$

where ρ is the mass per unit of volume; u, v, w the component velocities through the fixed point P at time t; and p the fluid pressure.   Hence, possible non-periodic motions being omitted, $u\,dx + v\,dy + w\,dz$ is a complete differential; and, denoting it by $d\phi$, we have

$$C - p = \rho\left(\frac{d\phi}{dt} - V\right) \ldots\ldots\ldots\ldots(92).$$

57. To find the surface conditions,—first, since the pressure has a constant value, Π, at the free surface,

$$p = g\rho h + \Pi \text{ when } r = a \dots\dots\dots\dots\dots (93),$$

the variations of gravity depending on the variations of figure being of course neglected in the infinitely small term $g\rho h$. And, since $\frac{dh}{dt}$ is the radial component of the velocity at E, we have, when $r = a$,

$$\frac{x}{r}\frac{d\phi}{dx} + \frac{y}{r}\frac{d\phi}{dy} + \frac{z}{r}\frac{d\phi}{dz} = \frac{dh}{dt} \dots\dots\dots\dots\dots(94).$$

Now since, the fluid being incompressible, $\nabla^2\phi = 0$, ϕ may be expanded in a series of solid harmonic functions; let

$$\phi = \Sigma\Phi_i \left(\frac{r}{a}\right)^i \dots\dots\dots\dots\dots\dots(95),$$

where Φ_1, Φ_2,... are surface harmonics. Hence, as the successive terms are homogeneous functions of the coordinates (x, y, z), of degrees 1, 2, &c.,

$$\frac{x}{r}\frac{d\phi}{dx} + \frac{y}{r}\frac{d\phi}{dy} = \frac{z}{r}\frac{d\phi}{dz} = \frac{1}{r}\Sigma i\Phi_i \left(\frac{r}{a}\right)^i \dots\dots\dots\dots (96),$$

and therefore, by (88) and (94),

$$\frac{dS_i}{dt} = \frac{i}{a}\Phi_i \dots\dots\dots\dots\dots\dots\dots(97).$$

58. Eliminating p between (92) with $r = a$ and (93), substituting for V by (89) and (90), differentiating, substituting for $\frac{dS_i}{dt}$ by (97), and comparing harmonic terms of order i, we have

$$\frac{d^2\Phi_i}{dt^2} + \frac{g}{a}i\left(1 - \frac{3}{2i+1}\right)\Phi_i = 0 \dots\dots\dots(98);$$

of which the integral is

$$\Phi_i = A\cos\left\{t\sqrt{\frac{g}{a}i\left(1 - \frac{3}{2i+1}\right)} - E\right\} \dots\dots\dots(99).$$

Here A is a surface spherical harmonic function of the coordinates of E expressing the maximum value of Φ_i, and E is the epoch (Thomson and Tait's *Natural Philosophy*, Part I. § 53) of the

simple harmonic function of the time which we find to represent
Φ_i. Using this solution in (97) and (88), we see that if the surface
be normally displaced according to a spherical harmonic of order
i, and left to itself, the resulting motion gives rise to a simple
harmonic variation of the normal displacement, having for period

$$2\pi \sqrt{\frac{a}{g} \frac{2i+1}{2i(i-1)}},$$

that is, the period of a common pendulum of length $\dfrac{(2i+1)\,a}{2i\,(i-1)}$.
It is worthy of remark that the period of vibration thus calculated
is the same for the same density of liquid, whatever be the
dimensions of the globe.

For the case of $i = 2$, or an ellipsoidal deformation, the length
of the isochronous pendulum becomes $\frac{5}{4}a$, or one and a quarter
times the earth's radius, for a homogeneous liquid globe of the
same mass and diameter as the earth; and therefore for this case,
or for any homogeneous liquid globe of about $5\frac{1}{2}$ times the density
of water, the half-period is $47^{\mathrm{m}} 12^{\mathrm{s}}$, which is the result stated in
§ 3 of the paper "On the Rigidity of the Earth", preceding the
present (Art. XCV. above).

APPENDIX, §§ 59—71.—*General Theory of the Equilibrium of an
Elastic Solid.*

59.   Let a solid composed of matter fulfilling no condition of
isotropy in any part, and not homogeneous from part to part, be
given of any shape, unstrained, and let every point of its surface
be altered in position to a given distance in a given direction. It
is required to find the displacement of every point of its substance
in equilibrium. Let x, y, z be the coordinates of any particle, P,
of the substance in its undisturbed position, and $x+a$, $y+\beta$, $z+\gamma$
its coordinates when displaced in the manner specified; that is
to say, let a, β, γ be the components of the required displacement.
Then, if for brevity we put

$$A = \left(\frac{d\alpha}{dx}+1\right)^2 + \left(\frac{d\beta}{dx}\right)^2 + \left(\frac{d\gamma}{dx}\right)^2,$$

$$B = \left(\frac{d\alpha}{dy}\right)^2 + \left(\frac{d\beta}{dy}+1\right)^2 + \left(\frac{d\gamma}{dy}\right)^2,$$

$$C = \left(\frac{d\alpha}{dz}\right)^2 + \left(\frac{d\beta}{dz}\right)^2 + \left(\frac{d\gamma}{dz}+1\right)^2,$$

$$a = \frac{d\alpha}{dy}\frac{d\alpha}{dz} + \left(\frac{d\beta}{dy}+1\right)\frac{d\beta}{dz} + \frac{d\gamma}{dy}\left(\frac{d\gamma}{dz}+1\right),$$

$$b = \frac{d\alpha}{dz}\left(\frac{d\alpha}{dx}+1\right) + \frac{d\beta}{dz}\frac{d\beta}{dx} + \left(\frac{d\gamma}{dz}+1\right)\frac{d\gamma}{dx},$$

$$c = \left(\frac{d\alpha}{dx}+1\right)\frac{d\alpha}{dy} + \frac{d\beta}{dx}\left(\frac{d\beta}{dy}+1\right) + \frac{d\gamma}{dx}\frac{d\gamma}{dy}$$

$$\dots (100);$$

these six quantities A, B, C, a, b, c, as is known*, thoroughly determine the strain experienced by the substance infinitely near the particle P (irrespectively of any rotation it may experience) in the following manner :—

60.   Let ξ, η, ζ be the undisturbed coordinates of a particle infinitely near P, relatively to axes through P parallel to those of x, y, z respectively; and let ξ_i, η_i, ζ_i be the coordinates, relative still to axes through P, when the solid is in its strained condition. Then

$$\xi_i^2 + \eta_i^2 + \zeta_i^2 = A\xi^2 + B\eta^2 + C\zeta^2 + 2a\eta\zeta + 2b\zeta\xi + 2c\xi\eta \dots(101);$$

and therefore all particles which in the strained state lie on a spherical surface

$$\xi_i^2 + \eta_i^2 + \zeta_i^2 = r_i^2,$$

are, in the unstrained state, on the ellipsoidal surface,

$$A\xi^2 + B\eta^2 + C\zeta^2 + 2a\eta\zeta + 2b\zeta\xi + 2c\xi\eta = r_i^2.$$

This, as is well known†, completely defines the homogeneous strain of the matter in the neighbourhood of P.

61.   Hence the thermo-dynamic principles by which, in a paper "On the Thermo-elastic Properties of Matter" in the first Number of the *Quarterly Mathematical Journal* (April 1855), [Art. XLVIII., Part VII. above; Vol. I. p. 291], Green's dynamical theory of elastic solids was demonstrated as part of the modern

---

* Thomson and Tait's *Natural Philosophy*, Part I. § 190 (e) and § 181 (5).

† Thomson and Tait's *Natural Philosophy*, Part I. §§ 155—165. See Art. XCII. above.

dynamical theory of heat, show that if $w . dx dy dz$ denote the work required to alter an infinitely small undisturbed volume, $dx dy dz$, of the solid, into its disturbed condition, when its temperature is kept constant, we must have

$$w = f(A, B, C, a, b, c) \dots\dots\dots\dots\dots (102),$$

where f denotes a positive function of the six elements, which vanishes when $A - 1$, $B - 1$, $C - 1$, a, b, c each vanish. And if W denote the whole work required to produce the change actually experienced by the whole solid, we have

$$W = \iiint w \, dx \, dy \, dz \dots\dots\dots\dots\dots (103),$$

where the triple integral is extended through the space occupied by the undisturbed solid.

62.   The position assumed by every particle in the interior of the solid will be such as to make this a minimum, subject to the condition that every particle of the surface takes the position given to it, this being the elementary condition of stable equilibrium.   Hence, by the method of variation,

$$\delta W = \iiint \delta w \, dx \, dy \, dz = 0 \dots\dots\dots\dots\dots (104).$$

But, exhibiting only terms depending on $\delta \alpha$, we have

$$\delta w = \left\{ 2 \frac{dw}{dA} \left(\frac{d\alpha}{dx} + 1 \right) + \frac{dw}{db} \frac{d\alpha}{dz} + \frac{dw}{dc} \frac{d\alpha}{dy} \right\} \frac{d\delta\alpha}{dx}$$

$$+ \left\{ 2 \frac{dw}{dB} \frac{d\alpha}{dy} + \frac{dw}{da} \frac{d\alpha}{dz} + \frac{dw}{dc} \left(\frac{d\alpha}{dx} + 1 \right) \right\} \frac{d\delta\alpha}{dy}$$

$$+ \left\{ 2 \frac{dw}{dC} \frac{d\alpha}{dz} + \frac{dw}{da} \frac{d\alpha}{dy} + \frac{dw}{db} \left(\frac{d\alpha}{dx} + 1 \right) \right\} \frac{d\delta\alpha}{dz}$$

$$+ \&c.$$

Hence, integrating by parts, and observing that $\delta\alpha$, $\delta\beta$, $\delta\gamma$ vanish at the limiting surface, we have

$$\delta W = - \iiint dx \, dy \, dz \left\{ \left(\frac{dP}{dx} + \frac{dQ}{dy} + \frac{dR}{dz} \right) \delta\alpha + \&c. \right\} \dots (105),$$

where for brevity P, Q, R denote the factors of $d\delta\alpha/dx$, $d\delta\alpha/dy$, $d\delta\alpha/dz$ respectively, in the preceding expression. In order that δW may vanish, the factors of $\delta\alpha$, $\delta\beta$, $\delta\gamma$ in the expression now found for it must each vanish; and hence we have, as the equations of equilibrium,

$$
\begin{aligned}
& \frac{d}{dx}\left\{ 2\frac{dw}{dA}\left(\frac{d\alpha}{dx}+1\right) + \frac{dw}{db}\frac{d\alpha}{dz} + \frac{dw}{dc}\frac{d\alpha}{dy}\right\} \\
& + \frac{d}{dy}\left\{ 2\frac{dw}{dB}\frac{d\alpha}{dy} + \frac{dw}{da}\frac{d\alpha}{dz} + \frac{dw}{dc}\left(\frac{d\alpha}{dx}+1\right)\right\} \\
& + \frac{d}{dz}\left\{ 2\frac{dw}{dC}\frac{d\alpha}{dz} + \frac{dw}{da}\frac{d\alpha}{dy} + \frac{dw}{db}\left(\frac{d\alpha}{dx}+1\right)\right\} = 0, \\
& \qquad\qquad \&c., \&c.
\end{aligned}
\right\} \dots(106),
$$

of which the second and third, not exhibited, may be written down merely by attending to the symmetry.

63. From the property of w that it is necessarily positive when there is any strain, it follows that there must be some distribution of strain through the interior which shall make $\iiint w\, dx\, dy\, dz$ *the least possible*, subject to the prescribed surface condition, and therefore that the solution of equations (106), subject to this condition, is possible. If, whatever be the nature of the solid as to difference of elasticity in different directions in any part, and as to heterogeneousness from part to part, and whatever be the extent of the change of form and dimensions to which it is subjected, there cannot be any internal configuration of unstable equilibrium, or consequently any but one of stable equilibrium, with the prescribed surface displacement and no disturbing force on the interior, then, besides being always positive, w must be such a function of A, B, &c. that there can be only one solution of the equations. This is obviously the case when the unstrained solid is homogeneous.

64. It is easy to include, in a general investigation similar to the preceding, the effects of any force on the interior substance, such as we have considered particularly for a spherical shell, of homogeneous isotropic matter, in §§ 35—46 above. It is also easy to adapt the general investigation to superficial data of *force*, instead of displacement.

65. Whatever be the general form of the function f for any part of the substance, since it is always positive it cannot change in sign when $A-1$, $B-1$, $C-1$, a, b, c have their signs changed; and therefore for infinitely small values of these quantities it

must be a homogeneous quadratic function of them with constant coefficients. (And it may be useful to observe that for all values of the variables A, B, &c., it must therefore be expressible in the same form, with varying coefficients, each of which is always finite, for all values of the variables.) Thus, for infinitely small strains, we have Green's theory of elastic solids, founded on a homogeneous quadratic function of the components of strain, expressing the work required to produce it. Putting

$$A - 1 = 2e, \quad B - 1 = 2f, \quad C - 1 = 2g \dots\dots\dots\dots(107),$$

and denoting by $\tfrac{1}{2}(e, e)$, $\tfrac{1}{2}(f, f)$, $\dots (e, f)$, $\dots (e, a)$, $\dots$ the coefficients, we have

$$
\left.
\begin{aligned}
w = \tfrac{1}{2}\,\{ (e, e)\,e^2 + (f, f)\,f^2 &+ (g, g)\,g^2 + (a, a)\,a^2 + (b, b)\,b^2 + (c, c)\,c^2\} \\
+ (e, f)\,ef &+ (e, g)\,eg + (e, a)\,ea + (e, b)\,eb + (e, c)\,ec \\
&+ (f, g)\,fg + (f, a)\,fa + (f, b)\,fb + (f, c)\,fc \\
&\qquad + (g, a)\,ga + (g, b)\,gb + (g, c)\,gc \\
&\qquad\qquad + (a, b)\,ab + (a, c)\,ac \\
&\qquad\qquad\qquad + (b, c)\,bc
\end{aligned}
\right\} (108).
$$

The one essential condition among the coefficients is that w is necessarily positive, i.e. reducible to six squares with each a positive or zero coefficient. The twenty-one coefficients in this expression constitute the twenty-one coefficients of elasticity, which Green first showed to be proper and essential for a complete theory of the dynamics of an elastic solid subjected to infinitely small strains.

66.   When the strains are infinitely small, the products

$$\frac{dw}{dA}\frac{da}{dx}, \quad \frac{dw}{db}\frac{da}{dz}, \quad \&c.$$

are each infinitely small, of the second order. We therefore omit them; and then, attending to (107), we reduce (106) to

$$
\left.
\begin{aligned}
\frac{d}{dx}\frac{dw}{de} + \frac{d}{dy}\frac{dw}{dc} + \frac{d}{dz}\frac{dw}{db} = 0, \\
\frac{d}{dx}\frac{dw}{dc} + \frac{d}{dy}\frac{dw}{df} + \frac{d}{dz}\frac{dw}{da} = 0, \\
\frac{d}{dx}\frac{dw}{db} + \frac{d}{dy}\frac{dw}{da} + \frac{d}{dz}\frac{dw}{dg} = 0
\end{aligned}
\right\} \dots\dots\dots (109),
$$

which are the equations of interior equilibrium. Attending to (108) we see that $dw/de \dots dw/da \dots$ are linear functions of e, f, g, a, b, c the components of strain. Writing out one of them as an example, we have

$$\frac{dw}{de} = (e, e)\,e + (e, f)\,f + (e, g)\,g + (e, a)\,a + (e, b)\,b + (e, c)\,c\dots(110).$$

And α, β, γ denoting, as before, the component displacements of any interior particle, P, from its undisturbed position (x, y, z), we have, by (107) and (100),

$$\left. \begin{array}{l} e = \dfrac{d\alpha}{dx}, \quad f = \dfrac{d\beta}{dy}, \quad g = \dfrac{d\gamma}{dz}, \\[2mm] a = \dfrac{d\beta}{dz} + \dfrac{d\gamma}{dy}, \quad b = \dfrac{d\gamma}{dx} + \dfrac{d\alpha}{dz}, \quad c = \dfrac{d\alpha}{dy} + \dfrac{d\beta}{dx}. \end{array} \right\} \dots\dots(111).$$

It is to be observed that the coefficients $(e, e)\,(e, f)$, &c. will be in general functions of (x, y, z), but will be each constant when the unstrained solid is homogeneous.

67. It is now easy to prove directly, for the case of infinitely small strains, that the solution of the equations of interior equilibrium, whether for a heterogeneous or a homogeneous solid, subject to the prescribed surface condition, is unique. For let α, β, γ be components of displacement fulfilling the equations, and let α', β', γ' denote any other functions of (x, y, z) having the same surface values as α, β, γ, and let $e', f', \dots, w'$ denote functions depending on them in the same way as $e, f, \dots, w$ depend on α, β, γ. Thus, by Taylor's theorem,

$$w' - w = \frac{dw}{de}\,(e' - e) + \frac{dw}{df}\,(f' - f) + \frac{dw}{dg}\,(g' - g)$$
$$+ \frac{dw}{da}\,(a' - a) + \frac{dw}{db}\,(b' - b) + \frac{dw}{dc}\,(c' - c) + H,$$

where H denotes the same homogeneous quadratic function of $e' - e$, &c. that w is of e, &c. If for $e' - e$, &c. we substitute their values by (111), this becomes

$$w' - w = \frac{dw}{de}\frac{d\,(\alpha' - \alpha)}{dx} + \frac{dw}{db}\frac{d\,(\alpha' - \alpha)}{dz} + \frac{dw}{dc}\frac{d\,(\alpha' - \alpha)}{dy} + \text{&c.} + H.$$

Multiplying by $dx\,dy\,dz$, integrating by parts, observing that $\alpha' - \alpha$, $\beta' - \beta$, $\gamma' - \gamma$ vanish at the bounding surface, and taking account of (109), we find simply

$$\iiint (w' - w)\,dx\,dy\,dz = \iiint H\,dx\,dy\,dz \dots\dots\dots\dots(112).$$

But H is essentially positive. Therefore every other interior condition than that specified by (α, β, γ), provided only it has the
same bounding surface, requires a greater amount of work than
w to produce it: and the excess is equal to the work that would
be required to produce, from a state of no displacement, such a
displacement as superimposed on (α, β, γ) would produce the other.
And inasmuch as (α, β, γ) fulfil only the conditions of satisfying
(110) and having the given surface values, it follows that no other
than one solution can fulfil these conditions.

68.   But (as has been remarked by Professor Stokes to the
author) when the surface data are of force, not of displacement,
or when force acts from without, on the interior substance of the
body, the solution is not in general unique, and there may be
configurations of unstable equilibrium, even with infinitely small
displacement.   For instance, let part of the body be composed of
a steel bar magnet; and let a magnet be held outside in the same
line, and with a pole of the same name in its end nearest to one
end of the inner magnet. The equilibrium will be unstable, and
there will be positions of stable equilibrium with the inner bar
slightly inclined to the line of the outer bar, unless the rigidity of
the rest of the body exceed a certain limit.

69.   Recurring to the general problem, in which the strains
are not supposed infinitely small, we see that, if the solid is
isotropic in every part, the function of A, B, C, a, b, c which
expresses w must be merely a function of the roots of the equation[*]

$$(A - \zeta^2)(B - \zeta^2)(C - \zeta^2) - a^2(A - \zeta^2) - b^2(B - \zeta^2) - c^2(C - \zeta^2)$$
$$+ 2abc = 0 \dots (113)$$

which (that is the positive values of ζ) are the ratios of elongation
along the principal axes of the strain-ellipsoid. It is unnecessary
here to enter on the analytical expression of this condition. For
the case of $A - 1$, $B - 1$, $C - 1$, a, b, c, each infinitely small, it
obviously requires that

$$\left.\begin{array}{l} (e,e)=(f,f)=(g,g)\,;\;(f,g)=(g,e)=(e,f)\,;\;(a,a)=(b,b)=(c,c)\,; \\[4pt] (e,a)=(f,b)=(g,c)=0\,;\;(b,c)=(c,a)=(a,b)=0\,; \\[4pt] \text{and} \\[4pt] (e,b)=(e,c)=(f,c)=(f,a)=(g,a)=(g,b)=0 \end{array}\right\} (114).$$

---

[*] Thomson and Tait's *Natural Philosophy*, Part I. § 181 (11). See Art. XCII. pp.
95, &c. above.

Thus the twenty-one coefficients are reduced to three—

(e, e), which we may denote by the single letter $\mathfrak{A}$,

(f, g),   „   „   „   „ $\mathfrak{B}$

(a, a),   „   „   „   „ n.

It is clear that this is necessary and sufficient for ensuring *cubic isotropy*—that is to say, perfect equality of elastic properties with reference to the three rectangular directions OX, OY, OZ. But for *spherical isotropy*, or complete isotropy with reference to all directions through the substance, it is further necessary that

$$\mathfrak{A} - \mathfrak{B} = 2n \ldots\ldots\ldots\ldots\ldots\ldots (115),$$

as is easily proved analytically by turning two of the axes of co-ordinates in their own plane through 45°; or geometrically by examining the nature of the strain represented by any one of the elements a, b, c (a "simple shear") and comparing it with the resultant of c, and $f = -e$ (which is also a simple shear). It is convenient now to put

$$\mathfrak{A} + \mathfrak{B} = 2m; \text{ so that } \mathfrak{A} = m + n, \; \mathfrak{B} = m - n \ldots\ldots (116);$$

and thus the expression for the potential energy per unit of volume becomes

$$2w = (m+n)(e^2+f^2+g^2) + 2(m-n)(fg+gh+ef) + n(a^2+b^2+c^2)$$
$$\ldots\ldots (117).$$

Using this in (108), and substituting for e, f, g, a, b, c their values by (111), we find immediately, for the equations of internal equilibrium, equations the same as (1) of § 2.

70.   To find the mutual force exerted across any surface within the solid, as expressed by (26) of § 22, we have clearly, by considering the work done respectively by P, Q, R, S, T, U (§ 21) on any infinitely small change of figure or dimensions in the solid,

$$P = \frac{dw}{de}, \quad Q = \frac{dw}{df}, \quad R = \frac{dw}{dg}, \quad S = \frac{dw}{da}, \quad T = \frac{dw}{db}, \quad U = \frac{dw}{dc} \; .. (118).$$

Hence, for an isotropic solid, (117) gives the expression (25) of § 21, which we have used above.

71.   To interpret the coefficients m and n in connexion with elementary ideas as to the elasticity of the solid; first let

$$a = b = c = 0, \text{ and } e = f = g = \tfrac{1}{3}\delta;$$

in other words, let the substance experience a uniform dilatation, in all directions, producing an expansion of volume from 1 to $1 + \delta$. In this case (117) becomes

$$w = \tfrac{1}{2}\left(m - \tfrac{1}{3}n\right)\delta^2;$$

and we have $\dfrac{dw}{d\delta} = \left(m - \tfrac{1}{3}n\right)\delta.$

Hence $\left(m - \tfrac{1}{3}n\right)\delta$ is the normal force per unit area of its surface required to keep any portion of the solid expanded to the amount specified by δ. Thus $\left(m - \tfrac{1}{3}n\right)$ measures the elastic force called out by, or the elastic resistance against, change of volume: and viewed as a *coefficient of elasticity*, it may be called the *elasticity of volume*. What is commonly called the " compressibility" is measured by $1/\left(m - \tfrac{1}{3}n\right)$.

And let next $e = f = g = b = c = 0$; which gives

$$w = \tfrac{1}{2}na^2; \text{ and, by (118), } S = na.$$

This shows that the tangential force per unit area required to produce an infinitely small shear*, amounting to a, is na. Hence n measures the innate power with which the body resists change of shape, and returns to its original shape when force has been applied to change it; that is to say, it measures *the rigidity* of the substance.

[Note added, December 1863.]

Since this paper was communicated to the Royal Society, the author has found that the solution of the most difficult of the problems dealt with in it, which is the determination of the effect produced on a spherical shell by a prescribed application of force to its outer and inner surfaces, had previously been given by Lamé in a paper published in *Liouville's Journal* for 1854, under the title " Mémoire sur l'Équilibre l'élasticité des enveloppes sphériques." In the same paper Lamé shows how to take into account the effect of internal force, but does not solve the problem thus presented except for the simple cases of uniform gravity and of centrifugal force. The form in which analysis has been applied in the present paper is very different from that chosen by Lamé (who uses throughout polar coordinates); but the principles are essentially the same, being merely those of spherical harmonic analysis, applied to problems presenting peculiar and novel difficulties.

---

* Thomson and Tait's *Natural Philosophy*, Part I. § 171 [Art. xcii. § 44 above].

(*Proceedings of the Royal Society of Edinburgh.* Read July 1 and 15, 1889.)

§ 1. THE scientific world is practically unanimous in believing that all tangible or palpable matter, molar matter as we may call it, consists of groups of mutually interacting atoms or molecules. This molecular constitution of matter is essentially a deviation from homogeneousness of substance, and apparent homogeneousness of molar matter can only be homogeneousness in the aggregate. " A body is called homogeneous when any two equal and similar parts of it, with corresponding lines parallel and turned towards the same parts, are undistinguishable from one another by any difference in quality*."   I now add that unless the " part " of the body referred to consists of an enormously great number of molecules, this statement is essentially the definition of crystalline structure.   It is, indeed, very difficult to imagine equilibrium, static or kinetic, in an irregular random crowd of molecules. Such a crowd might be a liquid,—I can scarcely see how it could be a solid.   It seems, therefore, that a homogeneous isotropic solid is but an isotropically macled crystal ; that is to say, a solid composed of crystalline portions having their crystalline axes or lines of symmetry distributed with random equality in all directions.   The proved highly perfect optical isotropy of the glass of object-glasses of great refracting telescopes, and of good glass prisms, seems to demonstrate that the ultimate molecular structure is fine-grained enough to let there be homogeneous crystalline portions, which contain very large numbers of molecules while their extent throughout space is very small in comparison with the wave-length of light.

§ 2. An ideal skeleton or framework for a homogeneous assemblage of bodies, or of material systems of any kind, or of qualities or properties of any kind, distributed periodically throughout space, is defined and explained in § 45 (*a*) to (*j*)

---

* Thomson and Tait's *Treatise on Natural Philosophy*, new edition, Vol. I. Part II. §§ 675—678; or *Elements of Natural Philosophy*, §§ 646—649 [Art. XCII. § 38 above].

below, substantially taken from Bravais' doctrine of homogeneous assemblages, which we may look upon as the grammar of
molecular construction.

SPACE-PERIODIC PARTITIONING (§§ 3—13).

§ 3. Given a homogeneous assemblage of points: let it be
required to partition all space accordingly. The thing to be done is
concisely defined in the second sentence of § 6 below.

§ 4. The problem is clearly indeterminate. Here is a solution
which has obvious relation to Brewster's kaleidoscope and the
corresponding doctrine of electric images, and which may be important in respect to Vortex Theory for a crystal or ether. From
P, a point of the given assemblage, draw a line, PN, of any length
in any direction, provided only that N is not a point of the assemblage of P's. Do the same relatively to every other of the
P-assemblage. We thus have a homogeneous assemblage of
double points, PN. Let Q be any point in space, and let Σ denote summation for all the PN's. Let $\phi(D)$ be a function which
decreases as D increases from 0 to ∞. The equation

$$\Sigma\left[\phi(QP) - \phi(QN)\right] = 0,$$

expresses a locus for Q which partitions space periodically, and
divides each periodic portion into two cells containing respectively
an N and a P. Every cell containing an N is a parallel pervert
(footnote on § 45 a below) of every cell containing a P. That this
is true we see by drawing any straight line to equal distances in
opposite directions through the point midway between N and P.
Its ends are similarly related, one of them to all the N's; the other
to all the P's.

§ 5. Here is a perfectly general solution. Around any one
of the points P describe a closed surface S, of which the greatest
distance from P is less than that of P's nearest neighbour. Describe an equal, homochirally similar, and same-ways oriented
surface around every other point P. None of these surfaces cuts
or touches any other. Expand all of them simultaneously, equally,
and without altering shape or orientation, till one of them touches
another. All corresponding pairs of the surfaces touch simultaneously at corresponding points. Continue the expansion,
annulling in each case the mutually enclosed portions of the

expanding surfaces, and substituting the portion of fixed surface
traced, or left behind, by the expanding line of mutual intersection.   This portion of surface we shall call (after my brother,
Professor James Thomson) an interface.   Follow the same rule
when another, another, and another contact takes places.   When
the borders of two of the growing interfaces thus traced meet
and begin to intersect, annul their projecting portions, so that
the intersection and what is left of the expansion of its previous
border now constitute the boundary of the interface.   Continue
the process until fresh growing intersections of interfaces are
formed, and the ends of these growing intersections meet, and at
last nothing is left of the expanded original surfaces, and therefore
nothing of space is left unenclosed by the cells—polyhedrons of
interfaces—thus constructed.

§ 6.   The interfaces formed in § 5 are generally curved, but
as we shall see (§ 7), may be plane, and are so in particular cases
of special interest.   In every case each cell contains one, and
only one, of the P's; there is no interstitial space between
them; they are all equal, homochirally similar, and con-orientational.

§ 7.   If the initiating surface, S, of § 5 is a polyhedron of plane
facets, the periodic partition to which it leads is in polyhedrons of
plane facets.   So it is also if the initiating surface is any ellipsoid
with P for centre.

§ 8.   Let S be a sphere.   The partitional polyhedron, to which
it leads, is the dodekahedron obtained by drawing planes through
the middle points of the lines between P and its twelve next-
neighbours, perpendicular to these lines.

§ 9.   If S is an ellipsoid similar to and con-orientational with
that determined in § 47 below, the partitional polyhedron to which
it leads is the rhomboidal dodekahedron to which the rhombic
dodekahedron of § 21 below is converted by the homogeneous
strain of § 46.   In this case the whole number of contacts of the
expanding surfaces (§ 5) is twelve, and they all take place simultaneously.

§ 10.   If the assemblage becomes equilateral, the partitional
dodekahedrons of §§ 8, 9 become, each of them, the rhombic dodekahedron of § 21.

§ 11. If S is an ellipsoid, having conjugate diameters along lines from P to other three points of the assemblage, and of magnitudes proportional to the distances from P to the nearest points in these lines, the partitional polyhedron to which it leads is a parallelepiped.

§ 12. If the three points chosen are nearest neighbours of P (§ 45i below), we are led to the *best conditioned* (or least oblique) of all the infinity of parallelepipedal partitions possible. This is the most obvious and the best known of the periodic partitions of space.

§ 13. Taking the parallelepipedal partitioning of § 11, let P' be the farthest corner from P, so that PP' is the longest diagonal of the parallelepiped. Let PA, PB, PC be conterminous edges and $A'P', B'P', C'P'$ their opposites conterminous in P'. Draw the planes $ABC, A'B'C'$. We thus divide the parallelepiped into three parts—an octohedron $ABCA'B'C'$; and two tetrahedrons, $PABC, P'A'B'C'$, which are parallel mutual perverts (footnote on § 45 a below). This grouping of eight points of a homogeneous assemblage is, as we shall see later, important in the dynamics of molecular structure, or at all events in Boscovich's theory.*

ON BOSCOVICH'S THEORY (§§ 14—44 and §§ 62—71).

§ 14. Without accepting Boscovich's fundamental doctrine that the ultimate atoms of matter are points endowed each with inertia and with mutual attractions or repulsions dependent on mutual distances, and that all the properties of matter are due to equilibrium of these forces, and to motions, or changes of motion, produced by them when they are not balanced; we can learn something towards an understanding of the real molecular structure of matter, and of some of its thermodynamic properties, by consideration of the static and kinetic problems which it suggests. Hooke's exhibition of the forms of crystals by piles

---

* Theoria Philosophiæ Naturalis redacta ad unicam legem virium in natura existentium, auctore P. Rogerio Josepho Boscovich, Societatis Jesu, nunc ab ipso perpolita, et aucta, ac a plurimis præcedentium editionum mendis expurgata. Editio Veneta prima ipso auctore præsente, et corrigente. Venetiis, MDCCLXIII. Ex Typographia Remondiniana superiorum permissu, ac privilegio.

of globes, Navier's and Poisson's theory of the elasticity of solids, Maxwell's and Clausius' work in the kinetic theory of gases, and Tait's more recent work on the same subject—all developments of Boscovich's theory pure and simple—amply justify this statement.

§ 15. Boscovich made it an essential in his theory that at the smallest distances there is repulsion, and at greater distances attraction; ending with infinite repulsion at infinitely small distance, and with attraction according to Newtonian law for all distances for which this law has been proved. He suggested numerous transitions from attraction to repulsion, which he illustrated graphically by a curve,—the celebrated Boscovich curve,—to explain cohesion, mutual pressure between bodies in contact, chemical affinity, and all possible properties of matter— except heat, which he regarded as a sulphureous essence or virtue. It seems now wonderful that, after so clearly stating his fundamental postulate which included inertia, he did not see intermolecular motion as a necessary consequence of it, and so discover the kinetic theory of heat for solids, liquids, and gases; and that he only *used* his inertia of the atoms to explain the known phenomena of the inertia of palpable masses, or assemblages of very large numbers of atoms.

§ 16. It is also wonderful how much towards explaining the crystallography and elasticity of solids, and the thermo-elastic properties of solids, liquids, and gases, we find without assuming more than one transition from attraction to repulsion. Suppose, for instance, the mutual force between two atoms to be zero for all distances exceeding a certain distance, I, which we shall call the radius of the sphere of influence; to be repulsive when the distance between them is $< \zeta$; zero when it is $= \zeta$; and attractive when it is $> \zeta$: and consider the equilibrium of groups of atoms under these conditions.

A group of two would be in equilibrium at distance ζ; and only at this distance. This equilibrium is stable.

A group of three would be in stable equilibrium at the corners of an equilateral triangle of sides ζ; and only in this configuration. There is no other configuration of equilibrium except with the three in one line. There is one, and there may be more than one, configuration of unstable equilibrium, of the three atoms in one line.

§ 17. The only configuration of stable equilibrium of four atoms is at the corners of an equilateral tetrahedron of edges ζ. There is one, and there may be more than one, configuration of unstable equilibrium of each of the following descriptions:—

(1) Three atoms at the corners of an equilateral triangle, and one at its centre.

(2) The four atoms at the corners of a square.

(3) The four atoms in one line.

There is no other configuration of equilibrium of four atoms, subject to the conditions stated above as to mutual force.

Important questions as to the equilibrium of groups of five, six, or greater finite numbers, of atoms occur, but must be deferred. The Boscovichian foundation for the elasticity of solids with no inter-molecular vibrations is the subject of §§ 62—71 below. A few preliminary remarks here may be useful.

§ 18. Every infinite homogeneous assemblage* of Boscovich atoms is in equilibrium. So, therefore, is every finite homogeneous assemblage, provided that extraneous forces be applied to all within influential distance of the frontier, equal to the forces which a homogeneous continuation of the assemblage through influential distance beyond the frontier, would exert on them. The investigation of these extraneous forces for any given homogeneous assemblage of single atoms—or of groups of atoms as explained below—constitutes the Boscovich equilibrium-theory of elastic solids.

§ 19. To investigate the equilibrium of a homogeneous assemblage of two or more atoms, imagine, in a homogeneous assemblage of groups of i atoms, all the atoms except one held fixed. This one experiences zero resultant force from all the points corresponding to itself in the whole assemblage, since it and they constitute a homogeneous assemblage of single points. Hence it must experience zero resultant force also from all the other $i-1$ assemblages of single points. This condition, fulfilled for each one of the atoms of the compound molecule, clearly

---

* "*Homogeneous assemblage of points, or of groups of points, or of bodies, or of systems of bodies,*" is an expression which needs no definition, because it speaks for itself unambiguously. The geometrical subject of homogeneous assemblages is treated with perfect simplicity and generality by Bravais, in the *Journal de l'École Polytechnique*, tome xix., cahier xxxiii. pp. 1–128 (Paris, 1850).

suffices for the equilibrium of the assemblage, whether the constituent atoms of the compound molecule are similar or dissimilar.

§ 20.   When all the atoms are similar—that is to say, when the mutual force is the same for the same distance between every pair—it might be supposed that a homogeneous assemblage, to be in equilibrium, must be of single points; but this is not true, as we see synthetically, without reference to the question of stability, by the following examples of homogeneous assemblages of symmetrical groups of points, with the condition of equilibrium for each when the mutual forces act.

§ 21.   *Preliminary.*—Consider an equilateral* homogeneous assemblage of single points, O, O', &c.   Bisect every line between nearest neighbours by a plane perpendicular to it.   These planes divide space into rhombic dodekahedrons.   Let A_1OA_5, A_2OA_6, A_3OA_7, A_4OA_8, be the diagonals through the eight trihedral angles of the dodekahedron inclosing O, and let $2a$ be the length of each.   Place atoms Q_1, Q_5, Q_2, Q_6, Q_3, Q_7, Q_4, Q_8, on these lines, at equal distances, r, from O; and do likewise for every other point, O', O'', &c., of the infinite homogeneous assemblage.   We thus have, around each point A, four atoms, Q, Q', Q'', Q''', contributed by the four dodekahedrons of which trihedral angles are contiguous in A, and fill the space around A.   The distance of each of these atoms from A is $a - r$.

§ 22.   Suppose, now, r to be very small.   Mutual repulsions of the atoms of the groups of eight around the points O will preponderate.   But suppose $a - r$ to be very small; mutual repulsions of the atoms of the groups of four around the points A will preponderate.   Hence for some value of r between zero and a, there will be equilibrium.   There may, according to the law of force, be more than one value of r between zero and a giving equilibrium; but whatever be the law of force, there is one value of r giving *stable* equilibrium, supposing the atoms to be constrained to the lines OA, and the distances r to be constrainedly equal.   It is clear from the symmetries around O and around A, that

---

* This means such an assemblage as that of the centres of equal globes piled homogeneously, as in the ordinary triangular-based, or square-based, or oblong-rectangle-based, pyramids of round shot, or of billiard-balls.

neither of these constraints is necessary for mere equilibrium; but without them the equilibrium might be unstable. Thus we have found a homogeneous equilateral distribution of 8-atom groups in equilibrium. Similarly, by placing atoms on the three diagonals, B_1OB_4, B_2OB_5, B_3OB_6, through the six tetrahedral angles of the dodekahedron around O, we find a homogeneous equilateral distribution of 6-atom groups in equilibrium.

§ 23.   Place, now, an atom at each point O. The equilibrium will be disturbed in each case, but there will be equilibrium with a different value of r (still between zero and a). Thus we have 9-atom groups and 7-atom groups.

§ 24.   Thus, in all, we have found homogeneous distributions of 6-atom, of 7-atom, of 8-atom, and of 9-atom groups, each in equilibrium. Without stopping to look for more complex groups, or for 5-atom or 4-atom groups, we find a homogeneous distribution of 3-atom* groups in equilibrium by placing an atom at every point O, and at each of the eight points A_1, A_5, A_2, A_6, A_3, A_7, A_4, A_8. There are four obvious ways of seeing this, found by choosing one or other of the four diagonals through trihedral angles referred to in § 21. Take, for example, A_1OA_5, and its congeners for all the dodekahedrons. These triplets include all the A's. (Compare § 25 below.)

§ 25.   Lastly, choosing A_2, A_3, A_4, so that the angles A_1OA_2, A_1OA_3, A_1OA_4, are each obtuse†, we make a homogeneous assemblage of 2-atom‡ groups in equilibrium by placing atoms at O, A_1, A_2, A_3, A_4. There are four obvious ways (compare § 24 above) of seeing this as an assemblage of di-atomic groups, one of which is as follows :—Choose A_1 and O as one pair. Through A_2, A_3, A_4, draw lines same-wards parallel to A_1O and each equal to A_1O. Their ends lie at the centres of neighbouring dodekahedrons, which pair with A_2, A_3, A_4 respectively.

§ 26.   For the Boscovich theory of the elasticity of solids, the consideration of this homogeneous assemblage of double atoms is

---

* This is the assemblage described in the footnote on § 69 below.

† This also makes A_2OA_3, A_2OA_4, and A_3OA_4 each obtuse. Each of these six obtuse angles is equal to $180° - \cos^{-1}(1/3)$.

‡ This is the assemblage described in § 69 below, and used in §§ 67, 68, 70.

very important. Remark that every O is at the centre of an equilateral tetrahedron of four A's; and every A is at the centre of an
equal and similar, but contrary-ways oriented, tetrahedron of O's.
The corners of each of these tetrahedrons are respectively A, and
three of its twelve nearest A-neighbours; and O and three of its
twelve nearest O-neighbours. By aid of an illustrative model
showing four of the one set of tetrahedrons with their corner
atoms painted blue, and one tetrahedron of atoms in their centres
painted red, the mathematical theory which had been communicated to the Royal Society of Edinburgh, was illustrated to Section
A of the British Association at its recent meeting in Newcastle.

§ 27. In this theory* it is shown that in an elastic solid
constituted by a single homogeneous assemblage of Boscovich
atoms, there are in general two different rigidities, n, n_1, and one
bulk-modulus, k; between which there is essentially the relation

$$3k = 3n + 2n_1,$$

whatever be the law of force. Here n_1 denotes what are called
the diagonal rigidities, and n the facial rigidities relative to the
primitive cube of § 53 below. By facial and diagonal rigidities
relative to any given cube I mean rigidities defined in the usual
manner†, one of them according to shearing parallel to any face
of the cube, the other according to shearing in planes parallel to
any plane-diagonal of the cube.

§ 28. A remarkable result of my mathematical investigation
is, that the facial rigidity, relatively to the primitive cube of § 52,
is double the diagonal rigidity in the case in which each atom
experiences force only from its twelve nearest neighbours. The
law of force may be so adjusted as to make $n_1 = n$; and in this
case we have $3k = 5n$, which is Poisson's relation. But no such
relation is obligatory when the elastic solid consists of a homogeneous assemblage of double, or triple, or multiple Boscovich
atoms. On the contrary, any arbitrarily chosen values may be
given to the bulk-modulus and to the rigidity, by proper adjust-
ment of the law of force, even though we take nothing more
complex than the homogeneous assemblage of double Boscovich
atoms above described.

---

* See §§ 62—71 below.

† Thomson and Tait's *Natural Philosophy*, 2nd ed., Vol. I. Part II. § 680; also
reprint of *Mathematical and Physical Papers*, Vol. III. Art. XCII. Part I.

Boscovichian Kinetic Theory of Crystals, Liquids, and Gases.

§ 29. The most interesting and important part of the subject, the kinetic, must, for want of time, be but slightly touched in the present communication. I hope to enter on it more fully in a future communication to the Royal Society of Edinburgh.

§ 30. To avoid circumlocutions, I shall call any velocity *moderate*, which is comparable with the maximum velocity acquired by two atoms attracting one another from rest, at distance I. It is the velocity that in the circumstances each would have when their distance becomes diminished to ζ When I speak of atoms or groups moving "rapidly," I mean that the velocities are moderate as thus defined.

§ 31. Let us consider what would follow if we had given at any time, scattered randomly but equably all through space, simple Boscovich atoms moving with velocities randomly equal in all directions. As we are supposing the masses of all the atoms equal, we may call the mass of each unity: thus $\Sigma\frac{1}{2}v^2$ for all the atoms in any part of space at any time, is the total of their kinetic energy. Both the number of atoms and their total energy we shall suppose to be equal in all very large equal volumes.

§ 32. The result of a collision between two atoms is essentially the same as that of the collision of two equal balls supposed simply repellent at contact, as in the elementary kinetic theory of gases as worked out by Maxwell and Tait*; but the size of the balls that would give the same result depends, for each collision, very complexly on the law of force, and on the velocities and lines of motion of the atoms before the collision. As long as there is no case of collision between more than two atoms, the average energy of the free atoms at any time, and the law of the distribution of energy among the multitude in their free paths between collisions, is not affected by this complication, and is the same as if the atoms were equal hard globes merely repellent at contact. It is only when the results of unequal distributions of density, of energy, or of components of momentum, are to be traced, and the

* Maxwell, *Philosophical Magazine*, 1860, and *Philosophical Transactions*, 1867 and 1878; Tait, "On the Foundations of the Kinetic Theory of Gases," *Trans. Roy. Soc. Edin.*, Vol. xxxiii., read May 14 and December 6, 1886, and January 7, 1887.

laws of the relation of pressure to density, or of thermal conduction, or of viscosity are to be investigated, that we can take
into account the law of force, and can find differences from what
the results would be if we had merely the hard equal balls to deal
with.

§ 33. But now suppose, while two atoms are in collision, a
third to come within their influential distance, so that three shall
be in collision at the same time. All three *may* go clear, or two
of them may remain in collision, or in other words, fall into
combination, and *one* go free. It is scarcely possible that all three
can remain in collision—that is to say, can combine. It will certainly be a very rare incident that they remain for any considerable
time in collision; but I cannot prove that the case may not occur
in which none will go free, and the three will remain in combination.

§ 34. If the initially-given velocities are very great, the general
result, even of triple collisions, will be to leave the individual
atoms free. The comparatively rare double atoms resulting from
triple collisions, and the still rarer triplets, will be liable to be
separated again into single atoms by all fresh collisions. This is
the case of a perfect monatomic gas, at a temperature much higher
than the Andrews' critical point.

§ 35. But if the originally-given velocity be exceedingly small,
the result of exceedingly nearly every triple collision will be to
form a combination of at least two of the three colliding atoms.
Immediately after the collision by which it was formed, each
doublet will generally have considerable relative motion of its two
atoms; that is to say, the two will describe orbits round their
common centre of inertia: or, in the extreme case of no moment
of momentum round this point, they will oscillate relatively to
their centre of inertia to and fro in a straight line; the centre of
inertia itself generally having a considerable velocity. Still supposing the average velocities of the free atoms to be very small,
and their number to be very great in comparison with that of the
double atoms, we now see that the general effect of the collisions
between double and single atoms must be to diminish the energies
of the relative and absolute motions of the constituents of the
doublets, and so reduce the doublets more and more nearly to

the condition of pairs of atoms in relative equilibrium (§ 16 above), at distance ζ asunder, with centre of inertia of each pair moving very slowly through space.

§ 36. But now consider the effect of a collision between two doublets each with little or no intestine commotion before the collision, and with its centre of inertia moving very slowly through space. The case in which the same description would be applicable to the four atoms after the collision, whether in the same pairs or in interchanged pairs, would be exceedingly rare. So also would be the case of the four atoms remaining combined. The result in exceedingly nearly every case would be a triplet with considerable intestine commotion, and its centre of inertia moving rapidly through space, and a single atom moving rapidly through space. The general tendency of subsequent collisions between these rapidly-moving triplets and single atoms, with the multitude of slowly-moving single atoms throughout space, would be to diminish the energy of the intestine commotions of the triplets, and of the motions of the centres of inertia, both of the triplets and of the single atoms, reducing each triplet to very nearly the condition of equilibrium (§ 16 above) at the corners of an equilateral triangle of side ζ with a slow translatory motion through space.

§ 37. By similar dynamical considerations we see that the general tendency of collisions between doublets and triplets, or between triplets and triplets, must be to form quartets, quintets, and sextets of atoms; and that when such groups, carrying away large kinetic energies from the generative collisions, subsequently collide with slowly-moving single atoms, the general tendency must be to diminish their kinetic energies, and reduce them more and more nearly to groups in one or other configuration of equilibrium, with slow motion of their centres of inertia through space.

§ 38. But now consider a collision between a slowly-moving triplet or quartet or more-multiple group, and a slowly-moving single atom. Even with the triplet the case will not be rare in which the single atom will remain in combination, and the result yielded be a quartet having considerable intestine commotion, and moving slowly through space. In collisions between a quartet and a single atom, the case will be relatively less rare, and with a quintet and single atom, still less rare for a single atom to remain in combination, and form a quintet or a sextet.

§ 39. If groups of large numbers of atoms in equilibrium, or slowly vibrating, have been thus formed, or are given ready formed, with single atoms slowly moving in the space all around them, each single atom colliding with a group will very frequently remain in the group; and in virtue of the exhaustion of potential energy thus effected the vibrational energy of the group will be slightly augmented. But in not rare cases either the single atom which collided, or one of the atoms of the group in the neighbour-hood of the collision, will be driven off, and generally with much greater velocity than the colliding atom had before the collision. Thus the average kinetic energy of vibration per atom of the group may be kept constant, while the group is gaining by the accession to it of more and more single atoms from without. But the exhaustion of potential energy due to the greater number falling into, than being thrown out from, the group would cause an augmentation of kinetic energy in the surrounding atmosphere of free atoms. To obviate this, let the atmosphere around the group be contained in a finite closed vessel, which, when left to itself, repels each atom that comes near enough to it, and sends it back inwards with unchanged energy. Now let portions of this bounding surface be movable, and let them be so moved by proper external appliances, that work shall be done upon them by the impinging atoms to just such a degree as to keep the average kinetic energy of the free atoms constant. We have thus a Boscovichian realization of a crystal of ice (hoar-frost) or other substance growing by condensation of a surrounding atmosphere of the same substance. The process in nature requires the ab-straction of what is called the latent heat of the vapour to allow it to condense. This in our Boscovichian system is performed by the arrangement for letting work be done outwards by the moving parts of the boundary.

§ 40. Even if there were no surrounding atmosphere of single moving atoms, our group, unless quite free from intestine com-motion, would occasionally throw off an atom in virtue of the chance concurrence of different sets of component vibrations at some of the outlying atoms. Now let there be just enough of atoms moving about in the space around the group to cause as many fallings-in as throwings-out of atoms, and with just enough of kinetic energy to neither gain nor lose energy in the surrounding atmosphere through these changes. This will also cause the average kinetic energy of the group to remain constant. Thus we have a crystal surrounded by an atmosphere of vapour at its own temperature, and at the proper temperature to cause neither condensation of the vapour nor evaporation of the solid.

§ 41. Now by somehow applying force to the atoms of the group increase their vibrational energy. We must, by introducing atoms from the boundary, increase the density of the atmosphere around it to cause as many atoms to enter the group as are thrown off from it. Continue this process until the inter-atomic oscillations in the group become so great that the atoms begin to pass from one configuration of equilibrium to another, and back ; as, for instance, the two configurations of § 46 (footnote) below. The group may still retain its form as a solid, and something of its rigidity as a solid.

§ 42. Now reverse the operations at the boundary so as to diminish the inter-atomic oscillatory energy of the group. The atoms *may* fall back into their previous positions of equilibrium. But they may *not;* and instead they may fall into another configuration more readily taken in a settlement from internal agitation than the previous configuration which was arrived at by growth from the boundary. This (with true molecules of matter instead of the ideal Boscovich atoms) seems to me, without doubt, the explanation of Madan's* beautiful discovery regarding chlorate of potash, and the change of crystalline structure by which Lord Rayleigh† has shown that the optical phenomena presented in it are to be explained. Virtually the same view to explain other changes of crystalline structure by differences of temperature or applications of pressure seems to have been given by M. Mallard‡, who is quoted by Madan in the article above referred to. In a future communication to the

---

* "On the Effect of Heat in Changing the Structure of Crystals of Potassium Chlorate," *Nature*, May 20, 1886.

† *Philosophical Magazine*, 1888.

‡ *Bulletin de la Société Minéralogique*, 1882, and December 1885.

Royal Society I hope to include considerations regarding the effect of inter-atomic forces and motions in guiding to one or other of the two configurations described in § 54 and footnote on § 46 below.

§ 43.   Once more communicate and continue communicating energy to the group by forces applied directly to its constituent atoms, and, at the same time, keep introducing fresh atoms from the outer boundary into the atmosphere surrounding the group to prevent the number of atoms in the group from diminishing. The intestine commotion will become so great that all configurations of equilibrium are utterly departed from, but still the atom is surrounded by neighbours well within the region of its attractive influence (a shell bounded by two concentric surfaces of radius I and ζ respectively) and constantly crossing and recrossing the spherical surface of radius ζ, or into and out of the sphere of repulsive force.   If the region of attractive force be sufficiently thick, and the augmentation of the repulsive force from zero towards infinity be sufficiently rapid, it is clear that our original group which was a crystal and is now fluid will remain more dense than the surrounding atmosphere of free atoms until we have imparted to the group far more of energy than was required to dislodge its constituent atoms from configurations of equilibrium.   There then is a mass of liquid surrounded by an atmosphere of its vapour, and in thermal equilibrium with the vapour if we cease the action on its atoms by which we imparted energy to it.   A little farther consideration would no doubt give us the virtual surface-tension of the liquid exactly according to Laplace's theory of capillary attraction; but we must not pause over this at present.

§ 44.   Recommence applying forces to the atoms of the group, now liquid, and introducing fresh atoms into the surrounding atmosphere.   The density of the atmosphere becomes greater, while that of the group becomes less.   Go on till the two densities become equal: thus we reach the Caignard de la Tour and Andrews' critical point.   If we continue now imparting energy to our original group, or to any of the atoms of the assemblage, we simply have a homogeneous assemblage in a state of homogeneous intestine commotion all through; the

Boscovish realisation of a fluid raised higher and higher above its critical temperature.

ON MOLECULAR TACTICS OF CRYSTALS AND OF THE ARTIFICIAL TWINNING OF ICELAND SPAR (§§ 45—60).

§ 45. (a)......(j).   *Summary of Bravais' Doctrine of a Homogeneous Assemblage of Bodies.*

(a)   The bodies must be equal, similar, and homochiral*.

(b)   They must be all similarly oriented.

(c)   They must be so distanced mutually that any point in one of them, and the corresponding points in all the others form a homogeneous assemblage of points. If this condition is fulfilled for any one chosen point of one body, (a) and (b) imply it for any other ; and *vice versa* if this condition is fulfilled for three points of one body chosen arbitrarily but not in one line, (b) is a necessary consequence.

(d)   A homogeneous assemblage of points means, and cannot mean other than, an assemblage which presents the same aspect and the same absolute orientation when viewed from different points of the assemblage. Some confusion of ideas has been introduced by leaving the generalised simplicity of Bravais, and considering an assemblage of double points, or triple points, or

---

* This will be more easily and not less thoroughly understood from illustrations than from a definition in general terms. Of an externally symmetrical man, the two hands are *allochirally* similar. Either is the *pervert* of the other; or they are *mutual perverts*. Two men of exactly equal and similar external figures would be allochirally similar if *one* holds out his right hand and *the other* his left; homochirally similar if each holds out his right hand, or each his left. (We ignore at present the monochiral anti-symmetry of one heart on one side; of interior structure of intestinal canal *not* in the plane bisecting the exterior symmetric figure, &c. &c.) Looking to § (i) below, we see two tetrahedrons, $OPQR$, $OP'Q'R'$, which are equal, and allochirally similar, being parallel perverts, either of the other, or parallel mutual perverts. From every point P of a body or group of points, draw a line through any one point O, and produce to P', making $OP' = PO$. The group of points (P') is a parallel pervert of the group (P). The groups (P) and (P') are parallel mutual perverts. Turn (P') 180° round any line OK. In the position thus reached, it is the image of (P) in a plane mirror through O, perpendicular to OK. In their present positions they are mutual perverts inverted relatively to the line OK. Mutual perverts are allochirally similar.

quadruple points, without noticing its being resolvable into two, or three, or four similar homogeneous assemblages of single points.

(e) *Rows of Points in a Homogeneous Assemblage.*—Through any two points of the assemblage draw a straight line, and produce it indefinitely in both directions. All points on this line at intervals successively equal to the distance between the two chosen points, are points of the assemblage. The interval between each point and the next to it on either side in the line is called by Bravais the *parameter* of the row.

(f) *Planes of Points* ("réseaux") *in a Homogeneous Assemblage.*—Take at random any three points of the group. The case of there being other points of the assemblage on the sides or within the area of the triangle of the chosen points may be excluded. Along the line of each side of the triangle produced in both directions, mark off in succession lengths equal to the side, and through each division draw parallels to the other two sides. The plane of the triangle extended indefinitely in all directions is thus divided into equal and homochirally similar triangles turned alternately in opposite directions. At every angle of each of these triangles a point of the assemblage is

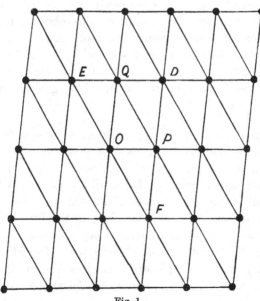

Fig. 1.

found. No point of the assemblage is to be found elsewhere in the same plane. Fig. 1 shows a homogeneous distribution of points in a plane. In the diagram they are joined by lines, determinately chosen according to § (*i*), so that all the angles of triangles formed by them are acute. Closely related to this triangular arrangement are three others. One of these is obtained by omitting PQ and its parallels and taking instead the other diagonal OD of the parallelogram $QOPD$ and drawing parallels to it through all the points. The two others are obtained similarly by omitting OQ and taking instead the other diagonal PE of the parallelogram $QPOE$; and by omitting OP and taking instead the other diagonal QF of the parallelogram $PQOF$.

(*g*) All the points of the assemblage lie in equidistant planes parallel to the plane of § (*f*); similarly placed at the angles of triangles equal, similar, and similarly oriented to the triangles of § (*f*). The distance between each of these planes, and the next plane to it, is easily proved to be equal to the reciprocal of the product of twice the area of the triangle into the number of points per unit volume. In fig. 2 the points $PQOP'Q'$ and their congeners represent a homogeneous distribution in one plane. The orthogonal projection on this plane of the points in the two nearest parallel planes are represented respectively by R and its congeners, black dots (•), and by R' and its congeners, white dots (◦) Thus explained, the diagram (fig. 2) is a complete specification of the whole homogeneous assemblage throughout space.

(*h*) *Tetrahedronal Grouping* —Choose any one of the triangles OPQ, and any point S in the nearest plane of points on either side of it; and imagine a tetrahedron of which these $OPQS$ are the four corner points. By similarly dealing with all the triangles of all the planes, con-orientational with the first chosen triangle, and the points corresponding to the first chosen point in the neighbouring plane, we form a homogeneous assemblage of equal homochirally similar, samely oriented, tetrahedrons. Thus, for example, take the triangle FGH which is con-orientational with QOP. The tetrahedron on the base FGH corresponding to $SQOP$ is $RFGH$.

Each point of the distribution is the common corner point of eight of those tetrahedrons; of which the twelve edges meeting

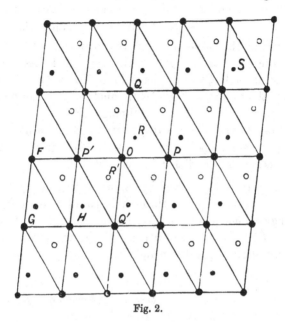

Fig. 2.

in it lie in the lines of six rows of points which intersect in that point.

(i) *Best conditioned Tetrahedronal Grouping. No Obtuse Angles.*—Instead of choosing our first two points and our first triangle at random, take any point O and its nearest neighbour on either side, P; and its next-nearest neighbour Q on the side making the angle QOP acute. The two other angles of this triangle are obviously, as Bravais remarks, acute. The only other way of thus finding *best conditioned* triangles is by taking O's other nearest neighbour, P', and its other next-nearest, Q'. The triangles $Q'OP'$ and QOP are equal, homochirally similar, and oppositely oriented; and thus we find the only other possible best conditioned triangular grouping. Every other triangle of the points in the same plane, having none of the points within its area, has, as Bravais remarks, an obtuse angle. Consider now the nearest parallel plane of points on one side of the plane of QOP. Let R and its con- geners (• black dots) be the orthogonal projections of its points on the plane of QOP. Let R' and its congeners (∘ white dots) be the projections of the points of the nearest parallel plane on the other side of QOP. These projections will be situated relatively to the triangle $P'OQ'$ and its congeners as are the former projections (• black dots) relatively to the triangle QOP.

R being, of the projections on the plane of POQ of all the points of the two parallel planes, the one which lies within the area of the triangle QOP, we have in $OPQR$ a best conditioned tetrahedronal grouping. $OP'Q'R'$ is another and the only other best conditioned tetrahedronal grouping. It is a parallel pervert of $OPQR$ [see footnote on § 45 (a) above]. Hence a homogeneous assemblage of single points is essentially free from monochiral anti-symmetry; or it is dichirally symmetrical.

(j) The tetrahedron found by taking, with O, P, Q, any other point than R in the plane through it parallel to QOP, has an obtuse angle along one, or obtuse angles along two, of its three edges, OP, PQ, QO: and so with O, P', Q', and any other point than R' in the other parallel plane.

Closest Packing of one Homogeneous* Assemblage of Equal and Similar Globes or Ellipsoids.

§ 46.   Take our tetrahedron $OPQR$, and by homogeneous distortional strain convert it into an equilateral tetrahedron $ABCD$,

---

* There is another closest packing of globes or ellipsoids which has the same density as, and might without careful attention be mistaken for, the closest *homogeneous* packing. For simplicity think only of globes, and take a plane covered with globes touching one another in equilateral triangular order. Look at the accompanying diagram, fig. 6 of § (55) below, and see that there are two ways of placing a second layer on the first to continue the formation of an assemblage. The globes of the second layer may be placed, all of them over the black dots (•) or all of them over the white dots (○). But having once chosen the position of the second layer there is no more freedom to choose in adding on layer after layer *if we are to make a single homogeneous assemblage*. Of the two positions which might be chosen for the third layer we must choose the one in which the globes are *not* over the globes of the first layer. The position of the fourth layer must be the one of which the globes are *not* over the globes of the second layer, but *are* over those of the first layer, and so on.

If on the contrary we place the globes of the third layer over the globes of the first, the globes of the fourth layer over those of the second and so on, we have a peculiar and symmetrical grouping which was first, so far as I know, described by Mr William Barlow (*Nature*, December 20 and 27, 1883). This grouping is not one homogeneous assemblage. It consists of two homogeneous

of equal volume. Take four globes, of diameters equal to the edges of this tetrahedron and place them with their centres at its corner points A, B, C, D. Alter this assemblage of globes by homogeneous strain till their centres, $ABCD$, become again the corner points of the original tetrahedron, $OPQR$. The globes have now become ellipsoids. Dealing thus with the whole original homogeneous assemblage of points, we find a closest packed homogeneous distribution of equal and similar ellipsoids through space.

§ 47. To find every possible closest packed homogeneous assemblage of given equal and similar ellipsoids, take a tetrahedron of four equal globes. Choose any three mutually perpendicular directions, and, by elongations and shrinkages of the group parallel to these directions, convert each globe into an ellipsoid equal and similar to the given ellipsoid. Every possible configuration of closest homogeneous packing of the given ellipsoids is clearly to be thus found; and is specified in terms of three independent variables,—the three orientational coordinates, relative to the equilateral tetrahedron of the system of rectangular lines.

§ 48. In §§ 46, 47 we have a solution of the problem *given four points, O, P, Q, R, not in one plane, to place con-orientationally four equal and similar ellipsoids with their centres at the four points, and the surface of every one touching the surface of each of the three others.* From it we have the following perfectly simple construction for the answer. Bisect OP, OQ, OR, in F, G, H, and QR, RP, PQ, in F', G', H', and join FF', GG', HH'. These three lines meet in one point S. The planes GSH, HSF, FSG are parallel to conjugate diametral planes of the required ellipsoids. These ellipsoids touch one another in the points F, F', G, G', H, H'. To construct them, first make four parallelepipeds, having a common corner at S, and their half-edges which meet in S, and their centres, as follows:—

assemblages, one of them constituted by the first, third, fifth, seventh, &c., layers; the other the second, fourth, sixth, eighth, &c., layers. The consideration of this peculiar mode of grouping may be of great interest in the dynamical investigations connected with the subject of the present communication, and, as Barlow has pointed out, may be of great importance in the theory of natural crystalline structure. I must however leave it for the present.

Half-edges.	Centres.		Half-edges.	Centres.
SF			*SH'*	
SG	} *O*		*SF''*	} *Q*
SA			*SG*	
SG'			*SF''*	
SH'	} *P*		*SG'*	} *R*
SF			*SH*	

Inscribe within the twelve edges of each parallelepiped an ellipsoid, touching them at their middle points. This construction is interesting as showing, in the middle points of the twelve edges of the parallelepiped, the twelve points of contact of the ellipsoid with its twelve next neighbours.

The ellipsoid touching the twelve edges is, it need scarcely be remarked, similar to the inscribed ellipsoid touching the surfaces, but of $\sqrt{2}$ times the linear dimensions.

§ 49. To understand the configuration of a closely packed homogeneous assemblage of ellipsoids, it is convenient to consider the assemblage of globes to which it is reduced by strain (geometrical distortion), in § 46. The assemblage of ellipsoids has all characteristic features the same, except the inequalities of lines and angles involved in the distortional transition from one configuration to the other.

§ 50. In the close homogeneous assemblage of globes, we may first remark, that each globe is touched by its neighbours, at twelve points, being the points in which its surface is cut by diameters parallel to the six edges of the tetrahedron. If we place a number of small globes (boys' marbles, or billiard balls), on a table in close triangular order, and three as close as they can be together above them, we see nine of the twelve points of contact on the ball below the middle of the triangle of these three; six points on the circle in which it is cut by a horizontal plane through its centre, and three symmetrically ranged on a small circle above it. The other ends of the diameters* through

---

* In the compound assemblage of two homogeneous assemblages described in the preceding footnote, there are twelve points of contact on each globe, of which nine are placed as those described in the text for the homogeneous single assemblage, and the remaining three are not "at the other ends of the diameters" as described in the text, but are at the opposite points of the small circle on which lie the ends of the diameters referred to.

these three are the remaining three of the twelve. Or if we join
the upper three by great circles, making a spherical triangle of

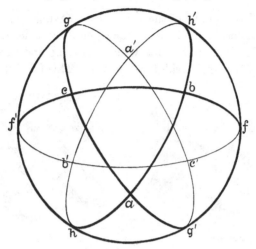

Fig. 3.

60° side, and complete these circles, they make another spherical
triangle of 60° side, whose angular points are the lower three of
the twelve contact points. The three great circles thus drawn cut
the horizontal great circle in the first six points. Thus we see
that the twelve points are the intersections of four great circles,
which divide the spherical surface into eight equilateral triangles,
and six squares; all with arcs of 60° for boundaries. Fig. 3 shows
an orthogonal projection of these circles on the plane of one of
them; each an ellipse whose minor axis is ⅓ of its major axis.
The eight equilateral spherical triangles are *abc, ahg', bfh', cgf',
a'b'c', a'h'g, b'f'h, c'g'f.* The six squares are *bcgh', cahf', abfg',
b'c'g'h, c'a'h'f, a'b'f'g.*

§ 51. Draw planes through the centre of the sphere, parallel
to the pairs of planes of the angular points of the eight spherical
triangles; these are four planes, *the* four planes in which the
assemblage is found in close triangular order. They are parallel
to the sides of the tetrahedron *ABCD.*

§ 52. Draw planes through the centre of the sphere, parallel
to the pairs of planes of the angular points of the six spherical
squares; these are three planes, *the* three planes, in which the
assemblage is found in square order. They are parallel to the

pairs (AB,CD), (AC,BD), (AD,BC) of the edges of the tetrahedron; and are mutually orthogonal.

§ 53. Take a cube of the assemblage, having its sides parallel
to the planes of § 51. It will present on every side, arrangement
of the globes in square order, with *rows* along and parallel to the
diagonals of the square sides of the cube. This I call the primitive

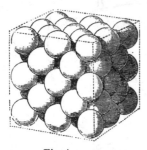

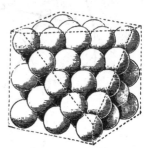

Fig. 4.   Fig. 5.

cube of a homogeneous assemblage of closely packed globes. It
is seen in fig. 4 taken from a paper published in *Nature* (Dec. 20,
1883), by Mr Barlow who, so far as I know, was the first to show
explicitly, a cubic part of the close-packed homogeneous assemblage
of equal globes.

§ 54. Bevel the corners of the primitive cube perpendicularly
to its four line-diagonals as shown for one only of the corners
bevelled in fig. 5, which also is taken from Mr Barlow's paper.
We thus get eight equilateral triangular facets, each showing close
triangular grouping of the globes appearing in it. The four pairs
of planes of these facets are, of course, parallel to the four faces of
the tetrahedron, $ABCD$. If we make the bevelling of each corner
deep enough, nothing is left of the cube but a regular octohedron,
whose eight faces are parallel to the four faces of the tetrahedron.

§ 55. If in building a triangular pyramid we commence with
globes in close triangular order on a horizontal plane, and place
the second layer above it over the white dots (○) of the diagram
(fig. 6), the third layer over the inner triangle of black dots (•),
and the fourth a single globe over the centre of the diagram, we
build up precisely the portion bevelled off the primitive cube in
§ 54. Thus we have a triangular pyramid whose three sides are

isosceles right-angled triangles meeting at right angles along the
three slant edges. The globes in these three faces are in square

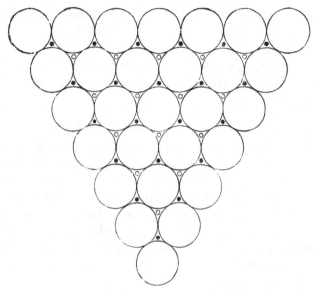

Fig. 6.

order. The lines of globes in contact in these faces are parallel
and perpendicular to the bounding edges of the base. In the
pyramid corresponding to the actual diagram, or any other with
an odd number of globes in each edge of the base, there are three
lines of globes in contact along the lines bisecting the three vertical
angles of the sides of the pyramid and ending in a single crowning
globe.

§ 56. If instead of building the second layer as in § 55, we
place a second layer over all the black dots (•), a third layer
over all the white dots (∘), a fourth layer over centres of
globes of the first layer, a fifth over black dots (•) again; a
sixth over white dots (∘), and the last a single globe as in
§ 55, we make an ordinary triangular pyramid having three
equilateral triangles for its slant sides and a fourth for base;
and having the globes arranged in equilateral triangular order
not only in the base as in § 55, but also in each of the three slant
sides.

§ 57. The ordinary square pyramid of globes has for its base
the same square order structure as the slant sides of the triangular pyramid of § 55, while its four slant sides have the same
equilateral triangular structure as each of the three slant sides
and the base of the pyramid of § 56. If we divide the ordinary
square pyramid into four parts by two diagonal vertical planes
through its centre, and turn one of these parts over till it rests
on its triangular slant side it becomes the triangular pyramid
of § 55.

§ 58. In considering Baumhauer's splendid discovery of the
artificial twinning of Iceland spar, by means of a knife, published
about 22 years ago, soon after Reusch's fundamental discovery
(1867) of the artificial twinning of Iceland spar by pressure, I
endeavoured to picture to myself the molecular tactics called into
play in the wonderful change of shape thus produced. It was
necessary first to suppose known the molecular arrangement in
the natural crystal. Two distinct hypotheses presented themselves, each perfectly definite; and it seems certain that the
structure is one or other of these two.

Hypothesis (1). Imagine an equilateral tetrahedron of a close
packed homogeneous assemblage of globes. To avoid circumlocution let one of its faces rest on a horizontal plane. Let the
whole system be shrunk homogeneously in lines perpendicular
to this plane till the originally acute trihedral angle of the
triangular pyramid of globes becomes the obtuse trihedral angle
of the rhomb of Iceland spar. The shrinkage ratio required to
do this would be exactly $\sqrt{8}$ to 1 if the inclination of each slant
face to the base were exactly 45°* in the triangular pyramid
obtained by truncating the obtuse trihedral angle of Iceland spar
perpendicularly to the "axis" (or line equally inclined to the
three edges meeting in the trihedral angle).

Hence if, instead of globes to begin with we take oblate
ellipsoids of revolution, each having its equatorial diameter
$\sqrt{8} (= 2\cdot83)$ times its polar axis, and make a pyramid of them

* At ordinary temperatures the angle is 44° 36'·6 (Phillips, Brooke, and
Miller's *Mineralogy*, § 407); and at temperature 300° it is almost exactly 45°.
Huyghens must have taken it as exactly 45°, as he gave $\sqrt{8}$ for the ratio of
the equatorial to the polar diameter in the statement of his hypothesis.

by laying a number of them flat on a horizontal plane and putting them together and building others up on them according to the rule of § 56, we have an obviously conceivable structure for Iceland spar. This is Hypothesis (1).

This Hypothesis I now find was given 200 years ago by Huyghens in his *Traité de la Lumière* (Leyden, 1690), and independently by Wollaston in the Bakerian Lecture for 1812, *Philosophical Transactions Royal Society* for the year 1813, Part I., but with priority attributed to Huyghens. I had thought of it independently, but did not feel altogether satisfied with it, in the first place because of the great internal commotion which it would imply in the tactics of Baumhauer's twinning. Then it occurred to me to think of the subject thus. It seems as if the æolotropic quality of Iceland spar, according to which there are differences of quality for directional actions along and perpendicular to the shortest line-diagonal of the rhomb, may be naturally supposed to depend on the rhomb not being a cube; and that the change from a cube to the Iceland spar rhomb should be looked to as the cause of the æolotropy. If this is so we must begin with a cube which is isotropic in respect to its four line-diagonals. This is the case with the cube described in § 53, but it is not the case with the cube which we find if in the shrinkage* of Hypothesis (1) we pause at the stage in which the acute trihedral angle of the equilateral tetrahedron is rectangular on its way to becoming obtuse; on the contrary, in this configuration each globe is an oblate with equatorial diameter twice as long as the polar axis. Hence I have been led to think it probable that the molecular structure of Iceland spar is not that of Hypothesis (1), but is as follows.

§ 59. *Hypothesis* (2).—Take a primitive cube § 53 of the assemblage and distort it by shrinkage along any one of its four line-diagonals, with (for simplicity) no change of length in

---

* The shrinkages to pass from the equilateral triangular pyramid to the pyramid with rectangular vertex and to the triangular pyramid for Iceland spar, will be understood in a moment by remarking that the tangents of inclinations of slant sides to base in the three cases are respectively $\sqrt{8}$, $\sqrt{2}$, and 1; and therefore the distances of vertex from base are as these numbers, the base being unchanged in the simple shrinkage specified in the text.

directions perpendicular to it. This reduces each globe to an
oblate ellipsoid of revolution, and the cube to a rhomb, which
is the rhomb of Iceland spar if the shrinkage-ratio is √2 to 1.

§ 60   Let *FG* (fig. 7) be one edge and *G* one of the two
obtuse trihedral angles of a rhomb of Iceland spar; and *F′*, *G′*

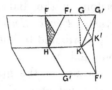

Fig. 7.

the corners opposite to *F*, *G* (so that *G′G* is the optic axis). Let
HKK′H′ be a diagonal plane parallel to *FG* and *F′G′*; *HK*, *H′K′*
being parallel to *FG*, *F′G′*. Now consider rows of oblates parallel
to *KK′* and *HH′*. The oblates are in contact at the ends of
equatorial diameters in the lines of these rows. Turn the oblates of each row round the line of the row, in the half of the
assemblage above *HKK′H′* (supposing this plane horizontal and
G to the right) all through equal angles against the motion
of the hands of a watch till their equators become horizontal.
The assemblage of centres shears to the right (with rotation *in
the direction* of the hands of a watch), *FG* moving rightwards,
till the angle between *FG* and the end face through *F* becomes
a right angle. Simultaneously with this shearing motion there
is a shrinkage of the assemblage in the direction perpendicular
to the plane *HKK′H′*, entailed by the fact of the equatorial
planes of the oblates turning from their primary inclined positions,
with equatorial planes perpendicular to the optic axis, to their
present horizontal positions. This is most readily seen by confining attention to the single row of oblates which initially had
their centres and points of mutual contact in the short diagonal
GF′ of the right-hand end face. The shrinkage of the assemblage
perpendicular to the plane *HKK′H′* implies elongation in lines
parallel to *FG*, because the volume remains constant, and there
is clearly neither elongation nor shrinkage perpendicular to the
plane of the diagram. Now to fit the tactics of Baumhauer's
twinning by the knife, we must have no change of dimensions
of the assemblage in the plane *HKK′H′*. Hence while the

turning and shearing motions described above are taking place, there must be a continual elongation of the substance of each oblate perpendicular to this plane, and shrinkage parallel to *FG**, to just such an extent as to prevent the centre of each oblate from coming nearer to the plane *HKK'H'*, but instead to cause all the centres to move in lines parallel to *FG*. The oblates are now no longer figures of revolution but are ellipsoids with three unequal axes: the shortest, vertical; the longest, perpendicular to the plane of the diagram; and the mean axis parallel to *FG*. To complete the process, proceed as follows:—

§ 61. Turn the oblates farther on in the same direction (opposite to the motion of the hands of a watch, as that in which they were turned in § 60), and through the same angle; and while, in consequence, the assemblage of centres shears to the right, give to the substance of each oblate a gradual shrinkage perpendicular to the plane *HKK'H'* and elongation parallel to the line *FG*, so as to cause the rightward shearing motion of the assemblage of centres to be still exactly parallel to the initial position of the line *FG*. The whole movement of which the first half has been described in § 60, and the second half in § 61, constitutes exactly what is done in Baumhauer's artificial twinning of an end portion of a prism of Iceland spar, by a knife applied at *F*, with its edge perpendicular to the plane of the diagram, and pressed against the edge *FG* of the obtuse angle between the two upper faces of the prism before and behind the plane of the diagram.

ON THE EQUILIBRIUM OF A HOMOGENEOUS ASSEMBLAGE OF MUTUALLY ATTRACTING POINTS (§§ 62—71).

§ 62. The chief object of this communication is to find the simplest possible way of realising, by means of an assemblage of

---

* Perhaps the simplest way of looking at the affair is found by considering that the elliptic section of each ellipsoid in the plane *HKK'H'* must remain constant; and so also must the horizontal and vertical axes of the elliptic section in the plane of the diagram. Hence, while the principal axes of the elliptic section turn in the manner described in §§ 60, 61, the ellipse itself must remain inscribed in a constant rectangle of vertical and horizontal sides in the plane of the diagram, while the third axis of the ellipsoid, which is perpendicular to the plane of the diagram, remains constant.

points acting upon one another with forces in the lines joining them, and depending merely on the lengths of the joining lines, an elastic solid which shall not be subject to Poisson's restriction of the bulk-modulus to be exactly $\frac{3}{5}$ of the rigidity-modulus; but which may on the contrary have, with given rigidity, any magnitude of bulk-modulus through the whole range from $-\frac{2}{3}$ of the rigidity to $+\infty$, shown to be imaginable by Green. That the thing can be done I showed in my Baltimore Lectures (1884), and I gave an easily conceived although a somewhat complex way of doing it. I now find that the next-to-the-simplest-possible mode of arranging an assemblage of points to produce an elastic solid realises Green's ideal; while the very simplest possible is restricted by Poisson's limitation.

§ 63. The simplest possible arrangement of points to make a homogeneous elastic solid, is a single homogeneous assemblage as defined in § 45 a—d above. In the first place, for simplicity we shall suppose it to be elastically isotropic, or as nearly isotropic as we can make it.

§ 64. To make the solid as nearly as may be isotropic, the unstrained equilibrium distribution must be the equilateral homogeneous assemblage of § 21 above. Consider now a finite assemblage containing a very great number of points thus distributed. To take the very simplest possible case, let there be no force exerted between others than nearest neighbours. For the case of equilibrium, no force acts from without on any of the points, whether on the boundary or in the interior; and therefore clearly there is no mutual action between any of the points according to our present supposition of forces between nearest neighbours only. Suppose now the assemblage to be in equilibrium under the influence of forces acting on points in the boundary, giving rise to infinitesimal deviations from the equilateral homogeneous grouping. Instead of zero force in each shortest distance, there will now be a force which, for stability of equilibrium, must be pull or thrust, according as the distance is greater or less than that which we had in the zero-equilibrium. Thus if, to help ideas, we look to a Boscovich curve, the distance between nearest neighbours for zero-equilibrium, which for brevity we shall call ζ, must be a point in which the curve cuts the line of abscissas with slope corresponding to repulsions for less distances and attractions for greater, and shows zero-force for all distances not less than $\zeta \sqrt{2}$.

§ 65.   To investigate moduluses of elasticity, we must suppose the forces applied from without to the points on the boundary to be such as to produce homogeneous strain throughout the assemblage.   The working out of this statical problem to be given in a future communication, shows that the solid so constituted is not elastically isotropic; but that, on the contrary, it has essentially two different rigidities.   It is in fact a cubical isotropic body with its two rigidities (article "Elasticity," *Encyclopædia Britannica*, ninth edition, or Article XCII., Part I., above) not equal.   An extension of the investigation to include the supposition of forces not only between nearest neighbours, but between nearest and next-nearest neighbours and none farther, gives of course the two rigidities generally not equal; but it allows them to be equalised by a certain definite relation between forces and variations of forces at the two distances ζ and $\zeta \sqrt{2}$.   Imposing this condition, we have elastic isotropy ; and I find the compressibility to be essentially $\frac{8}{3}$ of the rigidity.   The solid thus constituted is therefore subject to Poisson's restriction ; and it will no doubt be found that this restriction is valid for any single equilibrated homogeneous distribution of points, with mutual forces according to Boscovich, and sphere of influence not limited to nearest and next-nearest neighbours, but extending to any large, not infinite, number of times the distance between nearest neighbours.

§ 66.   Having thus failed to produce a solid free from Poisson's restriction, go back to the very simplest case, and try for another way of leaving its simplicity by which we may succeed. Try first to realise an incompressible elastic solid.   When this is done we shall see, by an inevitably obvious modification, how to give any degree of compressibility we please without changing the rigidity, and so to realise an elastic solid with any given positive rigidity, and any given positive or negative bulk-modulus (stable without any surface constraint, only when the bulk-modulus is positive).

§ 67.   To aid conception, make a tetrahedronal model of six equal straight rods, jointed at the angular points in which three meet, each having longitudinal elasticity with perfect anti-flexural rigidity. These constitute merely an ideal materialisation of the connection assumed in the Boscovich attractions and repulsions. A very telling *realisation* of the system thus imagined is made by taking six equal and similar bent bows and jointing their ends together by threes. The jointing might be done accurately by a ball and double socket mechanism of an obvious kind, but it would not be worth the doing. A rough arrangement of six bows of bent steel wire, merely linked together by hooking an end of one into rings on the ends of two others, may be made in a few minutes; and even its defects are not unhelpful towards a vivid understanding of our subject. We have now an element of elastic solid which clearly has an essentially definite ratio of compressibility to reciprocal of either of the rigidities (§ 27 above), each being inversely proportional to the stiffness of the bows. Now we can obviously make this solid incompressible if we take a boss jointed to four equal tie-struts, and joint their free ends to the four corners of the tetrahedron; and we do not alter either of the rigidities if the length of each tie-strut is equal to distance from centre to corners of the unstressed tetrahedron. If the tie-struts are shorter than this, their effect is clearly to augment the rigidities; if longer, to diminish the rigidities. The mathematical investigation proves that it diminishes the greater of the rigidities more than it diminishes the less, and that before it annuls the less it equalises the greater to it.

§ 68. If for the present we confine our attention to the case of the tie-struts longer than the un-strained distance from centre to corners, simple struts will serve; springs, such as bent bows, capable of giving thrust as well as pull along the sides of the tetrahedron, are not needed; mere india-rubber elastic filaments will serve instead, or ordinary spiral springs, and all the end-jointings become much simplified. A realised model accompanies this communication.

§ 69. The model being completed, we have two simple homogeneous Bravais assemblages of points; reds and blues, as we shall call them for brevity; so placed that each blue is in the centre of a tetrahedron of reds, and each red in the centre of a tetrahedron of blues. The other tetrahedronal groupings (Molecular Tactics, §§ 45, 60) being considered, each tetrahedron of reds is vacant of blue, and each tetrahedron of blues is vacant of reds*.

§ 70. Imagine the springs removed and the struts left; but now all properly jointed by fours of ends with perfect frictionless ball-and-socket triple-joints. We have a perfectly non-rigid three-dimensional skeleton frame-work, analogous to idealised plane netting consisting of stiff straight sides of hexagons perfectly jointed in threes of ends. [Compare Art. C., § 2, below.]

§ 71. Leaving mechanism now, return to the purely ideal mutually attracting points of Boscovich; and, as a simple example suppose mutual forces to be zero at all distances exceeding something between ζ and $\zeta\sqrt{2}$.

Let the group be placed at rest in simple equilateral homogeneous distribution :—shortest distance ζ. It will be in stable equilibrium, constituting a solid with the compressibility, and the two rigidities referred to in § 27 above. Condense it to a certain degree to be found by measurements made on the Boscovich curve, and it may become unstable. Let there be some means of consuming energy, or carrying away energy; and it will fall into a stable allotropic condition. The Boscovich curve may be such that this condition is the configuration of absolute minimum energy; and may be such that this configuration is the double homogeneous assemblage of reds and blues described above. Though marked red and blue, to avoid circumlocutions, these points are equal and similar in all qualities.

The mathematical investigation must be deferred for a future communication, when I hope to give it with some further developments.

---

* An interesting structure is suggested by adding another homogeneous assemblage, marked green; giving a green in the centre of each hitherto vacant tetrahedron of reds. It is the same assemblage of triplets as that described in § 24 above. It does not (as long as we have mere jointed struts of constant length between the greens and reds) modify our rigidity-modulus, nor otherwise help us at present, so, having inevitably noticed it, we leave it.

ART. XCVIII. [January 13, 1888]. FIVE APPLICATIONS OF FOURIER'S LAW OF DIFFUSION, ILLUSTRATED BY A DIAGRAM OF CURVES WITH ABSOLUTE NUMERICAL VALUES.

[Report of British Association Bath Meeting, September 1888.]

    I.   Motion of a viscous fluid.

   II.   Closed electric currents within a homogeneous conductor *.

  III.   Heat.

  IV.   Substances in solution.

   V.   Electric potential in the conductor of a submarine cable when electromagnetic inertia can be neglected †.

1. Fourier's now well-known analysis of what he calls the " Linear motion of heat " is applicable to every case of diffusion in which the substance concerned is in the same condition at all points of any one plane parallel to a given plane. The differential equation of diffusion ‡, for the case of constant diffusivity κ, is

$$\frac{dv}{dt} = \kappa \frac{d^2v}{dx^2},$$

where v denotes the "*quality*" at time t and at distance x from a fixed plane of reference. This equation stated in words is as follows :—rate of augmentation of the "*quality*," per unit of time, is equal to the diffusivity multiplied into the rate of augmentation

* This subject is essentially the "electromagnetic induction" of Faraday. It is essentially different from the conduction of electricity through a solid investigated by Ohm in his celebrated paper "Die Galvanische Kette mathematisch bearbeitet," Berlin, 1827: translated in Taylor's *Scientific Memoirs*, Vol. II. Part VIII.; "The Galvanic Circuit investigated Mathematically" by Dr G. S. Ohm. In Ohm's work electromagnetic induction is not taken into account, nor does any idea of an electric analogue to inertia appear. The electromotive force considered is simply that due to the difference of electrostatic potential in different parts of the circuit, unsatisfactorily, and even not accurately, explained by what, speaking in his pre-Greenian time he called "the electroscopic force of the body," and defined or explained as "the force with which the electroscope is repelled or attracted by the body"; the electroscope being "a second movable body of invariable electric condition."

† This subject belongs to the Ohmian electric diffusion pure and simple, worked out by aid of Green's theory of the capacity of a Leyden jar (see Art. LXXIII. Vol. II. above).

‡ See *Mathematical and Physical Papers*, Vol. II. Art. LXXII.

per unit of space of the rate of augmentation per unit of space of the "*quality.*"

The meaning of the word "*quality*" here depends on the subject of the diffusion which may be any one of the five cases referred to in the title above.

2. If the subject is motion of a viscous fluid the "*quality*" is any one of three components of the velocity, relative to rect-angular rectilineal coordinates. But in order that Fourier's diffusional law may be applicable we must either have the motion very slow, according to the special definition of slowness in Article XCIX. § 11 below: or the motion must be such that the velocity is the same for all points in the same stream-line, and would continue to be steadily so if viscosity were annulled at any instant. This condition is satisfied in laminar flow, and more generally in every case in which the stream-lines are parallel straight lines. It is also satisfied in the still more general case of stream-lines coaxal circles, with velocity the same at all points at the same distance from the axis. Our present illustration however is confined to the case of laminar flow, to which Fourier's diffusional laws for what he calls "Linear Motion" (as explained in § 1 above) are obviously applicable without any limitation to the greatness of the velocity in any part of the fluid considered (though with conceivably a reservation in respect to the question of stability*). In this case the "*quality*" is simply fluid velocity.

3. If the subject is electric current, with stream-lines parallel straight lines, the "*quality*" is simply current-density, that is to say strength of current per unit of area traversed by the current. The perfect mathematical† analogy between the electric motion thus defined, and the corresponding motion of a viscous fluid defined in § 2 above was accentuated by Mr Oliver Heaviside, in the *Electrician*, July 12, 1884; and in the following words in the *Philosophical Magazine* for 1886 second half-year p. 135: "Water in a round pipe is started from rest and set into a state of steady motion by the sudden and continued application of

---

* See "Stability of Fluid motion," § 28: *Philosophical Magazine*, August, 1887.

† It is essentially a mathematical analogy only; in the same sense as the relation between the "Uniform motion of heat" and the mathematical theory of electricity, which I gave in the *Cambridge Mathematical Journal* 46 years ago, and which now constitutes the first article of my "Collected Papers on Electrostatics and Magnetism," is a merely mathematical analogy.

a steady longitudinal dragging or shearing-force applied to *its boundary*. This analogue is useful because every one is familiar with the setting of water in motion by friction on its boundary, transmitted inward by viscosity." Mr Heaviside well calls this analogue "useful." It is indeed a very valuable analogy not merely in respect to philosophical considerations of electricity, ether, and ponderable matter, but as facilitating many important estimates, particularly some relating to telephonic conductors and conductors for electric lighting on the alternate-current system. In a non-magnetic metal, the diffusivity for electric current is the electric resistivity divided by 4π: in a magnetic metal it is the electric resistivity divided by 4π times the magnetic permeability. The diffusivity of fluid for viscous motion is its viscosity divided by its density. (See Art. XCIX. § 11 below.)

4. If the subject is heat, as in Fourier's original development of the theory of diffusion the "*quality*" is temperature. The thermal diffusivity of a substance is its thermal conductivity divided by its thermal capacity per unit bulk.

5. If the subject is diffusion of matter, the "*quality*" is density of the matter diffused or deviation of density from some mean or standard density considered. It is to Fick, 33 years ago Demonstrator of Anatomy and now Professor of Physiology, in the University of Zurich, that we owe this application of Fourier's diffusional theory, so vitally important in physiological chemistry and physics and so valuable in natural philosophy generally. When the substance through which the diffusion takes place is fluid a very complicated but practically important subject is presented if the fluid be stirred. The exceedingly rapid progress of the diffusion produced by vigorous up-and-down stirring, causing to be done in half a minute the diffusional work which would require years or centuries if the fluid were quiescent, is easily explained; and the explanation is illustrated by the diagram of curves § 7 below with the time-values given for sugar and common salt. Look at curve No. 1 and think of the corresponding curve with vertical ordinates diminished in the ratio of 1 to 40. The corresponding diffusion would take place for sugar in 11 seconds and for salt in 3·4 seconds. The case so represented would correspond to a streaky distribution of brine and water or of

syrup and water, in which portions of greatest and least salinity
or saccharinity are within half a millimetre of one another.
This is just the condition which we see, in virtue of the difference
of optic refractivity produced by difference of salinity or of
saccharinity when we stir a tumbler of water with a quantity
of undissolved sugar or salt on its bottom. If water be poured
very gently on a quantity of sugar or salt in the bottom of a
tumbler with violent stirring up guarded against by a spoon,
the now almost extinct Scottish species called "toddy ladle"
being the best form, or better still a little wooden disc which
will float up with the water; and if the tumbler be left to
itself undisturbed for two or three weeks, the condition at the
end of 17×10^5 seconds (20 days) for the case of sugar, or $5\cdot4 \times 10^5$
seconds (6 days) for salt, will be that represented by No. 10 curve
in the diagram.

6. If the subject be electricity in a submarine cable, the
"*quality*" is electric potential at any point of the insulated conductor. It is only if the cable were a straight line that x
would be (as defined above) distance from a fixed plane: but
the cable need not be laid along a straight line; and the proper
definition of x for the application of Fourier's formula to a submarine cable is the distance along the cable from any point of
reference (one end of the cable for example) to any point of
the cable. For this case the diffusivity is equal to the conductance* of unit of length of its conductor, reckoned in electrostatic
units, divided by the electrostatic capacity of the conductor per
unit length insulated as it is in gutta percha with its outer surface
wet with sea-water, which, in the circumstances, is to be regarded
as a perfect conductor. For demonstration of this proposition see
Art. LXXIII., Vol. II. above; and for various examples see Table
XIII. of Art. XCII., Part II., Vol. III. page 227, above.

* "Conductance," a valuable word introduced by Mr Oliver Heaviside, means
the reciprocal of "resistance," in respect to electric currents. It is also available
in respect to thermal conduction. It is what I, not so well, called conducting
power. Whether for heat or electricity, the conductance of unit length of a rod or
wire is equal to the conductivity of its substance multiplied by the area of its cross-
section. The electric conductance of any length of wire, or fine thread, reckoned
in electrostatic measure, is a velocity: and its electric resistance, reckoned in
electromagnetic measure, is a velocity. Explanations of these reckonings are given
in my *Popular Lectures*, Vol. I. pp. 123—126, and 436—443

DIAGRAM SHOWING PROGRESS OF LAMINAR DIFFUSION.

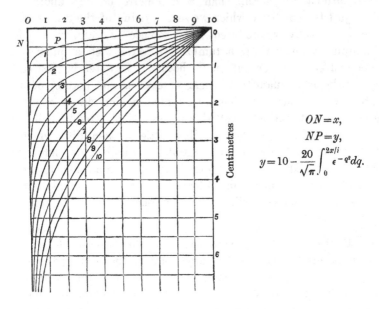

$$ON = x,$$
$$NP = y,$$
$$y = 10 - \frac{20}{\sqrt{\pi}} \int_0^{2x/i} \epsilon^{-q^2} dq.$$

7. EXPLANATION OF DIAGRAM SHOWING PROGRESS OF LAMINAR DIFFUSION.

In each curve

$$NP = 10 \left(1 - \frac{2}{\sqrt{\pi}} \int_0^{2x/i} dq \epsilon^{-q^2} \right),$$

where x denotes the number of centimetres in ON, and $i \, (= 4\sqrt{\kappa t})$ the "curve-number." The curves are drawn directly from the values of the integral given in Table III. appended to De Morgan's article "On the Theory of Probabilities," *Encyclopædia Metropolitana*, Vol. II. pp. 483—484, and reproduced in full at the end of the present article. No. 2 curve is simply a graphic representation of the Table. The horizontal ordinates are reckoned in an arbitrary unit: the vertical ordinates are the absolute distances referred to in the following statement, and are numbered in centimetres. Each vertical ordinate of No. i curve is i times that of No. 1 for the same horizontal ordinate. The time till the condition represented by No. i curve is attained is i^2 times the time till No. 1.

The constant "quality" at initiational surface being called 10, NP denotes the

"Quality"... $\begin{cases} \text{at distance } = ON \text{ from initiational surface,} \\ \text{and at time equal in seconds to ["curve-number"]}^2 \\ \text{divided by sixteen times the diffusivity in} \\ \text{square centimetres per second.} \end{cases}$

Subject of Diffusion.	"*Quality*" represented by NP.
Motion of a viscous fluid.	Velocity.
Closed electric currents within a homogeneous conductor.	Current-density.
Heat.	Temperature.
Substance in solution.	Quantity of substance per unit of volume.
Electricity in the conductor of a submarine cable.	Electric potential.

EXAMPLES.

"Curve-number."	Time in seconds.	Case of Diffusion.
1	27100	Zinc sulphate through water.
1	25700	Copper sulphate through water.
1	17100	Sugar through water.
1	5390	Common salt through water.
5	1200	Heat through wood.
5	118	Laminar motion of water at 10° Cent.
5	29·5	Laminar motion of air.
5	6·94	Heat through iron.
5	·60	Electric current through iron of magnetic permeability 300, and electric resistivity 9800.
5	1·324	Heat through copper.
		Electric current within a homogeneous non-magnetic conductor...
10	·0488	Copper,
10	·00403	„ „ Lead,
10	·00378	„ „ German silver,
10	·00231	„ „ Platinoid.
200,000,000	·086	Electric potential in the Direct U. S. Atlantic Cable.

APPENDIX.

[Being extract (Table III.) from De Morgan's article, "Theory of Probabilities," in *Encyclopædia Metropolitana*, Vol. II. (1845), pp. 483, 484.]

Values of $\dfrac{2}{\sqrt{\pi}}\displaystyle\int_0^q \epsilon^{-q^2}\,dq = f(q)$ for intervals of q, each $= \cdot01$,

from $q = 0$ to $q = 2$.

q	$f(q)$	Δ	Δ^2	q	$f(q)$	Δ	Δ^2
0·00	0·00000 00	1128 33	22	0·45	0·47548 18	917 37	8 40
0·01	0·01128 33	1128 11	45	0·46	0·48465 55	908 97	8 51
0·02	0·02256 44	1127 66	67	0·47	0·49374 52	900 46	8 61
0·03	0·03384 10	1126 99	90	0·48	0·50274 98	891 85	8 69
0·04	0·04511 09	1126 09	1 12	0·49	0·51166 83	883 16	8 78
0·05	0·05637 18	1124 97	1 35	0·50	0·52049 99	874 38	8 88
0·06	0·06762 15	1123 62	1 58	0·51	0·52924 37	865 50	8 96
0·07	0·07885 77	1122 04	1 79	0·52	0·53789 87	856 54	9 03
0·08	0·09007 81	1120 25	2 01	0·53	0·54646 41	847 51	9 10
0·09	0·10128 06	1118 24	2 24	0·54	0·55493 92	838 41	9 17
0·10	0·11246 30	1116 00	2 46	0·55	0·56332 33	829 24	9 23
0·11	0·12362 30	1113 54	2 67	0·56	0·57161 57	820 01	9 30
0·12	0·13475 84	1110 87	2 88	0·57	0·57981 58	810 71	9 35
0·13	0·14586 71	1107 99	3 10	0·58	0·58792 29	801 36	9 40
0·14	0·15694 70	1104 89	3 31	0·59	0·59593 65	791 96	9 45
0·15	0·16799 59	1101 58	3 52	0·60	0·60385 61	782 51	9 49
0·16	0·17901 17	1098 06	3 72	0·61	0·61168 12	773 02	9 53
0·17	0·18999 23	1094 34	3 93	0·62	0·61941 14	763 49	9 55
0·18	0·20093 57	1090 41	4 14	0·63	0·62704 63	753 94	9 59
0·19	0·21183 98	1086 27	4 34	0·64	0·63458 57	744 35	9 62
0·20	0·22270 25	1081 93	4 53	0·65	0·64202 92	734 73	9 63
0·21	0·23352 18	1077 40	4 73	0·66	0·64937 65	725 10	9 65
0·22	0·24429 58	1072 67	4 92	0·67	0·65662 75	715 45	9 66
0·23	0·25502 25	1067 75	5 12	0·68	0·66378 20	705 79	9 68
0·24	0·26570 00	1062 63	5 29	0·69	0·67083 99	696 11	9 67
0·25	0·27632 63	1057 34	5 49	0·70	0·67780 10	686 44	9 68
0·26	0·28689 97	1051 85	5 67	0·71	0·68466 54	676 76	9 68
0·27	0·29741 82	1046 18	5 84	0·72	0·69143 30	667 08	9 66
0·28	0·30788 00	1040 34	6 01	0·73	0·69810 38	657 42	9 66
0·29	0·31828 34	1034 33	6 19	0·74	0·70467 80	647 76	9 65
0·30	0·32862 67	1028 14	6 36	0·75	0·71115 56	638 11	9 62
0·31	0·33890 81	1021 78	6 52	0·76	0·71753 67	628 49	9 61
0·32	0·34912 59	1015 26	6 67	0·77	0·72382 16	618 88	9 57
0·33	0·35927 85	1008 59	6 84	0·78	0·73001 04	609 31	9 56
0·34	0·36936 44	1001 75	6 98	0·79	0·73610 35	599 75	9 52
0·35	0·37938 19	994 77	7 14	0·80	0·74210 10	590 23	9 48
0·36	0·38932 96	987 63	7 29	0·81	0·74800 33	580 75	9 45
0·37	0·39920 59	980 34	7 42	0·82	0·75381 08	571 30	9 41
0·38	0·40900 93	972 92	7 55	0·83	0·75952 38	561 89	9 36
0·39	0·41873 85	965 37	7 69	0·84	0·76514 27	552 53	9 31
0·40	0·42839 22	957 68	7 82	0·85	0·77066 80	543 22	9 26
0·41	0·43796 90	949 86	7 95	0·86	0·77610 02	533 96	9 21
0·42	0·44746 76	941 91	8 07	0·87	0·78143 98	524 75	9 16
0·43	0·45688 67	933 84	8 17	0·88	0·78668 73	515 59	9 09
0·44	0·46622 51	925 67	8 30	0·89	0·79184 32	506 50	9 04

q	$f(q)$	Δ	Δ^2	q	$f(q)$	Δ	Δ^2
0·90	0·79690 82	497 46	8 97	1·45	0·95969 50	135 85	3 91
0·91	0·80188 28	488 49	8 91	1·46	0·96105 35	131 94	3 82
0·92	0·80676 77	479 58	8 83	1·47	0·96237 29	128 12	3 74
0·93	0·81156 35	470 75	8 77	1·48	0·96365 41	124 38	3 65
0·94	0·81627 10	461 98	8 70	1·49	0·96489 79	120 73	3 57
0·95	0·82089 08	453 28	8 61	1·50	0·96610 52	117 16	3 49
0·96	0·82542 36	444 67	8 55	1·51	0·96727 68	113 67	3 40
0·97	0·82987 03	436 12	8 46	1·52	0·96841 35	110 27	3 32
0·98	0·83423 15	427 66	8 39	1·53	0·96951 62	106 95	3 25
0·99	0·83850 81	419 27	8 30	1·54	0·97058 57	103 70	3 16
1·00	0·84270 08	410 97	8 22	1·55	0·97162 27	100 54	3 09
1·01	0·84681 05	402 75	8 13	1·56	0·97262 81	97 45	3 01
1·02	0·85083 80	394 62	8 05	1·57	0·97360 26	94 44	2 94
1·03	0·85478 42	386 57	7 96	1·58	0·97454 70	91 50	2 86
1·04	0·85864 99	378 61	7 86	1·59	0·97546 20	88 64	2 79
1·05	0·86243 60	370 75	7 77	1·60	0·97634 84	85 85	2 73
1·06	0·86614 35	362 97	7 68	1·61	0·97720 69	83 12	2 64
1·07	0·86977 32	355 29	7 60	1·62	0·97803 81	80 48	2 59
1·08	0·87332 61	347 69	7 49	1·63	0·97884 29	77 89	2 51
1·09	0·87680 30	340 20	7 40	1·64	0·97962 18	75 38	2 45
1·10	0·88020 50	332 80	7 31	1·65	0·98037 56	72 93	2 38
1·11	0·88353 30	325 49	7 21	1·66	0·98110 49	70 55	2 31
1·12	0·88678 79	318 28	7 12	1·67	0·98181 04	68 24	2 26
1·13	0·88997 07	311 16	7 01	1·68	0·98249 28	65 98	2 20
1·14	0·89308 23	304 15	6 91	1·69	0·98315 26	63 78	2 12
1·15	0·89612 38	297 24	6 82	1·70	0·98379 04	61 66	2 08
1·16	0·89909 62	290 42	6 72	1·71	0·98440 70	59 58	2 01
1·17	0·90200 04	283 70	6 61	1·72	0·98500 28	57 57	1 96
1·18	0·90483 74	277 09	6 52	1·73	0·98557 85	55 61	1 90
1·19	0·90760 83	270 57	6 42	1·74	0·98613 46	53 71	1 85
1·20	0·91031 40	264 15	6 31	1·75	0·98667 17	51 86	1 79
1·21	0·91295 55	257 84	6 22	1·76	0·98719 03	50 07	1 75
1·22	0·91553 39	251 62	6 11	1·77	0·98769 10	48 32	1 68
1·23	0·91805 01	245 51	6 02	1·78	0·98817 42	46 64	1 65
1·24	0·92050 52	239 49	5 91	1·79	0·98864 06	44 99	1 59
1·25	0·92290 01	233 58	5 81	1·80	0·98909 05	43 40	1 54
1·26	0·92523 59	227 77	5 71	1·81	0·98952 45	41 86	1 50
1·27	0·92751 36	222 06	5 61	1·82	0·98994 31	40 36	1 44
1·28	0·92973 42	216 45	5 52	1·83	0·99034 67	38 92	1 41
1·29	0·93189 87	210 93	5 41	1·84	0·99073 59	37 51	1 36
1·30	0·93400 80	205 52	5 32	1·85	0·99111 10	36 15	1 33
1·31	0·93606 32	200 20	5 22	1·86	0·99147 25	34 82	1 27
1·32	0·93806 52	194 98	5 11	1·87	0·99182 07	33 55	1 24
1·33	0·94001 50	189 87	5 02	1·88	0·99215 62	32 31	1 20
1·34	0·94191 37	184 85	4 93	1·89	0·99247 93	31 11	1 16
1·35	0·94376 22	179 92	4 82	1·90	0·99279 04	29 95	1 12
1·36	0·94556 14	175 10	4 74	1·91	0·99308 99	28 83	1 08
1·37	0·94731 24	170 36	4 63	1·92	0·99337 82	27 75	1 06
1·38	0·94901 60	165 73	4 55	1·93	0·99365 57	26 69	1 01
1·39	0·95067 33	161 18	4 45	1·94	0·99392 26	25 68	99
1·40	0·95228 51	156 73	4 35	1·95	0·99417 94	24 69	95
1·41	0·95385 24	152 38	4 27	1·96	0·99442 63	23 74	91
1·42	0·95537 62	148 11	4 18	1·97	0·99466 37	22 83	89
1·43	0·95685 73	143 93	4 09	1·98	0·99489 20	21 94	85
1·44	0·95829 66	139 84	3 99	1·99	0·99511 14	21 09	82
				2·00	0·99532 23		

[Now published for the first time, May, 1890.]

§§ 1—11. *Viscous liquid.*

1. *Stress required to produce change of shape.* Stokes assumed
the stress to be in simple proportion to the speed of the change
of shape, as basis for his mathematical theory. Poiseuille's experiments on the flow of water through capillary tubes amply
confirm * this assumption, for water, and we have good reason
for believing that it is very near to the truth for all ordinary
fluids; with however possibly some need for correction to take
into account residual effects of previous conditions, especially in
the case of extremely viscous fluids such as treacle, or thick oil. But
for our present purely mathematical analogy we shall, and without
further question, use simply the law of simple proportionality,
which implies of course undisturbed superposition of stresses and
corresponding speeds and modes of change of shape.

2. (*Definitions.*) A simple shearing infinitesimal distortion,
or as it is also called a simple distortion, is a distortion in which
there are equal elongations and shortenings in two mutually perpendicular lines, and therefore, (the substance being incompressible) neither elongation nor shortening in the line perpendicular
to both. The numerical reckoning of a simple distortion † is the
difference from a right angle, reckoned in radian, which the distortion produces on the angle between lines bisecting the angles
between the lines of greatest elongation and greatest shortening.
This reckoning is easily proved to be equal to the greatest elongation, or equal to the greatest shortening. The *speed* of a simple
distortional change of shape is the angular velocity at which the

---

* See Art. xcii. Part i. above, p. 21, § 29, and insert Poiseuille's name there
accordingly.

† See Art. xcii. Part i. above, p. 33, § 43, and p. 91, example (4).

angle between two planes bisecting the angles between the lines
of greatest elongation and shortening varies; or it is the rate
per unit of time, of the greatest lengthening, or of the greatest
shortening : or it is the relative velocity of either of those planes
and a parallel plane at unit distance.   This last statement, which
for many purposes is the most convenient definition of speed of
change of shape, corresponds to the ordinary idea of "shearing"
motion.

3. The *viscosity** of a liquid (or the numerical reckoning
or measure of the viscous quality) is defined as the tangential
force per unit of area, in either of the mutually perpendicular
planes of zero-elongation, of a simple distortion, required to
produce change of shape at unit speed.   It is proved below (§ 4)
to be equal to (and it might be defined as) *one third of* the normal
pull per unit area required to produce unit speed of elongation;
when, upon pressure equal in all directions, we superimpose normal
pull in one line or perpendicular to one set of parallel planes.

4. To prove this proposition, look to pp. 35, 36 above (Art.
XCII. Part. I. Elasticity, §§ 45, 46) and, taking $k = \infty$ to enforce
constancy of volume, (incompressibility,) repeat the investigations
there given; but with *speed of simple distortion*, or *speed of
shearing*, substituted for "simple distortion," or "shear."   Then,
n will be the viscosity according to the definition of § 3 of the
present article, and by Art. XCII. Part I. § 46 (3) we find $P/3n$, for
the rate of elongation in the direction of the pull, (which is
equivalent to the proposition of § 4 to be proved;) and $P/6n$
for the rate of contraction in every direction perpendicular to
the pull, when a viscous liquid is pulled homogeneously in parallel
lines with a force amounting to P per unit of area perpendicular
to the direction of the pull.

---

* There is a curiously illogical tendency to introduce the word "coefficient" in
connection with numerical reckoning of properties of matter, which seems to have
originated in the too-long tolerated expression "coefficient of friction"; and which
has given us "coefficients of elasticity", the "coefficient of compressibility", the
"coefficient of rigidity", "coefficient of viscosity", "coefficient of magnetic permeability".   In each case the designation "coefficient of" is mathematically
vicious; and to the non-mathematical mind it is a mystery of circumlocution.
Frictionality is much shorter than "coefficient of friction", and might, I think,
advantageously replace it in dynamical and mechanical language.

5. Let now pulls P, Q, R, each reckoned per unit of area, be applied in three directions mutually at right angles, and let e, f, g be the resulting rates of elongation in these directions respectively. We have

$$e=(2P-Q-R)/6n; \quad f=(2Q-R-P)/6n; \quad g=(2R-P-Q)/6n \dots\dots(1);$$

or $\qquad e=(P+p)/2n; \quad f=(Q+p)/2n; \quad g=(R+p)/2n \dots\dots\dots(2),$

where $\qquad\qquad\qquad p=-\tfrac{1}{3}(P+Q+R)\dots\dots\dots\dots\dots\dots(3).$

6. Now let u, v, w be the velocity-components at any point (x, y, z) in a viscous liquid. The rates of elongation in the x, y, z directions (denoted as in § 6 by e, f, g): and the rates of shearing,

$\qquad a$ parallel to y in plane yx, or parallel to z in plane zx,

$\qquad\quad b \qquad$ „ $\qquad$ „ z „ $\quad$ „ $\quad zy,$ „ $\qquad$ „ $\qquad$ „ x „ $\quad$ „ $\quad xy,$

and $\quad c \qquad$ „ $\qquad$ „ x „ $\quad$ „ $\quad xz,$ „ $\qquad$ „ $\qquad$ „ y „ $\quad$ „ $\quad yz,$

are given by the following equations:—

$$e=\frac{du}{dx}, \qquad f=\frac{dv}{dy}, \qquad g=\frac{dw}{dz} \dots\dots\dots\dots(4),$$

$$a=\frac{dv}{dz}+\frac{dw}{dy}, \quad b=\frac{dw}{dx}+\frac{du}{dz}, \quad c=\frac{du}{dy}+\frac{dv}{dx}\dots\dots(5).$$

7. Consider now an infinitesimal parallelepiped $\delta x \, \delta y \, \delta z$, of the fluid, having its centre at (x, y, z). It experiences normal and tangential tractions, on its three pairs of faces, according to the following schedule:—

Faces acted on	Tractions parallel to		
	x	y	z
$\delta y \, \delta z$	$-p+2n\dfrac{du}{dx}$	$n\left(\dfrac{dv}{dx}+\dfrac{du}{dy}\right)$	$n\left(\dfrac{du}{dz}+\dfrac{dw}{dx}\right)$
$\delta z \, \delta x$	$n\left(\dfrac{dv}{dx}+\dfrac{du}{dy}\right)$	$-p+2n\dfrac{dv}{dy}$	$n\left(\dfrac{dw}{dy}+\dfrac{dv}{dz}\right)$
$\delta x \, \delta y$	$n\left(\dfrac{du}{dz}+\dfrac{dw}{dx}\right)$	$n\left(\dfrac{dw}{dy}+\dfrac{dv}{dz}\right)$	$-p+2n\dfrac{dw}{dz}$

$$(6).$$

To find the resultant force on the matter within the parallelepiped $\delta x \, \delta y \, \delta z$ due to the nine pairs of normal and tangential tractions, exerted on its three pairs of faces, by the surrounding matter, remark that they differ from (6) by the proper infinitesimal

differences due to the circumstance that the coordinates of the
centres of its six faces are

$$x \pm \tfrac{1}{2}\delta x, \quad y \pm \tfrac{1}{2}\delta y, \quad z \pm \tfrac{1}{2}\delta z.$$

Consider first the x-components. Of these we have a pair, in
opposite directions and infinitely nearly equal, on the parallel faces
$\delta y\, \delta z$, contributing to the required x-component their difference,
which is

$$\frac{d}{dx}\left(-p + 2n\frac{du}{dx}\right)\delta x\, \delta y\, \delta z.$$

This and the x-contributions from the pairs of faces $\delta z\, \delta x$,
and $\delta x\, \delta y$, similarly reckoned, give for the whole x-component

$$\left\{-\frac{dp}{dx} + n\left[2\frac{d}{dx}\frac{du}{dx} + \frac{d}{dy}\left(\frac{dv}{dx}+\frac{du}{dy}\right) + \frac{d}{dz}\left(\frac{du}{dz}+\frac{dw}{dx}\right)\right]\right\}\delta x\delta y\delta z \dots(7).$$

Now the fluid being incompressible, we have

$$\frac{du}{dx} + \frac{dv}{dy} + \frac{dw}{dz} = 0 \dots\dots\dots\dots\dots \ (8).$$

Modifying (7) accordingly we find

$$\left\{-\frac{dp}{dx} + n\left(\frac{d^2u}{dx^2} + \frac{d^2u}{dy^2} + \frac{d^2u}{dz^2}\right)\right\}\delta x\delta y\delta z \dots\dots\dots(9),$$

for the x-component of the resultant of tractional forces on the
matter within the parallelepiped $\delta x\delta y\delta z$. Hence if we denote by
F, G, H, the components of the resultant force *per unit of volume*
due to viscous action, and resistance to condensation (p), in the
fluid in the neighbourhood of x, y, z we have

$$F = n\nabla^2 u - \frac{dp}{dx}; \quad G = n\nabla^2 v - \frac{dp}{dy}; \quad H = n\nabla^2 w - \frac{dp}{dz}\dots(10);$$

where ∇^2 denotes $\qquad \dfrac{d^2}{dx^2} + \dfrac{d^2}{dy^2} + \dfrac{d^2}{dz^2} \dots\dots\dots\dots\dots\dots\dots(11).$

8. Let now X, Y, Z denote components of bodily force, if
any there is, by which is meant force (as, for example, gravity)
exerted from a distance on the fluid by other matter, and not force
transmitted through the fluid such as the force due to viscosity or
the force due to fluid pressure. The components of the whole
force per unit of volume, acting on the fluid at x, y, z at any
instant, are accordingly $F + X$, &c., or, by (10),

$$n\nabla^2 u - \frac{dp}{dx} + X; \quad n\nabla^2 v - \frac{dp}{dy} + Y; \quad n\nabla^2 w - \frac{dp}{dz} + Z \dots(12);$$

and these, each divided by the density of the fluid, must be equal
to its acceleration at (x, y, z).

9. Now by the well-known kinematics of elementary hydrokinetics we have

$$\left.\begin{aligned}
\frac{Du}{dt} &= \frac{du}{dt} + \frac{\partial u}{dt}\\
\frac{Dv}{dt} &= \frac{dv}{dt} + \frac{\partial v}{dt}\\
\frac{Dw}{dt} &= \frac{dw}{dt} + \frac{\partial w}{dt}
\end{aligned}\right\} \dots\dots\dots\dots(13);$$

where D/dt denotes rate of variation per unit of time of any
attribute (such as the temperature, or the velocity, or a velocity-
component) of one and the same portion of fluid in its actual
motion, if it has any; d/dt denotes rate of variation per unit
of time of an attribute of the portion of the fluid which is at a
fixed point x, y, z at any instant; and ∂/dt denotes what the rate
of variation per unit of time of an attribute would be for one and
the same particle of fluid, if variation of this attribute at any
place fixed in space were, at the instant of our reckoning,
temporarily annulled. In other words ∂/dt denotes the variation
of an attribute due to the fact of its having different values at
different points of space, and the fact that the portion of the
fluid considered is moving from point to point. This definition,
put into symbols according to the notation of the differential
calculus, is

$$\left.\begin{aligned}
\frac{\partial u}{dt} &= u\frac{du}{dx} + v\frac{du}{dy} + w\frac{du}{dz}\\
\frac{\partial v}{dt} &= u\frac{dv}{dx} + v\frac{dv}{dy} + w\frac{dv}{dz}\\
\frac{\partial w}{dt} &= u\frac{dw}{dx} + v\frac{dw}{dy} + w\frac{dw}{dz}
\end{aligned}\right\} \dots\dots\dots\dots(14).$$

10. Taking now the force-components (12) and the corresponding components of acceleration (8), we find for the dynamical
equations of the motion:—

$$\left.\begin{aligned}
\frac{du}{dt} + \frac{\partial u}{dt} &= \kappa\nabla^2 u + \frac{1}{\rho}\left(X - \frac{dp}{dx}\right)\\
\frac{dv}{dt} + \frac{\partial v}{dt} &= \kappa\nabla^2 v + \frac{1}{\rho}\left(Y - \frac{dp}{dy}\right)\\
\frac{dw}{dt} + \frac{\partial w}{dt} &= \kappa\nabla^2 w + \frac{1}{\rho}\left(Z - \frac{dp}{dz}\right)
\end{aligned}\right\} \dots\dots\dots(15),$$

where ρ denotes the density of the fluid, and

$$\kappa = n/\rho \quad\dots\dots\dots\dots\dots\dots\dots(16).$$

11. Going back to (14) and (10) we see that if velocity were diminished in the ratio of n to 1, the first terms of the force-components F, G, H and the accelerational items du/dt, dv/dt, dw/dt would be diminished in the same ratio; while the other accelerational items $\partial u/dt$, $\partial v/dt$, $\partial w/dt$ are diminished in the ratio n^2 to 1. Hence the motion may be so slow that $\partial u/dt$, $\partial v/dt$, $\partial w/dt$ may be neglected in comparison with du/dt, dv/dt, dw/dt, and in comparison with $\kappa\nabla^2 u$, $\kappa\nabla^2 v$, $\kappa\nabla^2 w$; and when it is so the equations of motion become reduced to

$$\left.\begin{aligned}
\frac{du}{dt} &= \kappa\nabla^2 u + \frac{1}{\rho}\left(X - \frac{dp}{dx}\right) \\
\frac{dv}{dt} &= \kappa\nabla^2 v + \frac{1}{\rho}\left(Y - \frac{dp}{dy}\right) \\
\frac{dw}{dt} &= \kappa\nabla^2 w + \frac{1}{\rho}\left(Z - \frac{dp}{dz}\right)
\end{aligned}\right\} \text{(viscous liquid)} \quad\dots(17).$$

§§ 12—13. *Equilibrium or motion of an elastic solid.*

12. Going back to § 4, replace "simple distortion," or "shear" instead of the speed of the change there considered: and let n now denote, as in Art. XCII. Part I., § 46, rigidity-modulus, (instead of viscosity, as in § 4 of the present Article); and let u, v, w, denote components of displacement (instead of velocity-components as in §§ 5—11 of the present Article). §§ 5—8 and all their formulas are applicable with no other change than elastic solids substituted everywhere for "viscous liquid." §§ 9—11 are exclusively applicable to fluid motion. Instead of these we have only now to remark that each member of (12) must be zero for the equilibrium of an elastic solid and must be equal to the rate, per unit of time, of augmentation of momentum per unit volume of the solid for varying displacement from the point (x, y, z). Hence if ρ denote the mass of the solid per unit volume we have

$$\left.\begin{aligned}
\rho\frac{d^2u}{dt^2} &= n\nabla^2 u - \frac{dp}{dx} + X \\
\rho\frac{d^2v}{dt^2} &= n\nabla^2 v - \frac{dp}{dy} + Y \\
\rho\frac{d^2w}{dt^2} &= n\nabla^2 w - \frac{dp}{dz} + Z
\end{aligned}\right\} \text{(elastic solid)}\dots(18).$$

13. These equations are not confined to the case of an in-compressible solid, because if we do not take $k = \infty$ as in § 4, (3) of § 5 gives

$$p = -k\left(\frac{du}{dx} + \frac{dv}{dy} + \frac{dw}{dz}\right) \dots\dots\dots\dots\dots(19),$$

by which we may eliminate p from (18). The definition of k, called for brevity the *bulk-modulus*, is implied in (1) § 45 of Article XCII. Part I. above. For the present however we shall suppose $k = \infty$, which requires that

$$\frac{du}{dx} + \frac{dv}{dy} + \frac{dw}{dz} = 0 \text{ (8) of § 7, repeated} \dots\dots\dots(20).$$

With this and the three equations (18) we have four equations for the four unknown quantities u, v, w, p.

§§ 14—20. *Equilibrium or motion of an ideal substance called for brevity, Ether.*

14. What I am for the present calling *ether*, is an ideal substance useful for extending the " Mechanical representation of electric, magnetic, and galvanic forces," which constitutes Art. XXVII. Vol. I. of the present Reprint, having first appeared under date Nov. 28, 1846, in the 1847 volume of the *Cambridge and Dublin Mathematical Journal*. For the present I suppose it absolutely incompressible. It has no intrinsic rigidity (elastic resistance to change of shape); but it has a *quasi* rigidity depending on an inherent *quasi* elastic resistance to absolute rotation. This *quasi* rigidity may be called simply rigidity for brevity; but when it is to be distinguished from the known natural rigidity of an elastic solid it will be called gyrostatic rigidity.

15. Let $n \cdot 2\theta$ denote the amount of torque per unit volume required to balance any portion of ether rotated through an infinitesimal angle θ. The components of θ round three rectangular axes are, as stated in my old paper of 1846 and as given originally by Stokes in 1845 (see Vol. I. p. 112 of his Papers)

$$\frac{1}{2}\left(\frac{dw}{dy} - \frac{dv}{dz}\right), \quad \frac{1}{2}\left(\frac{du}{dz} - \frac{dw}{dx}\right), \quad \frac{1}{2}\left(\frac{dv}{dx} - \frac{du}{dy}\right)\dots\dots(21),$$

where u, v, w denote the components of infinitesimal displacement at any point x, y, z. These rotational components (21), multiplied by $2n$ express the corresponding components of the torque per unit volume.

16. Hence, if we denote by S, T, U, tangential forces, each reckoned per unit of area, on the three quartets of faces of an infinitesimal parallelepiped $\delta x \, \delta y \, \delta z$, according to the directions,

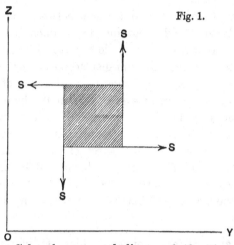

Fig. 1.

indicated for S by the annexed diagram* (fig. 1), and the symmetrically corresponding directions for T and U, we have,

$$\left. \begin{aligned} S &= n\left(\frac{dw}{dy} - \frac{dv}{dz}\right) \\ T &= n\left(\frac{du}{dz} - \frac{dw}{dx}\right) \\ U &= n\left(\frac{dv}{dx} - \frac{du}{dy}\right) \end{aligned} \right\} \quad\ldots\ldots\ldots\ldots\ldots\ldots (22).$$

Besides the tangential forces we may have equal pressures to any amount p on each of the twelve faces, and thus now instead of (6) above we have the following table of tractions on the faces of an infinitesimal parallepiped of ether.

Faces acted on	Tractions parallel to		
	x	y	z
$\delta y \, \delta z$	$-p$	$n\left(\dfrac{dv}{dx} - \dfrac{du}{dy}\right)$	$n\left(\dfrac{dw}{dx} - \dfrac{du}{dz}\right)$
$\delta z \, \delta x$	$n\left(\dfrac{du}{dy} - \dfrac{dv}{dx}\right)$	$-p$	$n\left(\dfrac{dw}{dy} - \dfrac{dv}{dz}\right)$
$\delta x \, \delta y$	$n\left(\dfrac{du}{dz} - \dfrac{dw}{dx}\right)$	$n\left(\dfrac{dv}{dz} - \dfrac{dw}{dy}\right)$	$-p$

$$\left. \right\} \ldots(23).$$

* Compare with diagram on p. 33 above.

Hence just as in §§ 7 and 12 above, we find exactly the same equations of motion (18) and (20) above, as we there found for an incompressible elastic solid.

17. What then is the difference between our ether, and jelly as for brevity and according to common usage we shall call the incompressible elastic solid ?   No difference whatever in respect to the equilibrium-displacement, or the motion, throughout any portion of homogeneous substance of either kind, if the position and motion of every point in the bounding surface of the portion considered are the same for the two.   But in respect to the traction on the bounding surface of a detached portion, and therefore also in respect of the interfacial relation between portions of the substance having different rigidities, there is an essential difference between the two, of vital importance for the inclusion of magnetic induction in our mechanical representation.

18. At an interface between jellies of different rigidities, the equality of tangential tractions requires that the direction of the tangential component of shearing distortion be the same on the two sides of the interface, and that the product of its magnitude into the rigidity be equal on the two sides.

The equality of normal tractions requires that $-p$ plus twice the rigidity into the normal component of extension be equal on the two sides of the interface.   These conditions we see immediately from the table of tractions (6) of § 7 above.

19. Similarly we find from table (23) above, the following interfacial conditions for the ether.—The equality of tangential components requires that the tangential components of rotation on the two sides of the interface have their axes coincident, and that the product of the rotation into the rigidity have always the same value on the two sides.

20. When the rigidity is equal on the two sides of an interface while there is discontinuity due to a difference of bodily force (§ 8 above), or of density and therefore of reaction against acceleration, on the two sides, all the interfacial conditions are the same for jelly and ether and may be mathematically expressed in the simplest manner by saying that p, and all the

nine differential coefficients of u, v, w, must have equal values on the two sides of the interface, as we see by the following considerations.

For simplicity let OX be perpendicular to the interface. The interfacial conditions are from (6) of § 7 above, for jelly,

$$\left. -p + 2n\frac{du}{dx} ; \quad n\left(\frac{dv}{dx} + \frac{du}{dy}\right); \quad n\left(\frac{du}{dz} + \frac{dw}{dx}\right) \right\} \dots (24),$$

continuous across the interface

and from (23) of § 16 above, for ether,

$$\left. -p ; \quad n\left(\frac{dv}{dx} - \frac{du}{dy}\right); \quad n\left(\frac{dw}{dx} - \frac{du}{dz}\right) \right\} \dots (25).$$

continuous across the interface

The continuity of u, v, w, requires that of the nine differential coefficients of u, v, w, the six with reference to y and z are equal on the two sides of the interface. Hence, and because

$$\frac{du}{dx} + \frac{dv}{dy} + \frac{dw}{dz} = 0 \quad (20), \text{ of § 13 repeated } \dots (26),$$

du/dx also has equal values on the two sides of the interface. Thus of the nine differential coefficients there only remain dv/dx and dw/dx which can be different on the two sides of the interface, and these also are necessarily equal when n is equal on the two sides, as we see for jelly by (24) and for ether by (25). Finally looking to the normal traction we see that the equality of du/dx secures for jelly by (24) above the equality of p when n is equal on the two sides of the interface; while by (25) for ether the condition is that p must be equal, whether the rigidity be equal on the two sides or not.

§§ 21—28. *Energy of stressed jelly or of stressed ether.*

21. Let bodily force be applied to the substance within any volume V, and surface tractions be applied to its surface, so as to produce, and to maintain in equilibrium any prescribed displacement, u, v, w, at any point x, y, z, of the substance: it is required to find the whole amount of work which must have been done by these forces. The total amount of work done by the supposed actual forces, is the same as that which would be done, if the body were ideally divided into an infinite number of

infinitely small parts, and surface tractions applied to each of these parts, so as to give it the distortion, in the case of the jelly, or the rotation in the case of the ether, which it actually has in the prescribed circumstances.

22. Taking the latter as the simpler case, first, we see at once that the work done on an infinitesimal portion is equal to the torque, due to half the actual rotation, which is the mean working couple ; multiplied into the amount of rotation, this being the angle through which the torque works. Hence if ϑ denote twice the rotation and n the rigidity, as in § 16, we have

$$\text{work per unit of volume} = \tfrac{1}{2}n\vartheta^2 \dots\dots\dots\dots(27),$$

or in terms of rectangular coordinates,

$$\tfrac{1}{2}n\left\{\left(\frac{dw}{dy}-\frac{dv}{dz}\right)^2+\left(\frac{du}{dz}-\frac{dw}{dx}\right)^2+\left(\frac{dv}{dx}-\frac{du}{dy}\right)^2\right\} \dots\dots(28).$$

23. To find corresponding expression for work done on the jelly, let O be the centre of an infinitesimal parallelepiped and consider, of its three pairs of faces, the displacements parallel respectively to x, y, z. These we find to be as shown in the following table.

| Faces | Displacements parallel to | | |
	x	y	z
$\delta y \delta z$	$\pm\tfrac{1}{2}\delta x\,\dfrac{du}{dx}$	$\pm\tfrac{1}{2}\delta x\,\dfrac{dv}{dx}$	$\pm\tfrac{1}{2}\delta x\,\dfrac{dw}{dx}$
$\delta z \delta x$	$\pm\tfrac{1}{2}\delta y\,\dfrac{du}{dy}$	$\pm\tfrac{1}{2}\delta y\,\dfrac{dv}{dy}$	$\pm\tfrac{1}{2}\delta y\,\dfrac{dw}{dy}$
$\delta x \delta y$	$\pm\tfrac{1}{2}\delta z\,\dfrac{du}{dz}$	$\pm\tfrac{1}{2}\delta z\,\dfrac{dv}{dz}$	$\pm\tfrac{1}{2}\delta z\,\dfrac{dw}{dz}$

$$\dots(29).$$

The amounts of work done on these displacements are to be calculated by multiplying by half the tractional forces shown in table (6) of § 7 above, remembering that these forces are in opposite directions on each pair of parallel faces. We thus find, for example, work done on pair of faces $\delta y\,\delta z$, by traction parallel to x,

$$= \delta x\,\delta y\,\delta z\,\frac{du}{dx}\times\frac{1}{2}\left(-\,p+2n\,\frac{du}{dx}\right),$$

work done on same faces by traction parallel to y,

$$= \delta x\, \delta y\, \delta z\, \frac{dv}{dx} \times \frac{1}{2}\left(\frac{dv}{dx} + \frac{du}{dy}\right),$$

and similarly other seven items of the total work. Summing these, remembering that $du/dx + dv/dy + dw/dz = 0$, because of incompressibility, which annuls the sum of terms with p as factor, and dividing by $\delta x\, \delta y\, \delta z$ to reduce to work down per unit of volume, we find

$$\tfrac{1}{2}n\left\{2\left(\frac{du^2}{dx^2} + \frac{dv^2}{dy^2} + \frac{dw^2}{dz^2}\right)\right\}$$

$$+ \tfrac{1}{2}n\left\{\left(\frac{dv}{dz} + \frac{dw}{dy}\right)^2 + \left(\frac{dw}{dx} + \frac{du}{dz}\right)^2 + \left(\frac{du}{dy} + \frac{dv}{dx}\right)^2\right\} \quad....(30).$$

This formula agrees with (7) of § 695 of Thomson and Tait's *Natural Philosophy*, for the case of $k = \infty$ and $k(e + f + g)^2 = 0$. Its first line expresses the work done in virtue of distortion consisting of stretchings and shortenings parallel to x, y, z. The second line expresses the work done in virtue of three shearings in planes respectively perpendicular to x, y, z.

24. Considering the fact that the equations of internal equilibrium or motion [(18) of § 12 above] are identical for jelly and ether, we see that if the boundary of our volume V (§ 21) of either substance, be held fixed, the work required to produce any specified internal displacement, must be the same for the two. Hence the volume integral of (30) for the whole of V, must be equal to the volume integral of (28), provided we have at boundary

$$u = 0, \quad v = 0, \quad w = 0 ....(31).$$

That it is so, we verify as follows :—

From the first line of (30) subtract

$$\tfrac{1}{2}n \times 2\left(\frac{du}{dx} + \frac{dv}{dy} + \frac{dw}{dz}\right)^2 \quad \text{(which is zero)}.$$

Thus instead of the first line of (30), we have

$$-\tfrac{1}{2}n \times 4\left(\frac{dv}{dy}\frac{dw}{dz} + \frac{dw}{dz}\frac{du}{dx} + \frac{du}{dx}\frac{dv}{dy}\right)..........(32).$$

Now by a well-known process of integration by parts, performed twice on each term, with (31) satisfied at the boundary, we find

$$\iiint dx\,dy\,dz \left(\frac{dv}{dy}\frac{dw}{dz} + \frac{dw}{dz}\frac{du}{dx} + \frac{du}{dx}\frac{dv}{dy} \right)$$

$$= \iiint dx\,dy\,dz \left(\frac{dv}{dz}\frac{dw}{dy} + \frac{dw}{dx}\frac{du}{dz} + \frac{du}{dy}\frac{dv}{dx} \right) \ldots(33).$$

Taking now the volume integral of (30) with its first line changed to (32), using (33), and taking the two lines together, we find for the total work required to produce the supposed displacement of the *jelly*,

$$\tfrac{1}{2}n \iiint dx\,dy\,dz \left\{ \left(\frac{dw}{dy} - \frac{dv}{dz}\right)^2 + \left(\frac{du}{dz} - \frac{dw}{dx}\right)^2 + \left(\frac{dv}{dx} - \frac{du}{dy}\right)^2 \right\} \ldots(34),$$

which is the same as that given directly for *ether* by (28) above.

25. As an illustration, consider a case in which the given displacement is everywhere tangential to circles on coaxial cylindric circular surfaces, and equal at equal distances from the axis. Taking x, y in a plane perpendicular to the axis, we have

$$u = -\frac{qy}{r}, \quad v = \frac{qx}{r}, \quad w = 0, \ldots\ldots\ldots(35),$$

where r denotes $\sqrt{(x^2 + y^2)}$ and q the displacement at x, y, z, which is a function of r. We find,

$$\left.\begin{aligned}
&\frac{du}{dx} = -\frac{xy}{r}\frac{d}{dr}\left(\frac{q}{r}\right), \quad \frac{du}{dy} = -\frac{y^2}{r}\frac{d}{dr}\left(\frac{q}{r}\right) - \frac{q}{r}, \quad \frac{du}{dz} = 0 \\
&\frac{dv}{dx} = \frac{x^2}{r}\frac{d}{dr}\left(\frac{q}{r}\right) + \frac{q}{r}, \quad \frac{dv}{dy} = \frac{xy}{r}\frac{d}{dr}\left(\frac{q}{r}\right), \quad \text{and } \frac{dv}{dz} = 0
\end{aligned}\right\} \ldots(36).$$

Whence,
$$\frac{dv}{dx} + \frac{du}{dy} = \frac{x^2 - y^2}{r}\frac{d}{dr}\left(\frac{q}{r}\right) \ldots\ldots\ldots\ldots\ldots(37),$$

and
$$\frac{dv}{dx} - \frac{du}{dy} = r\frac{d}{dr}\left(\frac{q}{r}\right) + 2\frac{q}{r} \ldots\ldots\ldots\ldots\ldots(38).$$

The component of shearing parallel to x and y, is expressed by (37). Taking $y = 0$ we find, for the shearing perpendicular to r in any part of the solid,

$$r\frac{d}{dr}\left(\frac{q}{r}\right) \ldots\ldots\ldots\ldots\ldots\ldots(39).$$

Contrast this with the formula (38) which expresses twice the value of the rotation. Using (36) and (37) in (30) we find,

$$\frac{1}{2} n \times 4 \frac{x^2 y^2}{r^2} \left[\frac{d}{dr}\left(\frac{q}{r}\right) \right]^2 + \frac{1}{2} n \times \left[\frac{x^2 - y^2}{r} \frac{d}{dr}\left(\frac{q}{r}\right) \right]^2 \dots\dots(40).$$

Of this the first term expresses the contribution due to stretchings and shortenings in the directions x and y; and the second term the contribution due to shearing parallel to x and y. Taking the two terms together and modifying algebraically we find for the total amount,

$$\frac{1}{2} n \left[r \frac{d}{dr}\left(\frac{q}{r}\right) \right]^2 \dots\dots\dots(41),$$

which of course might have been derived direct from (39).

26. To find now, the total energy of displacement, in any cylindric portion of the substance of radius b, coaxal with the line of displacement, we must, for the jelly, multiply (41) by $2\pi r dr$ and integrate; and for the ether we must multiply the square of (38) by $\frac{1}{2}n \times 2\pi r dr$ and integrate; taking each integration between the limits $r = 0$ and $r = b$. Modifying the formula in the second case, by performing the integration in a part of it, we find, for the substance within a cylindric surface of radius a,

$$\text{energy of jelly} = \pi n \int_0^b r dr \left[r \frac{d}{dr}\left(\frac{q}{r}\right) \right]^2 \dots\dots (42),$$

$$\text{energy of ether} = \pi n \left\{ \int_0^b r dr \left[r \frac{d}{dr}\left(\frac{q}{r}\right) \right]^2 + 2 q_b^2 \right\} \dots(43),$$

where q_b denotes the value of q for $r = b$. In the case of zero displacement over the bounding surface, $q_b = 0$, and the energy of the jelly is equal to the energy of the ether, verifying our result of § 24 above.

27. As a sub-example let (in accordance with the notation of § 22 above)

$$q = \tfrac{1}{2}\vartheta r \text{ for } r < a \dots\dots(44),$$

$$q = \tfrac{1}{2}\vartheta \frac{a^2}{r} \text{ for } r > a \dots\dots(45).$$

By (39) and (38) we find,

$$\left. \begin{array}{l} 0 = \text{shearing perpendicular to } r; \\ \vartheta = 2 \times \text{rotation}; \end{array} \right\} \text{ for } r < a \dots(46).$$

$$\vartheta \frac{a^2}{r^2} = \text{shearing perpendicular to } r\,; \atop 0 = \text{rotation}\,;} \Bigg\} \text{ for } r > a \dots (47).$$

In this case (42) and (43) give, for the cylindric portion of radius a,

$$\text{energy of jelly} = 0\,; \quad \text{energy of ether} = \tfrac{1}{2}\pi n \vartheta^2 a^2 \dots (48).$$

Similarly by volume-integration, through space to infinity around the cylinder, we find,

$$\text{energy of jelly} = \tfrac{1}{2}\pi n \vartheta^2 a^2\,; \quad \text{energy of ether} = 0 \dots (49)\,;$$

and thus we see that the total energy is the same in the two cases, although its seats are different, being the external space for the jelly, and the internal space for the ether. This case, whether we take jelly or ether as the substance, represents perfectly the circumstances of the electro-magnetic action on the space inside and outside of an infinitely thin circular cylindric solenoid of electric current, according to the principle given in my old paper referred to in § 14 above.

28. A more elaborate illustration, easily worked out, to represent the case of a solenoid of finite length, and to show for the case of very great length, the continuity from uniform rotation in the interior at great distance from the ends, to irrotational circular displacement in the external space around the central parts of the solenoid, was part of the development which in Nov. 1846 I reserved " for a future paper." The whole doctrine of lines of magnetic force, and electro-magnetic solenoids, is now so well known, that I need scarcely now reserve these details for a future paper, and may safely leave them to be worked out by students of electromagnetism as exercises on the subject.

§§ 29—45. *Mechanical representation of the magnetic force of an electro-magnet.*

29. Imagine a piece of endless cord, in the shape of a circle, or of any other closed curve or polygon, to be imbedded in jelly, and a tangential force to be applied to this cord uniformly all round its circuit. To render our representation quite exact, let the substance of the cord be of exactly the same quality as the jelly in which it is imbedded; or simply imagine a portion of homogeneous continuous jelly, to take the place of the cord

which we imagined first to fix the ideas: and let the supposed uniform tangential force be applied directly to this circuital* portion of the ether uniformly all round its circuit. For the present we suppose the greatest transverse diameter of cross-section of the circuital volume to be small in comparison to the radius of curvature of the circuital line: in other words we suppose it to be the space occupied by a thin wire (which need not be of circular section nor of uniform section, nor of uniform gauge, throughout its length), bent into any shape with its two ends united, subject only to the condition that the radius of curvature of the bend must be everywhere great in comparison with the greatest diameter of the wire in any part.

30. The forces thus applied tangentially all round an endless line of the jelly produces a tangential drag on the jelly all around, and causes displacement and distortion less and less at greater and greater distances, becoming nil only at infinitely great distances. The rotatory displacement, or as we shortly call it the rotation, at any point of the jelly, caused by the supposed circuital force, is equal to half the magnetic force at the corresponding point in the neighbourhood of a conducting wire, taking the place of our tangentially applied force and having an electric current steadily maintained through it.

This is the "mechanical representation" of electro-magnetic force due to a closed circuit, deduced from the expression (III.) given for the magnetic force due to an infinitesimal element of a circuit in Art. XXVII., Vol. I., already referred to. It is in-

---

* After much consideration I have adopted this word, though it is not found in the dictionaries. Instead of it I should have said *annular*, were it not that this would unduly limit the idea to be conveyed. *Annular* would describe perfectly a "toroidal" or "anchor-ring" shape, even though considerably deviating from the circular forms of aperture and cross-section, but it could scarcely convey to the mind the idea of an ordinary helix of wire, with its two ends united, whether outside or inside the helix: and still less could it convey the idea of a single endless cord in the form of some complex knot, or of an ordinary piece of knitting with the ends of the thread united: all which configurations are included in the general designation of a circuital portion of space or a circuital piece of matter. In connection with Riemann's geometrical doctrine of multiple continuity, the "circuital" portion of space here considered, of however complicated configuration, has duplex continuity. A piece of matter of any shape with a single hole bored through it, presents the simplest case of a circuital piece of matter and is continuous in species with an ordinary ring of any proportions.

teresting to know that it is applicable not merely to all cases
of coils of wire in electro-magnetic instruments, but to the most
complex piece of knitting, or knotting or weaving, referred to
in the footnote above, with one thread only or any number of
threads, provided only that no thread has an end.

31.   To learn to understand it perfectly however, think first
of the very simplest case,—a thin circular ring with an electric
current somehow maintained through it, an arrangement called

Fig. 2.

a "ring electro-magnet" (*Brit. Association*, Belfast, 1852, *Electrostatics and Magnetism*, Art. 35). Its lines of magnetic force are represented in the accompanying diagram, which was first given in my paper on "Vortex Motion" (*Trans. R. S. E.*, 1869), to represent the lines of motion of a liquid circulating irrotationally through a circular ring. The line *OZ* is perpendicular to the plane of the ring, through its centre, and the diagram shows the lines of fluid motion, in one plane through this axis. Imagine now the smallest oval (approximately a circle) seen with its centre in *OX*, to be a cross-section of a solid circular ring, imbedded in the jelly, and let an infinitesimal rotation be given to this ring round its axis, *OZ*. The jelly dragged round with it will be infinitesimally rotated, in every part. Every part of *OZ* is clearly an axis of the rotation of the jelly around it, and at every part of this line the rotation is in the same direction as that given to the solid ring. Looking at any one of the completed oval curves in the diagram, for instance the largest complete one, (marked ·3), we can readily understand that at the part of the line nearest to the point *O*, the rotation of the jelly is in the same direction as that of the ring. Following this line out to the farthest point, we see how the axis of molecular rotation gradually turns through an angle of 180°, until we come to the most remote part of the oval from *O* where the rotation is round an axis parallel to the axis of the ring, but in a direction opposite to that of the rotation of the ring. Now instead of a thin ring, imagine a thin open circular cylinder of rigid material, of any length very small or very great, imbedded in the jelly and an infinitesimal rotation given to it: we thus have, when the cylinder is very long and the distances of its ends very great from the part of the jelly considered whether outside or inside the cylinder, the case which was fully investigated in § 26 above. We can now readily understand, whether with reference to lines of electro-magnetic force, or with reference to lines of rotational axes in jelly, the continuity of species between the ring electro-magnet and the bar electro-magnet. But the bar electro-magnet of which the mechanical representation is given by a thin *rigid* cylinder imbedded in jelly and infinitesimally rotated, is as we shall see presently, one in which the surface-density of the current round the cylinder diminishes towards each end according to law discoverable only by transcendental analysis.

32.   Whether we consider the thin ring or the cylinder, let us now suppose its material to be jelly of the same rigidity as, and continuous with, the jelly which we are supposing to occupy all space: and let us suppose a force perpendicular to the axis of the cylinder or ring, to be uniformly applied at every point of this portion of the jelly.

The displacement at every point of the substance, whether in the infinite portion not acted on by force, or in the finite portion acted on by force, is calculated synthetically from (III.) of Art. XXVII., Vol. I. above, without any mathematical *analysis*. This is true not only for a ring of circular section, but for one of any sectional figure whatever; look for example at the annexed diagram, representing a cross-section with an angular projection

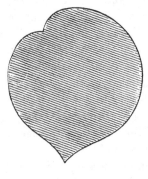

Fig. 3.

and an angular groove.   The synthetical calculation will clearly show a less displacement of the jelly in the neighbourhood of the projecting edge, and a greater displacement in the neighbourhood of the angular groove, than over the rest of the circumference of the portion on which the force acts*.   In the case in

* If the portion acted on were rigid, a highly transcendental problem of analysis would have to be solved, to compel the displacement of the jelly to be the same as that of the boundary of the rigid ring, over the surface of meeting of the two substances.   This problem has no special interest unless of a purely mathematical kind, and I refer to it only to show the significance of the condition (§§ 29—32 above), that the circuital portion of the jelly acted on by force has the same rigidity as all the surrounding jelly within influential distance.

which the cross-section is circular, the applied force will clearly give the greatest displacement at or near the centre of the circle, and the plane circle of the undisturbed cross-section will become infinitesimally convex in the direction in which it is urged by the force (which, be it remembered, is in each part of the ring perpendicular to the cross-section). It will be proved in § 34 below that the convexity* in the case of parallel straight rods, acted on by equal contrari-wards forces, is the same in all parts of the cross-section, whatever be the shape of its boundary.

The subject of these §§ 32 and 33 is the mechanical representation of an electric current flowing through a conductor of whatever shape of cross-section, for example that of the annexed diagram. We have hitherto supposed the force uniform in all parts of the cross-section; and it is on this supposition, that the convexity produced by it is uniform in the case of the straight rods. But if the forces in different parts of the cross-section be different, the electric-current system which is represented is that of the unequal distribution of current in a conductor composed of different metals laid together parallel, so as to constitute one conductor of different conductivity in different parts of its cross-section, but of the same conductivity in corresponding parts of different cross-sections. When, through a conductor thus constituted, a continuous current is maintained by a dynamo or a voltaic battery, the current-density across different parts of the cross-section is in direct proportion to the electric conductivity, provided the radius of curvature of the circuit is large in comparison with diameter of cross-section.

33. In §§ 31—32 we have considered a circular circuit. This has the simplicity of perfect homogeneousness all round the circuit; but on the other hand it presents a problem of which the solution, even in its simplest case, of a very thin ring of circular cross-section can only be expressed in terms of elliptic transcendents. We shall now consider a mathematically much easier

---

* By "the convexity of a surface" I mean the sum of its curvatures in mutually perpendicular normal cross-sections (the algebraic sum, of course, when the two curvatures are in opposite directions). In the present case the surface deviates very little from the plane cross-section of the ring, and any two planes perpendicular to it and to one another are approximately enough the " two mutually perpendicular normal sections " of the definition.

case,—forces of equal total amounts applied longitudinally in opposite di ections to two parallel cylindric or prismatic portions of an infin te homogeneous jelly: the electric-current analogue is, *equal and opposite currents in two straight parallel conductors.* Everything said in § 32 is applicable without change to this case; and conversely the results of the simple mathematical treatment which we can now apply are available to demonstrate the proposition as to convexity, which was stated in § 32.

34. Take OZ parallel to the portions of the jelly of which the forces are applied; so that with the notation of § 12,

$$u = 0, \quad v = 0, \quad X = 0, \quad Y = 0, \text{ and } dp/dz = 0.$$

As we are at present concerned with equilibrium $d^2w/dt^2 = 0$, and thus (18) reduces to

$$n\nabla^2 w + Z = 0 \dots\dots\dots\dots\dots\dots(50);$$

and in the present case we have

$$\nabla^2 w = \frac{d^2 w}{dx^2} + \frac{d^2 w}{dy^2} \dots\dots\dots\dots\dots(51).$$

This shows that $\nabla^2 w$ is equal to the convexity of the disturbed cross-section, and therefore (50) proves all the statements of § 32 regarding convexity.

35. By (50) we see according to Poisson's well-known theorem (Thomson and Tait's *Natural Philosophy*, 2nd Edition, § 494) that the displacement, w, is equal to the potential of an ideal distribution of positive or negative matter, of density numerically equal to $Z/4\pi n$. The displacement of any point of the jelly is therefore calculable by mere summation without other mathematical work, whatever be the forms of cross-sections of the two cylindric portions of the jelly acted on; or generally whatever be the whole distribution of the force, supposing only that Z is given for every pair of values of x and y. The condition that the total amounts of force in the two directions are equal (mathematically expressed by $\int_{-\infty}^{\infty}\int_{-\infty}^{\infty} dx\, dy\, Z = 0$), is imposed, because if it were not fulfilled, w would be infinitely great.

36. Let now force be applied arbitrarily at every point of an infinite homogeneous elastic solid, which for a moment we may

suppose to be not incompressible. Let X, Y, Z denote its components, reckoned per unit of volume, at any point x, y, z. Supposing X, Y, Z to be arbitrarily given functions of x, y, z, we can find u, v, w synthetically (by mere addition) just as we solved the simpler problem of § 35 above. To prove this, take the equations of equilibrium, (18) and (19) of § 13 above, with the accelerational terms of (18) omitted. Taking $\dfrac{d}{dx}$ of the first, $\dfrac{d}{dy}$ of the second, $\dfrac{d}{dz}$ of the third of (18), and using (19), we find

$$\left(\frac{n}{k}+1\right)\nabla^2 p = \frac{dX}{dx}+\frac{dY}{dy}+\frac{dZ}{dz} \dots\dots\dots\dots(52).$$

Hence by Poisson's theorem (Thomson and Tait, § 494), we see that p is equal to the potential of an ideal distribution of matter of density equal to

$$-\frac{k}{n+k}\left(\frac{dX}{dx}+\frac{dY}{dy}+\frac{dZ}{dz}\right)/4\pi \dots\dots\dots\dots(53).$$

Thus determinately we find p by purely synthetical process (mere addition). Then using this value in equation (18) we find by another application of Poisson's theorem u, v, w determinately.

37. We have an exceedingly important case if we take

$$\frac{dX}{dx}+\frac{dY}{dy}+\frac{dZ}{dz}=0 \dots\dots\dots\dots\dots(54)^*.$$

A distribution of bodily force on matter (solid or fluid) continuously occupying space, I call a circuital forcive† if it fulfils this condition (54): because the forcive described in § 29 above is a particular case of it, and it on the other hand may be regarded as constituted of an infinite number of forcives applied along closed curves‡ according to the condition stated in § 29.

---

* See footnote on *Electrostatics and Magnetism*, § 513.

† Forcive is a word which has been introduced by my brother, Prof. James Thomson, to denote quite generally any system of forces such as, for example, a force or a number of forces acting on a rigid body, or on any system of particles, or any distribution of bodily or surface force on an elastic or fluid body.

‡ An example of a circuital forcive is, the triangle of forces or the polygon of forces, in elementary statics, by which I mean forces numerically equal to the sides of the triangle or polygon and applied in the lines of the sides, not as sometimes meant, forces applied in parallel lines through one point (Thomson and Tait's

38. For the case in which the given forcive is circuital, (52) shows that $p = 0$ through all space and therefore equations (18) become

$$n\nabla^2 u + X = 0, \quad n\nabla^2 v + Y = 0, \quad n\nabla^2 w + Z = 0 \ldots\ldots (55).$$

Hence in this case as in the sub-case in § 35 above, u, v, w are equal to the potentials of distributions of matter, having their densities respectively equal to $X/4\pi n$, $Y/4\pi n$, $Z/4\pi n$; and it is interesting to remark that the elastic solid need not be incompressible; the displacement at every point being the same, whatever its compressibility may be, as if its resistance to compression were infinite.

39. The completed mechanical representation of an electromagnet consisting of any distribution whatever of closed electric currents, is as follows:—Let X, Y, Z denote components of electric current per unit of area and F, G, H the components of the magnetic force at (x, y, z) due to it: we have

$$F = \frac{dv}{dz} - \frac{dw}{dy}, \quad G = \frac{dw}{dx} - \frac{du}{dz}, \quad H = \frac{du}{dy} - \frac{dv}{dx} \ldots (56),$$

where u, v, w are determined by equations (55) of § 38 from X, Y, Z, supposed given.

40. From (56), with (20) we find

$$\frac{dG}{dz} - \frac{dH}{dy} = -\nabla^2 u, \quad \frac{dH}{dx} - \frac{dF}{dz} = -\nabla^2 v, \quad \frac{dF}{dy} - \frac{dG}{dx} = -\nabla^2 w \ldots (57),$$

whence by (55),

$$\frac{dG}{dz} - \frac{dH}{dy} = \frac{X}{n}, \quad \frac{dH}{dx} - \frac{dF}{dz} = \frac{Y}{n}, \quad \frac{dF}{dy} - \frac{dG}{dx} = \frac{Z}{n} \ldots (58).$$

Hence if an infinite homogeneous solid (compressible or incompressible) be in some part or parts acted on by a circuital forcive or circuital forcives, we have throughout every part unacted on by force,

$$F = \frac{d\chi}{dx}, \quad G = \frac{d\chi}{dy}, \quad H = \frac{d\chi}{dz} \ldots\ldots\ldots (59),$$

Elements of Natural Philosophy, §§ 411, 416). The polygon of forces (which need not be a plane polygon) applied to a rigid body, is reducible to a single couple, whose component round any axis is equal to twice the area of the orthogonal projection of polygon on a plane perpendicular to that axis.

where χ denotes a function of x, y, z, which obviously must be of the "many valued" class (such as $\tan^{-1} y/x$, or generally the apparent area of a circuit, as in § 43 below). In § 43 the following rule for finding it will be proved. Divide all the space acted on by circuital forcives or forcives into infinitely thin linear portions separated by ideal interfaces, each constituted by endless lines of force; so that each of these linear portions is endless, and may for brevity be called a ring. It is convenient to consider the cross-section of each ring as being an infinitesimal quadrilateral or an infinitesimal hexagon. The defining condition (54) of the circuital forcive is equivalent to a statement*, that for different points of each one of the rings the cross-sectional area varies inversely as the force per unit of volume. Perform the partitioning so that the cross-sectional areas are inversely proportional to the forces across them. Let q be the force per unit of volume, multiplied by the cross-sectional area in any part of the solid which is acted on by force, and n the rigidity of the solid; then $\chi = (q/4\pi \mathrm{n}) \times$ the sum of the apparent areas as seen from P of all the rings into which the part of the solid acted on by force is ideally divided.

The function χ thus determined is, in the electro-magnetic analogue, the magnetic potential at (x, y, z) of the electric current system (X, Y, Z).

§§ 41—43. *Synthesis of a circuital forcive from a single force applied through a space comprised within an infinitely small distance from a point in an incompressible elastic solid (jelly).*

Let P be the force, l, m, n the direction cosines of OK, its direction. The solution (III.) of my old paper† (Vol. I. Art. XXVII.) gives for the displacement at (x, y, z),

$$\left. \begin{aligned} u &= C\left(-\frac{1}{2}\frac{d}{dx}\frac{lx + my + nz}{r} + \frac{l}{r} \right) \\ v &= C\left(-\frac{1}{2}\frac{d}{dy}\frac{lx + my + nz}{r} + \frac{m}{r} \right) \\ w &= C\left(-\frac{1}{2}\frac{d}{dz}\frac{lx + my + n}{r} + \frac{n}{r} \right) \end{aligned} \right\} \dots\dots\dots(60),$$

---

* See footnote on *Electrostatics and Magnetism*, § 513.

† See also Thomson and Tait, § 731. In the reference to "Mathematical and Physical Papers," "Stokes" is written inadvertently instead of "Thomson."

where C denotes a coefficient which must be determined to make the total force applied equal to the given amount P. A ready and easily understood way of doing this is to consider an infinite cylinder of any finite radius, b, having its axis along OK. The jelly within this cylinder drags the jelly outside it with tangential traction, parallel to the axis of the cylinder, diminishing in intensity with distance from O in either direction, and is equal in total amount to P. To work this out in the easiest way take OK along OX. We have, $l = 1$, $m = 0$, $n = 0$, and (56) give

$$u = \frac{1}{2} C \left(\frac{x^2}{r^3} + \frac{1}{r} \right) \Bigg\rbrace$$
$$v = \frac{1}{2} C \frac{xy}{r^3} \quad\quad\quad \Bigg\rbrace \quad \dots\dots\dots\dots\dots(61).$$
$$w = \frac{1}{2} C \frac{xz}{r^3} \quad\quad\quad \Bigg\rbrace$$

The tangential drag parallel to OX, in the plane parallel to ZOX through any point (x, y, z), is as follows,

$$- \text{n} \left(\frac{du}{dy} + \frac{dv}{dx} \right) = \text{n}C \frac{3x^2 y}{r^5} ,$$

where n denotes the rigidity.

Putting $y = b$, and $z = 0$, in this we see that the tangential drag produced by the jelly within, upon the jelly external to the cylindric surface of radius r and axis OX, per unit length of the cylinder, is

$$2\pi b \text{n}C \frac{3bx^2}{(b^2 + x^2)^{5/2}} \quad \dots\dots\dots\dots\dots(62).$$

Hence we have,

$$P = 6\pi b^2 \text{n}C \int_{-\infty}^{\infty} \frac{x^2 dx}{(b^2 + x^2)^{5/2}} = 4\pi \text{n}C \dots\dots\dots\dots(63),$$

whence,

$$C = \frac{P}{4\pi \text{n}} \quad \dots\dots\dots\dots\dots\dots(64).$$

42. Let us now find the displacements due to a circuital forcive applied to an infinitely thin endless line, or ring, $ABQDA$, of the jelly. Let ξ, η, ζ be the coordinates of any point, Q, in this line, and ds an infinitesimal element of its length through Q. Then if q denote the force per unit of length of the circuit, so that the force on the element ds is qds, we may substitute this for P in (64); and the components of displacement at x, y, z

due to this force will be given by (60) with $x - \xi$, $y - \eta$, $z - \zeta$, substituted for x, y, z. Thus taking first one component only, we find

$$\text{contribution to } u = \frac{q\,ds}{4\pi \mathrm{n}}\left[\frac{1}{2}\left(l\frac{d}{d\xi} + m\frac{d}{d\eta} + n\frac{d}{d\zeta}\right)\frac{x - \xi}{r} + \frac{l}{r}\right] \quad \ldots(65)\,;$$

the form of this expression being found from (60) by remarking that

$$-\frac{d}{dx}\frac{x - \xi}{r} = \frac{d}{d\xi}\frac{x - \xi}{r}\,, \qquad -\frac{d}{dx}\frac{y - \eta}{r} = \frac{d}{d\eta}\frac{x - \xi}{r}\,,$$

$$-\frac{d}{dx}\frac{z - \zeta}{r} = \frac{d}{d\zeta}\frac{x - \xi}{r}\,.$$

Remembering now that l, m, n are the direction-cosines of the length ds at the point (ξ, η, ζ) of the circuit, we see that, the first chief part in the second member of (65) disappears in integration round the circuit $ABQDA$, and we find for the total of u, and similarly for those of v and w,

$$u = \frac{q}{4\pi \mathrm{n}}\int ds\,\frac{l}{r}\,, \qquad v = \frac{q}{4\pi \mathrm{n}}\int ds\,\frac{m}{r}\,, \qquad w = \frac{q}{4\pi \mathrm{n}}\int ds\,\frac{n}{r}\,\ldots(66),$$

where $\int ds$ denotes integration round the whole circuit $ABQDA$.

43. Hence by (56) we find

$$F = \frac{q}{4\pi \mathrm{n}}\int ds\,\frac{n(y - \eta) - m(z - \zeta)}{r^3}\,,$$

$$G = \frac{q}{4\pi \mathrm{n}}\int ds\,\frac{l(z - \zeta) - n(x - \xi)}{r^3}\,,$$

$$H = \frac{q}{4\pi \mathrm{n}}\int ds\,\frac{m(x - \xi) - l(y - \eta)}{r^3}\,\ldots\ldots\ldots\ldots\ldots(67).$$

Now from these we verify immediately that

$$F\,dx + G\,dy + H\,dz$$

is a complete differential; and for its integral χ according to notation of (59), we find

$$\chi = \int_{\infty}^{x} dx\,F = \frac{q}{4\pi \mathrm{n}}\int ds\,\frac{[n(y - \eta) - m(z - \zeta)]\,[(x - \xi) - r]}{[(y - \eta)^2 + (z - \zeta)^2]\,r}\,\ldots(68).$$

Calling the points (x, y, z) and (ξ, η, ζ), P and Q respectively, consider the conical surface traced by carrying Q all round the circuit $ABQDA$, and think of the spherical area enclosed by it on a spherical surface of unit radius, having P for its centre. By elementary geometry of three dimensions, the formula after $\int$ in (68) is seen to be equal to the portion of the area contained between the element of periphery corresponding to ds and planes through its extremities and PX, a line through P parallel to x. Hence the complete integral is equal to the whole of the spherical area considered, and therefore, denoting this area by Ω, we have

$$\chi = \frac{q\Omega}{4\pi n} \dots\dots\dots\dots\dots\dots\dots(69).$$

Ω is called the apparent area of the circuit as seen from P. This is the promised proof, of the rule for magnetic potential, and its analogue in the theory of elastic solids which was given in § 40.

44. We have now completed in §§ 29—43, the development which in my short article of forty-three years ago, "I reserved for a future paper." Everything in this development is applicable indifferently to our ideal substance which we have called 'ether', and to an ordinary incompressible elastic solid (jelly). Why not then be satisfied with the ordinary solid for our mechanical representation? The peculiar effect on lines of magnetic force due to "inductive magnetisation," was not alluded to in my article of 1847. It might be imagined that these effects could be included in the analogy by giving different rigidities to the jelly in different parts, to correspond to their different magnetic permeabilities*. But this is not so; and it was the marvellous exigencies of an attempt to include inductive magnetisation†, in the mechanical representation, that compelled the assumption of a quasi-elastic force, depending on absolute rotation, and not otherwise on distortion; and this compelled the introduction of the new ideal substance which we have been calling 'ether'. The interfacial condition to be satisfied at the interface between two substances of different permeabilities in a magnetic field, is that the normal components of the magnetic force are equal on the two sides of the interface, while the tangential components are to one another as the magnetic permeabilities of the substance on the two sides: the term "magnetic force" being here used according to what I have

---

* See *Electrostatics and Magnetism*, § 628.

† See Art. CII. § 19 below.

called the electro-magnetic definition*. The magnetic force
being in our analogy the rotation of the jelly, or ether, we see
by § 20 above that the proper interfacial condition between substances of different rigidity (n), is not fulfilled.by the jelly, and is
fulfilled by the ether.

45. We can now understand perfectly in our mechanical representation, the enormously greater energy of a bar electro-magnet
with a soft iron core, than that of an equal and similar bar electro-
magnet with the same strength of current but with no soft iron
core. In the place of the soft iron core we must suppose ether
of vastly less rigidity than that of the ether through the rest of
space, whether copper or air. If we suppose the copper which
carries the electric current to lie in a thin cylindric shell of
mean radius a, the mechanical analogue, for the case of no soft
iron core, is the fully investigated case in §§ 26—28. Now, to
represent the case of a soft iron core of permeability 300, suppose
the value of n for the ether in the space corresponding to the
soft iron core, to be 1/300 of its value elsewhere, and let the
circuital forcive be the same as that in the former case. Remark
that, within the cylinder and within the space around outside
it, at distances from the axis small in comparison with distances
from the ends, the equilibrium of the ether depends simply on
the balancing of the sum of torques resisting the rotation of
the ether, within the cylindric shell by the circuital forcive in
the shell, there being no rotation, and therefore no contribution
to the forcive balanced, in the ether outside the shell, except
in the neighbourhoods of the ends. Hence in this case, the
rotation, and therefore the energy of the ether within the shell,
must be 300 times what it was in the former case, except near
the ends. In the neighbourhoods of the ends we should have
for the ether, the transcendental problem to solve which Green
attacked and solved for magnetic induction, by a confessedly
imperfect approximation in § 17† of his now celebrated essay

---

* That is to say, the magnetic force in the air in an ideal crevasse, perpendicular
to the lines of magnetisation. To avoid circumlocutions I shall always use the
term "magnetic force" according to this definition, unless in any case it is
expressly stated, that the "polar definition," or the magnetic force in air in an
ideal crevasse tangential to lines of magnetisation is meant. See *Electrostatics and
Magnetism*, footnotes on §§ 516 and 517, and end of postscript to § 517.

† See Green's *Math. Papers*, p. 107.

on the application of mathematical analysis, to the theories of electricity and magnetism.

Contrast this with the circumstances when we have 'jelly' instead of 'ether'. There being no distortion (except in the neighbourhoods of the ends) within the cylinder, the greater or less rigidity of the jelly there makes no difference to the result. We might in fact suppose the jelly removed and a hollow space left within the cylinder. The irrotational motion through the jelly outside, except near the ends, would be the same as if the whole space were filled with homogeneous material. The seat of the energy is altogether outside the cylinder as was seen in (§ 27) above. The equilibrium of the jelly in these circumstances is the balancing of the circuital forcive in the shell, by forcive of elastic resistance to distortion throughout the surrounding jelly, with its irrotational displacement in the parts surrounding the middle of the cylinder, and its rotational displacement corresponding to the lines of magnetic force in the neighbourhoods of the ends.

46. It need scarcely be said that the 'ether' which we have assumed, is a merely ideal substance. It seems to me highly probable however, that the assumed dependence of its forcive on absolute rotation, is at all events analogous to the truth of real ether. Even in the simple assumption of § 14 to which we were forced by the consideration alone of magnetic susceptibility, we tacitly adopted a property which goes straight towards explaining Stokes' theory of aberration*, which is, that the earth and the other heavenly bodies give only irrotational motion to the ether by their motions through it. This in fact is the case for the infinitesimal motions of smooth hard solids or plastic solids filling vesicular hollows in our ether. Generally our 'ether,' whether extending to infinity in all directions, and having vesicular or tubular hollows, or a finite portion of it given with a boundary of any shape, provided that only normal pressure act on the boundary; takes precisely the same motion for any given motion of the boundary, as does a frictionless incompressible liquid in the same space, showing the same motion of boundary. I need scarcely remind persons who know hydrokinetics, that the velocity of any part of the sub-

---

* Stokes, *Math. and Phys. Papers*, Vol. i. pp. 124, 153—156.

stance is always that which for normal component velocity, given at every part of the boundary, has minimum* kinetic energy.

47. Hitherto our representation has been purely static. An obvious kinetic extension is expressed by the equations of motion (18) of § 12 above to any case in which the forcive (X, Y, Z) is applied to a limited portion of the jelly or ether, and is a periodic function of the time. We thus have simply *the undulatory theory of light*, as an inevitable consequence of believing that the displacement of elastic solid by which, in my old paper, I gave merely a "*representation*" of the electric currents and the corresponding magnetic forces, is a reality. But to give anything like a satisfactory material realisation of Maxwell's electro-magnetic theory of light, it is necessary to show *electro-static force* in relation to the forcive (X, Y, Z) of my formulas; to explain the generation of heat according to Ohm's law in virtue of the action of this forcive when it causes an electric current to flow through a conductor; and to show how it is that the velocity of light *in ether* is equal to, or perhaps we should rather say, *is*, the number of electro-static units in the electro-magnetic unit of electric quantity. All this essentially involves the consideration of ponderable matter permeated by, or imbedded in ether, and a *tertium quid* which we may call electricity, a fluid go-between, serving to transmit force between ponderable matter and ether and to cause by its flow the molecular motions of ponderable matter which we call heat. I see no way of suggesting properties of matter, of electricity, or of ether, by which all this, or any more than a very slight approach to it, can be done, and I think we must feel at present that the triple alliance, ether, electricity, and ponderable matter is rather a result of our want of knowledge, and of capacity to imagine beyond the limited present horizon of physical science, than a reality of nature.

48. The following article is a contribution towards showing how a jointed framework subject to gyrostatic domination can fulfil the law of forcive which, without molecular hypothesis, and without any other suggestion as to ultimate cause, is assumed in § 14 for our ether. See also Art. CII. §§ 20—27.

* See Thomson and Tait, Vol. I. § 317, Ex. 3.

ART. C.   ON A GYROSTATIC ADYNAMIC CONSTITUTION FOR
'ETHER.'

§§ 1...6   Translated from *Comptes Rendus*, Sept. 16, 1889.
§§ 7...15 From *Proceedings Royal Society of Edinburgh*, Mar. 17, 1890.

1. CONSIDER the double assemblage of the red and blue atoms
of § 69 of Art. XCVII. above.   Annul all the forces of attraction
and of repulsion between the atoms.   Join each red to its blue
neighbour by a rigid bar, as in the little model which I submitted
to the Academy in my last communication.   We shall thus have,
abutting on each red atom and on each blue, four bars making
between them obtuse angles, each equal to $\pi - \cos^{-1}\frac{1}{3}$.

2. Let us suppose that each atom be a little sphere, instead
of being a point; that each bar is provided at its extremities with
spherical caps (as in § 70 of Art. XCVII.), rigidly fixed to it, and
kept in contact with the surface of the spheres by proper guards,
leaving the caps free to slide upon the spherical surfaces.   We
shall thus have realised an articulated molecular structure, which
in aggregate constitutes a perfect incompressible quasi-liquid.
The deformations must be infinitely small, and such deformations
imply diminutions of volume, infinitely small and of the second
order, or proportional to their squares, which we may neglect.   It
is because of this limitation that we have not a perfect incompressible liquid, without the qualification 'quasi."   But this
limitation does not alter at all the perfection of our ether, so
far as concerns its fitness to transmit luminous waves.

3. Now to give to our structure the quasi-elasticity which it
requires in order to produce the luminous waves, let us attach to
each bar a gyrostatic pair composed of two Foucault gyroscopes,
mounted according to the following instructions.

4. Instead of a simple bar, let us take a bar of which the
central part, for a third of its length for example, is composed of
two rings in planes perpendicular to one another.   Let the centre
of each ring, and a diameter of each ring, be in the line of the bar.
Let the two rings be the exterior rings of gyroscopes, and let the
axes of the interior rings be mounted perpendicularly to the line

of the bar.   Let us now place the interior rings, with their planes
in those of the exterior rings, and consequently with the axes of
their flywheels in the line of the bar.   Let us give speeds of
rotation, equal, but in opposite directions, to the two flywheels.

5.   The gyrostatic pair thus constituted (that is to say, thus
constructed and thus energised) has the singular property of
requiring a Poinsot couple to be applied to the bar in order to
hold it at rest in any position inclined to the position in which it
was given.   The moment of this couple, L, remains sensibly constant until the axes of the flywheels have turned through considerable angles from their original direction in the primitive line of
the bar; and is given by the following formula which is easily
demonstrated by the theory of the gyroscope,

$$L = \frac{(mk^2\omega)^2}{\mu}\, i,$$

i meaning the angle, supposed infinitely small, between the length
of the bar in its deviated position and in its primitive position, m
meaning the mass of one of the flywheels, mk^2 meaning its moment
of inertia, ω meaning its angular velocity, μ meaning the moment
of inertia about the axis of the pivots of the interior ring, of the
entire mass (ring and flywheel) which they support.

6.   Our jointed structure, with the bars placed between the
black and white atoms, carrying the gyrostatic pairs, is not now
as formerly without rigidity; but it has an altogether peculiar
rigidity, which is not like that of ordinary elastic solids, of which
the forces of elasticity depend simply on the deformations which
they suffer.   On the contrary, its forces depend directly on the
absolute rotations of the bars and only depend on the deformations,
because these are kinematic consequences of the rotations of the
bars.   This relation of the quasi-elastic forces with absolute rotation, is just that which we require for the ether, and especially to
explain the phenomena of electro-dynamics and magnetism.

7.   The structure thus constituted, though it has some interest
as showing a special kind of quasi-solid elasticity, due to rotation
of matter having no other properties but rigidity and inertia,
does not fulfil exactly the conditions of Art. XCIX., § 14.   The
irrotational distortion of the substance or structure, regarded as

30—2

a homogeneous assemblage of double points, involves essentially
rotations of some of the connecting bars, and therefore requires
a balancing forcive.   For the 'ether' of Art. XCIX. no forcive
must be needed to produce any irrotational deformation: and any
displacement whether merely rotational, or rotational and de-
formational, must require a constant couple in simple proportion
to the rotation and round the same axis.   In a communication
to the Royal Society of Edinburgh of a year ago, I stated the
problem of constructing a jointed model under gyrostatic domination to fulfil the condition of having no rigidity against irrotational
deformations, and of resisting rotation, or rotational deformation,
with quasi-elastic forcive in simple proportion to rotation.   I
gave a solution, illustrated by a model, for the case of points all
in one plane; but I did not then see any very simple three-
dimensional solution.   After many unavailing efforts, I have
recently found the following.

8.   Take six fine straight rods and six straight tubes all of
the same length, the internal diameter of the tubes exactly equal
to the external diameter of the rods.   Join all the twelve together with ends to one point P.   Mechanically this might be
done (but it would not be worth the doing), by a ball-and-twelve-
socket mechanism.   The condition to be fulfilled is simply that
the axes of the six rods and of the six tubes all pass through
one point P.   Make a vast number of such clusters of six tubes
and six rods, and, to begin with, place their jointed ends so as
to constitute an equilateral homogeneous assemblage of points
P, P', ... each connected to its twelve nearest neighbours by a
rod of one sliding into a tube of the other.   This assemblage of
points we shall call our primary assemblage.   The mechanical connections between them do not impose any constraint: each point
of the assemblage may be moved arbitrarily in any direction,
while all the others are at rest.   The mechanical connections exist
merely for the sake of providing us with rigid lines joining the
points, or more properly rigid cylindric surfaces having their axes
in the joining lines.   Make now a rigid frame G of three rods
fixed together at right angles to one another through one point O.
Place it with its three bars in contact with the three pairs of
rigid sides of any tetrahedron

$$(PP', \ P''P'''), \ (PP'', \ P'''P'), \ (PP''', \ P''P'),$$

of our primary assemblage. Place similarly other similar rigid frames G, G', &c., on the edges of all the tetrahedrons congener (Art. XCVII., § 13) to the one first chosen, the points O, O', $O'O''$ &c. form a second homogeneous assemblage, related to the assemblage of P's just as the reds are related to the blues in Art. XCVII., § 69.

9. The position of the frame G, that is to say its orientation and the position of its centre O (six disposables) is completely determined by the four points P, P', P'', P'''. (Thomson and Tait's *Natural Philosphy*, § 198, and *Elements*, § 168.) If its bars were allowed to break away from contact with the three pairs of edges of the tetrahedrons, we might choose as its six coordinates, the six distances of its three bars from the three pairs of edges; but we suppose it to be constrained to preserve these contacts. And now let any one of the points P, P', P'', P''' or all of them be moved in any manner, the position of the frame G is always fully determinate. This is illustrated by a model accompanying the present communication, showing a single tetrahedron of the primary assemblage and a single G frame. The edges of the tetrahedron are of copper wires sliding into glass tubes. The wires and tubes are provided with an eye or staple respectively, through which a ring passes to hold three ends together at the corners. Two of the rings have two glass tubes and one copper wire linked on each, while the other two rings have each two copper wires and one glass tube.

10. Returning now to our multitudinous assemblage, let it be displaced by stretchings of all the edges parallel to PP' with no rotation of PP' or $P''P'''$. This constitutes a homogeneous irrotational deformation of the primary assemblage. The frames G, G', &c. experience merely translatory motions without any rotation, as we see readily by confining our attention to G and the tetrahedron PP', $P''P'''$. Consider similarly five other displacements by stretchings parallel to the five other edges of the tetrahedron. Any infinitely small homogeneous deformation of the primary assemblage (§ 8 above), may be determinately resolved into six such simple stretchings, and any infinitely small rotational deformations may be produced by the superposition of a rotation without deformation, upon the irrotational

deformation.   Hence an infinitely small homogeneous deformation of the primary assemblage without rotation, produces only translatory motion, no rotation of the G frames: and any infinitely small homogeneous displacement whatever of the primary assemblage, produces a rotation of each frame equal to, and round the same axis as, its own rotational component.

11.   It now only remains to give irrotational stability to the G frames.   This may be done by mounting gyrostats properly upon them according to the principle stated in §§ 3—5 above and Art. CII. §§ 21—26 below.   Three gyrostats would suffice but twelve may be taken for symmetry and for avoidance of any resultant moment of momentum of all the rotators mounted on one frame.   Instead of ordinary gyrostats with rigid flywheels we may take liquid gyrostats as described below, § 12, and so make one very small step towards abolishing the crude mechanism of flywheels and axles and oiled pivots.   But I chose the liquid gyrostat at present merely because it is more easily described.

12.   Imagine a hollow anchor ring, or tore, that is to say an endless circular tube of circular cross-section.   Perforate it in the line of a diameter and fix into it tubes to guard the perforations as shown in the accompanying diagram.   Fill it with frictionless liquid, and give the liquid irrotational circulatory

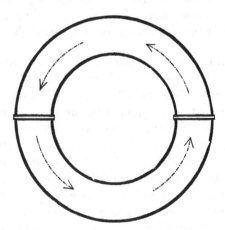

motion as indicated by the arrow heads in the diagram.   This arrangement constitutes the hydrokinetic substitute for our mechanical flywheel. Mount it on a stiff diametral rod passing through the perforations, and it becomes the mounted gyrostat, or Foucault gyroscope, required for our model. Looking back to §§ 3—4 above we see how much its use would have simplified and shortened the descriptions there given, which however was given purposely as they were because they describe real mechanism by which the exigences of our model can be practically realised in a very interesting and instructive manner, as may be seen in Art. CII., §§ 21—23 below.

13. Let XOX', YOY', ZOZ' be the three bars of the G frame: mount upon each of them four of our liquid gyrostats, those on XOX' being placed as follows and the others correspondingly. Of the four rings mounted on XX' two are to be placed in the plane of YY', XX', the other two in the plane of ZZ', XX'. The circuital fluid motions are to be in opposite directions in each pair.

14. The gyrostatic principle stated in § 5 above, applied to our G frame, with the twelve liquid gyrostats thus mounted on it, shows that if, from the position in which it was given with all the rings at rest, it be turned through an infinitesimal angle i round any axis, it requires, in order to hold it at rest in this altered position, a couple in simple proportion to i; and that this couple remains sensibly constant, as long as the planes of all the gyrostats have only changed by very small angles from parallelism to their original directions. Hence with this limitation as to time our primary homogeneous assemblage of points controlled by the gyrostatically dominated frames G, G' &c. fulfils exactly the condition stated for the ideal ether of § 14 of Art. XCIX. If the velocity of the motion of the liquid in each gyrostat be infinitely great, the system exerts infinite resistance against rotation round any axis; and if the bars and tubes constituting the edges of the tetrahedron, and the bars of the G frames are all perfectly rigid, the primary assemblage is incapable of rotation or of rotational deformation: but if there is some degree of elastic flexural yielding in the edges of the tetrahedron, or in the bars of the G frame, or in all of them, the primary assemblage fulfils the definition of gyrostatic rigidity of § 14 Art. XCIX. without any limit as to time, that is to say with perfect durability of its quasi-elastic rigidity.

15.   A homogeneous assemblage of points with gyrostatic quasi rigidity conferred upon it in the manner described in §§ 8—14 would, if constructed on a sufficiently small scale, transmit vibrations of light exactly as does the ether of nature: and it would be incapable of transmitting condensational-rarefactional waves, because it is absolutely devoid of resistance to condensation and rarefaction.   It is in fact, a mechanical realisation of the medium to which I was led one and a half years ago*, from Green's original theory, by purely optical reasons, in endeavouring to explain results of observation regarding the refraction and reflection of light.

* *Philosophical Magazine*, Nov. 1888, On the reflection and refraction of light, by Sir W. Thomson.

ART. CI. ON AN ACCIDENTAL ILLUSTRATION OF THE SHALLOW-
NESS OF A TRANSIENT ELECTRIC CURRENT IN AN IRON BAR.

[Read March 17, 1890.]

1. AFTER the recent meeting of the British Association at
Newcastle, Lord Armstrong, in showing me the appliances by
which his house at Cragside is lighted electrically by water-power,
told me of a very wonderful incident which he had recently experienced. A bar of steel, which he was holding in his hand,
was allowed accidentally to come in contact with the two poles
of a dynamo in action. He instantly perceived a painful sensation
of burning, and let the bar drop. He found his hand or fingers,
where it had touched the bar, severely blistered. The bar itself
was found immediately afterwards to be quite cold, or not perceptibly hot. This was a very marvellous incident. It proved
(1) the outer surface of the steel to have been intensely heated;
(2) that not enough of heat was generated to sensibly warm the
whole bar. The explanation, of course, was to be found in the
known laws of diffusion of electric currents, through non-magnetic conductors, considered in connection with the effect of
magnetic susceptibility of unknown amount and law, in conductors
of steel or iron.

2. Lord Armstrong's accidental experiment seemed to me such
a very instructive illustration of fundamental principles of electro-
magnetic induction, that I wrote to him asking his permission to
communicate it to the Royal Society of Edinburgh, at the same
time inquiring as to some details. In reply I immediately received a letter, of date 7th March, kindly giving the desired
permission, and containing the following very interesting statement :—

"I send you, by parcel post, the steel (not iron) bar which I
"held when it accidentally short-circuited the current. You will
"observe two little hollows*, which were burnt out of the metal

* The distance between the hollows is 15½ cms., the bar is about a foot long, and
its diameter is 14 mm.

" at the instant of contact, and these mark the distance between
" the points of contact. The bar was held by my fingers midway
" between these two marks, and the burns were inflicted at the
" places where my fingers touched the metal. The sudden pain
" caused me to dash the bar instantaneously to the ground, and
" an attendant *immediately* picked it up and found it quite cold.
" Three of my fingers and my thumb were blistered, and had the
" injuries not been immediately treated by an expert who
" happened to be present, they would probably have developed
" into troublesome sores ; as it was, my arm had to be carried in
" a sling during the first day, and I was not able to hold a pen
" with comfort for many days afterwards. There was a great
" blaze of light from the two points of metallic contact but the
" flame could not possibly have got under my fingers where they
" touched the metal and were burnt. If the flame had done
" the injury, it would have taken effect upon the exposed parts
" of my hand, but nothing was scorched except the skin at the
" points of grasp.
    " The dynamo was of Crompton's pattern, with compound
" winding. The speed was about 1300 revolutions per minute.
" The dynamo was not employed in charging batteries, but in
" the direct lighting of incandescent lamps. The duty of the
" dynamo at the time would be about 85 amperes and the
" potential 103 volts. No check in the dynamo was perceived,
" nor was it likely to be observable, seeing that the momentum
" of the revolving parts would be enormously powerful to over-
" come any momentary increase of resistance. There being two
" dynamos on one axis, both in motion, though only one was doing
" work, besides the turbine-wheel and the impinging jets, there
" would be a collective momentum of great energy for a momen-
" tary effort."

    3. The requisites for working out fully the theory of transient
or periodic currents in conductors of any form, are included in
Maxwell's fundamental equations of electro-magnetic induction,
and are given explicitly for straight cylindric conductors of non-
magnetic material in §§ 685—689 of his great work. Lord
Rayleigh in his paper " On the Self-Induction and Resistance
of Straight Conductors" (*Phil. Mag.*, May 1886) gives explicitly
the proper formulas for transient or periodic currents, in straight

cylindric rods of iron, on the supposition of constant magnetic susceptibility. The details of this highly interesting and important branch of the subject have also been investigated by Heaviside in a very comprehensive manner. The tendency of periodically alternating currents to be condensed in the outer part of a cylindric conductor, while the current may be exceedingly feeble or quite insensible in the central parts, was discussed and explained by Lord Rayleigh in p. 388 of his article already referred to, and its aggravation in an iron conductor specially pointed out. The same considerations show that a transient current resulting from the application, for a very short time, of the electromotive force of a voltaic battery, or of electro-magnetic induction acting not directly on* the cylindric conductor considered, but on a conductor such as the inductor of a dynamo of any kind, momentarily in circuit with it, is only skin deep if the duration of the electromotive force be but short enough; and that the depth to which the current penetrates in a given very short time is much smaller for iron than for copper. This is certainly the explanation of Lord Armstrong's wonderful experiment.

4. To find something towards a mathematical solution for the increasing current at any instant during the electric contact of Lord Armstrong's experiment, it is convenient first to solve the problem of finding the subsidence of current initially given in a circuit of two very long parallel bars connected by end bridges, or in a circuit of one long bar insulated within a conducting sheath except at its ends, which are in metallic connection with the sheath.

5. In all cases of electric currents given in parallel straight lines, and left to subside†, without any other electromotive force

---

* This caution is introduced to avoid leading any reader into an error into which I myself fell in the text, and corrected in footnotes, of an article in the *Philosophical Magazine* for March 1890, "On the Time-Integral of a Transient Electro-Magnetically Induced Current."

† The thermal analogue for a varying or constant electromotive force applied by a voltaic battery or dynamo substituted for one of the end-bridges is positive and negative sources of heat applied at the interfaces between the thermal analogues of electric conductor and electric insulator. The quantity of heat generated per unit of area per unit of time, at any point of either interface in the thermal analogue, is equal to the rate of variation per unit of length along the electric conductor of the

than that of their mutual electro-magnetic induction, the thermal
analogy is exceedingly convenient. For electric conductivity, c,
we have thermal capacity divided by 4π; for magnetic permeability, the reciprocal of thermal conductivity; and for current-
density, temperature multiplied by thermal capacity. Thus, if
two infinitely long straight parallel conducting bars, separated
by insulating material, be given with equal currents in opposite
directions through them, and left to themselves, we have precisely
the same mathematical problem to solve as if in every line of the
thermal analogue, we had initial temperature multiplied by
thermal capacity given equal to the current-density in the corresponding line of the electro-magnetic problem, and the system left
to itself, with the positive and negative temperatures in the two
bars subsiding towards zero.

6. The thermal analogue for the insulating material of the
electro-magnetic problem is an ideal medium of zero thermal
capacity. Thus in process of equalisation of temperature we have
diffusion of heat through the substance of each bar, according to
Fourier's original use of the term diffusion; while in the ideal
medium taking the place of the electric insulator, we have merely
conduction of heat, without any diffusion properly so called, that
is to say, without any excess of heat conducted out of, above heat
conducted into, any portion of the medium.

7. If ρ denote the temperature at time t, in the thermal
analogue, at any point, P; $4\pi c$ the thermal capacity per unit of
volume; $1/\varpi$ the thermal conductivity of the analogue to either
of the electric conducting bars; and $1/\varpi'$ the thermal conductivity
of the analogue to the insulating medium : the equations express-

electrostatic force in the insulator in contact with it in the electric analogue.
Remark that, in the electric system the potential is uniform over each normal
section of either conductor, and that the variation of potential within each conductor per unit of distance along its length, that is to say, the component electrostatic force, in the direction of the length, is exceedingly small in comparison with
the component electrostatic force perpendicular to the length, at any place in the
insulator except close to the ends metallically connected by a bridge. The equations in the text are unchanged, except the second interfacial condition; it becomes

$$\sigma + \frac{1}{\varpi}\left[\frac{d\rho}{d\nu}\right] = \frac{1}{\varpi'}\left[\frac{d\rho}{d\nu}\right]',$$

where σ denotes the quantity of heat generated per unit of time in the source

ing all the conditions of the problem are

$$4\pi c \frac{d\rho}{dt} = \frac{1}{\varpi}\left(\frac{d^2\rho}{dx^2} + \frac{d^2\rho}{dy^2}\right), \ [P, \text{ in either bar}],$$

$$0 = \frac{d^2\rho}{dx^2} + \frac{d^2\rho}{dy^2}, \ [P, \text{ in the analogue to the insulating medium}]:$$

$$\left.\begin{array}{l} [\rho] = [\rho]', \\ \dfrac{1}{\varpi}\left[\dfrac{d\rho}{d\nu}\right] = \dfrac{1}{\varpi'}\left[\dfrac{d\rho}{d\nu}\right]' \end{array}\right\} [P, \text{ in the interface}];$$

where $d/d\nu$ denotes rate of variation per unit length in the direction of the normal, at any point of the interface: and [ ], and [ ]′ denote values infinitely near the interface outside and inside respectively.

8. In the particular case of one of the bars of circular cross-section, and the other a hollow circular cylinder surrounding it coaxially, the problem becomes greatly simplified. It becomes still farther so if we suppose the electric conductivity, or in the thermal analogue the thermal capacity, of this outside sheath to be infinitely great. In this last case we have identically the same mathematical problem as that regarding a heated cylinder left to cool, which was presented and fully solved by Fourier in the sixth chapter of his great work (*Théorie analytique de la Chaleur*). Instead of the bodily thermal conductivity divided by surface emissivity of Fourier's problem, essentially a line which we shall denote by λ, we have in the electro-magnetic problem,

$$\lambda = \frac{\varpi'}{\varpi} a \log \frac{b}{a},$$

when ϖ and ϖ' are the magnetic permeabilities of the conducting rod and insulating medium around it, and a and b the radii of the cross-sections of the rod, and of the inner surface of the enclosing conductor.

9. Not considering for the present the interesting case suggested by Heaviside of an insulating medium composed of soft iron filings imbedded in wax or other ordinary insulating solid, we have practically $\varpi' = 1$, whether the insulator be air or any ordinary insulating solid or liquid.

10. Consider now two cases—a copper rod and a steel or iron rod, each of the same diameter, 1·4 cm. as Lord Armstrong's steel rod, and suppose, for example, b equal to ten times a, we have

$$\lambda = 2 \cdot 3a \doteqdot 1 \cdot 6 \text{ cm. for copper};$$
$$\lambda = \varpi^{-1} \,.\, 2 \cdot 3a = \varpi^{-1} \,.\, 1 \cdot 6 \text{ cm. for steel or iron.}$$

If b, instead of being 7 cm., were 70 cm. or 700 cm., λ would be only doubled or tripled; on the other hand, if b were very small in comparison with a, $\lambda \doteqdot b - a$. Excluding this case, we see that for copper λ is greater than a, or not incomparably less than a. We thus have a very fine and a very easy example for working out numerical results by Fourier's solution.

11. On the other hand, for an iron or steel rod we have for ϖ some large number, possibly about 300; or if the currents and therefore the magnetic forces concerned are very small, we may, according to Lord Rayleigh's important experimental investigation on the subject of magnetic induction* by very small magnetic forces, have ϖ as small as 80; on the other hand, for moderately strong currents and correspondingly high electromotive forces, we may have ϖ greater than 3000†. We shall take it as 300 merely by way of example and illustration, but as the permeability varies enormously with the amount of the magnetising force, and in a manner desperately complicated by magnetic retentiveness, hysteresis according to Ewing's designation, no accurate mathematical investigation is practicable with only our present knowledge of the requisite data for the diffusion of electric currents through an iron or steel conductor.

12. Taking for iron or steel $\varpi = 300$, and, as above, $a = \cdot 7$, $b = 7$, we find $\lambda = 1/130$ of a, or $1/187$ of a centimetre. Now, because λ is so small a fraction as $1/130$ of the radius of the rod, we see that the current-density at the surface (or surface-temperature in the thermal analogue) drops nearly to zero, while there is still but a relatively small diminution of current density (or temperature in the thermal analogue) farther in from the surface than a distance of $1/10$ of the radius. Hence, for the roughly approxi-

---

* This collocation of words illustrates the exceeding inconvenience of Maxwell's use of "magnetic induction" to designate the magnetic force in an air-crevasse perpendicular to the lines of magnetisation in magnetised steel or soft iron.

† Rowland for one specimen of iron found the magnetic permeability as high as 3595 for magnetising force 1·317 (see *Phil. Mag.*; Aug. 1873).

mate investigation with which we must be content, we may be satisfied with the very simplifying supposition of $\lambda = 0$.

13.   If we take 10000 c.g.s., or square centimetres per second, as the resistivity* of steel or iron, we must divide this by $300 \times 4\pi$ to find the diffusivity for electric current (thermal diffusivity, or conductivity divided by thermal capacity of unit volume, in the thermal analogue), which therefore is 2·7 square centimetres per second.   This is only about 15 times the thermal diffusivity of heat in iron (which is ·18 of a square centimetre per second)   Hence in 1/4300 of a second (if $\lambda = 0$), the state of things as regards falling off of the strength of current towards the final zero, at different distances from the surface, would be that represented by number 0·1 curve of my diagram of laminar diffusion†.   That is to say, the diffusion curve would be curve number 1 with its vertical ordinates reduced to 1/10, or curve number 10 with its vertical ordinates reduced to 1/100. Now by number 1 curve we see that at 1/2 centimetre from the initiational surface (supposed plane) the amount of the falling off is 16 per cent. of the whole.   Hence in our iron or steel rod (on the supposition $\lambda = 0$) the current at 1/4300 of a second from the beginning would have fallen off by 16 per cent. of its given amount.   Thus we judge that during the first 1/4000 of a second the effect of the cylindric curvature is but slight, and the diffusion follows sensibly the law of plane laminar diffusion.

14.   The supposition of a circular cylindric sheath of infinite electric conductivity, coaxial with the rod considered, and separated from it by the insulating material, which we have adopted for the sake of simplicity and definiteness, may be departed from, and instead we may substitute any conductor parallel to the rod considered, provided that the distance between the two is a considerable multiple of the greatest diameter of either.   In virtue of this proviso, the distribution of current-density is necessarily but very little disturbed from the equality at equal distances from the axis of the rod, which was provided

---

* This seems to me a much better word than specific resistance to denote the resistance per centimetre of length of a bar of a square centimetre of cross-section of any substance.   The resistivity of Lord Armstrong's steel bar, I have found by measurement to be about 14000 c.g.s.

† See *British Association Report*, Bath, 1888, p. 571; or Art. xcviii. § 7 of the present Volume.

for by the supposition of a cylindric sheath. If the second conductor is of the same diameter and of the same material as the
first, and placed at a distance $2a$ from it, the expression for λ will
be the same as that given above.

15. Now, instead of the great simplicity of a current, generated (no matter how), and given initially as a steady current,
through the circuit of two long parallel conductors and the end-
bridges between them; suppose the conductors and one end-bridge
to be given with no current, and let a voltaic battery be suddenly
applied instead of the other end-bridge. If the difference of
potentials maintained by the latter between the ends to which
it is applied be absolutely constant, the rise of the current through
the parallel conductors from its initial zero to its ultimate steady
amount will follow *nearly* the same law as the fall from the initial
steady current to the final zero in our former simple case: *exactly*
the same law of the quantity of positive electricity on one conductor, and of negative electricity on the other, called forth according to electro-static law, in virtue of the gradient of potential,
is nothing in comparison with the quantity flowing through the
circuit. The quantity of electricity required for the static electrification is not negligible in a large variety of telegraphic and
telephonic problems*. In the ocean cable problem it is of
paramount importance; and the electro-magnetic induction with
which we are now occupied is negligible. In shorter cables and
high-speed action both the electro-static and electro-magnetic
induction must be taken into account, and both have been very
practically taken into account by Heaviside in the mathematical
theory of the telephone.

16. In a vast variety of laboratory experimental arrangements
the currents required for the static charges are quite negligible.
In such a case as that of Lord Armstrong's experiment the quantity
of electricity required for producing or changing the electrification
of every part of the circuits or of the conductors concerned is
clearly quite insensible in comparison with the quantity which
must have flowed through the bar to produce the observed heating
effect.

17. But although we are not troubled with any difficulty in
respect to electro-static charge, we have in Lord Armstrong's case

* *Papers*, Vol. II. Arts. 72—77 and 80—83.

circumstances of such extreme complexity that it is of no use to attempt to work out a complete mathematical theory. It seems probable, however, that the solution indicated above, and represented by my diffusion diagram, fairly illustrates the circumstances of the current which actually flowed through the steel bar, though scarcely with any approach to quantitative correctness. At all events we have a very striking illustration of what really took place, and ample explanation of the intensity and suddenness of the effect perceived by Lord Armstrong, by working out the result numerically of what would take place if a difference of potentials of 100 volts were suddenly instituted, and forcedly maintained constant during one or two or three ten-thousandths of a second between two points of the bar $15\frac{1}{2}$ centimetres asunder. This with any reasonable assumption as to the magnetic permeability of the iron or steel bar, and its diffusivity for electric currents, is easily done on the supposition $\lambda = 0$, and the solution conveniently represented by the diffusion diagram.

18. We must not, however, suppose that the difference of potentials between the two points of the steel bar touched by the main electrodes of the dynamo was in reality constant at 103 volts, even for so long a time as two or three ten-thousandths of a second. What was really constant must have been the strength of the current through the electrodes leading from the complex circuit of dynamo-armature and of shunt and series coils of the electromagnet, to the external permanent electric lighting circuit and temporary circuit through the steel bar. The difference of potentials between the two points of the steel touched must have been at first 103 volts, and must have fallen very rapidly, while the current which it produced in the steel rose from 0 to 85 amps, against ohmic resistance sinking from infinity towards ·00037 of an ohm (this being the actual resistance of Lord Armstrong's bar, to a current running full-bore through it, as I have found by measurement).

19. The immense quasi-inertia of each partial circuit within the dynamo forbids the supposition that there can have been any great augmentation of the outgoing current during the few hundredths of a second of the short-circuiting by the steel bar;

and, probably with no practical error, we may suppose that current to have been constant during the whole time. Hence, at each instant the electric lighting circuit must have lost just as much current as that which was passing through the steel bar. Hence, considering the smallness of the quasi-inertia of the electric lighting circuit, the 85 amperes through it before the accident must, after two or three ten-thousandths of a second, have been very nearly annulled: and, therefore, very nearly a constant current of 85 amperes must have passed for the rest of the time through the outer skin of the steel bar*. We have thus no

---

* This suggests an interesting and happily an easy problem regarding electro-magnetic induction in rectilinear electric currents in a conductor surrounded by an insulator. Let the electromotive action, whatever its kind, be so regulated that the integral amount of current crossing the normal section of the conductor is kept constant. The mathematical statement of this condition according to the notation of § 7 above, is

$$\frac{d}{dt} \int\int dA \, \rho = 0,$$

where $\int\int dA$ denotes surface integration over the cross section of the conductor. From this, by the first equation in § 7, we have

$$\int\int dA \left(\frac{d^2\rho}{dx^2} + \frac{d^2\rho}{dy^2} \right) = 0.$$

Now, as is well known, and very easily proved, we have in every case

$$\int\int dA \left(\frac{d^2\rho}{dx^2} + \frac{d^2\rho}{dy^2} \right) = \int ds \, \frac{d\rho}{d\nu},$$

where $\int ds$ denotes integration all round the border of the cross-section. Hence, the condition for constant total amount of current is simply

$$\int ds \, \frac{d\rho}{d\nu} = 0.$$

For the case of circular cross-section with uniform electric conductivity in all parts of it; and with the circuit-completing conductor either a coaxial cylindric sheath, or a conductor of any form whatever, provided only that no part of it is near enough to the considered part of the given conductor to sensibly disturb the distribution of the current through its circular cross-section from being of equal current-density at equal distances from the axis, the condition for constancy of total current becomes simply,

$$\frac{d\rho}{dr} = 0,$$

at the boundary of the conductor, where r denotes distance from the axis. The full numerical solution of this problem from the instantaneous commencement of a current of given total strength (which must necessarily be in the very outer skin, and must require an infinite current-density for the first instant) through the whole time until the current becomes as nearly as may be uniform throughout the cross-section, is particularly easy but must be reserved for a future communication. It is identi-

difficulty in understanding that there should have been amply sufficient current through an exceedingly thin shell of the bar to produce very suddenly the high temperature of the surface which Lord Armstrong perceived, and yet that the total amount of heat generated was insufficient to heat the bar to any sensible degree after the second or two required for the thermal diffusion (diffusivity ·18 of a sq. cm. per sec.), to spread it nearly uniformly through the body of the bar. The heat lost outside the bar by surface emissivity (which is about 1/4000 of a gramme water thermal unit per second per sq. cm. of surface per degree of excess) would be quite ineffective to considerably diminish the whole quantity in the time required for diffusion to nearly equal temperature throughout the bar. If the dynamo had been doing no work externally at the time of the accident, the time required to get up a strong enough current out of the dynamo to produce much heating effect would have been very much longer than it was. The result to Lord Armstrong might not have been very noticeably different from what it was, but the attendant's fingers would have been burned also.

cal with the following particular case of Fourier's thermal problem: Let a given quantity of heat be initially distributed uniformly through an infinitely thin surface-layer of a solid cylinder, coated with an impermeable surface varnish; it is required to find, for any subsequent time, the temperature at any distance inwards from the surface of the cylinder.

ART. CII. ETHER, ELECTRICITY, AND PONDERABLE MATTER.

[Part of the Presidential Address to the Institution of Electrical Engineers,
Delivered January 10, 1889.]

1. THE demand for something like a mechanical explanation
of electrical phenomena is not new, but it is growing in intensity
every year. The proceedings of recent meetings of the British
Association—and especially of the last meeting of the British
Association—illustrate the growing desire to know something
below the surface; to know something of the internal relations
connected with the wonderful manifestations of force and energy
which are put before us in the action of the magnet, in the
working even of a common electrical machine, and in electro-
magnetic phenomena. The Addresses of Past-Presidents of the
parent Societies of Telegraph Engineers, and Telegraph-Engineers
and Electricians, illustrate also the growing desire to know something of the molecular theory or the dynamical theory of
electricity and magnetism. Mr Preece, in his Address of 1880,
pointed out how Maxwell had shown the velocity of light to be
related to electricity in such a way that we can scarcely doubt
but that the propagation of electro-magnetic disturbance through
space, which we have every reason to believe does exist—which,
in fact, from known laws we may say certainly does exist—is
effected with a velocity equal to that of light, and that the
propagation of electrical disturbance and of light may be identical.
In support of these remarks, Mr Preece alluded to the disturbances at the sun's surface and the simultaneous magnetic
disturbances which had been observed in the telegraphs and in
other operations of an electro-magnetic character on the surface of
the earth.

2. In 1883 Mr Willoughby Smith described shortly some
experiments, very beautiful and very instructive, with which he
was then engaged. Those experiments demonstrated and illustrated the screening effect of sheets of different kinds of metal
upon electro-magnetic and electrostatic inductions. Electric induction, simply, we may say, because we begin to fail to distinguish between electrostatic induction and electro-magnetic

induction. In Willoughby Smith's subsequent work he gave an exceedingly beautiful set of experimental investigations of the screening effect of lead, copper, and iron, of which, as I have said, a slight sketch was given in his Presidential Address. A little earlier the subject was mathematically worked out with great power by several mathematicians, but perhaps most notably by Horace Lamb. I feel it almost invidious to mention names when there are so many thorough workers who touched upon the same subject very closely. Charles Niven *, three years earlier than Lamb†, went through very much the same kind of work—in fact, obtained the same solutions of some important and interesting problems regarding electric currents in spherical conductors. I specially mention Lamb's name because the subject of screening is more particularly developed in his mathematical paper.

3. In the memorable Presidential Address of Professor Hughes, another allied branch of electro-magnetic induction was very admirably illustrated by experiments which are now more or less familiar to us all, but which have been of an immensely suggestive and stimulating character, both to mathematicians and to experimental workers. The very criticisms by mathematicians upon some of the experiments and modes of statement by Professor Hughes have, with Professor Hughes's own experiments, given a large body of electric knowledge and electromagnetic knowledge which, without such stimulus and such mathematical and experimental scrutiny as it has led to, might have been wanting for many a year.

4. One of the earliest problems in which electric induction had to be considered was that of the submarine telegraph. The subject of induction in telegraph wires presented itself in a peculiarly perplexing way to the first workers in that department. There was the general knowledge of electro-magnetic induction between two wires, which had been worked out by Faraday and Henry in a very full manner. That was the only kind of induction which was thought of by some of the pioneers

* "On the Induction of Electric Currents in Infinite Plates and Spherical Shells," *Phil. Trans. Roy. Soc.*, 1881, p. 307 (read Jan. 29th, 1880).
† "On Electrical Motions in a Spherical Conductor," *Phil. Trans. Roy. Soc.*, 1883, p. 519 (read April 5th, 1883).

of submarine telegraphy. Another kind of induction was more
thought of by others. That was the electrostatic induction due
to the Leyden-jar charge of the insulated wire. Faraday, in this
department, as in the other department, was the origin of nearly
all that we now know. He explained in a very beautiful and
clear way the electrostatic charge of the submarine cable, and
showed how the electricity conducted through the cable from one
end, to give what of potential is necessary in the middle or the
other end of the cable—in the middle of the cable for any mode
of working, at the other end of the cable for modes of working in
which the other end is insulated—gave rise to the Leyden-jar
charge. He pointed out (without going into any of these details,
however) the doctrine of the conduction of electricity through
the wire to supply the Leyden-jar charge which the wire must
have, in the course of working, in order to be raised to the
difference of potential from the earth required to cause the signal
current to pass through it. Cromwell Varley made very important advances in that direction. At the meeting of the
British Association in 1854, at Liverpool, he brought forward
some important developments of Faraday's doctrine. And then
came on the great Atlantic Cable question. I always remember
how that question came upon me. I see in Professor Stokes's
presence with us this evening a reminder of the circumstances.
I was hurriedly leaving the meeting of the British Association,
when a son of Sir William Hamilton, of Dublin, was introduced
to me with an electrical question. I was obliged to run away
to get to a steamer by which I was bound to leave for Glasgow,
and I introduced him to Professor Stokes, who took up the subject
with a power which is inevitable when a scientific question is submitted to him. He wrote to me on the subject soon after that
time, and some correspondence between us passed, the result of
which was that a little mathematical theory was worked out,
which constituted, in fact, the basis of the theory of the working
of the submarine cable. In that theory, electro-magnetic induction was not taken into account at all. The leaving it out of
account was justified by the speed of signalling which the circumstances of a cable exceeding 200 or 300 miles in length dictated.
For a cable more than 200 or 300 miles long the speed of
working was essentially limited by these electrostatic considerations—limited so much that the electro-magnetic induction

certainly could have no sensible effect. But the possible speed
of working in a cable of 20 miles or 50 miles, or even 100 miles,
was so great that, in short lengths like that, the electro-magnetic
induction might well come into play. I worked out the subject
partially myself. I found it necessary to do so to satisfy myself
that the doctrine upon which the Atlantic Cable project, then
growing up, was ultimately founded, was thoroughly trustworthy,
I found it necessary to investigate the question of electro-
magnetic induction. This question was further forced upon me
by communications that I had with my friends, Lewis Gordon,
and the two brothers, Charles Wm. Siemens and Werner Siemens,
with reference to Mediterranean cables. It was imagined that
electro-magnetic induction alone was operative—that embarrassment in working through the submarine cable was due to
electro-magnetic induction alone. On its being demonstrated by
me that electro-magnetic induction could have no sensible effect
on the signalling through proposed Mediterranean cables, the
proposal to have two thin wires close together in order to obviate inductive embarrassment, was given up. Experiments in
Germany had shown considerable electro-magnetic induction on
short lengths of cable, and it had been supposed that there would
be embarrassment from this cause in the working of the cables,
which would be diminished by using wires very close together.
But this diminution of the electro-magnetic inductive influence
would produce a corresponding increase in the electrostatic inductive influence; and when it was pointed out that the electro-
magnetic inductive influence would be absolutely imperceptible
at the highest speeds of working of the proposed cables, and that
it would be the electrostatic induction which would limit the
speed, the idea of making them of thin twin wires—two pairs of
wires close together in metallic circuit—was given up, and the
present type of submarine cables was adopted.

5. But now it is very interesting to us to find that old
question revived. I had myself laid it aside in some corner of
my mind and in some slight corners of my note-books for forty
years. Within the last forty days I have really worked it out
to the uttermost, merely for my own satisfaction. But in the
meantime it had been worked out in a very complete manner
by Mr Oliver Heaviside; who has pointed out and accentuated

this result of his mathematical theory—that electro-magnetic induction is a positive benefit : it helps to carry the current. It is the same kind of benefit that mass is to a body shoved along against a viscous resistance. Suppose, for instance, you had a railway carriage travelling through a viscous fluid. Take a boat not floated but partially supported on wheels, so that when loaded more heavily it will not sink deeper in the fluid. Take a boat on wheels in a viscous fluid. We will shove off two boats with a certain velocity—the boats of the same shape ; but let one of them be loaded to ten times the mass of the other : it will take greater force to give it its impulse, but it will go further. That is Mr Heaviside's doctrine about electro-magnetic induction. It requires more electric force to produce a certain amount of current, but the current goes further. It is a very crude way in which I am putting it. I am not doing justice, of course, I know, to his statement in one short sentence. The whole question is treated by him in the most complete mathematical way. The effect of electro-magnetic induction and electrostatic induction taken together (and they cannot be separated) is fully worked out. One thing that was known of old is made a point of in Mr Heaviside's treatment of the cable problem—that is, the beneficial effect of leakage in respect to clearness of signals. Old telegraphists remember that. They always used to say three or four good leaks in a cable, *if they would but kindly remain constant, and not introduce extra trouble by earth currents,* would make the signalling more distinct. That used to be well known, and the reason used to be fairly well known ; mathematical theory had pointed it out. Now Mr Oliver Heaviside has taken up that subject again and included it in his work. It is a practical point of importance that the question of clearness of signals is not simply or even very importantly this—How much is the current attenuated at the remote end of the cable ? how much is the amplitude of the electric current in one mode of working, or of the variation of electric potential in another mode of working, altered in transmission through a thousand miles or two thousand miles of cable ? A certain range is given at the sending end ; and what is the range at the receiving end ? That is an important question, but it is not the most important question with reference to clearness of signalling ; in fact, we might almost say it is not an important question at all. It is not the smallness of the signals at the receiving end that is the real difficulty in a submarine cable just now at all; it is the running of one signal into another; it is the want of correspondingly definite distinctions of single signals or of a group of signals at the receiving end and at the sending end.

6. Now in the mathematical theory there are two things to be considered in respect to the distortion (as Heaviside called it) of the signals in passing through the cable. One thing to be considered is the retardation of phase; another is the diminution of amplitude. If the retardation of phase were the same for alternating currents of all periods, then this retardation of the phase would be of no consequence whatever—it could not diminish the distinctness at all. Again, if the diminution of the amplitude were precisely in the same proportion for alternating currents of all periods, then when we come to make non-periodic signals we should find that the signals would be transmitted with perfect sharpness. If we compare the transmission of electric signals through a wire with the transmission of sound through air, we have in the course of transmissions of sound through air great attenuation by distance—inversely as the square of the distance, in fact—but the same for all notes; and, again, retardation of phase depending upon the velocity of sound, the same for all notes. The result is that speaking, and musical performances, and signals of all kinds in air, lose none of their clearness by distance. It is just a question whether at the very greatest distance at which a sound can be heard there is any want of clearness due to different attenuations of the different notes or of the different elements forming the compound sound, or to difference of retardations of phase. I must not occupy you too long with this subject, but it is one of large practical importance. Heaviside points out that electro-magnetic induction causes a less great difference in the attenuation of signals of different periods than there is without it; and that electro-magnetic induction (as we knew forty years ago) tends to reduce the retardation of phase to the same for all different notes—that is, to the retardation equal to what would depend on a velocity not very different from the velocity of light—if the signals have but sufficient frequency. That velocity was then and is still known as the velocity which is the conductance in electrostatic measure, and the resistance

in electro-magnetic measure of one and the same conductor. But its relationship to the velocity of light was brought out in a manner by Maxwell to make it really a part of theory which it never was before. Maxwell pointed out its application to the possible or probable explanation of electric effects by the influence of a medium, and showed that that medium—the medium whose motions constitute light—must be ether. Maxwell's "electro-magnetic theory of light" marks a stage of enormous importance in electro-magnetic doctrine, and I cannot doubt but that in electro-magnetic practice we shall derive great benefit from a pursuing of the theoretical ideas suggested by such considerations. In fact, Heaviside's way of looking at the submarine cable problem is just one instance of how the highest mathematical power of working and of judging as to physical applications, helps on the doctrine, and directs it into a practical channel.

7. The telephone—one of the added subjects of the Institution of which we are members—illustrates very splendidly these developments of the theory of the transmission of signals through the submarine cable. The telephonic signals have, in fact, sufficient frequency to make electro-magnetic induction very sensibly influential. The frequencies in telephony correspond to from 250 periods per second, up to four, or five, or six times that; being the frequencies involved in speaking in the human voice—tenor and soprano—and in the quality of the voice as affected by the over-tones. I say frequencies of from 250 per second to 1,000 or 1,500 periods per second, are concerned in the fundamental notes, and in the characterising over-tones, of the sounds transmitted by the telephone. Now there seems no doubt but that the clearness of the telephone through great distances is to a large degree due to the circumstance which Heaviside has pointed out—that we have much less of difference of attenuation and difference of phasal retardation, for different notes, with the actual frequencies of the notes in sounds transmitted through the telephone wire, and the practical dimensions of the telephone wire, than we should have without electro-magnetic induction.

8. Leaving all questions of submarine telegraphy, and of telegraphy or telephony, in which—whether from the greatness

of the distance through which the communications are made, or
smallness of distance between insulated conductor and sheath,
or between the twin wires when insulated metallic circuit is
used, the effects of electrostatic capacity give rise to sensible
difference of strength of current in different parts along the
length of the conductor—I wish to call your attention to the
differences of current-density across different parts of the cross
section, which are produced when alternate currents are sent
through a wire.   Consider a copper wire, and a copper tube
surrounding it for return.   Or consider what is, after all, one
of the very simplest cases—two parallel copper wires.   If the
distance between them is a large multiple of the diameter of
each, as is generally the case in telegraphy and telephony, the
problem is the same as the problem of a single copper wire in
the centre of a cylindrical tube of infinitely conductive metal,
and of radius equal to the distance between the wires.   The
distribution of current within the solid conductor depends only
on the period of the alternations, and on the diameter and the
specific resistance of the metal; and is quite independent of
the surroundings, provided only they be symmetrical all round,
or provided, if the case be that of two parallel wires, the distance
between the two wires be a large multiple of the diameter of
each, so that the current in each is not sensibly disturbed, by
the influence of the other, from being arranged in co-axial
cylindric layers of equal current-density.   For this problem the
mathematical theory gives us a remarkably interesting and very
useful practical result; and I really, in proposing to speak upon
such a very abstruse subject as " Electricity, Ether, and Ponderable Matter," wish to try to give one little piece of practical
information to-night.   Here is the solution, expressed in a
formula, and a table of numerical results calculated from it.
[The demonstration is given in §§ 29—35 below.]

9.   Alternate Currents through a Straight Conductor of Round
Rod of Non-Magnetic Material.

Let σ denote the specific resistance in square centimetres per
second (or the "specific resistance C.G.S.");

a „   the radius of the wire;

$R\,(S)$ denote the value of $\sigma l \div \pi a^2$ [or the resistance (in centimetres per second) of any length (l) of the wire, with steady current through it];

$R\,(N)$ „   the effective ohmic resistance * of the same length (l), with alternate current of N periods per second through it.

$c\,(N)$ „   the current-density at distance r from the axis, and at time t.

$C\,(N)$ „   the current-density in the axis at time t.

We have $c\,(N) = C\,(N)\,(\text{ber } q \cos \theta - \text{bei } q \sin \theta)$,

where q denotes $\left(2\pi \sqrt{\dfrac{2N}{\sigma}}\right) r$;

„ θ „ $(2\pi N)\,t$;

„   ber and bei denote two functions defined as follows:—

$$\text{ber } q = 1 - \frac{q^4}{2^2\,4^2} + \frac{q^8}{2^2\,4^2\,6^2\,8^2} - \&\text{c.}\ ;$$

$$\text{bei } q = \frac{q^2}{2^2} - \frac{q^6}{2^2\,4^2\,6^2} + \&\text{c.}$$

And if p denote the value of q, with $r = a$, we have

$$\frac{R\,(N)}{R\,(S)} = \tfrac{1}{2}\,p\,\frac{\text{ber } p\ \text{bei}'\,p - \text{bei } p\ \text{ber}'\,p}{(\text{ber}'\,p)^2 + (\text{bei}'\,p)^2}\,,$$

where the accents denote differential coefficients.

The following table of numerical results has been calculated for me by Mr Magnus Maclean, official assistant to the Professor of Natural Philosophy in the University of Glasgow:—

---

* This expression I have introduced to designate (in contradistinction to Mr Heaviside's *impedance*) the coefficient by which the time-average of the square of the total current must be multiplied to find the time-average of the work done in maintaining the current, or of the dynamical value of the heat which the current generates, per unit of time, in the conductor. See § 29 below.

q	ber q	bei q	ber' q	bei' q	$\dfrac{\text{bei}'\,\text{ber} - \text{ber}'\,\text{bei}}{\text{ber}'^2 + \text{bei}'^2}$	$\tfrac{1}{2}q \times \dfrac{\text{bei}'\,\text{ber} - \text{ber}'\,\text{bei}}{\text{ber}'^2 + \text{bei}'^2}$
0·0	1·0000	0·0000	0·0000	0·0000	∞	1·0000
0·5	·999	0·0625	0·0078	0·24992	4·0000	1·0000
1·0	·9844	·2496	·062446	·499947	2·00014	1·0001
1·5	·9211	·5576	·210011	·730251	1·3678	1·0258
2·0	·7517	·9723	·4931	·9170	1·0805	1·0805
2·5	·3999	1·4571	·9436	·9983	·9398	1·1747
3·0	−·2214	1·9376	1·5698	·8805	·8787	1·3180
3·5	−1·1936	2·2833	2·3361	·4353	·8526	1·4920
4·0	−2·5634	2·2927	3·1347	−·4911	·8389	1·6778
4·5	−4·2991	1·6859	3·7537	−2·0526	·8279	1·8628
5·0	−6·2301	·1160	3·8442	−4·3538	·8172	2·0430
5·5	−7·9735	−2·7902	2·9070	−7·3729	·8069	2·2190
6·0	−8·8584	−7·3348	·2931	−10·8462	·7979	2·3937
8·0	−20·9739	−35·0167	−38·2944	−7·6615	·7739	3·0956
10·0	−138·8405	−56·3704	−51·373	−135·23	·7588	3·7940
15·0	−2969·79	2952·33	86·648	−4089·2	·7431	5·5732
20·0	47583·7	11500·8	24325·1	41491·5	·7325	7·3250
∞					·7071*	

* Or, $\sqrt{\tfrac{1}{2}}$: see §§ 29—35 below.

For copper we have $\sigma = 1{,}610$ square centimetres per second. Hence, with $N = 80$ we find

$$q = 1\cdot 98r \doteqdot 2r\,;$$

thus in respect to the ohmic resistance of the whole wire, we may for copper take the column headed q as the diameter of the wire, and in respect to the distribution of the current through the wire (expressed by the ber bei formula above) we may take q the diameter of the cylindric shell in which the current-density is to be calculated.

10.   Take, for example, 80 periods per second as the frequency —that is about what is adopted in the alternate-current system of distribution for electric light; at all events in one great system I know, the Grosvenor Gallery installation, that is the frequency of the period; and I believe it is pretty much the same gene-rally.   First, consider copper wire of 1 centimetre diameter: the ohmic effective resistance is greater than for steady current through the same wire, but only about $\frac{1}{100}$ per cent. greater. Take, now, copper wire of $1\frac{1}{2}$ centimetres diameter: the ohmic effective resistance is $2\frac{1}{2}$ per cent. greater than the resistance for steady current.   Next, take copper wire of 2 centimetres diameter: the ohmic resistance is 8 per cent. more for the alternating current than for the steady current.   In round copper rod of 4 centimetres diameter, the ohmic resistance is 68 per cent. more for the 80 periods per second alternate currents than for steady currents.   In round copper bar of 10 centimetres diameter, the ohmic resistance is $3\cdot 8$ times what it would be for the steady current.   In a solid copper cylinder of 100 centimetres diameter the ohmic resistance is 35 times greater than for steady currents. From 10 centimetres diameter upwards the ohmic effective con-ductance—that is, the reciprocal of the ohmic effective resistance —increases scarcely more than as the diameter simply, and not as the square of the diameter.   The conductance for steady currents is as the square of the diameter for all sizes.   The effective conductance for alternate currents follows a law which can only be expressed by aid of Fourier-Bessel functions till we get to very great diameters.   When we get to so great a diameter that the shell, or outer portion, of the wire into which the current is practically confined, is moderate or small in pro-portion to the diameter of the wire, then, for diameters exceeding

that, you can all see perfectly without calculation, that the conductance is in simple proportion to the circumference, and therefore in simple proportion to the diameter. This very imperfect
explanation of the results may give some idea which, I think, is
of rather an interesting and important kind, but the figures speak
for themselves With quadruple frequency, the same figures apply
to wires of half diameter. There we get the telephone problem.
Four times 80 is 320, which is among the frequencies for telephonic notes; and for the 320 frequency, take the figures I
have given, but with half the linear magnitudes. Thus, for
instance, for copper wire of 1 centimetre diameter, transmitting
musical notes of 320 periods per second, the ohmic resistance is
8 per cent. greater than the resistance for steady currents; for
a copper wire 2 centimetres diameter, and frequency of musical
note 320 per second, the ohmic resistance is 68 per cent. greater
than the resistance for steady currents; and so on.

11. In respect to electro-magnetic theory; we have a very
fine analogy with viscous fluid motion, which has been obvious,
more or less, from the time the known laws of electro-magnetic
induction were put into formulæ in the beautiful manner in which
Maxwell put them,—we have a very fine analogy, I say, with the
diffusion of laminar motion into a viscous fluid, and its analogue
in the diffusion of heat by conduction through a solid, first
pointed out by Professor Stokes. The actions concerned in the
distribution of alternating electric current through a conductor
such as copper, and the distribution of the motion of water in
a viscous fluid disturbed by periodical tangential motions of its
surface, follow identically the same law. Mr Heaviside, referring
to this, has well said that this analogy is very useful, because we
can see the motions in a viscous fluid, and understand them, and
picture them to our minds, while it is much more difficult to
fancy we see the distribution of electric current in a wire.
Take now definitively, this analogy for the distribution of electric
current in a round copper wire through which alternate currents
of electricity are sent. Take a viscous fluid in a tube, in place of
the conductor: move the tube to and fro with a regular
alternating motion—a simple harmonic motion. In order that
we may fulfil at all approximately what I am speaking of, the
length of the tube must be very great in comparison with the

diameter, and the place in which we consider the motion of the fluid must be at a distance of many diameters from the ends, which we may suppose to be closed by frictionless pistons, limiting the fluid at its two ends. In the first place, if the fluid were not viscous—if it were perfectly liquid—you might move the tube to and fro, but the fluid inside of it would remain at rest. Water, however, would move; oil would move; the more viscous the fluid is, the more liable it would be to experience motion in that way. Now there is a perfect analogy between the alternating motion of the fluid transmitted inwards from the surface, and the distribution of the electric current in a wire through which the effect of the alternating current machine is being conveyed.

12. Another very interesting analogy in which exactly the same law holds, is the change of temperature of a conducting solid, due to variations of external temperature* Imagine a column of rock or stone or metal, and let the atmosphere around it be periodically varied in temperature: the law of the inwards progress of changes of temperature, the law of the maximums and minimums and zeros of temperature, is identical with the law of the corresponding features of electric currents and of viscous fluid motion. In each case we have a propagation inwards, with diminishing amplitude. In each case the rate of diminution of amplitude corresponds to the retardation of phase according to exactly the same law. I need not attempt at this time to state the law—mathematicians know it perfectly well †.

13. Now take another case. Here the thermal analogy absolutely fails us, but the fluid motion analogy still holds. Take a tube of fluid and give it an alternating motion—a periodically varying motion round its axis which gives a tangential drag to the fluid in the inside. Now you can all see that the inwards penetration of the tangential drag, if the alternations of the motion be very quick, will follow the same law for the to and fro motion of the cylinder and for the rotatory motion of the cylinder. The question is this, Does the variation penetrate sensibly to a large distance in, from the outside or not? If, for

* See foot-note on § 5, Art. CI. of the present Volume.
† See § 11 of Art. XCIII.

example, it penetrates in only the one-hundredth of the radius, then it is obvious that we shall have sensibly the same law of penetration inwards for the disturbance, whether for the case of the rotatory motion of the cylinder round its axis, or of longitudinally to and fro motion. Exactly the same thing holds with reference to electro-magnetic induction. The one case of electro-magnetic induction that I mentioned first is the most important, being the telegraph and telephone case; but another very interesting case, and not at all without practical importance, is the penetration of induced currents into a copper or other metallic core within a solenoid. Take a common helix or solenoid: send an alternating current through its coil—you know what it does. It produces alternating magnetic force, with lines of force parallel to the axis, in the interior of the solenoid. But alternating longitudinal magnetic force, in the copper bar, induces electric currents in circles perpendicular to the direction of the force. Thus we have currents induced, as you all very well know, in a solid metal core of a solenoid. A metallic core other than iron, is a subject for investigation of an exceedingly easy kind. The Fourier-Bessel functions come in here just as they do in the other cases in which we are concerned with circular cylinders. If we have, instead of copper, an iron core, we must take into account its inductive magnetisation. This presents no *mathematical* difficulty if we suppose the magnetic susceptibility constant; and the same law of amplitudes and phasal retardation holds as for copper or other non-magnetic metal. The difficulties, both experimental and mathematical, to take into account, are the enormous differences of the inductive quality of iron with different degrees of magnetisation, and with reversals of magnetisation. The great complications of the inductive effects on account of the "magnetic friction" in the iron, introduce corresponding complications in the theory of the induced currents, and they are complications of a kind that are very formidable.

14. Now I can only just go on to say two or three words about an extension of that viscous fluid theory that allows us to take into account all that goes on both in air and in metal, and in different metals, whether in contact with one another or separated by air. For illustration, consider our two simple

cases—parallel wires with alternating currents through them, and the cylinder rotated with a periodic motion of rotation alternately in opposite directions.   The analogy is simply this : To represent different metals, densities of fluid in simple proportion to the electric conductivities must be taken; the viscosity must be the same in all.   The representative of an insulator in this analogy is a massless fluid.   By "massless" I mean devoid of inertia— perhaps I ought to say an "inertialess" fluid, because people attach other ideas to "mass" sometimes than "inertia," but in the strictest dynamical language "mass" is taken as the measure of inertia.   An inertialess viscous fluid must take the place of air or other non-conductor; a viscous fluid of a certain density, but the same degree of viscosity, must take the place of lead.   A fluid of twelve times the density of lead would take the place of copper, the conductivity of copper being, say, twelve times the conductivity of lead.

15.   Time does not allow me to pursue the subject further in the way of illustration at present, but I must return to the second case later on, because I am going to speak of iron and rotation.

Now, with reference to the electrostatic effect, the hopeless— I must not say "hopeless:" that is too large a word; we are never without hope in science—I was going to use another word, "despair"—well, I feel it desperately difficult ; I feel the pro- bability of my seeing the solution of it *is* hopeless.   To merely introduce into the analogy electrostatic effect is very simple. Simply imagine an interface between the two fluids, and give it such stiffness against change of shape as is required to cause it to fulfil the conditions which electrostatic knowledge and our knowledge of the laws of electric and electro-magnetic influence, dictate to us.   I say, put in at the interface the requisite normal force, and you can extend the analogy to include the complete problem of the submarine cable, in which electro-magnetic and electrostatic induction are both taken into account.   But it is only by putting in, and in an arbitrary manner, a force at the surface to fulfil the requisite conditions, that we can complete the analogy.

16.   The analogy I have just sketched cannot be considered as being in any respect a physical analogy.   In it the analogue to

electric current is not velocity of the liquid; it is not the molecular rotation of the liquid. It is the rotation of a second liquid whose translational velocity is, at every point, equal to and in the direction of the axis of the rotation of the first liquid. This is too difficult a subject to explain fully at present: but I may illustrate it by an example. Take, in the viscous fluid analogue, what corresponds to the steady current in a wire. Think of the tube with viscous fluid and pistons as before. At one end of the tube press a piston in with a uniform motion, continued long enough to cause the fluid throughout the tube to come to a state of steady motion. In the neighbourhood of the piston the motion is disturbed by the rigidity of the piston; but go to a distance of ten or twenty diameters from the piston, and the motion of the fluid takes a perfectly regular character, [*Illustrating on blackboard as Fig.* 1.] Suppose *that* to be the inner surface of the tube. This dotted line represents a portion of the liquid which at one time is plane. A little later, while the fluid in contact with the containing surface remains unmoved, in the doctrine of viscous fluid as given by Stokes there is absolutely no slip at the containing surface. This portion of the fluid which was plane becomes the paraboloid of revolution, which you see shown in axial section in the diagram. The velocity of the fluid is nothing at the bounding surface, and it is a maximum at the centre. Well, we have two functions derivable from the consideration of that distribution of velocity. The first is the rate of shearing of the fluid; the second is the rate of change per unit change of distance from the axis of the rate of shearing*. The rate of shearing represented graphically is equal to the tangent of the inclination (T P N) of this curve to the transverse surface, the inclination of the curve being the angle which is represented by the letter i. Now the rate of change from point to point of the rate of shearing is the analogue to the strength of the current, and that is uniform.

---

* Suppose the parabola in the drawing to represent the fluid which lay along the dotted line a unit of time earlier. The distance of P from the dotted line is equal to the velocity of the fluid (u) at the distance (r) of P from the axis. We have $u = c(a^2 - r^2)$, where a denotes the radius of the tube and ca^2 the fluid velocity along its axis. We have $\dfrac{-du}{dr} = 2cr$, which is the rate of shearing; and $\dfrac{-d^2u}{dr^2} = 2c$, which is our representative of the electric-current density. The whole strength of the electric current is $2c . \pi a^2$

So in this analogy of a viscous fluid forced through a tube, we have not the fluid velocity equal to the electric current, but something else, quite intelligible; and the reason for it, in our analogy, is clear enough. But there is something interesting, perhaps, in this idea—that we have a super-subtle mathematical definition of electric current which is not fluid velocity. Well, now, perhaps

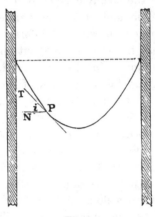

Fig. 1.

some one will say, " Had not we better get an analogy in which a " fluid velocity is equal to the velocity of the electric flow ?" Well, I do not say whether we had better do so or had better not, but we do not, otherwise than in the way I have defined, get the working analogy; and there is an advantage in this analogy. It gives us a motion of which the rotation is the magnetic force. This is what Maxwell calls " vector-potential." I think it is an unhappily chosen name; but Maxwell's use of the thing is most happy and most instructive, as it seems to me. Maxwell does not translate this into realities of motion, but he puts down in his formulas, as the foundation from which one step leads to magnetic force and the next step to electric current, something which, translated into realities of motion, gives us a motion of which the rotation is the magnetic force; and here it seems to me that if we are ever to have a real theory it must be founded upon this view. The hand of the clock warns me time is going on so rapidly that we must leave this viscous liquid analogy absolutely unfinished. Perhaps it is well to be obliged to leave it now, because the more we look at it the less we like it, if we wish to see and to like a true mechanical explanation of electro-magnetism. The work is done in the wrong place. Throughout the liquid, work is done and heat generated in proportion to the square of the rate of shearing. In the electric reality no heat is generated in the surrounding insulator; and the work is done and heat generated uniformly throughout the conductor. In the viscous analogue we have work done and heat generated in the massless fluid taking the place of the insulator. We must discredit that absolutely; but the reason for judging the analogy worth so much notice as even it has had to-night, is that it is a perfect mathematical working analogy, and an exceedingly useful and instructive kind of analogy, and a very potent one to help us in guessing out, and in thinking out, and estimating results in practical problems of electro-magnetic induction in dynamos and in alternate-current machines, and in telephones and in electric instruments of great varieties of shape and mutual relations.

17. But now there is another line of thought in connection with this subject, and that is the elastic solid idea. Will you allow me to read a very short statement which was published in the *Cambridge and Dublin Mathematical Journal* for the year 1847*? It is dated Glasgow University, November 28th, 1846. It was written after I had been twenty-eight days at work in my professorship, and it is as follows:—"Mr Faraday, "in the 11th series of his 'Experimental Researches on Elec- "'tricity,' has set forth a theory of electrostatical induction which "suggests the idea that there may be a problem in the theory of "elastic solids corresponding to every problem connected with "the distribution of electricity on conductors or with the forces "of attraction and repulsion exercised by electrified bodies. The "clue to a similar representation of magnetic and galvanic forces "is afforded by Mr Faraday's recent discovery of the affection "with reference to polarised light, of transparent solids subjected "to magnetic or electro-magnetic forces. I have thus been led "to find three distinct particular solutions of the equations of "equilibrium of an elastic solid, of which one expresses a "state of distortion, such that the absolute displacement of a

* *Papers*, Vol. I. Art. XXVII.

"particle in any part of the solid represents the resultant
"attraction at this point produced by an electrified body. An-
"other gives a state of the solid in which each element has a
"certain resultant angular displacement, representing in mag-
"nitude and direction the force at this point produced by a
"magnetic body; and the third represents in a similar manner
"the forces produced by any portion of a galvanic wire; the
"directions of the force in the latter cases being given by the
"axes of the resultant rotations impressed upon the elements
"of the solid." Then come the mathematics, in three pages,
and then comes the last sentence : "I should exceed my present
"limits were I to enter into a special examination of the states
"of a solid body representing various problems in electricity,
"magnetism, and galvanism, which must, therefore, be reserved
"for a future paper." As to this last sentence, I can say now,
what I said forty-two years ago—"*must be reserved for a future
"paper* !*" I may add that I have been considering the subject
for forty-two years—night and day for forty-two years. I do
not mean all of every day and all of every night; I do not mean
some of each day and some of each night; but the subject has
been on my mind all these years. I have been trying, many
days and many nights, to find an explanation, but have not
found it.

18. Let there be an elastic solid body of exceedingly small
density, and let there be a tubular portion of it porous, but
with the same aggregate rigidity as that of the continuous
elastic matter round it. Let the pores be filled with a dense
viscous fluid, and let this fluid be forced, by aid of a piston
or otherwise, to move through the tube. The pull of the fluid
upon the porous solid will produce static rotational displacement exactly proportional to the continued rotatory motion
which we had in the case of the viscous fluid. Some of the
most interesting practical problems of electro-magnetic induction
can be dynamically realised, as it were, in model, by following
out this idea; in fact, if we had nothing but electricity and
ether, the thing would be done. If it were not for the gross
ponderable matter that we are forced to consider, I should be

* The paper which was "future" in 1846 and 1889, appears at last in §§ 29—43
of Art. xcix. of the present Volume.

perfectly satisfied with the problem of electro-magnetic induction, by taking the electricity as a viscous fluid, and ether an elastic solid, porous in some places, and continuous or non-porous elsewhere.

19. Now, if you will pardon me, though it is very late for introducing another topic on which to speak. I shall confine myself to one, and that is magnetism. I must return to the rotational case. Imagine this (Fig. 2 or Fig. 3) to be the section

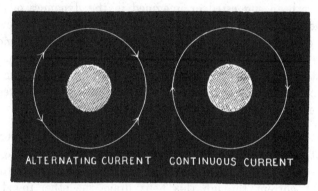

ALTERNATING CURRENT   CONTINUOUS CURRENT

Fig. 3.   Fig. 2.

of an ordinary helix or solenoid with a solid copper core. Imagine a continuous electric current (Fig. 2) or an alternating electric current (Fig. 3) of electricity sent through the solenoid, shown in section by the outer circle. Whatever the current of electricity may be, I believe *this* is a reality: *it does pull the ether round* within the solenoid. I do not think this is a dream of electro-magnetic theory; difficult as the idea is, I believe it to be a reality. Whatever ether is, we move through it—the earth moves through it. Astronomers and opticians do not cry out and make their lives miserable because of the aberration of light. Fresnel and Professor Stokes have done all that man, up to the 9th of January, 1889, has been able to do, to explain the dynamics of the aberration of light. It may be not beyond man's range to complete the solution—how the earth can tear through this elastic solid ether and yet the waves of light be propagated through it as they are. The aberration of light is still an absolute mystery. Yet people who deal with optics and

astronomy are not expected to be miserable for life because
they have that difficulty ever before them. Well, are we to be
absolutely unhappy because, while we see a mobile wire caused,
in virtue of an electric current through it, to move by electro-
magnetic force, we cannot see any possibility of explaining how
a medium capable of the "magnetic stress" can allow it to
move? After all, great as the mystery there is, there is a
mystery greater than that. An act of free-will is, with reference
to the laws of matter, a greater mystery than anything that
has ever been suggested or imagined in the dynamics of ether,
and electro-magnetism, and light. Somehow or other, however it
is, the ether is pulled round, the ether does get a turning motion
in the interior of a solenoid; somehow or other the electric
current through the surrounding wire, does give a turning motion
to ether in our supposed copper core and in the air between it
and the wire through which the current is flowing.

20. But now for the iron. And now, instead of an alternating
current through the helix, take a constant current through it.
What can it do? One thing or the other it does: either a
constant current through this helix drags the ether round and
round inside, or it drags it round to a certain angle proportionate to the strength of the electric current, and brings it to
static equilibrium so turned. It does either one or other of
those things. Now, how *can* iron differ, in the principle of the
interfacial law, from copper? Our interfacial law depending on
equal viscosities is quite clear, but when you introduce iron you
introduce an interfacial difference depending on rotation, without anything that could possibly be a cause of any viscous action,
or a cause of any elastic action. Elastic action (unless of compression or rarefaction, and these are not of our present subject)
requires distortion. You have no elasticity of an incompressible
elastic solid without distortion. Now if, by applying a tangential
force all round the space within a cylinder, you keep turning the
circumference, you will keep turning the contents. Ultimately
the whole fluid within will go round with the same angular
velocity as the circumference in contact with the cylinder. Thus
our viscous fluid analogue works out perfectly well for the magnetic force within a solenoid having any non-magnetic material
within it, and illustrates the fact that it is the same for conducting

and non-conducting matter. But with iron the case is something quite different. Our viscous fluid analogue is called on to give us a greater permanent rotational velocity, or a greater static rotational displacement, in the space occupied by iron in the magnetic analogue, than in this surrounding space! Thus the primary phenomenon of the magnetisation of a bar of iron within a helix, absolutely leaves us behind, cuts the ground from under us, both as to our viscous fluid analogy and our elastic solid analogy. If it is to be a fluid going round and round, we must have an action between the portions of fluid on the two sides of the interface, depending, not on distortion, but on rotation. Or if we take our elastic solid analogue, we must have static equilibrium of the elastic cylinder, with the inner part turned through a greater angle than the rotational part of the displacement of the surrounding matter. An irrotational circular displacement of the outer part added to this, procures fulfilment of the *no slip* condition at the interface. Hence we must have an arrangement of matter in which a constant torque produces a constant angular displacement in a body, and does not produce continued rotation. The only thing that can do that is an inherent rotation existing in the molecules of matter. This seems the only thing that can do it, and this *can* do it certainly. But consider this—that the gyrostat shows us the thing done; and I will just conclude, if you will allow me, with a simple gyrostatic experiment—a very well-known old gyrostatic experiment—which I bring before you because I want to accentuate the application of it.

21. I am going to show by this illustration with reference to the idea of a medium, a medium which has the properties of an incompressible fluid, and no rigidity except what is given to it gyrostatically. Here is, so to speak, a molecular skeleton that can give us such a fluid—a set of rigid squares with their neighbouring corners joined by endless flexible inextensible threads, running frictionlessly through holes in the corners, or round pulleys mounted in the corners (Fig. 4). Here is a model thus constructed—sixteen rigid squares and nine endless cord segments connecting the corners in this pattern, forming a kind of web. Now, if we take an ordinary cloth web, and pull it in different directions: in the direction of the warp and the direction

of the woof, you cannot stretch it ; but at 45 degrees from the

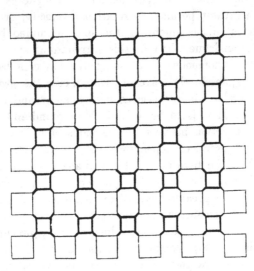

Fig. 4.

warp and woof you can stretch it very freely. We all understand
that. You know how the surgeons take advantage of it in their
diagonally cut bandages. Now here is a web which is equally
easily stretchable in all directions, and yet which is of constant
area—a constant area for infinitesimal displacements, not a constant area for very great displacements. The circumference of
each rigid and of each flexible square is given. Well, now, if you
infinitesimally alter the square into a not-square rectangle, or into
a rhombus, the area remains sensibly unchanged. The first change
of the area is a diminution in whatever direction you stretch it ;
but that is proportional to the square of the strain, so that you
may say, in language of infinitesimals, the area is unchanged.
The constancy of the periphery, then, of each of these figures
gives rise to and entails the condition of an approximate constancy
of the area. Here, then, we have in this skeleton a two-
dimensional working model of a medium which is unchangeable
in area, but is freely extensible in any direction, provided you
allow it to shrink proportionately in the perpendicular direction.
Well, now, let us put a gyrostat into each of those squares
(Fig. 5), and you have all that is wanted to fulfil the strange—

almost inconceivable—condition for a dynamical model of electro-magnetic induction in iron which I have put before you. I will

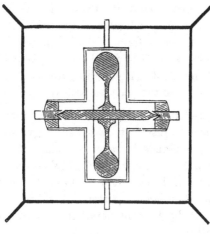

Fig. 5.

just make an experiment illustrating that, if it is not occupying too much time. [*Sir William Thomson then spun the gyrostat.*] I turn azimuthally the square frame by which I hold it—first in one direction, and the red end of the bearing of the axle of the fly-wheel turns up; I turn the other way, and up comes the blue end. The gyrostat is mounted in a square frame, as you see, which I hold in my hand. The rigid case bearing the axle of the fly-wheel is, as you see, free to turn round the axis of these trunnions, mounted horizontally on bearings in opposite sides of the square frame which I hold in my hand. The axis of these trunnions is perpendicular to the axis of the fly-wheel. I shall walk round and round to right, to keep the red side up; I walk round to the left, and it keeps the blue side up. It is a curiously interesting experiment. There are three little objects on a tray, as it were. Imagine this to be a butler's tray, with wine-glasses on it represented by these india-rubber corks. As long as I turn ever so little to my left all goes well; if I go straight forward, it is doubtful; but if I turn by an infinitesimal angle to my right, over it goes and everything falls off it.

22.   Look, now, at the gyrostat resting in this position on its trunnions, the axes of the trunnions and of the fly-wheel being

both at present horizontal. The outer square frame seems immovable in azimuth. When I apply a couple tending to move
it in azimuth it does not move. It does not move in azimuth
till the gyrostat turns round its trunnion axis and brings its
fly-wheel axis to be perpendicular to the plane in which I am
trying to turn the square frame. And I must apply a couple
whose time-integral is equal to double the moment of momentum
of the fly-wheel, before I can get the gyrostat from the position
with the blue end up, to the position with the red end up.

23. This closed brass case, with a rapidly rotating fly-wheel
mounted on bearings inside it, is called a gyrostat because in
virtue of rotation it stands, however you place it, with any of its
edges resting on a hard, smooth table. You see, place it as I
will, it cannot fall. If I place it with its centre of gravity above
the supporting point, it stands at rest. With its centre of gravity
not vertically over the bearing point, it goes round in azimuth,
but it does not fall.

24. Now imagine mounted in each one of the rigid squares
of this web a gyrostat exactly as this one is in the square frame
which I hold in my hand. If the fly-wheel speed be great enough,
each of those rigid squares is practically immovable in azimuth.
I do not say it is immovable, but I say you may make it practically
immovable by making the velocity of the fly-wheel sufficiently
great.

25. Thus we have a skeleton model of a special elastic solid
with a structure essentially involving a gyrostatic contribution to
rigidity. Now do not imagine that a structure of this kind, gross
as it is, is necessarily uninstructive. Look at the structures of
living things; think of all we have to explain in electricity and
magnetism; and allow, at least, that there must be some kind of
structure in the ultimate molecules of conductors, non-conductors,
magnetic bodies, and non-magnetic bodies, by which their wonderful properties now known to us, but not explained, are to be
explained. We cannot suppose all dead matter to be without
form and void, and without any structure; its molecules must
have some shape; they must have some relation to one another.

26. So that I do not admit that it is merely playing at theory,
but it is helping our minds to think of possibilities, if by a model,

however rough and impracticable, we show that a structure can be produced which is an incompressible frictionless liquid when no gyrostatic arrangement is in it, and which acquires a peculiar rotational elasticity or rigidity as the effect of introducing the gyrostats into these squares. Imagine a corresponding model in three dimensions, with rigid cubes instead of the rigid squares which you see in the model before you. Instead of the endless flexible cords which you see, you may imagine elastic threads stretched between neighbouring corners of the cubes. In each cube mount three gyrostats, with their trunnion axes perpendicular to the three pairs of its faces. The gyrostatic domination thus provided, causes the cubes to be practically immovable in rotation, but leaves them perfectly free to take translatory motion. There you have a body, then, that you could not distinguish from an ordinary elastic solid in respect to any irrotational distortion, or in respect to translational motion of the whole, but which if you try to turn it, will resist. It will not be immovable in respect of the turning, but it will be balanced by a constant couple with a constant degree of rotatory displacement. [See Art. C. § 5, above.] Thus upon this solid, the effect of a constant couple is not to produce continued rotation, but to produce and balance a constant displacement; and that balance might last for any time, however long, if the rotational moment of momentum of the fly-wheels is but great enough.

27. Now, lastly, I should just explain briefly that this rotational rigidity of ether must be equal in copper and all other non-magnetic metals, and in air and other non-conductors; but that it must be enormously less in iron. These conditions fulfil exactly what we want for the relation of ether between air and iron inside the helix of an electro-magnet*. But, alas! we are only led on to inscrutable difficulties. How much does our elastic solid go towards the explanation, when in the very fundamental fact of the mutual motions by which electro-magnetic forces are made manifest to us, we have a force as of a strained solid between the bodies (magnets or wires) whose motions revealed to Œrsted and Ampère the existence of electro-magnetic force? Why is it that those strains do not simply balance themselves in

---

* See § 45, Art. xcix. of the present Volume.

the solid? How can there be a solid capable of giving rise to that wonderful condition which we have in the air between the poles of an electro-magnet—for instance, such that a piece of copper will fall down through it at the rate of, perhaps, a quarter of a centimetre per second? Look on the subject as engineers, and think of the "strength of materials" wanted for ether in air, with the molecules of the air itself tearing through it in all directions at speeds averaging 500 metres per second, or more or less according to temperature. Think of the forces, amounting to 110 kilogrammes weight per square centimetre, with which two bars of iron magnetized to 1,700 C.G.S., with faces separated by a thin space of air, and with Ewing's 46,000 C.G.S. of magnetic force in the air around the bars, are urged towards one another. How can it be that these prodigious forces are developed in ether, an elastic solid, and yet ponderable bodies be perfectly free to move through that solid? Now I simply say, all that has been done to think out this subject merely gives us a dynamical theory on one part of it. I have absolutely—not ignored, because I have spoken of it two or three times—but I have left out in the cold, the electrostatic part, the thing we knew first. Our first love was electrostatics. That is absolutely left out in the cold; we do not touch it. We do not get near to explaining the mutual force between two electrified bodies, in any of these illustrations or attempted explanations; we do not even get near the mutual attraction between the iron of an electro-magnet, or the steel of a permanent magnet, and its armature or keeper; we do not get near to explaining the possibility of the motions of the bodies that demonstrate the forces. We only try to explain for a quiescent system of conductors and insulators, the variable distributions of electric currents which from mathematical theory and experimental observation we know to exist.

28. And here, I am afraid, I must end by saying that the difficulties are so great in the way of forming anything like a comprehensive theory, that we cannot even imagine a finger-post pointing to a way that can lead us towards the explanation. That is not putting it too strongly. I only say we cannot now imagine it. But this time next year,—this time ten years,—this time one hundred years,—probably it will be just as easy as we think it is to understand that glass of water, which seems now so plain and

simple. I cannot doubt but that these things, which now seem to us so mysterious, will be no mysteries at all; that the scales will fall from our eyes; that we shall learn to look on things in a different way—when that which is now a difficulty will be the only common-sense and intelligible way of looking at the subject.

I ask you to pardon me for leading you up to so impotent a conclusion as that we really know nothing below the surface of this grand subject which constitutes the province of the Institution of Electrical Engineers.

29—35. [Added May 13, 1890.] *Effective ohmic resistance.*

29. The definition of this expression given in the foot-note on § 9 above, is applicable to any portion B of a conductor having a zone, Z, of its surface in contact with insulating material all round, and the two thus separated parts S, S' of its surface in contact with other conducting matter through which electricity flows so as to traverse it either with constant or with periodically varying flow. Let

σ be the resistivity at any point, P;

dB, an infinitesimal element of bulk at P;

ρ, the current density at time t, and place P;

dA, an infinitesimal element of area of any surface Q edged by the zone;

q the normal component of ρ, at dA;

γ, the total current across Q;

w the rate of total work at time t.

We have

$$w = \iiint dB \sigma \rho^2 \dots\dots\dots\dots\dots(1),$$

and

$$\gamma = \iint dA q \dots\dots\dots\dots\dots(2),$$

where $\iiint dB$ denotes integration through the whole volume, B, and $\iint dA$ integration over the whole surface, Q.

30. To include every case of periodically varying current, let $\rho =$ resultant of

$$\begin{cases} P_0 + P_1 \cos \omega t + P_1' \sin \omega t + P_2 \cos 2\omega t + P_2' \sin 2\omega t + \&c. \\ Q_0 + Q_1 \cos \omega t + Q_1' \sin \omega t + Q_2 \cos 2\omega t + Q_2' \sin 2\omega t + \&c. \\ R_0 + R_1 \cos \omega t + R_1' \sin \omega t + R_2 \cos 2\omega t + R_2' \sin 2\omega t + \&c. \end{cases} \dots(3),$$

where P_0, Q_0, R_0, P_1, &c. denote functions of (x, y, z). We find time-average of

$$\rho^2 = P_0^2 + Q_0^2 + R_0^2 + \tfrac{1}{2}(P_1^2 + P_1'^2 + Q_1^2 + Q_1'^2 + P_2^2 + P_2'^2 + \ldots)\ldots(4).$$

Hence

time aver. of $w = \iiint dB\sigma \{P_0^2 + Q_0^2 + R_0^2 + \tfrac{1}{2}(P_1^2 + \&\text{c.})\}\ldots(5).$

Again, let

$$q = G_0 + G_1 \cos \omega t + G_1' \sin \omega t + G_2 \cos 2\omega t + G_2' \sin 2\omega t + \&\text{c.}\ldots(6).$$

We find, from (2),

time-average of $\gamma^2 = (\iint dA\, G_0)^2$

$$+ \tfrac{1}{2}\{(\iint dA\, G_1)^2 + (\iint dA\, G_1')^2 + (\iint dA\, G_2)^2 + \&\text{c.}\}\ldots\ldots(7).$$

The required "effective ohmic resistance," as defined in the footnote on § 9 above, is the quotient of (5) divided by (7).

31. In the important practical case of rectilineal flow in a straight conductor, of whatever form of cross-section, these formulas become greatly simplified. Thus if we take OZ parallel to the length of the conductor, the first two lines of (3) vanish, and if we take, for the surface Q, a plane cross-section perpendicular to the length, we have

$q = \rho$; and therefore $G_0 = R_0$, $G_1 = R_1$, $G_1' = R_1'$, &c.$\ldots\ldots(8)$;

and each of these quantities is a function of (x, y). For the very important particular case of purely "alternate" current according to the simple harmonic law, all the G's vanish except G_1 and G_1'; and, if we take for B the bulk between plane cross-sections with a length l of conductor between them;

time-average of $w = \tfrac{1}{2}l \iint dA \cdot \sigma (G^2 + G'^2)\ldots\ldots\ldots(9),$

and time-average of $\gamma^2 = \tfrac{1}{2}\{(\iint dA\, G)^2 + (\iint dA\, G')^2\}\ldots\ldots\ldots(10).$

32. Going back to (3) and (8), we now have

$$\rho = G \cos \omega t + G' \sin \omega t \ldots\ldots\ldots\ldots\ldots(11);$$

and, by Maxwell's fundamental equations of electro-magnetic induction, we have, as in Art. CI. § 7 above

$$\frac{d\rho}{dt} = \kappa \nabla^2 \rho \ldots\ldots\ldots\ldots\ldots\ldots\ldots(12);$$

where $\kappa = \dfrac{\sigma}{4\pi\varpi}$, and ∇^2 denotes $\dfrac{d^2}{dx^2} + \dfrac{d^2}{dy^2}$ $\ldots\ldots\ldots(13).$

Hence by (11),

$$G = -\frac{\kappa}{\omega} \nabla^2 G' ; \text{ and } G' = \frac{\kappa}{\omega} \nabla^2 G \ldots\ldots\ldots(14).$$

Hence, and by well known integrations, with the notation of the foot-note on § 19 of Art. CI., above,

$$\iint dA\, G = -\frac{\kappa}{\omega}\iint dA\,.\,\nabla^2 G' = -\int ds\,\frac{dG'}{d\nu}$$
$$\left.\vphantom{\int}\right\}\quad\dots\dots(15),$$
and similarly
$$\iint dA\, G' = \frac{\kappa}{\omega}\int ds\,\frac{dG}{d\nu}$$

and

$$\iint dA\,(G'^2 + G^2) = \frac{\kappa}{\omega}\iint dA\,(G'\nabla^2 G - G\nabla^2 G')$$
$$= \frac{\kappa}{\omega}\int ds\left(G'\frac{dG}{d\nu} - G\frac{dG'}{d\nu}\right)\dots\dots\dots(16).$$

By (11), (8), (2), and (15) we find, as the expression for the total current at time t,

$$\gamma = \frac{\kappa}{\omega}\left(-\cos\omega t\int ds\,\frac{dG'}{d\nu} + \sin\omega t\int ds\,\frac{dG}{d\nu}\right)\dots\dots(17);$$

whence,

$$\text{time-average of }\gamma^2 = \tfrac{1}{2}\left(\frac{\kappa}{\omega}\right)^2\left\{\left(\int ds\,\frac{dG'}{d\nu}\right)^2 + \left(\int ds\,\frac{dG}{d\nu}\right)^2\right\}\dots\dots(18).$$

And by (9) and (16),

$$\text{time-average }w = \tfrac{1}{2}\sigma l\,\frac{\kappa}{\omega}\int ds\left(G'\frac{dG}{d\nu} - G\frac{dG'}{d\nu}\right)\dots\dots(19).$$

Lastly, dividing (19) by (18), we find effective ohmic resist-

ance
$$= \sigma l\,\frac{\omega}{\kappa}\,\frac{\displaystyle\int ds\left(G'\frac{dG}{d\nu} - G\frac{dG'}{d\nu}\right)}{\displaystyle\left(\int ds\,\frac{dG'}{d\nu}\right)^2 + \left(\int ds\,\frac{dG}{d\nu}\right)^2}\dots\dots\dots(20).$$

33.   For the case of a conductor of circular cross-section (of radius a), with symmetrical surroundings, or with any surroundings such as to render G and G' functions of distance, r, from the centre, we have $\int ds = 2\pi a$, and $d/d\nu = (d/dr)_{r=a}$.

Hence (20) becomes

$$\frac{1}{2\pi a}\sigma l\,\frac{\omega}{\kappa}\left\{\frac{G'\dfrac{dG}{dr} - G\dfrac{dG'}{dr}}{\left(\dfrac{dG'}{dr}\right)^2 + \left(\dfrac{dG}{dr}\right)^2}\right\}_{r=a}\dots\dots\dots(21).$$

The solution of (14) in ascending powers of r, used in (11), and in (21) divided by $\sigma l/\pi a^2$, yields the expressions given in § 9 above. The calculations for the Table were of course made by aid of the semi-convergent series in descending powers of r, for values of r so large as to render the convergence of the converging series inconveniently slow. The designations *ber q* and *bei q* were given because three letters are convenient (after the pattern of *sin, cos, tan,* &c.) for designating determinate functions, and the series so designated are respectively the real part and ι^{-1} of the imaginary part, of the Fourier-Bessel function $J_0(\iota q)$.

34. For a conductor of any form of cross-section, the greater the frequency of the alternation, the less is the depth inwards from the boundary, within which there is any sensible intensity of current. That this must be so we see by taking ω exceedingly great in equations (14). The determinate solution of these equations for great or small values of ω, with proper determining boundary conditions for practical problems, is an exceedingly interesting and practically important question for mathematical treatment. For the present consider the case of ω so great that the thickness of the shell of sensible current is very small in comparison with the radius of curvature at the most curved part of the cross-section, and in comparison with the distance from the nearest centre of curvature of any other conductor or conductors insulated from it. Let OX be the inward normal through any point O of the bounding surface. For distances from O considerable in comparison with the radius of curvature at this point, but large in comparison with the thickness of the shell of sensible current, G and G' do not vary sensibly with distance from OX: that is to say they are functions of x; and therefore (14) become

$$G = -\frac{\kappa}{\omega}\frac{d^2G'}{dx^2} \; ; \;\; G' = \frac{\kappa}{\omega}\frac{d^2G}{dx^2} \dots\dots\dots\dots\dots(22).$$

The appropriate solution of these equations, that is to say *the* solution for simple harmonic alternation in period $2\pi/\omega$ with amplitude diminishing as x increases, is

$$G = g\cos\left(x\sqrt{\tfrac{\omega}{2\kappa}}\right)\epsilon^{-x\sqrt{\tfrac{\omega}{2\kappa}}}, \;\; G' = g\sin\left(x\sqrt{\tfrac{\omega}{2\kappa}}\right)\epsilon^{-x\sqrt{\tfrac{\omega}{2\kappa}}}\dots(23),$$

where g denotes an arbitrary constant; and using these in (11) we have

$$\rho = g\cos\left(x\sqrt{\tfrac{\omega}{2\kappa}} - \omega t\right)\epsilon^{-x\sqrt{\tfrac{\omega}{2\kappa}}}\dots\dots\dots\dots(24).$$

This is Fourier's solution for the variation of underground temperature due to a simple-harmonic variation of surface-temperature, which was stated verbally in Art. XCIII. § 11, above.

With (23) for G and G' we find

$$G \frac{dG'}{dx} - G' \frac{dG}{dx} = g^2 \sqrt{\frac{\omega}{2\kappa}} \epsilon^{-x \sqrt{\frac{\omega}{2\kappa}}} \quad \ldots \ldots \ldots \ldots (25);$$

and from this, remarking that $d/d\nu = -d/dx$, and that $x = 0$ at the surface of the conductor; and that g, though a constant with reference to x, is not generally constant round the circumference of the conductor, we have

$$\text{time-average of } w = \tfrac{1}{2}\sigma l \sqrt{\frac{\kappa}{2\omega}} \int ds g^2 \quad \ldots \ldots (26).$$

Similarly by (18) we find

$$\text{time-average of } \gamma^2 = \frac{\kappa}{2\omega} \left(\int ds g \right)^2 \quad \ldots \ldots \ldots (27).$$

Lastly, dividing (26) by (27) we find

$$\text{effective ohmic resistance} = \sigma l \sqrt{\frac{\omega}{2\kappa}} \frac{\int ds g^2}{\left(\int ds g \right)^2} \quad \ldots \ldots (28).$$

35. When the surroundings (the neighbouring conductors insulated from the given conductor) are such that g is constant all round the cross-section, (28) becomes, if we denote the circumference by c,

$$\sigma l \sqrt{\frac{\omega}{2\kappa} \frac{1}{c}}, \text{ or } \sigma l / c \sqrt{\frac{2\kappa}{\omega}} \quad \ldots \ldots \ldots \ldots (29),$$

which shows that the effective ohmic resistance of the given conductor for a simple-harmonic alternating current of period $2\pi/\omega$, is the simple ohmic resistance for constant current through an outer shell of thickness $\sqrt{(2\kappa/\omega)}$ of the given conductor supposed insulated from the conducting matter within. I therefore call $\sqrt{(2\kappa/\omega)}$ the mhoic effective thickness, or simply the mhoic thickness, of the alternate current stratum. With $\omega = 503$, or frequency 80 periods per second, it is 2·48 cm. for lead, ·714 cm. for copper; and would be 104 cm. for iron if its diffusivity for electric currents were 2·7 square centimetres per second as calculated from $\varpi = 300$ in § 13 of Art. CI. above.

For the case of circular cross-section of radius a, we have $c = 2\pi a$; and (29) divided by $\sigma l / \pi a^2$ becomes

$$\frac{1}{\sqrt{2}} \cdot \frac{1}{2} a \sqrt{\frac{\omega}{\kappa}},$$

which proves the last entry of the Table of § 9, above.

ART. CIII. PROFESSOR TAIT'S EXPERIMENTAL RESULTS REGARD-
ING THE COMPRESSIBILITY OF FRESH WATER AND SEA WATER
AT DIFFERENT TEMPERATURES ; COMPRESSIBILITIES AT SINGLE
TEMPERATURES OF MERCURY, OF GLASS*, AND OF WATER WITH
DIFFERENT PROPORTIONS OF COMMON SALT IN SOLUTION;
BEING AN EXTRACT FROM HIS CONTRIBUTION TO THE REPORT
ON THE SCIENTIFIC RESULTS OF THE VOYAGE OF H.M.S.
CHALLENGER, VOL. II. PHYSICS AND CHEMISTRY, PART IV.
(PUBLISHED IN 1888).

THE pressures employed in the experiments ranged from
150 to 450 atm., so that results given below for higher or lower
pressures [and enclosed in square brackets] are extrapolated.
A similar remark applies to temperature, the range experimentally treated for water and for sea-water being only 0° to 15° C.
Also it has been stated that the recording indices are liable to
be washed down the tube, to a small extent, during the relief of
pressure, so that the results given are probably a little too *small.*

Compressibility of Mercury, per atmosphere, 0·0000036†

,,   ,, Glass, . . . 0·0000026

* Tait's determinations of the Compressibility of Glass were made by direct
observation of the diminution of length of a glass rod subjected to pressure equal in
all directions (water pressure from 150 to 450 atmos), according to a method first
described by J. Y. Buchanan in the *Transactions of the Royal Society of Edinburgh*
for 1880, pp. 589—598. They are interesting as having been thus obtained directly,
instead of by the very indirect method of calculation of observations of Young's
Modulus and the Rigidity Modulus as described above in §§ 45—47 of Art. XCII.

† Tait appends to the description of his own determination of the compressibility of Mercury the following remarks on the discordant results obtained by
experimenters who came after Regnault: "It is well to remember that though
" Grassi, working with Regnault's apparatus, gave as the compressibility of mercury

0·00000295,

" which Amaury and Descamps afterwards reduced to

0·00000187,

" the master himself had previously assigned the value

0·00000352.

" Had Grassi's result been correct, I should have got only about half the displace-
" ments observed: had that of Amaury and Descamps been correct the apparent
" compressibility would have had the *opposite sign* to that I obtained, so that the
" index would not have been displaced."

[Bulk Moduluses.]

In Atmospheres	In Grms. p. Sq. Cm.	In Dynes p. Sq. Cm.
278000	286×10^6	281×10^9
385000	397×10^6	389×10^9

Average compressibility of fresh water :—

[At low pressures $(520 \ - \ 3{\cdot}55t \ + \ {\cdot}03t^2)\,10^{-7}]$

For 1 ton $= 152{\cdot}3$ atm.   504   3·60   ·04

   2 ,, $= 304{\cdot}6$,,   490   3·65   ·05

   3 ,, $= 456{\cdot}9$,,   478   3·70   ·06

The term independent of t (the compressibility at $0°$ C.) is of the form $10^{-7}\,(520 - 17p + p^2)$,

where the unit of p is 152·3 atm. (one ton-weight per sq. in.). This must not be extended in application much beyond $p = 3$, for there is no warrant, experimental or other, for the minimum which it would give at $p = 8{\cdot}5$.

The point of minimum compressibility of fresh water is probably about $60°$ C. at atmospheric pressure, but is lowered by increase of pressure.

As an *approximation* through the whole range of the experiments we have the formula :—

$$\frac{0{\cdot}00186}{36 + p}\left(1 - \frac{3t}{400} + \frac{t^2}{10,000}\right);$$

while the following formula exactly represents the average of all the experimental results at each temperature and pressure :—

$$[(520 - 17p + p^2) - (3{\cdot}55 + {\cdot}05\,.\,p)\,t + ({\cdot}03 + {\cdot}01\,.\,p)\,t^2]\,.\,10^{-7}.$$

Average compressibility of sea-water (about 0·92 of that of fresh water) :—

[At low pressures $(481 \ - \ 3{\cdot}40t \ + \ {\cdot}03t^2)\,.\,10^{-7}]$

For 1 ton   462   3·20   ·04

   2 ,,   447·5   3·05   ·05

   3 ,,   437·5   2·95   ·05

Term independent of t :—

$$10^{-7}\,(481 - 21{\cdot}25p + 2{\cdot}25p^2).$$

Approximate formula :—

$$\frac{0{\cdot}00179}{38 + p}\left(1 - \frac{t}{150} + \frac{t^2}{10,000}\right).$$

Minimum compressibility point, probably about $56°$ C. at atmospheric pressure, is lowered by increase of pressure.

Average compressibility of solutions of NaCl for the first p tons of additional pressure, at 0° C.:—

$$\frac{0\cdot00186}{36 + p + s},$$

where s of NaCl is dissolved in 100 of water.

Note the remarkable resemblance between this and the formula for the average compressibility of fresh water at 0° C and $p + s$ tons of additional pressure.

Various parts of the investigation seem to favour Laplace's view that there is a large molecular pressure in liquids. In the text it has been suggested, in accordance with a formula of the Kinetic Theory of Gases, that in water this may amount to about 36 tons-weight on the square inch. In a similar way it would appear that the molecular pressure in salt solutions is greater than that in water by an amount directly proportional to the quantity of salt added.

Six miles of sea, at 10° C. throughout, are reduced in depth 620 feet by compression. At 0° C. the amount would be about 663 feet, or a furlong. (This quantity varies nearly as the square of the depth.) Hence the pressure at a depth of 6 miles is nearly 1000 atmospheres.

The maximum-density point of water is lowered about 3° C. by 150 atm. of additional pressure.

From the heat developed by compression of water I obtained a lowering of 3° C. per ton-weight per square inch.

From the ratio of the volumes of water (under atmospheric pressure) at 0° C. and 4° C., given by Despretz, combined with my results as to the compressibility, I found 3°·17 C.:—and by direct experiment (a modified form of that of Hope) 2°·7 C. The circumstances of this experiment make it certain that the last result is too small.

Thus, at ordinary temperatures, the expansibility of water is increased by the application of pressure.

In consequence, the heat developed by sudden compression of water at temperatures above 4° C. increases in a higher ratio than the pressure applied; and water under 4° C. may be heated by the sudden application of sufficient pressure.

The maximum density coincides with the freezing-point at −2°·4 C., under a pressure of 2·14 tons.

Art. CIV. Velocities of Waves of different character;
and corresponding Moduluses in cases of Waves due
to Elasticity: being Appendix to Art. XCII., § 51.

1. Velocities, Lengths, and Periods, of Deep-water* Waves.

(Gravity) $g = 981$ centimetres per second per second.

(Surface-Tension) $T = \cdot075$ grammes weight per centimetre
$= 73\cdot6$ dynes per centimetre.

Case A. Gravitational; when $l > 30$ cms., and therefore
surface-tension may be neglected

$$V = \sqrt{\frac{gl}{2\pi}}; \quad P = \sqrt{\frac{2\pi l}{g}}.$$

Velocity		Length in feet	Period in seconds
In Nautical miles per hour†	In feet per second		
2	3·38	2·23	·659
3	5·07	5·01	·989
4	6·76	8·91	1·32
5	8·45	13·9	1·65
6	10·1	19·9	1·97
7	11·8	27·2	2·30
8	13·5	35·6	2·63
9	15·2	45·1	2·97
10	16·9	55·7	3·30
11	18·6	67·5	3·63
12	20·2	79·6	3·94
13	22·0	94·4	4·29
14	23·6	109	4·60
15	25·4	126	4·96
16	27·0	142	5·27
17	28·7	161	5·60
18	30·4	180	5·93
19	32·1	201	6·26
20	33·8	223	6·59
22	37·2	270	7·26
24	40·4	318	7·88
26	44·0	378	8·58
28	47·2	435	9·21
30	50·8	501	9·91
35	59·2	684	11·5
40	67·6	891	13·2

* Rigorously this means water infinitely deep; but if the depth be as small
as half the wave-length, the velocity is less by only $e^{-2\pi}$, or 1/535, of what it
is in infinitely deep water. Hence the error is less than 1/5 per cent. on the
velocity if we calculate for infinitely deep water, when the depth is anything
more than half the wave-length.

† The Nautical Mile is taken, according to the English Admiralty Rule, as
6086 feet, being the mean length of one minute of longitude at the Equator.

Case B. Gravitational-Cohesional*, when $l < 30$ centimetres, and $> 1.72^2/30$, or ·0986 of a centimetre.

$$V = \sqrt{\left(\frac{lg}{2\pi} + \frac{2\pi T}{l}\right)}; \quad P = \frac{l}{V}.$$

Velocity		Wave-length in centimetres	Period in seconds
In Nautical miles per hour	In centimetres per second		
·446	23 (min^m)	1·72	·0750
·452	23·3	{ 2·0 { 1·45	·0860 ·0622
·485	25·0	{ 3·0 { ·963	·120 ·0385
·528	27·2	{ 4·0 { ·723	·147 ·0265
·573	29·5	{ 5·0 { ·578	·169 ·0196
·779	40·1	{ 10·0 { ·289†	·249 ·00721†
1·33	68·5	{ 30·0 { ·096†	·438 ·00140†

Case C. Cohesional*; when $l < $ ·0986 of a centimetre and therefore gravity may be neglected.

$$V = \sqrt{\frac{2\pi T}{l}}; \quad P = \sqrt{\frac{l^3}{2\pi T}}.$$

Velocity in centimetres per second	Length† in centimetres	Period† in seconds
71·7	·090	·001256
96·2	·050	·000520
215	·010	·0000465

* For theoretical investigation of these waves, and experimental determination of the minimum wave-velocity of deep-water waves, see *Phil. Mag.* Nov. 1871. (Sir William Thomson, "Hydrokinetic Solutions and Observations.")

† These wave-lengths and periods are those of the standing vibrations, seen on

2. Velocity of "long waves" in water of given depth.

$$V = \sqrt{gD}.$$

Depth in Fathoms	Velocity in Nautical miles per hour
1	8·23
2	11·63
5	18·40
10	26·02
100	82·28

the surface of water in a thin wine-glass, or tumbler, or finger-glass, when caused to sound a note by sliding a wetted finger round the rim; but they are, in the most easily observed cases, of double the period of the exciting vibration of the glass, as was first noticed sixty years ago by Faraday ("On a Peculiar Class of Acoustical Figures," *Phil. Trans.* 1831; Appendix " On the Forms and States assumed by Fluids in contact with vibrating elastic surfaces;" republished in Faraday's Volume of " Experimental Researches in Chemistry and Physics"). This curious and surprising result is thoroughly explained in Lord Rayleigh's dynamical investigation of " Maintained Vibrations " (*Phil. Mag.* 1883, first half year) and is experimentally illustrated in his article "On the Crispations of a Fluid resting on a vibrating support" (*Phil. Mag.* 1833, second half year). The whole subject of Tables B and C has been subjected to a very searching experimental examination by L. Matthiessen (Wiedemann's *Annalen*, 1889, Vol. XXXVIII.), in which such telling quantitative verifications of the dynamical theory are found, with values for the surface-tension of water, alcohol, sulphuric ether, bisulphide of carbon, and mercury, taken from generally accepted results of static capillary measurement, that conversely Matthiessen's measurements of wave-lengths for given short enough periods might be used for determining the surface-tensions of the liquids experimented on. This principle has been used by Prof. Michie Smith (Royal Society of Edinburgh, Mar. 17, 1890), to determine the surface-tension of mercury; and it may prove to be a useful method in many cases for the determination of the surface-tension of this and other liquids; as for instance water with its surface either pure, or coated with molecularly thin layers of oil as in Lord Rayleigh's interesting and important investigation communicated to the Royal Society of London, March 27, 1890. (See *Proc. R. S.* for that date, or Sir W. Thomson's *Popular Lectures and Addresses*, Vol. I., Second edition, where Lord Rayleigh's paper is reprinted by permission.)

3. Velocity of Elasticity-waves in Kilometres per second.

1 Kilometre = ·6214 English Statute mile

= ·5391 Nautical mile.

	Distortional $V = \sqrt{\dfrac{n}{\rho}}$	Condensational-rarefractional in infinite solid or fluid $V = \sqrt{\dfrac{k + \frac{4}{3}n}{\rho}}$	Longitudinal in free rod $V = \sqrt{\dfrac{M}{\rho}}$
Iron	3·18	5·72	5·08
Copper	2·24	5·06	3·72
Brass	2·15	4·38	3·53
Glass	2·82	4·95	4·48
Water...........	0	1·43	0
Ether...........	300,000	—	—

4. Moduluses in dynes per square centimetre.

	Rigidity n	Resistance to compression k	Young's M $= \dfrac{9nk}{3k+n}$	Resistance to simple longitudinal extension $k + \frac{4}{3}n$	Density ρ
Iron.........	770×10^9	1459×10^9	1964×10^9	2486×10^9	7·6
Copper......	447 ,,	1683 ,,	1231 ,,	2279 ,,	8·9
Brass.......	370 ,,	1042 ,,	997 ,,	1535 ,,	8·0
Glass........	239 ,,	415 ,,	601 ,,	734 ,,	3·0
Fresh water at 10°.....	0	20·5 ,,	0	20·5	1·0
Sea water at 10°.....	0	22·2 ,,	0	22·2	1·027
Ether	$9 \times 10^{20}\rho$				ρ
,, ...	$\equiv 0·09!$			unless $\rho > 10^{-22}$ *	

* See *Papers*, Vol. II., Art. LXVII. p. 32.

INDEX.

END OF VOL. III.

CAMBRIDGE: PRINTED BY C. J. CLAY, M.A. & SONS, AT THE UNIVERSITY PRESS.

Printed in the United States
By Bookmasters